BIOLOGY AND CONSERVATION OF

Martens, Sables, and Fishers

BIOLOGY AND CONSERVATION OF

Martens, Sables, and Fishers

A New Synthesis

EDITED BY

Keith B. Aubry, William J. Zielinski,
Martin G. Raphael, Gilbert Proulx,
and Steven W. Buskirk

COMSTOCK PUBLISHING ASSOCIATES a division of

CORNELL UNIVERSITY PRESS Ithaca and London

First published 2012 by Cornell University Press

Printed in the United States of America

Library of Congress Cataloging-in-Publication Data

Biology and conservation of martens, sables, and fishers : a new synthesis / edited by Keith B. Aubry ... [et al.].
p. cm.
Includes bibliographical references and index.
ISBN 978-0-8014-5088-4 (cloth : alk. paper)
1. Martes. 2. Martes—Ecology. 3. Wildlife conservation. I. Aubry, Keith Baker.
QL737.C25B516 2012
599.76'65—dc23 2012003137

Cornell University Press strives to use environmentally responsible suppliers and materials to the fullest extent possible in the publishing of its books. Such materials include vegetable-based, low-VOC inks and acid-free papers that are recycled, totally chlorine-free, or partly composed of nonwood fibers. For further information, visit our website at www.cornellpress.cornell.edu.

Cloth printing 10 9 8 7 6 5 4 3 2 1

Contents

Preface

Martens, sables, and fishers are iconic mesocarnivores that occur in North America, Europe, and Asia. In the ecosystems they occupy, which are generally but not always forested, these semiarboreal mustelids contribute to the functioning of healthy ecosystems (especially as predators), serve as indicators of structurally complex habitats, and provide economic benefits as furbearers. Despite their ecological and economic value, however, many marten, sable, and fisher populations are at risk of further decline or extirpation. We believe that the conservation of these populations will depend largely on the application of scientifically sound and practical programs for habitat and population management and public education. To facilitate the development of such programs, we recognized the need to synthesize the current state of knowledge on the genus *Martes*, and to develop a reliable basis for organizing interdisciplinary knowledge and identifying key elements to communicate to wildlife biologists, resource managers, and policy makers. This book is intended to provide the empirical foundation for meeting that need.

In 1994, the first synthesis of the current state of knowledge about the genus *Martes*, titled, *Martens, Sables, and Fishers: Biology and Conservation* and edited by S.W. Buskirk, A.S. Harestad, M.G. Raphael, and R.A. Powell, was published by Cornell University Press. That book was one of the products of the First International *Martes* Symposium, convened at the University of Wyoming in 1991. This first formal North American gathering of people with a particular interest in the genus *Martes* also led to the formation of the *Martes* Working Group, with the primary purpose of facilitating communication among people with a common interest in research, conservation, and management programs for *Martes* species. By the early 1990s, most species in the genus *Martes* had experienced range reductions or population declines, but very little was known about their biology, ecological relations, or conservation status. Thus, one of the objectives of the original synthesis book was to identify critical information gaps and promising hypotheses for future research on *Martes* species throughout the world.

Since that time, the *Martes* Working Group has overseen the publication of 3 additional books that were the proceedings of subsequent *Martes* symposia: (1) Proulx et al.'s (1997) Martes: *Taxonomy, Ecology, Techniques, and Management*, published by the Provincial Museum of Alberta; (2) Harrison et al.'s (2004) *Martens and Fishers* (Martes) *in Human-altered Environments: An International Perspective*, published by Springer Science+Business Media; and (3) Santos-Reis et al.'s (2006) Martes *in Carnivore Communities*, published by Alpha Wildlife Publications.

Over the years, many changes have occurred both in research objectives for *Martes* species and in public concerns about the conservation of their populations. Some resource management issues, such as trapping, became less important, and other considerations, such as the conservation of genetic stocks, the reintroduction of populations, and the sustainability of forested habitats, became more prominent; thus, during the past few decades, the nature and direction of research on *Martes* species have paralleled many of the socioeconomic changes that occurred during that time.

Since publication of the original synthesis volume in 1994, field and laboratory studies on these species have continued apace, and some key information gaps have been filled. Additionally, with the subsequent explosion in the use of genetic data to study wildlife populations as well as emerging concerns about the potential effects of global warming on many threatened or sensitive species (including most members of the genus *Martes*), many new focal areas for research on these species have emerged. Significant changes have occurred in both research emphases on *Martes* species and the investigative tools that can be used. Accordingly, this new synthesis provides an update to many of the primary topics included in the 1994 book, but it also includes many new subjects the original synthesis did not cover, including the use of genetic data in *Martes* research and conservation; pathogens, parasites, and the biogeography and coevolution of host-parasite systems; ecophysiological relations; multiscale analyses of habitat relationships; new developments in radiotelemetry techniques; occupancy modeling using noninvasive survey methods; reintroduction or augmentation of populations; use of habitat and viability models for conservation and restoration; bioregional conservation strategies; and the potential effects of global warming on the distribution and status of *Martes* populations. Thus, we emphasize that this book should not be viewed simply as an updating of the 1994 book; rather, it is an entirely new synthesis of the genus *Martes* that reflects the breadth of modern scientific investigations. As before, we have made a concerted effort to ensure that this new synthesis is truly international in scope, and that all *Martes* species are considered whenever possible.

To develop the structure and content of this book and identify key topics to include in each section, we began by carefully considering recent advances in research on *Martes* species, the development of new technologies, and current societal concerns regarding their conservation. We then invited the researchers whom we considered best qualified to serve as the lead authors for chapters in their areas of expertise; those who accepted our invitation to contribute a chapter to this book also agreed to give an oral presentation at the Fifth International *Martes* Symposium, convened at the University of Washington in September 2009. It is also important to understand, however, that this book is not the proceedings of the Fifth *Martes* Symposium. Many contributed papers were also presented at the Symposium, but we have limited the content of this book to the subset of invited, synthetic presentations that we identified while planning the symposium.

This book is divided into 5 major topic areas: (1) Evolution and Biogeography of the Genus *Martes*, (2) Biology and Management of *Martes* Populations, (3) Ecology and Management of Habitat for *Martes* Species, (4) Advances in Research Techniques for *Martes* Species, and (5) Conservation of *Martes* Populations. For each of the 5 sections, one of the editors directed an anonymous (at the reviewer's discretion) peer-review process for each chapter that included 2 to 5 referees whom we considered particularly well qualified to review each chapter. In much the same way that scientific journals conduct their review process, all chapters that were provisionally accepted by each editor were then submitted to the lead editor, who conducted the final review and acceptance process, including editorial guidance to improve the clarity and conciseness of each chapter and ensure that various usages and conventions were applied consistently among the chapters. A complete list of the referees who generously provided their time and expertise during the review process for all of the chapters we considered for inclusion in this book is presented in the Acknowledgments.

The genus *Martes* includes 8 formally recognized species, including the American marten (*M. americana*) and fisher (*M. pennanti*) in northern North America; the European pine marten (*M. martes*) and stone (beech, house) marten (*M. foina*) in Europe and south-central Asia; the sable (*M. zibellina*) in northern and eastern Asia; the Japanese marten (*M. melampus*) in Japan and the Korean Peninsula; the yellow-throated marten (*M. flavigula*) in southeast Asia; and the Nilgiri marten (*M. gwatkinsii*) in southern India. However, based on the detailed evaluation of genetic and morphometric variation in North American martens presented by Natalie Dawson and Joseph Cook in chapter 2, and previous work by other researchers, we also recognize (and urge taxonomic authorities to recognize) a ninth species in the genus *Martes*—the Pacific marten (*M. caurina*) of the western United States and Canada. Because Dawson and Cook's proposed taxonomic revision had not yet been evaluated critically by mammalian taxonomists, we did not require the authors of other chapters in this book to follow this taxonomy. Consequently, in some chapters, this taxon is given specific or subspecific status, whereas in others, it is considered a unique evolutionary clade whose taxonomic status remains uncertain. We hope that the publication of this book will help to resolve this long-standing controversy.

Despite important advances in our knowledge of the genus *Martes* since 1994, many significant knowledge gaps remain. For example, although it has been noted repeatedly that little was known about the yellow-throated marten in Southeast Asia, the Japanese marten in Japan and Korea, and the Nilgiri marten in southern India, the scarcity of reliable scientific information on all 3 of these species remains problematic. Also, in the future, much new research will be needed on the effects of climate change, human development, and industrial activities on the distribution and persistence of *Martes* populations. Data are also lacking on the population dynamics of *Martes* species in many habitat conditions, and on the factors that drive habitat selection at both very small (microhabitat) and large (landscape) scales. In general, much more research is needed to develop conservation programs that will maintain sustainable populations and habitats for martens, sables, and fishers.

We strongly believe that the *Martes* Working Group will continue to play a major role in bringing together wildlife researchers and managers from around the world to share information and identify new areas of research. Because martens, sables, and fishers live in sympatry with species that are taxonomically and ecologically similar (e.g., tayras and wolverines) or in competition for the same resources (e.g., genets), it will also be important for future research to include consideration of other carnivores, whose adaptations and life history strategies may help us to understand why *Martes* species behave as they do.

To produce this book, we brought together 62 scientists from 12 countries, who reviewed and synthesized thousands of scientific publications to produce the 20 syntheses included here. We sincerely hope that we have met our objective of providing wildlife biologists, resource managers, and policy makers with a comprehensive review of the biology and conservation of the genus *Martes* that can provide the scientific basis for efforts designed to maintain or enhance their populations and their habitats throughout the world.

Keith B. Aubry
William J. Zielinski
Martin G. Raphael
Gilbert Proulx
Steven W. Buskirk

Acknowledgments

We thank the many referees who participated in the peer-review process and generously shared their knowledge and expertise on the genus *Martes:*

Mike Badry, Wildlife Branch, British Columbia Ministry of Environment, Canada; **Eric Beever**, Northern Rocky Mountain Science Center, USDI U.S. Geological Survey, Montana, USA; **Jeff Bowman**, Wildlife Research and Development Section, Ontario Ministry of Natural Resources, Canada; **Mark Boyce**, Department of Biological Sciences, University of Alberta, Canada; **Carlos Carroll**, Klamath Center for Conservation Research, California, USA; **Fraser Corbould**, Peace/Williston Fish and Wildlife Compensation Program, British Columbia, Canada; **Jeff Dunk**, Department of Environmental Science and Management, Humboldt State University, California, USA; **Jennifer Frey**, Department of Fish, Wildlife, and Conservation Ecology, New Mexico State University, USA; **Kurt Galbreath**, Department of Biology, Western Washington University, USA; **Jon Gilbert**, Great Lakes Indian Fish and Wildlife Commission, Wisconsin, USA; **Russ Graham**, College of Earth and Mineral Sciences, Pennsylvania State University, USA; **Brad Griffith**, Alaska Cooperative Fish and Wildlife Research Unit, USDI U.S. Geological Survey, USA; **Dan Harrison**, Department of Wildlife Ecology, University of Maine, USA; **Kristofer Helgen**, National Museum of Natural History, Smithsonian Institution, Washington, D.C., USA; **Jan Herr**, Department of Biology and Environmental Science, University of Sussex, UK; **David Jachowski**, Department of Fisheries and Wildlife Sciences, University of Missouri, USA; **Kurt Jenkins**, Forest and Rangeland Ecosystem Science Center, USDI U.S. Geological Survey, Washington, USA; **Dave Jessup**, Marine Wildlife Veterinary Care and Research Center, California Department of Fish and Game, USA; **Doug Kelt**, Department of Wildlife, Fish, and Conservation Biology, University of California at Davis, USA; **Bill Krohn**, Maine Cooperative Fish and Wildlife Research Unit, USDI U.S. Geological Survey, USA (retired);

Chris Kyle, Department of Wildlife Genetics and Forensics, Trent University, Ontario, Canada; **John Litvaitis**, Department of Natural Resources and the Environment, University of New Hampshire, USA; **Robert Long**, Western Transportation Institute, Montana State University, USA; **Diane Macfarlane**, Pacific Southwest Region, USDA Forest Service, California, USA; **Pat Manley**, Pacific Southwest Research Station, USDA Forest Service, California, USA; **Andrew McAdam**, Department of Integrative Biology, University of Guelph, Ontario, Canada; **Kevin McKelvey**, Rocky Mountain Research Station, USDA Forest Service, Montana, USA; **Josh Millspaugh**, Department of Fisheries and Wildlife, University of Missouri, USA; **Garth Mowat**, British Columbia Ministry of Environment, Canada; **Barry Noon**, Department of Fish, Wildlife and Conservation Biology, Colorado State University, USA; **Allan O'Connell**, Patuxent Wildlife Research Center, USDI U.S. Geological Survey, Maryland, USA; **Kim Poole**, Aurora Wildlife Research, British Columbia, Canada; **Roger Powell**, Department of Biology, North Carolina State University, USA (retired); **Richard Reading**, Denver Zoological Foundation, Colorado, USA; **DeeAnn Reeder**, Biology Department, Bucknell University, Pennsylvania, USA; **Alexis Ribas Salvador**, Department of Health Microbiology and Parasitology, University of Barcelona, Spain; **Maria Santos**, Graduate Group in Ecology, University of California at Davis, USA; **Mike Schwartz**, Rocky Mountain Research Station, USDA Forest Service, Montana, USA; **Philip Seddon**, Department of Zoology, University of Otago, New Zealand; **Winston Smith**, Pacific Northwest Research Station, USDA Forest Service, California, USA (retired); **Wayne Spencer**, Conservation Biology Institute, California, USA; **Karen Stone**, Department of Biology, Southern Oregon University, USA; **Craig Thompson**, Pacific Southwest Research Station, USDA Forest Service, California, USA; **Ian Thompson**, Great Lakes Forestry Centre, Canadian Forest Service, Ontario, Canada; **Christina Vojta**, USDI U.S. Fish and Wildlife Service, Southwest Region, Arizona, USA; **Eric Walteri**, City College, City University of New York, USA; **Rich Weir**, Artemis Wildlife Consultants, British Columbia, Canada; **John Whitaker, Jr.**, Department of Biology, Indiana State University, USA; **Izabela Wierzbowska**, Institute of Environmental Sciences, Jagiellonian University, Poland; **John Withey**, School of Forest Resources, University of Washington, USA; **Mieczyslaw Wolsan**, Department of Paleozoology, Polish Academy of Sciences, Poland.

Financial support for the preparation and publication of this book was provided by Miranda Mockrin, Peter Stine, and Carlos Rodriguez-Franco of the USDA Forest Service, Research and Development, Washington, D.C., USA; John Laurence of the USDA Forest Service, Pacific Northwest Research Station, Land and Watershed Management Program, Portland, Oregon, USA; and Jamie Barbour of the USDA Forest Service, Pacific Northwest Research Station, Focused Science Delivery Program, Portland, Oregon, USA.

We extend special thanks to Cathy Raley, Yasmeen Sands, and Sandra Maverick of the USDA Forest Service, Pacific Northwest Research Station, Olympia, Washington, USA for the invaluable help they provided to the editors of this book. Cathy spent countless hours working directly with each author to resolve a multitude of issues relating to the figures, tables, and text; preparing the final version of each chapter for submission to Cornell University Press; and compiling the comprehensive literature cited section. Yasmeen prepared the index, and Sandra proofed the literature citations in each chapter.

Finally, we gratefully acknowledge the Pacific Northwest and Pacific Southwest Research Stations of the USDA Forest Service, Alpha Wildlife Research & Management, Ltd., and the University of Wyoming for their contributions to the development and publication of this book.

Contributing Authors

Alexei V. Abramov, Zoological Institute, Russian Academy of Sciences, St. Petersburg, Russia; e-mail: a.abramov@mail.ru and abramov@zin.ru

Jon M. Arnemo, Department of Forestry and Wildlife Management, Hedmark University College, Campus Evenstad, NO-2480 Koppang, Norway; e-mail: jon.arnemo@hihm.no

James A. Baldwin, USDA Forest Service, Pacific Southwest Research Station, P.O. Box 245, Berkeley, California 94701, USA; e-mail: jbaldwin@fs.fed.us

Jeff Bowman, Ontario Ministry of Natural Resources, Wildlife Research and Development Section, 2140 East Bank Drive, Peterborough, Ontario, K9J 7B8, Canada; e-mail: Jeff.Bowman@ontario.ca

Scott M. Brainerd, Alaska Department of Fish and Game, Division of Wildlife Conservation, 1300 College Road, Fairbanks, Alaska 99701, USA; and Norwegian University of Life Sciences, Department of Ecology and Natural Resource Management, P.O. Box 5003, NO-1432, Ås, Norway; e-mail: scott.brainerd@alaska.gov

Richard N. Brown, Humboldt State University, Wildlife Department, 1 Harpst Street, Arcata, California 95521, USA; e-mail: Richard.Brown@humboldt.edu

Steven W. Buskirk, Department of Zoology and Physiology, University of Wyoming, Laramie, Wyoming 82071, USA; e-mail: marten@uwyo.edu

Carlos Carroll, Klamath Center for Conservation Research, P.O. Box 104, Orleans, California 95556, USA; e-mail: carlos@klamathconservation.org

Joseph A. Cook, Department of Biology and Museum of Southwestern Biology, University of New Mexico, Albuquerque, New Mexico 87131, USA; e-mail: cookjose@unm.edu

Samuel A. Cushman, USDA Forest Service, Rocky Mountain Research Station, 2500 S. Pine Knoll Drive, Flagstaff, Arizona 86001, USA; e-mail: scushman@fs.fed.us

Natalie G. Dawson, Wilderness Institute, College of Forestry and Conservation, University of Montana, Missoula, Montana 59812, USA; e-mail: natalie.dawson@umontana.edu

John Fryxell, Department of Integrative Biology, University of Guelph, Guelph, Ontario, N1G 2W1, Canada; e-mail: jfryxell@uoguelph.ca

Mourad W. Gabriel, Integral Ecology Research Center, 102 Larson Heights Road, McKinleyville, California 95519, USA; and Department of Veterinary Genetics, University of California, Davis, California 95616, USA; e-mail: mwgabriel@ucdavis.edu

Jonathan H. Gilbert, Great Lakes Indian Fish and Wildlife Commission, P.O. Box 9, Odanah, Wisconsin 54806, USA; e-mail: jgilbert@glifwc.org

Evan H. Girvetz, The Nature Conservancy, 1917 First Avenue, Seattle, Washington 98101, USA; and School of Forest Resources, University of Washington, Box 352100, Seattle, Washington 98195, USA; e-mail: egirvetz@tnc.org

Rebecca A. Green, USDA Forest Service, Pacific Southwest Research Station, 2081 E. Sierra Avenue, Fresno, California 93710, USA; e-mail: regreen@ucdavis.edu

Daniel J. Harrison, Department of Wildlife Ecology, University of Maine, 5755 Nutting Hall, Orono, Maine 04469, USA; e-mail: harrison@maine.edu

J. Mark Higley, Hoopa Tribal Forestry, P.O. Box 368, Hoopa, California 95546, USA; e-mail: mhigley@hoopa-nsn.gov

Eric P. Hoberg, USDA Agricultural Research Service, U.S. National Parasite Collection, Animal Parasitic Diseases Laboratory, BARC East 1180, 10300 Baltimore Avenue, Beltsville, Maryland 20705, USA; e-mail: Eric.Hoberg@ars.usda.gov

Susan S. Hughes, Pacific Northwest National Laboratory, P.O. Box 999, MSIN K6–75, Richland, Washington 99352, USA; e-mail: Susan.Hughes@pnnl.gov

Neil R. Jordan, The Vincent Wildlife Trust, Eastnor, Ledbury, Herefordshire, UK; e-mail: enquiries@vwt.org.uk

Anson V.A. Koehler, Department of Zoology, University of Otago, P.O. Box 56, Dunedin, New Zealand; e-mail: anson76@gmail.com

William B. Krohn, USDI U.S. Geological Survey, Maine Cooperative Fish and Wildlife Research Unit, 5755 Nutting Hall, University of Maine, Orono, Maine 04469, USA (retired); e-mail: wkrohn@maine.edu

Joshua J. Lawler, School of Forest Resources, University of Washington, Box 352100, Seattle, Washington 98195, USA; e-mail: jlawler@u.washington.edu

Jeffrey C. Lewis, Washington Department of Fish and Wildlife, 600 Capitol Way N., Olympia, Washington 98501, USA; e-mail: lewisjcl@dfw.wa.gov

Eric C. Lofroth, British Columbia Ministry of Environment, P.O. Box 9358, 395 Waterfront Crescent, Victoria, British Columbia, V8W 9M1, Canada; e-mail: Eric.Lofroth@gov.bc.ca

Robert A. Long, Western Transportation Institute, Montana State University, 420 North Pearl Street, Suite 305, Ellensburg, Washington 98926, USA; e-mail: robert.long@coe.montana.edu

Paula MacKay, Western Transportation Institute, Montana State University, 420 North Pearl Street, Suite 305, Ellensburg, Washington 98926, USA; e-mail: paula.mackay@coe.montana.edu

Bruce G. Marcot, USDA Forest Service, Pacific Northwest Research Station, 620 S.E. Main Street, Suite 400, Portland, Oregon 97205, USA; e-mail: brucem@SpiritOne.com

Ryuichi Masuda, Department of Natural History Sciences, Faculty of Science, Hokkaido University, Sapporo, 060–0810, Japan; e-mail: masudary@ees.hokudai.ac.jp

Marina Mergey, Centre de Recherche et de Formation en Eco-éthologie, Université de Reims Champagne-Ardenne, 5 rue de la héronnière, 08240 Boult-aux-Bois, France; e-mail: marina.mergey@cerfe.com

Vladimir Monakhov, Institute of Plant and Animal Ecology, Russian Academy of Sciences, 8 Marta Street, Yekaterinburg, Russia; e-mail: monv@mail.ru

Takahiro Murakami, Shiretoko Museum, 49–2 Hon-machi, Shari-cho, Shari-gun, Hokkaido, Japan 099–4113; e-mail: murakami.ta@town.shari.hokkaido.jp

Anne-Mari Mustonen, University of Eastern Finland, Faculty of Science and Forestry, Department of Biology, P.O. Box 111, FI-80101, Joensuu, Finland; e-mail: Anne-Mari.Mustonen@uef.fi

Petteri Nieminen, University of Eastern Finland, Faculty of Science and Forestry, Department of Biology, and Faculty of Health Sciences, School of Medicine, Institute of Biomedicine/Anatomy, P.O. Box 111, FI-80101, Joensuu, Finland; e-mail: Petteri.Nieminen@uef.fi

Cino Pertoldi, Department of Biological Sciences, Ecology and Genetics, Aarhus University, Ny Munkegade, Building 1540, 8000 Århus C, Denmark; and Mammal Research Institute, Polish Academy of Sciences, Waszkiewicza 1c, 17–230 8 Białowieża, Poland; e-mail: dcino.pertoldi@biology.au.dk

Roger A. Powell, Department of Biology, North Carolina State University, Raleigh, North Carolina 55731, USA; e-mail: newf@ncsu.edu

Gilbert Proulx, Alpha Wildlife Research & Management Ltd., 229 Lilac Terrace, Sherwood Park, Alberta, T8H 1W3, Canada; e-mail: gproulx@alphawildlife.ca

Kathryn L. Purcell, USDA Forest Service, Pacific Southwest Research Station, 2081 E. Sierra Avenue, Fresno, California 93710, USA; e-mail: kpurcell@fs.fed.us

Catherine M. Raley, USDA Forest Service, Pacific Northwest Research Station, 3625 93rd Avenue SW, Olympia, Washington 98512, USA; e-mail: craley@fs.fed.us

Martin G. Raphael, USDA Forest Service, Pacific Northwest Research Station, 3625 93rd Avenue SW, Olympia, Washington 98512, USA; e-mail: mraphael@fs.fed.us

Luis M. Rosalino, Centro de Biologia Ambiental, Universidade de Lisboa, Faculdade de Ciências de Lisboa, Ed. C-2, 1749–016, Lisboa, Portugal; e-mail: lmrosalino@fc.ul.pt

Aritz Ruiz-González, Department of Zoology and Animal Cell Biology, Zoology Laboratory, Facultad de Farmacia, Universidad del País Vasco-Euskal Herriko Unibertsitatea, Paseo de la Universidad, 7, 01006 Vitoria-Gasteiz, Spain; e-mail: martes_martes99@yahoo.es

Hugh D. Safford, USDA Forest Service, Pacific Southwest Region, 1323 Club Drive, Vallejo, California 94592, USA; and Department of Environmental Science and Policy, University of California, Davis, California 95616, USA; e-mail: hughsafford@fs.fed.us

Margarida Santos-Reis, Universidade de Lisboa, Centro de Biologia Ambiental, Faculdade de Ciências, Campo Grande, Bloco C2–5° Piso, 1749–016, Lisboa, Portugal; e-mail: mmreis@fc.ul.pt

Joel Sauder, Idaho Department of Fish and Game, 3316 16th Street, Lewiston, Idaho 83501, USA; e-mail: joel.sauder@idfg.idaho.gov

Michael K. Schwartz, USDA Forest Service, Rocky Mountain Research Station, 800 E. Beckwith Avenue, Missoula, Montana 59801, USA; e-mail: mkschwartz@fs.fed.us

Andrew J. Shirk, University of Washington Climate Impacts Group, Box 355672, Seattle, Washington 98195, USA; e-mail: ashirk@uw.edu

Keith M. Slauson, USDA Forest Service, Pacific Southwest Research Station, 1700 Bayview Drive, Arcata, California 95521, USA; e-mail: kslauson@fs.fed.us

Brian G. Slough, 35 Cronkite Road, Whitehorse, Yukon Territory, YIA 2C6, Canada; e-mail: slough@northwestel.net

Wayne D. Spencer, Conservation Biology Institute, 815 Madison Avenue, San Diego, California 92116, USA; e-mail: wdspencer@consbio.org

Richard A. Sweitzer, Department of Environmental Science, Policy, and Management, University of California, Berkeley, California 94720, USA; e-mail: rasweitzer@berkeley.edu

Craig M. Thompson, USDA Forest Service, Pacific Southwest Research Station, 2081 E. Sierra Avenue, Fresno, California 93710, USA; e-mail: cthompson@fs.fed.us

Ian D. Thompson, Canadian Forest Service, 1219 Queen Street East, Sault Ste. Marie, Ontario, P6A 2E5, Canada; e-mail: Ian.Thompson@NRCan-RNCan.gc.ca

Richard L. Truex, USDA Forest Service, Rocky Mountain Region, 740 Simms Street, Golden, Colorado 80401, USA; e-mail: rtruex@fs.fed.us

Emilio Virgós, Departamento de Biología y Geología, Universidad Rey Juan Carlos, C/Tulipán s/n, 28933 Móstoles, Madrid, Spain; e-mail: emilio.virgos@urjc.es

Tzeidle N. Wasserman, School of Forestry, Northern Arizona University, Flagstaff, Arizona 86001, USA; e-mail: moonhowlin@yahoo.com

Greta M. Wengert, Integral Ecology Research Center, 102 Larson Heights Road, McKinleyville, California 95519, USA; and Department of Veterinary Genetics, University of California, Davis, California, 95616, USA; e-mail: gmwengert@ucdavis.edu

J. Scott Yaeger, USDI U.S. Fish and Wildlife Service, 1829 S. Oregon Street, Yreka, California 96097, USA; e-mail: Scott_Yaeger@fws.gov

Andrzej Zalewski, Mammal Research Institute, Polish Academy of Sciences, 17–230, Białowieża, Poland; e-mail: zalewski@zbs.bialowieza.pl

William J. Zielinski, USDA Forest Service, Pacific Southwest Research Station, 1700 Bayview Drive, Arcata, California 95521, USA; e-mail: bzielinski@fs.fed.us

Patrick A. Zollner, Department of Forestry and Natural Resources, Purdue University, West Lafayette, Indiana 47907, USA; e-mail: pzollner@purdue.edu

SECTION 1

Evolution and Biogeography of the Genus *Martes*

1

Synthesis of *Martes* Evolutionary History

SUSAN S. HUGHES

ABSTRACT

In this chapter, I synthesize recent information on the evolutionary history and biogeography of *Martes,* drawing on 4 lines of evidence: the fossil record, genetic analyses, *Martes* adaptations, and paleoclimatic information. Although genetic analyses generally support the fossil record, they have revised our understanding of some of the taxonomic relationships among this group. For example, *Gulo gulo* (the wolverine) and *Eira barbara* (the tayra) are more closely related to *Martes* than previously thought, and they should be included in the same lineage. *Martes lydekkeri* is likely not a direct ancestor of *M. flavigula,* as some researchers have suggested, nor is *M. vetus* an ancestor of *M. foina. Martes americana* and *M. melampus* probably split before *M. zibellina* arose, and diversification of the 2 North American marten lineages (*M. americana* and *M. caurina*) occurred after its arrival in North America. The first members of the genus appear to have evolved in western Eurasia in the middle Miocene and colonized North America repeatedly, although the last 2 dispersals, dated at 1.8 and 1.0 million years ago, included *M. pennanti,* as well as the ancestral form of *M. americana* and *M. caurina*; *G. gulo* also colonized North America during the Pleistocene. Initial diversification within the genus likely occurred in southeast Asia, and diversification of the true martens was strongly influenced by glacial events in the Pliocene and Pleistocene that created barriers to gene flow. Members of the genus also show greater adaptive plasticity than previously thought; their adaptation to boreal forest environments occurred late in their evolutionary history. Although genetic studies have further refined our knowledge of the phylogeny and evolutionary history of the genus, significant gaps remain that can be resolved only with a better understanding of the fossil record.

Table 1.1. Taxonomy of extant species in the genera *Martes, Eira, and Gulo* based on Wilson and Reeder (2005) and Nowak (1999)

Genus	Subgenus	Species	Distribution
Martes	*Pekania*	*M. pennanti*	Northern North America
	Charronia	*M. flavigula*	Asia
		M. gwatkinsii	Southern India
	Martes	*M. americana*	Northern North America
		M. foina	Europe/southwest Asia
		M. martes	Europe/western Palearctic
		M. melampus	Japan/Korea
		M. zibellina	Eastern Palearctic/northern Japan
Eira		*E. barbara*	South America
Gulo		*G. gulo*	Palearctic

Introduction

Much new information has come to light since Anderson (1994) reviewed *Martes* evolution, biogeography, and systematics. At that time, reconstructions of *Martes* phylogeny were based primarily on comparative studies of living and fossil taxa. Today, genetic analyses and more-nuanced studies of the behavior and physiology of *Martes* have provided new insights into this topic. In this chapter, I synthesize recent information on the phylogeny and evolutionary history of the genus, drawing on 4 lines of evidence: the fossil record, genetic analyses, *Martes* adaptations, and paleoclimatic information. I review and evaluate the *Martes* fossil record in light of new genetic data and discuss the evolutionary history and biogeography of *Martes* based on evolutionary adaptations and paleoclimatic data. I use Wilson and Reeder's (2005) taxonomic classification of *Martes* in this paper with the exception that the genus is subdivided into 3 subgenera: *Pekania* (fishers), *Charonnia* (yellow-throated martens), and *Martes* (true martens), following the classification of Anderson (1970, 1994) and others (Nowak 1999; Table 1.1). Where appropriate, I have also included *Gulo gulo* (the wolverine) and *Eira barbara* (the tayra) in my synthesis because of their close, and possibly congeneric, relationships with *Martes* (Koepfli et al. 2008).

What's New in the Fossil Record

A phylogenetic reconstruction of the genus *Martes* has not been attempted since Anderson's (1970, 1994) comparative studies of fossil and recent skeletal morphology. One notable feature of her phylogeny is how few fossil taxa were assigned to *Martes* ancestry (Anderson 1970, 1994): *M. lydekkeri*, as a possible ancestor of *M. flavigula*; *M. laevidens*, *M. wenzensis*, and *M. vetus* as probable ancestors of true martens; and *M. paleosinensis* and *M. diluviana* as possible ancestors of *M. pennanti* (Figure 1.1). The list became even smaller,

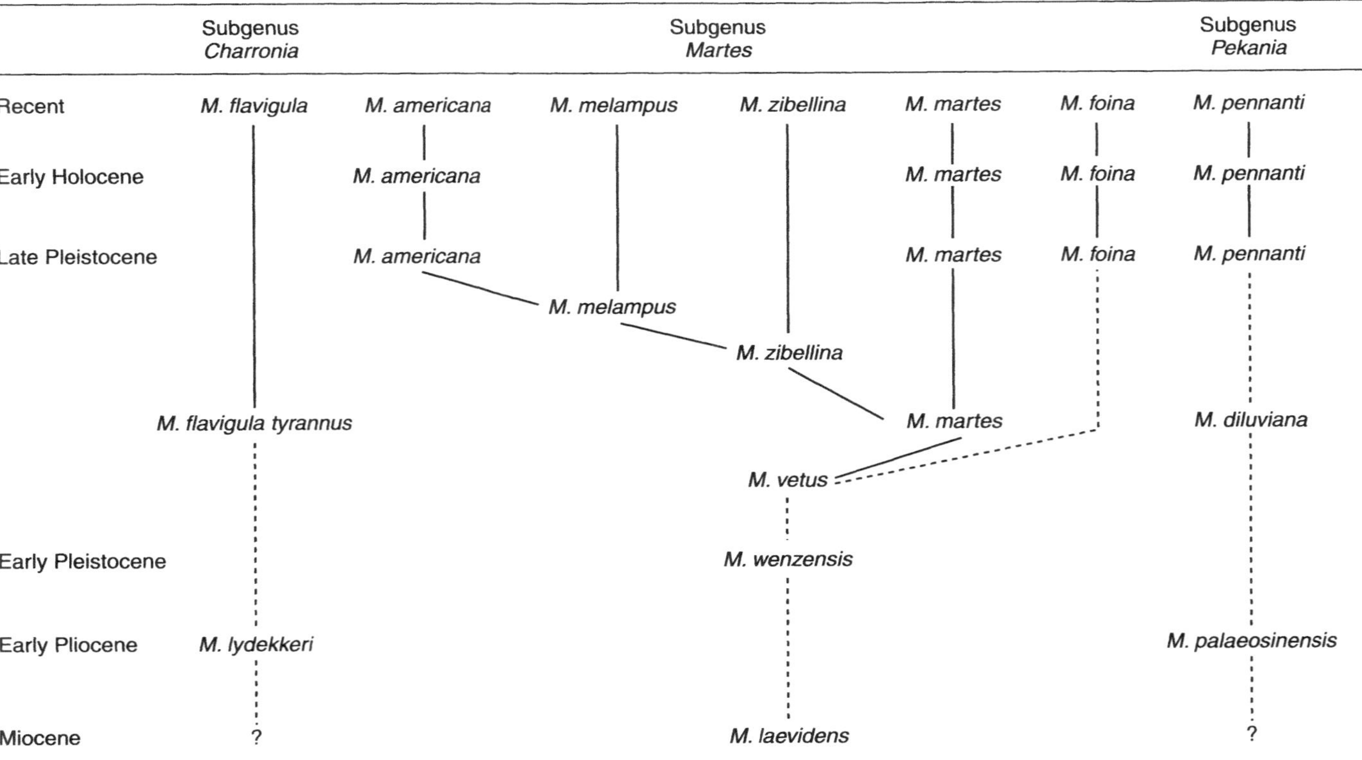

Figure 1.1. Phylogeny of the genus *Martes* based on skeletal morphology (from Anderson 1970, 1994). *Martes laevidens* has since been removed from the lineage (Sato et al. 2003). A solid line indicates ancestry; a dashed line, probable ancestry.

however, when Sato et al. (2003) demonstrated that *M. laevidens* is not a true marten because its suprameatal fossa is not completely ossified, a condition that predates the divergence of *Martes* (see Wolsan 1993a and 1999 for discussions of this feature).

The small number of *Martes* fossil ancestors can be partially explained by characteristics of the species and its recent evolutionary history. Skeletal material from small arboreal carnivores does not preserve well in the fossil record (Baskin 1998); the morphological diversity that characterizes extant species of *Martes* likely existed in the distant past, making it more difficult to assign ancestry (Wolsan et al. 1985; Wolsan 1988, 1989), and morphological differences are small within the genus because of its brief evolutionary history. Although the fossil record is limited, Anderson did note that other "*Martes*-like" fossils exist but are either poorly described in the literature, not accessible for study, or not sufficiently diagnostic to be included in the *Martes* family tree (Anderson 1970, 1994).

A review of the fossil literature today shows little change from these interpretations. I queried 4 online fossil databases to compile a list of identified *Martes* fossil remains throughout the world—the Neogene of the Old World (NOW) database at the University of Finland (Fortelius 2009), the online Paleobiology database (http://www.paleodb.org/cgi-bin/bridge.pl), the MIOMAP database at the University of California at Berkeley (Carrasco et al. 2005), and the FAUNMAP database at the Illinois State Museum (Faunmap Working Group 1994). Altogether, 36 extinct fossil species from the Neogene (Miocene and Pliocene) exist in these databases, of which only 18 appear to be recognized as valid species today (Table 1.2). Extant species emerge in the Pleistocene and post-Pleistocene. Recent trends in the literature indicate that some of the older fossil material has been reassigned and some new species, such as *M. khelifensis, M. lefkonensis,* and *M. nambianus*, have been identified. Most problematic is that much of the fossil material has not been compared with living forms (Hunt 1996), and the taxonomic position of most of this material is undetermined. Some have even claimed that the genus has become a "waste basket" for mustelids that are difficult to identify (Kaya et al. 2005). In-depth comparative study and a taxonomic revision of the fossil material are clearly needed.

In lieu of such a taxonomic revision, the "valid" taxa listed in Table 1.2 must serve as a proxy for evolutionary patterns in *Martes* ancestry. Plotting these taxa through time by geographic region (Figure 1.2) reveals that the oldest fossils assigned to *Martes* appear about 18 million years before present (myBP), and extend throughout the Miocene (23.0–5.3 myBP), Pliocene (5.3–2.6 myBP), and Pleistocene (2.6–0.012 myBP). The plot in Figure 1.2 is time-transgressive throughout Eurasia, with the oldest taxa appearing in the circum-Mediterranean region and central Europe. By the middle to late Miocene, *Martes*-like taxa appear in India and, slightly later, in China. Early Eu-

Table 1.2. Fossil records assigned to *Martes* dating from the Early Miocene to Late Pleistocene

Taxa	Country	Region[a]	Number of localities	Dates (myBP)	References	Comments	Valid taxa
M. andersonii	China	—	1	—	NOW; Anderson 1970	Conspecific with *paleosinensis* (Anderson 1970)	—
M. basilii	Spain	CM	3	8.2–7.1	NOW; Petter 1965; Roussiakis 2002	—	X
M. burdigaliensis	France, Spain	CE, CM	4	18.0–17.0	NOW; Ginsburg 2002; Azanza et al. 2004; Peigne et al. 2006	—	X
	Turkey	CM	1	11.2–9.0			
M. cadeoti	France	CE	1	18.2–15.2	NOW; Peigne et al. 2006	—	X
M. campestris	USA, South Dakota	—	?	Early Pliocene	Anderson 1970; Baskin 2005	Reassigned to *Sthenictis* (Baskin 2005:434)	—
M. collongensis	France	CE	1	18.0–15.2	NOW; Guerin and Mein 1971; Ginsburg 1999	—	X
M. crassa	China				Anderson 1970	Not *Martes*	—
M. delphinensis	France	CE	2	12.5–11.2; 18.0–15.2	NOW; Trouessart 1893; Peigne et al. 2006; Nagel et al. 2009	—	X
	Spain	CM	1	12.5–11.2			
M. filholi	Armenia	CE	1	3.0–1.8	NOW; Trouessart 1893; Anderson 1970; Peigne et al. 2006	Some identifications uncertain	X
	France	CE	3	12.5–11.2; 18.0–15.2			
	Germany	CE	1	12.5–11.2			
	Austria	CE	1	17.0–15.2			
M. foxi	USA, Kansas	NA	1	4.9–1.8	PB; Hibbard and Riggs 1949	—	X
M. gazini	USA, Oregon	—	1	Late Tertiary	PB; Hall 1931; Anderson 1970	Reassigned to *Plionictis* (Anderson 1970)	—
M. jaegari	Germany	—	1	11.2–9.5	NOW	No information	—

Table 1.2—*cont.*

Taxa	Country	Region[a]	Number of localities	Dates (myBP)	References	Comments	Valid taxa
M. khelifensis	Morocco	CM	1	13.8–11.6	PB; Wolsan et al. 1985	Identified at Beni Mellal	X
M. kinseyi	USA, Montana	—	1	Miocene	PB; Gidley 1927; Gazin 1934; Anderson 1970	Reassigned to *Plionictis* (Gazin 1934)	—
M. laevidens	Czechoslavakia	—	2	20.0–18.0	Anderson 1970; Wolsan 1999	Not *Martes* (Wolsan 1999)	—
	France	—	1	20.0–19.0			
	Germany	—	1	20.0–18.0			
	Spain	—	1	18.0–17.0			
M. lefkonensis	Greece	CM	1	7.1–4.2	NOW; Roussiakis 2002	—	X
M. leporinum	Romania	CE	?	Pontian?	Anderson 1970; Roussiakis 2002	Relationship unknown (Anderson 1970)	X
M. mellibula	Spain	CM	2	9.5–9.0	NOW; Agustí et al. 2003		X
M. munki	France	CE	7	18.0–15.2; 12.5–11.2	NOW; Ginsburg 2001; Peigne et al. 2006; Nagel et al. 2009	Well-known, some identifications uncertain	X
	Germany	CE	2	12.5–11.2; 18.0–17.0			
	Spain	CM	3	9.5; 12.5–11.2; 18.0–17.0			
M. nambianus	USA, New Mexico	NA	1	14.1–12.5	MIOMAP	—	X
M. oregonensis	USA, Oregon	NA	2	6.7–5.9	MIOMAP; Shotwell 1970	—	X
M. pachygnatha	China	—	1	1.8–1.2	NOW; Anderson 1970	Identification uncertain	—
M. parapennanti	USA, Maryland	—	1	—	PB; Anderson 1970	Reassigned to *M. diluviana* (Anderson 1970)	—

M. parviloba	USA, Nebraska, Kansas	—	1	Miocene	PB; Scott and Osborn 1887; Voorhies 1990	Reassigned to *Plionictis* (Matthew 1924 and Galbreath 1953); recombined with no evidence (Voorhies 1990)	—
M. pentelici	Greece, Pikermi	—	1	Pontian	Anderson 1970; Roussiakis 2002	Reassigned to *Sinictis* (Koufos 2006; Nakaya 1994)	—
M. pusilla	Germany, Sandelzhausen	—	1	17.0–15.0	Koufos 2008; Nagel et al. 2009	Reassigned to *Proputorius sansaniensis* (mustelid)	—
M. sainjoni	France, Spain	CE, CM	3	18.0–17.0	NOW; Ginsburg 2002	—	X
M. sansaniensis	Southern France	CM	1	15.2–12.5	NOW; Nakaya 1994; Ginsburg 2001; Peigne et al. 2006; Morlo et al. 2010	—	X
M. sinensis	China	—	1	0.6–0.01	NOW; Wan-bo 1979	No information	—
M. stirtoni	USA, Kansas	NA	1	12.0–10.0	MIOMAP, PB; Wilson 1968; Stevens and Stevens 2003	—	X
M. transitoria	France	—	1	12.5–11.2	NOW	—	—
M. woodwardi	Greece	CM	1	8.2–7.1	NOW; Anderson 1970; Nakaya 1994; Bernor et al. 1996; Roussiakis 2002; Koufos 2008	—	X
M. zdanski	China	—	?	Pliocene	Anderson 1970	Not *Martes* (Anderson 1970)	—

Sources: Data compiled from NOW (Neogene of the Old World; Fortelius 2009), MIOMAP (Carrasco et al. 2005), and Paleobiology (PB) databases
Note: Regions correspond to those in Figure 1.2
[a] CE = central Europe, CM = circum-Mediterranean, NA = North America

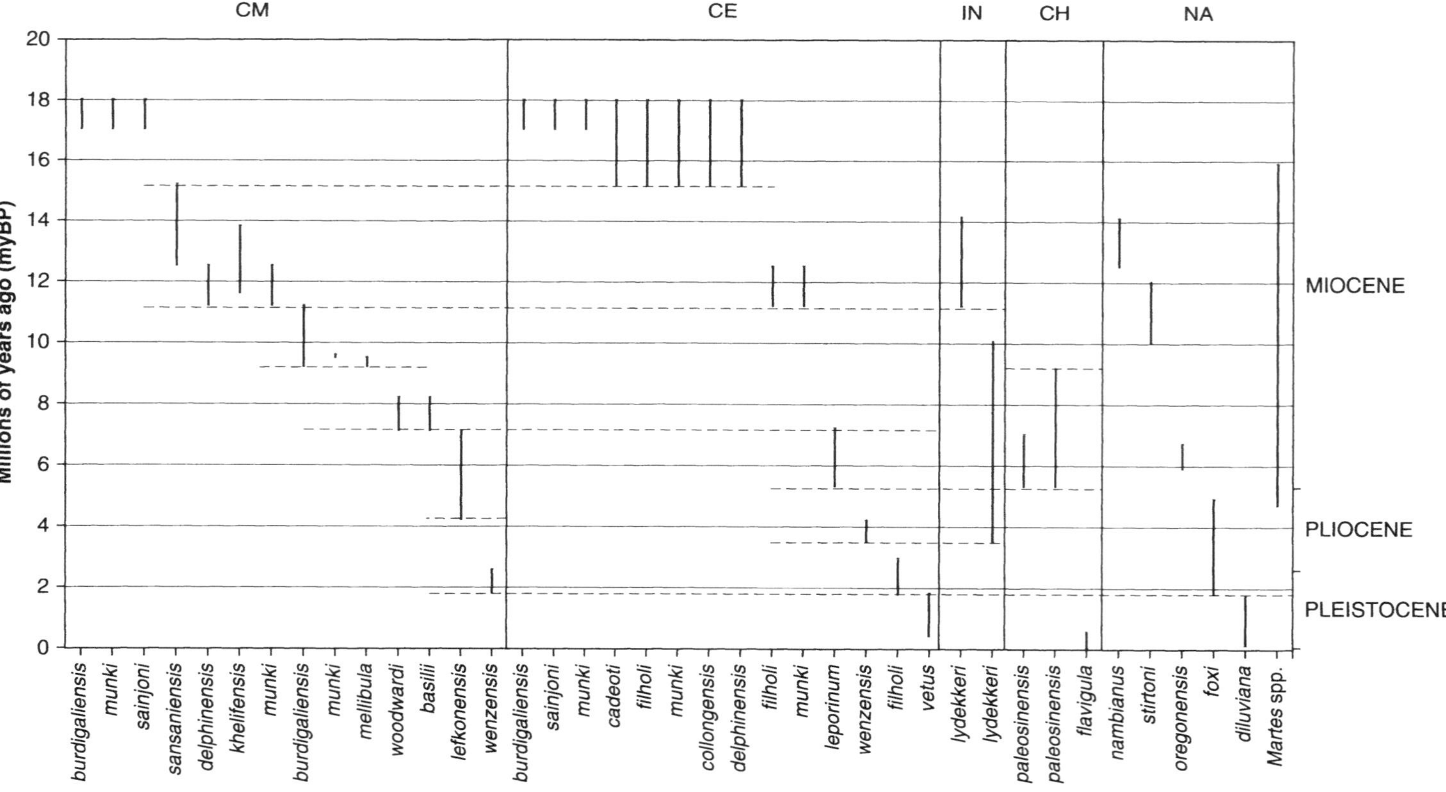

Figure 1.2. Plot of "valid" *Martes*-like fossils through time (myBP) and by geographic region (CM = circum-Mediterranean, CE = central Europe, IN = India, CH = China/Kazakstan, NA = North America). Data are from the databases of NOW (Fortelius 2009), MIOMAP (Carrasco et al. 2005), FAUNMAP (Faunmap Working Group 1994), and Paleobiology (see Table 1.2), and also include fossil taxa shown in Figure 1.1. Dashed horizontal lines mark apparent *Martes* turnover events. Some taxa are duplicated because they appear in different time periods and regions.

ropean *Martes* forms appear in North America by 16.0 myBP, the first of many dispersals from Eurasia to North America (Hunt 1996; Baskin 1998). The time-transgressive pattern of these data suggests that the earliest members of the genus evolved in Europe and the circum-Mediterranean region and dispersed eastward. After a hiatus of about 1 million years, *Martes* reappeared in the circum-Mediterranean region, ca. 14.0–5.0 myBP, and from then on *Martes* is well represented in the fossil record of that region. About 14.0 myBP, *Martes* first appeared in India and China. A turnover event occurred around 11.2 myBP that impacted all Eurasian species. A small turnover event occurred in the circum-Mediterranean region about 9.2 myBP, and after another turnover 7.2 myBP, *Martes* reappeared in central Europe. Turnover events that occurred 5.3 and 3.2 myBP impacted species in central Europe, Asia, and the Far East, whereas a turnover event ca. 4.1 myBP impacted western Eurasia. Both Eurasian and North American *Martes* were impacted by a turnover event that occurred 1.8 myBP.

The fossil record in North America during the middle to late Miocene is incomplete but shows that fossils identified as *Martes* appear as early as 16.0 myBP, and that the genus is well represented from then on. Turnover events in North America do not appear to correspond to turnover events in Eurasia, except the most recent one that occurred 1.8 myBP, and possibly one at the Miocene-Pliocene boundary. Turnovers within North American mustelids have been reported at 18.0–16.0, 10.0–9.0, and 5.0 myBP (Tedford et al. 1987; Baskin 1998) that correspond to turnovers seen in the limited North American *Martes* data shown in Figure 1.2.

Although I believe that Figure 1.2 accurately depicts the broad patterns in *Martes* evolutionary history, and roughly corresponds to other carnivore turnover events, the dataset is problematic in several respects. Dates are not completely accurate because they bracket geological deposits, rather than indicating dates for the fossils themselves; some taxa may be misidentified as *Martes*; and the sample is strongly biased toward western Europe, where most faunal research has been conducted.

The New Genetic Record

Genetic research over the last 20 years has revitalized the study of *Martes* evolutionary history. In addition to identifying phylogenetic relationships, genetic analyses provide date estimates for species divergence and help identify patterns of dispersal. A broad range of genetic studies have been conducted on a variety of topics, including phylogenetic relationships, species distributions, and introduced taxa (e.g., Hosoda et al. 1999; Drew et al. 2003; Kyle and Strobeck 2003; Small et al. 2003; Wisely et al. 2004), as well as broader phylogenetic studies that include all or subsets of the *Martes* taxa, often providing estimated dates of divergence (e.g., Bininda-Emonds et al.

Table 1.3. Divergence-date estimates for extant species in the genera *Martes*, *Eira*, and *Gulo* and subgenera in the genus *Martes* based on genetic data (myBP)

Speciation events	Diversification	Hosoda et al. (2000)	Marmi et al. (2004)	Koepfli et al. (2008)	Sato et al. (2009a)
I.	*Martes/Eira/Gulo* clade split from other mustelids	23.0–15.0	14.0–11.0	11.2	12.0
II.	*E. barbara*	N/A	N/A	7.0	N/A
III.	*Pekania*	14.0–10.0	7.6–6.6	6.8	6.0
	G. gulo	14.0–10.0	8.6–5.8	5.5	6.3
	Charronia	14.0–10.0	6.1–4.6	4.8	5.3
IV.	*M. foina*	8.0–5.0	3.1–2.2	2.8	3.8
V.	*M. americana*	4.0–2.0	1.8–0.9	1.8–1.6	2.8 *(martes)*
	M. melampus	4.0–2.0	1.0	1.8–1.6	1.8 *(americana)*
VI.	*M. zibellina* from *M. martes*; *M. caurina* from *M. americana*	1.0	1.0	1.0	1.0 *(melampus/ zibellina)*

1999; Hosoda et al. 2000; Sato et al. 2003; Marmi et al. 2004; Yonezawa et al. 2007; Koepfli et al. 2008). The reliability of molecular-based phylogenies, and especially divergence-date estimates, depend, among other factors, on the size of the sample, the genomic characters sampled, and the methodologies used. Studies have improved considerably over the years. One important limitation of divergence-date estimates is that they are calibrated to the fossil record and will only be as accurate as our contemporary understanding of that record.

Martes phylogeny and divergence-date estimates from 4 of the more recent and comprehensive genetic studies are compared in Table 1.3. These studies are not in complete agreement because of differences in both the samples and the methodologies used, but are consistent in a number of respects. First, all genetic research shows that *G. gulo* and *E. barbara* are much more closely related to *Martes* than originally thought. *Gulo gulo* represents an early split from the *Martes* lineage that occurred between 8.6 and 5.5 myBP, possibly from an ancestor that gave rise to both the *Pekania* and *G. gulo*, although the precise relationship between these 2 taxa and *E. barbara* needs further resolution (Carr and Hicks 1997; Hosoda et al. 2000; Koepfli et al. 2008). *Gulo gulo* is a direct descendant of *G. schlosseri*, with representatives found in both Europe and North America during the Pleistocene (Pasitschniak-Arts and Larivière 1995). *Eira barbara,* an inhabitant of Central and South American tropical forests, is also closely related to *Martes* and diverged from the lineage about 7.0 myBP (Koepfli et al. 2008). This genus may be represented by Eurasian fossil remains dating to the Pliocene (Presley 2000).

Genetic research also supports a history of punctuated equilibria, whereby *Martes* evolution has undergone periods of radiation and rapid diversification leading to new species (Hunt 1996; Werdelin 1996; Koepfli et al. 2008). The proposed number of radiations varies: Hosoda et al. (2000) recognized 5 radiations, Marmi et al. (2004), 6, and Koepfli et al. (2008), 2. All studies, however, identify 5 basic periods of *Martes* speciation (Table 1.3): (1) the basal divergence of *Martes* from other mustelids; (2) divergence of the 3 *Martes* subgenera, *G. gulo*, and *E. barbara*; (3) divergence of the first true marten, the ancestor of *M. foina*; (4) diversification of the Holarctic true martens; and (5) the recent split between *M. zibellina* and *M. martes* in northern Asia, and the *americana* and *caurina* lineages in North America or, alternatively, an earlier split by *M. americana* and *M. martes*, and a more recent split by *M. melampus* and *M. zibellina* (Sato et al. 2009a; Table 1.3).

Synthesis of the Fossil and Genetic Records

A comparison of the fossil and genetic records shows some discrepancies. Most genetic studies place the rise of the genus around 12.0 myBP, a much more recent date than was previously proposed and more recent than the fossil material assigned to the genus (Figure 1.2). Anderson suggested that the *Martes* lineage diverged sometime between 20.5 and 16.4 myBP; however, only results reported by Hosoda et al. (2000) support Anderson's estimate, which is considered too old (Sato et al. 2003). If the divergence of *Martes* occurred closer to 12.0 myBP, then many of the early *Martes* fossils (i.e., *M. burdigaliensis, M. munki, M. sainjoni, M. sansaniensis, M. collongensis, M. cadeoti, M. filholi, M. delphinensis, M. nambianus,* and the earliest *Martes* spp. in North America) are either misidentified or represent transitional forms (Figure 1.2).

The genetic analysis by Koepfli et al. (2008) is the only one that includes *E. barbara.* Their data show that *Eira* split from the *Martes* lineage, ca. 7 myBP and displays a close genetic relationship to *M. pennanti,* either as a separate clade or as successive lineages sister to a clade containing *G. gulo* and the remaining species of *Martes* (Koepfli et al. 2008).

Genetic studies generally agree on the pattern and timing of divergence of the 3 *Martes* subgenera and *G. gulo*. *Pekania* and *G. gulo* were the first to diverge, although disagreement exists on which split first, *G. gulo* (Marmi et al. 2004; Sato et al. 2009a) or *Pekania* (Koepfli et al. 2008), with dates ranging between 8.6 and 5.5 myBP. The divergence of the *Charronia* followed soon after, with dates centering around 5.0 myBP. The divergence estimates provided by Hosoda et al. (2000) are much older and not in agreement with recent studies (Table 1.3). The split of *Pekania*, ca. 7.0–6.0 myBP, is consistent with fossils of *M. paleosinensis* recovered from strata dated between 9.0 and 5.3 myBP in China (Fortelius 2009), although the genetic dates may slightly underestimate the date of divergence.

If the subgenus *Charronia* split off sometime after 6.1 myBP, as the genetic divergence dates suggest, then fossil remains of *M. lydekkeri* are too old to be in the direct line of *Charronia*, as Anderson (1970) suggested. *Martes lydekkeri* was first recovered from the Chinji Formation in northern India (Colbert 1935), which dates between 14.0 and 11.0 myBP (Behrensmeyer and Barry 2005). Recently, another specimen was identified in faunal assemblages from the Dhok Pathan Formation dating to the late Miocene (Ghaffar et al. 2004), revealing that this was a long-lived taxon in south-central Asia. This fossil taxon, positioned between the East and the West, may be an important key to understanding *Martes* evolutionary history.

Genetic studies generally agree on the divergence of true martens from other *Martes* between 4.0 and 3.0 myBP (Table 1.3). The earliest records of *M. foina* occur in Palestine and Iraq and date to the late Pleistocene (Fortelius et al. 2006). The genetic divergence dates are supported by the earliest fossil representative of true martens, *M. wenzensis*, which dates between 4.0 and 3.3 myBP in Poland (Czyzewska 1981; Wolsan 1988; Sato et al. 2003). This taxon may be ancestral to both *M. foina* and more recent true martens. The distribution of *M. wenzensis* suggests that true martens first evolved in central Europe and later dispersed to the Near East. Anderson (1970, 1994) and Kurtén (1968) attribute the ancestry of true martens to a more recent fossil, *M. vetus*, appearing in European faunal assemblages dating from 1.8 to 0.4 myBP. Anderson (1994) suggested that *M. vetus* was the morphological and ecological forerunner of the forest-adapted *M. martes* because of its geographic distribution. The genetic divergence dates indicate that *M. vetus* cannot be the ancestor of true martens, because this taxon appears too recently in the fossil record.

Both fossil and genetic data show a rapid diversification of the Holarctic true martens, *M. martes, M. melampus, M. zibellina*, and *M. americana* (Figure 1.1) around 2.0–1.8 myBP, that coincides with the appearance of *M. vetus*. The morphological data suggest a divergence sequence starting with *M. martes*, followed by *M. zibellina*, *M. melampus*, and *M. americana* (Anderson 1970). Anderson (1970) proposed that *M. melampus* split from *M. zibellina* after expansion into China.

The genetic phylogeny of the Holarctic true martens is not yet resolved because of the similarity of their genomes (Fulton and Strobeck 2006). Most studies show the early divergence of *M. americana* (Hosoda et al. 1997; Marmi et al. 2004) and a concurrent or slightly more recent divergence of *M. melampus* (Hosoda et al. 2000; Koepfli et al. 2008). In contrast, Sato et al. (2009a) show an early split of *M. martes*, followed by *M. americana*. Most studies place the divergence of *M. zibellina* last, either as a sister species of *M. melampus* (Sato et al. 2009a) or, more commonly, as a sister species of *M. martes* (Carr and Hicks 1997; Hosoda et al. 2000; Sato et al. 2003; Koepfli et al. 2008).

Genetic divergence-date estimates for the first Holarctic true martens range from 2.8 to 1.8 myBP, with the final split occurring 1.0 myBP (Table 1.3).

Only the results of Sato et al. (2009a) support Anderson's (1970) suggestion that *M. melampus* branched from *M. zibellina*; others point to a more recent split of *M. zibellina* from *M. martes* (Table 1.3).

The genetic data also support the long-held belief that both *M. pennanti* and *M. americana* colonized North America from Eurasia across the Bering Land Bridge. The approximate date for the arrival of *M. pennanti*, based on genetic data, is 1.8 myBP, which corresponds to the appearance of *M. diluviana*, the ancestor of *M. pennanti*, in at least 25 Irvingtonian fossil assemblages in the United States (Anderson 1970, 1994; see Graham and Graham 1994 for a list of fossils).

Both fossil and genetic data demonstrate that true martens colonized North America around 1.0 myBP. Fossils appear in a number of assemblages dating back as far as 0.8 myBP (Anderson 1970; Barnosky 2004). Anderson (1970) recognized 2 subgroups of American martens in North America, the *americana* group, distributed across Canada and eastern North America, and the *caurina* group, distributed along the Pacific coast. Because *caurina* bears a closer morphological resemblance to *M. zibellina*, Anderson (1970) suggested that the *americana* and *caurina* groups represented 2 separate colonizations. Genetic studies now show, however, that the *americana* and *caurina* clades evolved in North America from a single ancestor that arrived from northern Eurasia sometime before 1.0 myBP. In addition, there is compelling evidence from both morphological and genetic studies that these clades represent 2 distinct species: *M. americana* and *M. caurina* (Dawson and Cook, this volume).

Evolutionary History and Biogeography of *Martes* Revisited

New insights into the ancestry, behavior, and physiology of *Martes* allow for a general reinterpretation of its evolutionary history. The fossil and genetic records point to the rise of the genus in moist tropical forests of southwest Eurasia during the early and middle Miocene. From this point, extant members of the lineage (*Martes* spp., *G. gulo*, and *E. barbara*; Koepfli et al. 2008) evolved through a series of dispersals and speciations against a backdrop of cooling climates, sea level changes, mountain uplifts, aridification, and cyclical glacial events, all creating barriers to gene flow.

Before I discuss the evolutionary history of *Martes*, several adaptive features of the genus that are important for understanding this history require mentioning. First, the genus has great adaptive plasticity (Croiter and Brugal 2010) and occupies a range of habitat conditions (Clevenger 1994a; Proulx et al. 2004). Although forests are an important habitat for most extant members of the genus, recent studies show that some species (e.g., *M. foina*) spend little time in forests when other types of protective cover are available

(Clevenger 1994a; Proulx et al. 2004; Slauson et al. 2007). Members of the genus also appear to prefer mesic or moist environments (Hughes 2009), perhaps because they have a relatively high rate of evaporative loss in their water balance (Meshcherskii et al. 2003). This feature may be a carryover from a Miocene ancestor adapted to the humid tropical and subtropical forests covering southern Eurasia during the early and middle Miocene.

Miocene (23.0–5.3 myBP)

Genetic studies indicate that *Martes* diverged from other mustelids about 12.0 myBP, although the fossil record pushes this estimate back to 18.0 myBP. The earliest fossil forms are found in central Europe and the circum-Mediterranean region, suggesting that *Martes* first evolved there. After 15.2 myBP, fossil forms almost disappear from central Europe, continue in the circum-Mediterranean region, and appear for the first time in south-central and southeast Asia (Figure 1.2). Temperatures gradually cooled during this period, but most of southern Eurasia was covered with humid tropical forests (Werdelin 1996).

The fossil record indicates a major turnover event about 11.2 myBP that impacted all *Martes* across southern Eurasia. This turnover likely correlates with the Serravellian sea-lowering event that impacted many Eurasian mustelids (Bernor et al. 1996; Koepfli et al. 2008). The Mid-Vallesian crisis, a faunal turnover dating to 9.2 myBP (Bernor et al. 1996; Werdelin 1996; Koufos et al. 2005), affected early *Martes* forms in the circum-Mediterranean region. What precipitated this event is unknown, but the collision of the Indian Plate, which gave rise to the Himalayan Mountains, may have been a contributing factor (Barry et al. 2002). Only *M. lydekkeri* in northern India appears to have survived this event; shortly thereafter, *M. palaeosinensis* first appears in northern China. *Martes basilii* and *M. woodwardi* reappear in the circum-Mediterranean around 8.0 myBP (Figure 1.2).

The collision of the Indian subcontinent with Asia and the rising Himalayas precipitated major environmental changes in central Eurasia that likely isolated ancestral subgenera on both sides of the Eurasian continent. As the rising Himalayas blocked monsoonal flow, India and central Asia transitioned from a wet monsoonal forest before 8.5 myBP to a dry monsoonal forest by 7.0 myBP, with savanna in place by 6.0 myBP (Karanth 2003; Badgley et al. 2008). A similar pattern of increased aridity and seasonality in the eastern Mediterranean region resulted in a gradual shift from humid subtropical forests to open woodlands, and then savanna (Bernor et al. 1996; Koufos et al. 2005; Fortelius et al. 2006). The expansion of savanna environments drew steppe-adapted species from Asia into the region, ca. 8.4 myBP, where this faunal assemblage, the Pikerman fauna, occupied an extended bioprovince from the eastern Mediterranean basin to the Middle East (Bernor et al.

1996; Koufos et al. 2005). Meanwhile, subtropical forests persisted in western Europe and the Far East (Bernor et al. 1996; Fortelius et al. 2006). A faunal turnover event ca. 7.2 myBP in India and the circum-Mediterranean region, with continuity of forms in China, coincides with these climatic changes (Figure 1.2). These changes likely set the stage for the rise of *Pekania* and *Charronia* in southeast Asia (and possibly *G. gulo* and *E. barbara*).

The *Pekania*, represented by *M. palaeosinensis*, arose first in China, between 9.0 and 8.0 myBP, and expanded to Kazakhstan in central Asia by 7.2 myBP. A contemporary species was *M. lydekkeri*, whose remains have been found in strata at the base of the Himalayas (Figure 1.2). The ancestral form of *Charronia* diverged sometime between 6.0 and 5.0 myBP. Among the 3 *Martes* subgenera, only the *Charronia* has remained in southern Asia to the present day, inhabiting the humid subtropical forests of southern India and southeast Asia.

Available evidence indicates that *Martes* first appeared in North America about 16.0 myBP; *M. nambianus* was recovered from deposits in New Mexico dating to 14.0 myBP. If the genetic dates are correct, these early *Martes* specimens in North America are likely transitional forms. After *Martes* diverged 12.0 myBP, the genus is represented by a succession of fossils recovered from locations in the western United States (Carrasco et al. 2005). All Miocene forms of *Martes* are thought to have entered North America prior to the closing of the Bering Land Bridge ca. 5.5–5.4 myBP, and all became extinct (Hunt 1996; Baskin 1998), an assessment that is supported by the genetic data (Koepfli et al. 2008).

Pliocene (5.3–2.6 myBP)

Another faunal turnover occurred at the Miocene-Pliocene boundary (5.3 myBP) in southeast Asia and North America (Figure 1.2). This was followed by a return to warmer and wetter conditions and the expansion of forests across the northern hemisphere until the onset of the Northern Hemisphere Glaciations (NHG) event 3.2 myBP (Zachos et al. 2001; Koufos et al. 2005). The beginning of the Pliocene also marks the uplift of the Alps in central Europe, creating a barrier between northern and southern Europe (Koufos et al. 2005). At this time, *M. foxi* appears in North America, perhaps crossing over during a brief opening of the Bering Land Bridge (Marincovich and Gladenkov 1999). We know this taxon did not survive, because genetic analyses show that the ancestors of *M. americana*, *M. caurina*, and *M. pennanti* arrived much more recently in North America.

The fossil record indicates additional turnovers localized in Europe and the Far East 4.0 and 3.2 myBP, perhaps related to the onset of the NHG (Steininger and Wessely 1999; Zachos et al. 2001; Croiter and Brugal 2010). Two fossil forms appear in central Europe at this time, *M. wenzensis*, the first true marten, at 4.0 myBP in Poland, and *M. filholi,* possibly as early as 3.0 myBP. The disappearance of *M. wenzensis* in Poland 3.2 myBP corresponds to the onset of glacial conditions in Europe, although the species persisted in the circum-Mediterranean region until 2.0 myBP.

The first glacial advance dates to around 3.2 myBP, and such events would have forced a tropical or subtropical carnivore like *Martes* to retreat to glacial refugia. Repeated expansion and contraction of *Martes* habitat throughout this glacial period caused rapid radiations and subsequent isolation that led to new adaptations and speciation. Early in this process, a group of true martens, perhaps confined to forested refugia along water courses in dry montane regions of southwest Asia, evolved adaptations to drier and more open habitats, such as in the extant species *M. foina*. The earliest fossil evidence of *M. foina* appears in southwest Asia in the late Pleistocene. At the close of the Pleistocene, this taxon expanded across Turkey and the Caucasus, then rapidly colonized southern and central Europe, replacing *M. martes* in drier, more open habitats (Wolsan 1993b; Sommer and Benecke 2004).

Pleistocene (2.6–0.012 myBP)

With the onset of the NHG, climatic conditions fluctuated with increasing aridity, especially in the circum-Mediterranean region. During glacial advances, ice covered the northern European continent and most of the mountain ranges of central Europe, including the Cantabrias, Pyrenees, Alps, and Caucasus. Across central Europe and the circum-Mediterranean region, dry arctic steppe environments supported a faunal assemblage known as the Villafranchian fauna (Koufos et al. 2005).

Glacial advances forced the subtropical true martens into glacial refugia that were probably located in humid, forested valleys in southern Europe, for example, the Caucasus, Balkans, and eastern Mediterranean, and the peninsulas of Iberia, Italy, and Greece (Hewitt 1999). With glacial retreat, martens rapidly expanded back into central Europe. Isolation in refugia created genetic bottlenecks that ultimately brought about adaptive changes conferring greater success in cool temperate forests. This new adaptation allowed the radiation of true martens across northern Eurasia, ca. 1.8 myBP, at a time of increased humidity and the expansion of temperate forests across this region (Koufos et al. 2005). Subsequent glacial advances and other barriers led to further diversification within this lineage.

The fossil record and present distribution of true martens supports the above hypothesis. Today, *M. martes* inhabits the forests of northern and central Europe, extending east to the Yenisei River in Russia; *M. zibellina* replaces *M. martes* in the trans-Ural region and was distributed historically across the vast montane forests of northern Eurasia to China (Kurtén 1968; Anderson 1970); *M. americana* replaces *M. zibellina* in the boreal forests of northern North America; and *M. melampus* is endemic to the Japanese islands in southeast Asia, whereas *M. zibellina* occupies mainland China (Proulx et al. 2004). These distributional patterns suggest that *M. americana* likely diverged first, followed by *M. melampus* and *M. zibellina. Martes americana* arrived in North America sometime before 1.0 myBP. The earliest fossil evidence of *M. americana* is found in Porcupine Cave, Colorado, and dates between 1.0 and 0.6 myBP (Barnosky 2004).

The Pleistocene true martens of Europe are represented by *M. vetus*, which appears in the fossil record between 1.8 and 0.4 myBP (Wolsan 1993b). A larger form of *M. martes,* the modern European pine marten, appeared about 0.12 myBP (Kurtén 1968; Anderson 1970; Wolsan 1989, 1993b), with fossil records in Belgium, Bulgaria, Germany, England, France, Italy, Romania, Switzerland, and Scandinavia. The recent appearance of *M. martes* in Europe suggests that this species evolved elsewhere during the later Pleistocene and re-invaded Europe around 0.4 myBP. Because no *M. martes* fossils exist elsewhere, the fossil evidence points to more recent speciation, which agrees with the genetic record (Hosoda et al. 2000; Koepfli et al. 2008).

Martes melampus is represented by a late-Pleistocene fossil from northern China (Anderson 1970). If the earliest true marten evolved in the West, then the ancestor of *M. melampus* likely expanded across northern Eurasia, only to become isolated in China during a glacial event. The warm subtropical forests of the early and middle Pleistocene in northern China were replaced by dry steppe in the later Pleistocene (Dexin and Robbins 2000), which was apparently an effective barrier to subsequent gene flow.

After 1.0 myBP, northern Eurasia and North America experienced cycles of alternating glacial and temperate periods caused by rapid ice advances and retreats (Koufos et al. 2005). This created a pattern of north-south migrations in Europe, with the mountains of central Europe serving as a possible barrier to dispersal (Koufos et al. 2005).

The distribution of late-glacial and Holocene remains of *M. martes* suggests that this species occupied multiple refugia during the last glacial advance (Sommer and Benecke 2004); however, only 1 or 2 refugia are indicated by molecular study (Davison et al. 2001). Both types of data show rapid colonization of northern Europe by *M. martes* after the last glacial retreat.

Three of the 4 genetic studies shown in Figure 1.2 point to a recent split of *M. zibellina* from *M. vetus* or another *Martes* ancestor in northern or central Europe sometime around 1.0 myBP. The historical distribution of this taxon suggests that it expanded across northern Eurasia in the late Pleistocene. Fossils of *M. zibellina* in the Altai Mountains of north-central Asia date between 0.155 and 0.033 myBP (Fortelius 2009). *Martes zibellina* entered northern China in the late Pleistocene, and competition with the endemic *M. melampus* likely pushed the latter to coastal margins, where post-glacial sea level rise confined representatives of both species to the southern Japanese islands.

Two genetic studies suggest the possibility of introgression between the 2 species in the recent past (Kurose et al. 1999; Murakami et al. 2004), whereas the data of Sato et al. (2009a) show that the 2 species are sister taxa.

Shortly after North American colonization, a glacial advance isolated *M. americana* in eastern and western refugia that remained separated throughout the late Pleistocene. Based on its greater genotypic diversity, *M. caurina* was isolated in multiple glacial refugia along the west coast, whereas *M. americana* occupied a single refugium in the east (Carr and Hicks 1997; Stone and Cook 2002; Small et al. 2003; Dawson and Cook, this volume). As continental glaciers retreated northward at the end of the Pleistocene, *M. americana* expanded westward while *M. caurina* expanded northward and eastward, colonizing the mountains and temperate coastal forests of the Pacific Northwest (Graham and Graham 1994; Stone et al. 2002; Small et al. 2003; Dawson and Cook, this volume). Rising sea levels isolated some populations on coastal islands (Small et al. 2003). Small et al. (2003) suggested that *M. americana* arrived more recently in the Pacific Northwest and continues to expand, possibly outcompeting *M. caurina*. Hybridization between the 2 species has occurred on Kuiu Island in southeastern Alaska (Small et al. 2003) and in southern Montana (Wright 1953). Genetic data also suggest that the Newfoundland subspecies (*M. a. atrata*) was isolated from the main population by rising sea levels during an earlier glacial cycle (McGowan et al. 1999; Kyle and Strobeck 2003).

The noble marten (*M. a. nobilis*), an extinct subspecies identified by Anderson (1970, 1994), may represent an early radiation of *M. caurina* from a glacial refugium in northern California at the close of the Pleistocene (Hughes 2009). The noble marten appears to have evolved slightly different physiological and behavioral traits that allowed it to colonize open, lower-elevation habitats associated with forested riverine ecosystems in the American West (Hughes 2009). This adaptation is supported by the recent discovery of noble martens at Marmes Rockshelter in Washington state dating between 0.013 and 0.012 myBP (Lyman 2011).

The subgenus *Charronia* is represented by *M. flavigula* and its Pleistocene ancestor *M. f. tyrannus*, with fossil remains identified from several cave sites in China dating to 1.2 myBP (Fortelius 2009). *Martes gwatkinsii*, the endangered Nilgiri marten of the Western Ghats in south India, is genetically and morphologically similar to *M. flavigula* (Anderson 1970) with an estimated divergence date of 0.9 myBP (Bininda-Emonds et al. 1999). These results suggest that *Charronia*, adapted to the subtropical forests of southeast Asia, once had a wider distribution than it does today (Karanth 2003).

The subgenus *Pekania* disappeared from China in the early Pleistocene and reappeared as *M. diluviana* in North America ca. 1.8 myBP (Anderson 1970, 1994), having evolved in the eastern subarctic before colonizing North America (Carr and Hicks 1997). *Martes pennanti* is present in eastern North

American fossil assemblages dating to 0.030 myBP. Both the fossil and genetic records indicate the isolation of fishers in an eastern refugium south of the ice margin during the last glacial advance. Following glacial retreat, fishers expanded west toward the northern Pacific coast, where they radiated south along the spine of the Sierra Nevada Mountains (Kyle et al. 2001; Drew et al. 2003; Wisely et al. 2004).

Summary and Conclusions

This synthesis of *Martes* phylogeny and evolutionary history demonstrates that little comparative study has been conducted on *Martes* fossil material during the last 20 years, although fossils continue to be found. Our knowledge base has expanded considerably in other areas of *Martes* research, however. Genetic analyses have provided valuable information on *Martes* phylogeny and divergence dates, although the existing fossil record suggests that some divergence dates may be too recent, or that some of the early *Martes* fossils were misidentified. New insights on *Martes* evolutionary history derived from genetic research include the following: (1) *Gulo* and *Eira* are closely associated with the *Martes* lineage; (2) *M. lydekkeri* is likely not the ancestor of *Charronia*, as proposed previously; and (3) *M. vetus* is not the ancestor of *M. foina*.

Information derived from multiple sources provides a more nuanced understanding of *Martes* evolutionary history. The genus appears to have evolved initially in the humid subtropical forests of southwest Asia during the early and middle Miocene. Initial diversification of the genus appears to have occurred in southeast Asia with the rise of *Pekania* and *Charronia*. The true martens likely evolved in Europe. Diversification took place against a backdrop of cooling climates, sea level rising and lowering, uplift of major mountain barriers, and cyclical glacial events that created barriers to gene flow and fostered rapid evolutionary change. Late in their evolutionary history, true martens evolved characteristics that enabled them to persist in the cool, temperate forests of northern Eurasia, and they subsequently radiated out to other regions to occupy this niche. The plasticity of the genus has allowed it to persist in rapidly changing environments.

This synthesis reveals several directions for future research. Primarily, a re-examination of all Neogene fossil material is needed to fill gaps in the fossil record. Such information would help to refine genetic divergence dates and clarify the evolutionary history of the genus. Another important research need is study of the extant *Charronia: M. flavigula* and *M. gwatkinsii*. Very little is known about the habits, physiology, or ecological relations of these 2 species. This information would be particularly valuable because the *Charronia* may contain adaptations that are characteristic of the original *Martes* ancestor. Regardless of these research needs, great strides have been made during the last 20 years toward understanding the phylogeny, evolutionary history, and adaptation of *Martes*.

Acknowledgments

I thank Keith Aubry and the other organizers of the 5th International *Martes* Symposium for their encouragement in taking on this project. I thank Russell Graham of Pennsylvania State University, Keith Aubry, Corey Duberstein of Pacific Northwest National Laboratory, and an anonymous reviewer for valuable comments on an earlier draft of this chapter. I also thank the individuals and institutions who developed the fossil databases used in this study; these resources offer easy access to information that would otherwise be very difficult to acquire. Last, I offer a posthumous thanks to Elaine Anderson, whose identification of a noble marten from Mummy Cave in Wyoming ultimately led to the development of this chapter. The Pacific Northwest National Laboratory Information Release Number for this paper is PNNL-SA-73060.

2

Behind the Genes

Diversification of North American Martens (*Martes americana* and *M. caurina*)

NATALIE G. DAWSON AND JOSEPH A. COOK

ABSTRACT

Originally, 2 species of martens were described in North America: the American marten (*Martes americana*) and the Pacific marten (*M. caurina*). The description of intermediate forms suggested hybridization between the species, and they were reclassified as a single species, *M. americana*, containing 14 subspecies. Additional morphometric and phylogeographic studies have uncovered patterns of differentiation that support the recognition of 2 distinct species. Previous studies discovered genetic patterns associated with differential range expansions by each species after the retreat of glaciers during the last ice age in North America. Hybridization occurs between these closely related species, but phylogeographic studies based on a suite of molecular markers document the diversification and independent histories of these 2 species of martens in North America. This chapter summarizes results from both morphometric and phylogeographic studies of North American martens. Altogether, those studies strongly support the occurrence of 2 independent and evolutionarily distinct groups of martens in North America that are concordant with the original taxonomic descriptions of 2 different species. We conclude with a brief discussion regarding the use of taxonomic classifications for the management and conservation of North American martens.

Introduction

According to the taxonomic overview of Wozencraft (2005), the genus *Martes* includes 8 extant species worldwide (*M. americana, flavigula, foina, gwatkinsii, martes, melampus, pennanti,* and *zibellina*). Recent genetic studies (Stone and Cook 2002; Koepfli et al. 2008) suggest that this radiation may also include the wolverine (*Gulo gulo*) and, possibly, the tayra (*Eira barbara*). Detailed assessments of variation within the genus have highlighted other unresolved taxonomic problems, including the question of whether extant North American martens represent a single species, *Martes americana*, the American marten; or 2 species, *M. americana* and *M. caurina*, the Pacific marten.

Although *M. caurina* was considered a distinct species as early as 1890 by Merriam, Wright's (1953) subsequent morphometric study of martens within a hybrid zone resulted in the assignment of *M. caurina* as a subspecies within *M. americana*. Recent taxonomic summaries (e.g., Hall 1981; Baker et al. 2003; Wozencraft 2005) followed Wright's recommendations and included all extant martens in North America in a single species, *M. americana*, although several authors continued to refer to 2 distinct "groups" of martens in North America, the *americana* group and the *caurina* group (e.g., Hagmeier 1955; Anderson 1970; Clark et al. 1987).

In the past few decades, several studies based on molecular data independently concluded that 2 distinct species of martens exist in North America (e.g., Carr and Hicks 1997), returning to the earlier taxonomic designations of *M. americana* and *M. caurina*. Initially, genetic studies of North American martens were based on a single mitochondrial marker, but additional sets of molecular markers consistently identified 2 reciprocally monophyletic clades among North American martens (Stone and Cook 2002; Small et al. 2003; Dawson 2008). Those analyses also revealed distinct zoogeographic histories for the 2 marten groups and provided a much more robust empirical framework for exploring the evolutionary processes that led to their divergence.

Marten Taxonomy and Morphometrics in North America

Martes americana (Turton 1806) was first described on the basis of specimens from eastern North America. Rafinesque (1819: 82–83) described a distinct marten found in the "regions watered by the Missouri" as "very different from the common marten of Europe, Asia and America." In one of the early assessments of martens in western North America, Gray (1865) described *M. americana abietinoides* from the Selkirk Mountains in southern British Columbia, indicating that the species' range extended westward to include portions of the Rocky Mountains. Merriam (1890: 27–28) later described *Mustela (= Martes) caurina* based on a specimen from Grays Harbor County in western Washington, noting that "The skull of *Mustela caurina* differs from that of *M. americana* in the following particulars: The rostral portion is broader and shorter; the audital bullae are shorter and less inflated; the frontals are broader both interorbitally and postorbitally; the shelf of the palate is less produced behind the plane of the last molar; the first upper premolar is smaller and more crowded; the upper molars are larger . . . the last upper molar is not only larger, but has a much broader saddle." These differences are clearly evident when skulls from each group are compared directly (Figure 2.1). Merriam's

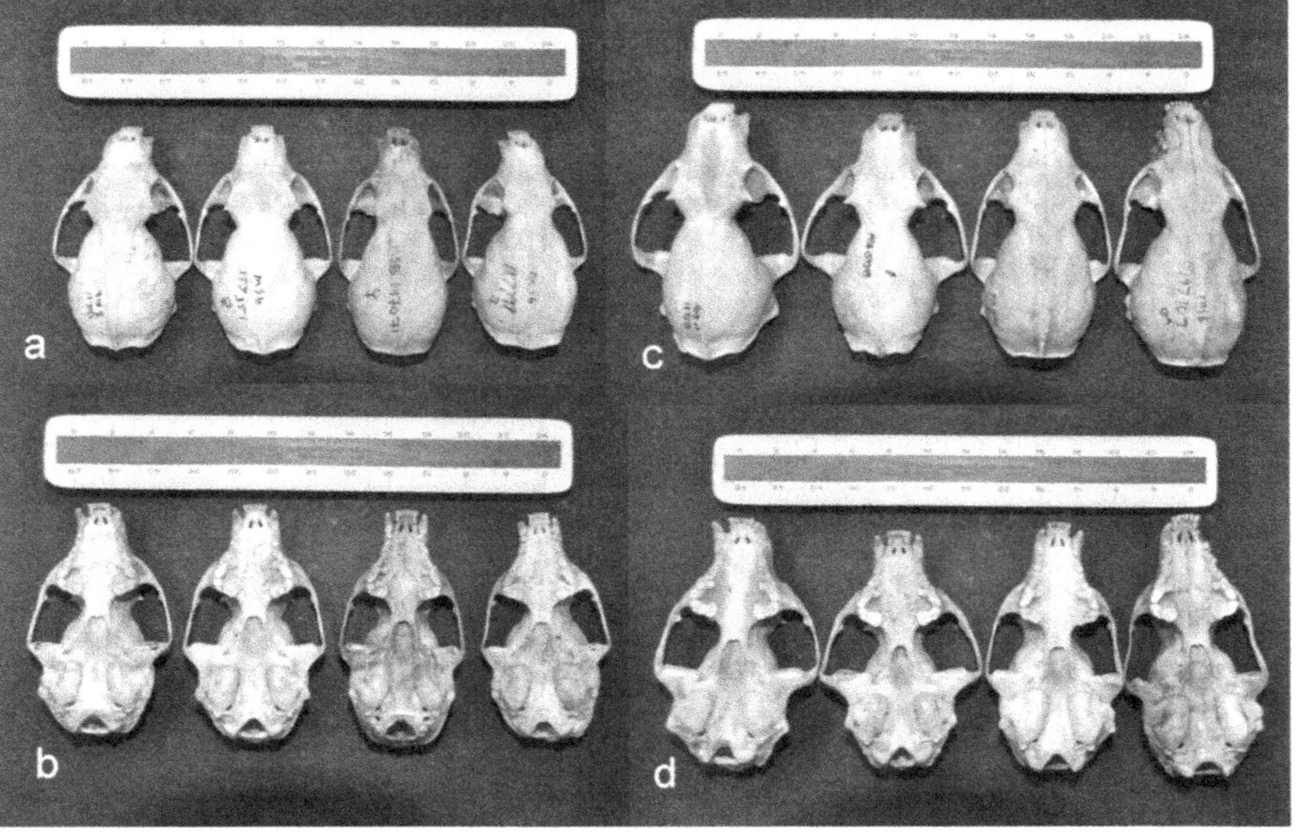

Figure 2.1. Dorsal and ventral views of adult skulls (Museum of Southwestern Biology) of *Martes americana* and *M. pacifica* from populations in northwestern North America. (a) Dorsal and (b) ventral views of 2 female *M. pacifi*ca (MSB 197756, British Columbia, Vancouver Island; MSB157351, British Columbia, Graham Island) and 2 female *M. americana* (MSB197071, Northwestern Territories, Vermillion Creek; MSB197799, British Columbia, Francois Lake). (c) Dorsal and (d) ventral views of 2 male *M. pacifica* (MSB 197229, Alaska, Admiralty Island; MSB157349, British Columbia, Moresby Island) and 2 male *M. americana* (MSB197101 and MSB197109, Northwestern Territories, Vermillion Creek).

(1890) species description includes 17 skull and 8 tooth measurements, as well as 25 ratios of cranial and dental measurements. Rhoads (1902: 456) examined 129 martens from throughout their range in North America, and his results supported the distinct cranial characteristics of *M. caurina* in an expanded species account: "As compared with skulls of *M. a. actuosa* from Mackenzie, B.A. [British America], and *M. americana* from north of Lake Superior. . . . The skull of *caurina* viewed from above is greatly broadened and flattened. . . . In conformity to this relative shortening and widening, *caurina* has a markedly wide brain-case and interorbital region, the postorbital processes being widely and bluntly developed and the frontals abruptly constricted behind them. . . . The lowness or flatness of the cranium is also marked."

Rhoads (1902) described a new subspecies, *M. caurina origenes*, from the mountains of southern Colorado, thereby extending the range of *M. caurina* eastward to the southern Rocky Mountains and raising the possibility of geographic contact or overlap with *M. americana*. Later mammalogists acknowledged the distinctive shape of *M. caurina* skulls in state-level monographs of western mammals, as they attempted to clarify the distribution of the 2 species based on the cranial characteristics put forth by Merriam (1890) and Rhoads (1902). Davis (1939) reported the occurrence of *M. caurina* in central and southern Idaho, and Dalquest (1948) reported that both species were present in Washington. Hall (1946) and Durrant (1952) considered the range of *M. caurina* to include Nevada and Utah, respectively. In general, *M. americana* was considered the eastern species of marten, with a range extending from the eastern edge of North America into the Great Plains region, whereas *M. caurina* was considered the western form, restricted to the Rocky Mountains and the west coast of North America. Davis (1939) hypothesized that the Rocky Mountains were a geographic barrier between the 2 species, clearly considering their ranges to be allopatric.

Another issue regarding the question of how many species of martens occur in North America concerns the extinct noble marten, *M. nobilis*. This Pleistocene species occupied western North America (Lyman 2011) and, although larger in size, has been allied with *M. caurina* (Anderson 1970; Youngman and Schueler 1991). Additional study of morphometric or genetic variation using ancient DNA analyses should help clarify the taxonomic status and paleoecology of this species (Hughes 2009).

For many biologists, however, the number of marten species occurring in North America was settled by a study of morphometric variation in martens conducted by Wright (1953) and based on 245 skulls and "several hundred" skins from northern Idaho, western Montana, and southern British Columbia. When Wright (1953) began his investigation, his objectives were to refine the distributional limits of the 2 species in the central Rocky Mountains, and to explore the consequences of potential contact between them. Wright's

(1953) findings, based on 13 standard cranial measurements, suggested contact with intergradation between the 2 species. For his assessment, Wright chose representative *M. americana* from 2 sites in southern British Columbia, and representative *M. caurina* from Idaho and the Sapphire Range in southern Montana. Martens from areas located between these "pure" populations were intermediate in size and shape. For his analyses, Wright (1953) classified samples from northern Idaho and near Whitefish, Montana, as *M. americana*, although the northern Idaho samples showed "some tendency" toward characteristics of *M. caurina*. He assigned specimens from Clearwater and Red Lodge counties in Montana to *M. caurina*, whereas specimens from the Swan, South Fork, and Sun River drainages were considered intermediate, based on 4 of the cranial measurements used in the original description of *M. caurina* (Merriam 1890): (1) length of auditory bullae, (2) length of inner lobe of the upper molar, (3) rostral breadth, and (4) height of the cranium (Wright 1953). Wright (1953) noted that the specimens of *M. americana* used by Merriam (1890) and Rhoads (1902) were both few in number and from distant locations (e.g., the Adirondack Mountains in New York). Wright (1953: 84) concluded that "since intergradation is occurring between populations previously regarded as distinct species, it appears that we are dealing with only a single species." He asserted that the occurrence of intermediate-sized specimens in the Swan, South Fork, and Sun river drainages of Montana (17 females and 15 males) reflected clinal variation. In accordance with biological thought at that time, Wright (1953) assumed that the occurrence of hybridization meant that these forms were conspecific; however, with the development of molecular techniques, hybridization between species is known to be far more common than was generally accepted when Wright published his findings (Hewitt 1989; Arnold 1997; Wayne and Brown 2001).

Wright (1953) also assessed variation in pelage color, but concluded that these characters were locally specific. For example, the bright orange throat patch and light coat color he identified as characteristic of martens from the Sapphire Mountains and Bitterroot Range in southern Montana, fall within the range of *M. caurina* and, coincidentally, are characteristic of island populations of *M. caurina* in southeast Alaska (R. Flynn, Alaska Department of Fish and Game, personal communication).

In his monograph on the taxonomy of North American mammals, Hall (1981) followed Wright's (1953) recommendation and included the *caurina* group as a subspecies within *M. americana*. Nonetheless, other researchers continued to acknowledge the existence of 2 morphologically distinct groups of martens in North America (Hagmeier 1955; Anderson 1970, 1994; Banfield 1974; Clark et al. 1987; Graham and Graham 1994). No comprehensive reevaluation of morphometric variation among North American martens has been conducted since Wright's study, although one researcher inadvertently indentified differences between the species based on samples collected along the North Pacific coast of North America. Nagorsen (1994) examined variation in body weights among island and mainland martens in the Pacific Northwest; in British Columbia, he found that island martens (*M. caurina*) were larger than mainland martens (*M. americana*). In contrast, martens from islands in the Alexander Archipelago (located north of the islands in British Columbia) were only weakly divergent from those of mainland southeast Alaska. Nagorsen's (1994) study included samples only from Baranof and Chichagof islands in the Alexander Archipelago, both of which contain populations of *M. americana* that were introduced from mainland sources. Comparing body weights between island (*M. caurina*) and mainland (*M. americana*) martens, Nagorsen (1994) found significant differences ($P < 0.01$) among all age classes and between the sexes. Thus, without knowing the history of these reintroductions, or the taxonomic status of martens in southeast Alaska or in the Queen Charlotte Islands, Nagorsen (1994) documented morphometric differences consistent with the recognition of *M. caurina* and *M. americana* as distinct species.

North American martens have been partitioned into as many as 14 subspecies, including *M. americana caurina* (Figure 2.2). With the exception of intensive studies of a few subspecies of conservation concern, such as *M. a. atrata* (Newfoundland Island endemic) and *M. a. humboldtensis* (Coast Range of northern California), there have been no recent comprehensive reviews of subspecific variation among North American martens. Limited sampling within 12 of the 14 recognized subspecies (1–3 samples from each) by Carr and Hicks (1997) precluded a rigorous assessment of mtDNA variation and subspecies limits.

Phylogeography of North American Martens

Phylogeographic analyses use genetic data to understand the historical processes that resulted in the current geographic distribution of a species. These methodologies have also been used to refine taxonomic nomenclature, especially when morphological differences are combined with assessments of genetic variation to define species (e.g., Rissler et al. 2006). The first phylogeographic studies of North American martens showed divergence between the *caurina* and *americana* groups, as reported in previous morphometric studies. Carr and Hicks (1997) used the maternally inherited mitochondrial cytochrome *b* gene to examine relationships among *Martes* species and, in particular, to investigate variation in North American martens. They found that pairwise sequence divergence (an index of base-pair differences in DNA sequences) between the *americana* group and the *caurina* group was as great as the sequence divergence among Palearctic species of *Martes* (*M. zibellina*, *M. martes*, *M. melampus*). Carr and Hicks (1997) included specimens from the Selkirk Mountains in British Columbia, as did Wright (1953), and found

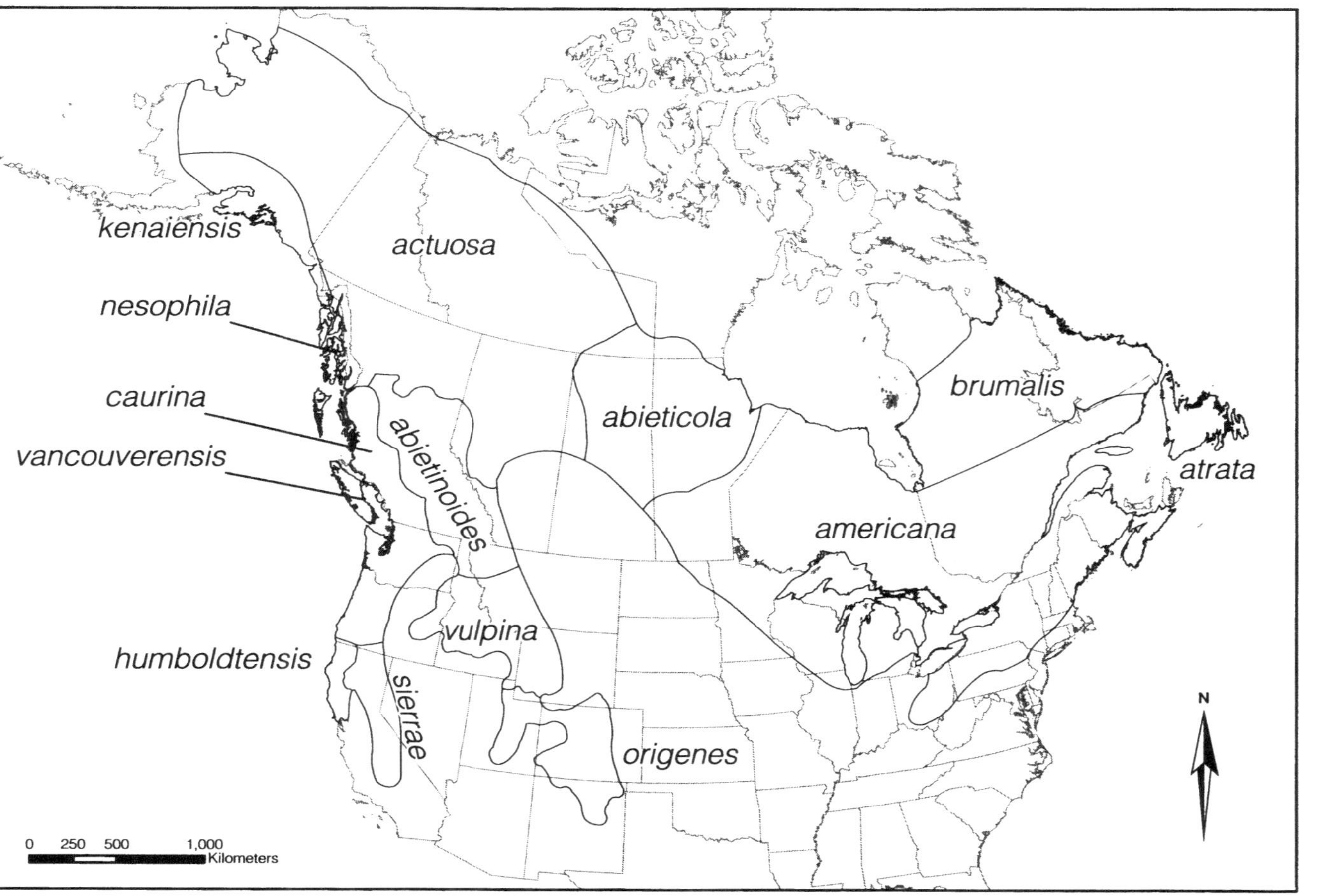

Figure 2.2. Geographic distribution of the 14 subspecies of *Martes americana* recognized by Hall (1981).

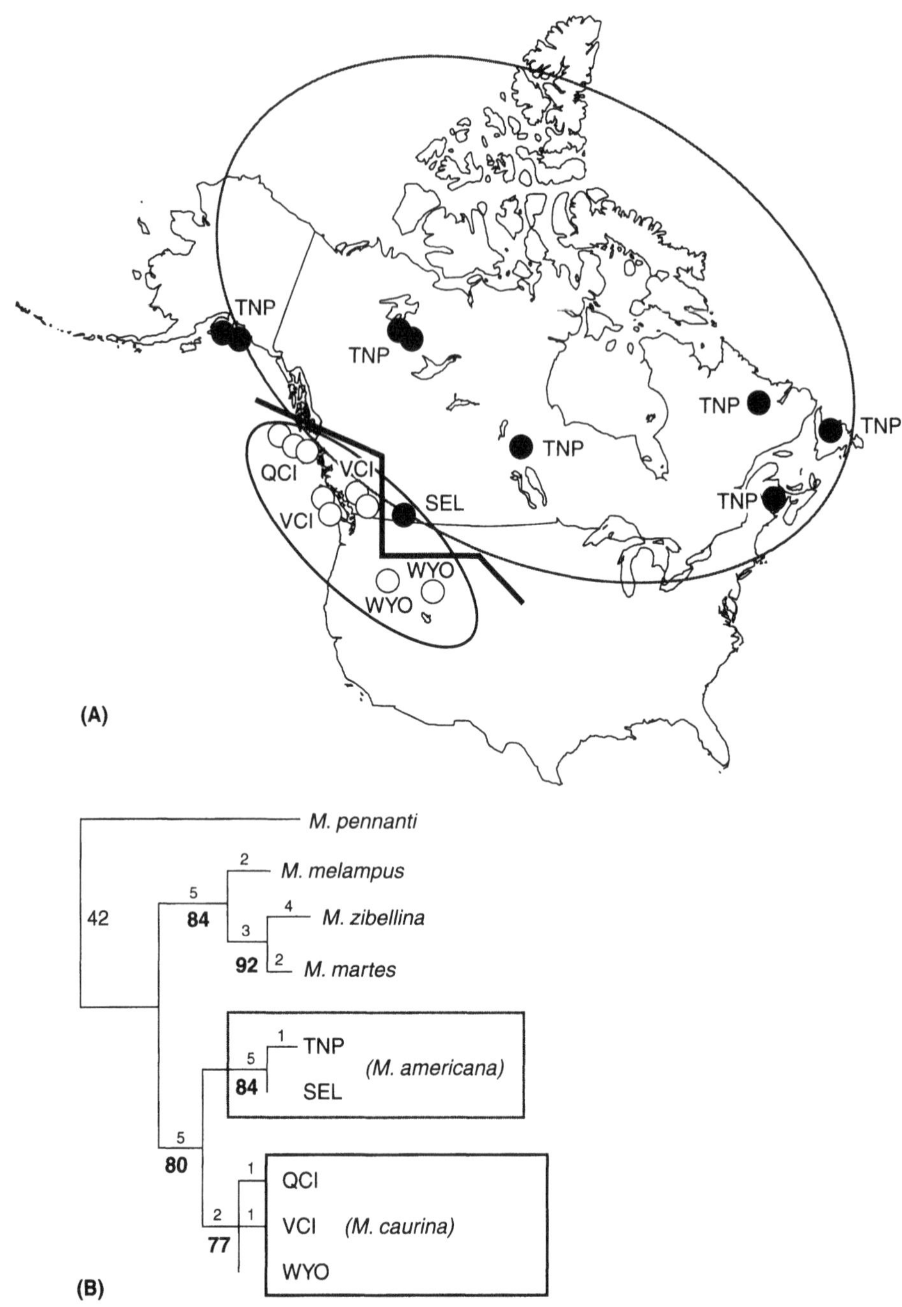

Figure 2.3. (A) Depicts the geographic boundary between *Martes americana* (light gray ellipse) and *M. caurina* (dark gray ellipse) based on cytochrome *b* sequence data. (B) Shows reciprocal monophyly between the *americana* and *caurina* clades, as well as the sequence divergence compared to other members of *Martes*. Both figures are adapted from Carr and Hicks (1997).

that those individuals were from the *americana* clade, consistent with the original taxonomic designation for this population (Gray 1865). Thus, the findings of Carr and Hicks (1997) provided the first evidence of genetic differentiation between the *americana* and *caurina* groups. Moreover, based on the phylogenetic species concept (Cracraft 1989), they concluded that these 2 groups represent distinct species (Figure 2.3).

Stone and Cook (2002) found that *M. caurina* and *M. americana* are reciprocally monophyletic sister taxa resulting from a single colonization event from Asia across the Bering Land Bridge into North America. Subsequent isolation into distinct eastern and western glacial refugia during the late Pleistocene resulted in divergence of the 2 groups of martens within North America. Stone et al. (2002) found that these taxa corresponded to previously recognized morphometric clusters, with the *americana* group extending from the east coast of North America (including Newfoundland populations) into northwestern North America, and the *caurina* group extending from the west coast of North America eastward into the Rocky Mountains. The divergence of these reciprocally monophyletic groups ranges between 2.5 and 3.0%, which corresponds to about 1 million years of independent evolution based on the estimated mutation rate of the cytochrome *b* gene in martens (Stone et al. 2002). Whereas the *americana* group comprised populations with little intraclade divergence and a signal of population expansion, the *caurina* group contained substantial subpopulation structure and no signal of recent expansion. That study also demonstrated a potential zone of contact similar to that described by Wright (1953). Although Stone et al. (2002) substantiated the conclusion drawn by Carr and Hicks (1997) that 2 distinct species of martens occur in North America, they did not formally revise marten taxonomy.

Small et al. (2003) then explored patterns of molecular variation between the *americana* and *caurina* clades using nuclear microsatellites, and largely corroborated findings from mtDNA analyses that there were 2 independent marten clades in North America (Figure 2.4). Small et al. (2003) described contact zones between the 2 clades in Montana and on Kuiu Island in southeast Alaska by identifying hybrid individuals using Bayesian assignment tests. Furthermore, their estimates of population structure indicated generally lower heterozygosity values, higher inbreeding coefficients, and numerous discrete alleles in the *caurina* group, whereas populations in the *americana* group were generally more genetically variable and shared more alleles across localities. Similarly, Kyle and Strobeck (2003) found little substructuring within *M. americana* populations in northern North America. These findings are consistent with the hypothesized northwestward expansion of *M. americana* during glacial retreat, as they followed expanding boreal forest conditions northward.

Small et al. (2003) also shed new light on the underlying taxonomic framework for Wright's (1953) study. All the populations used by Wright to compare

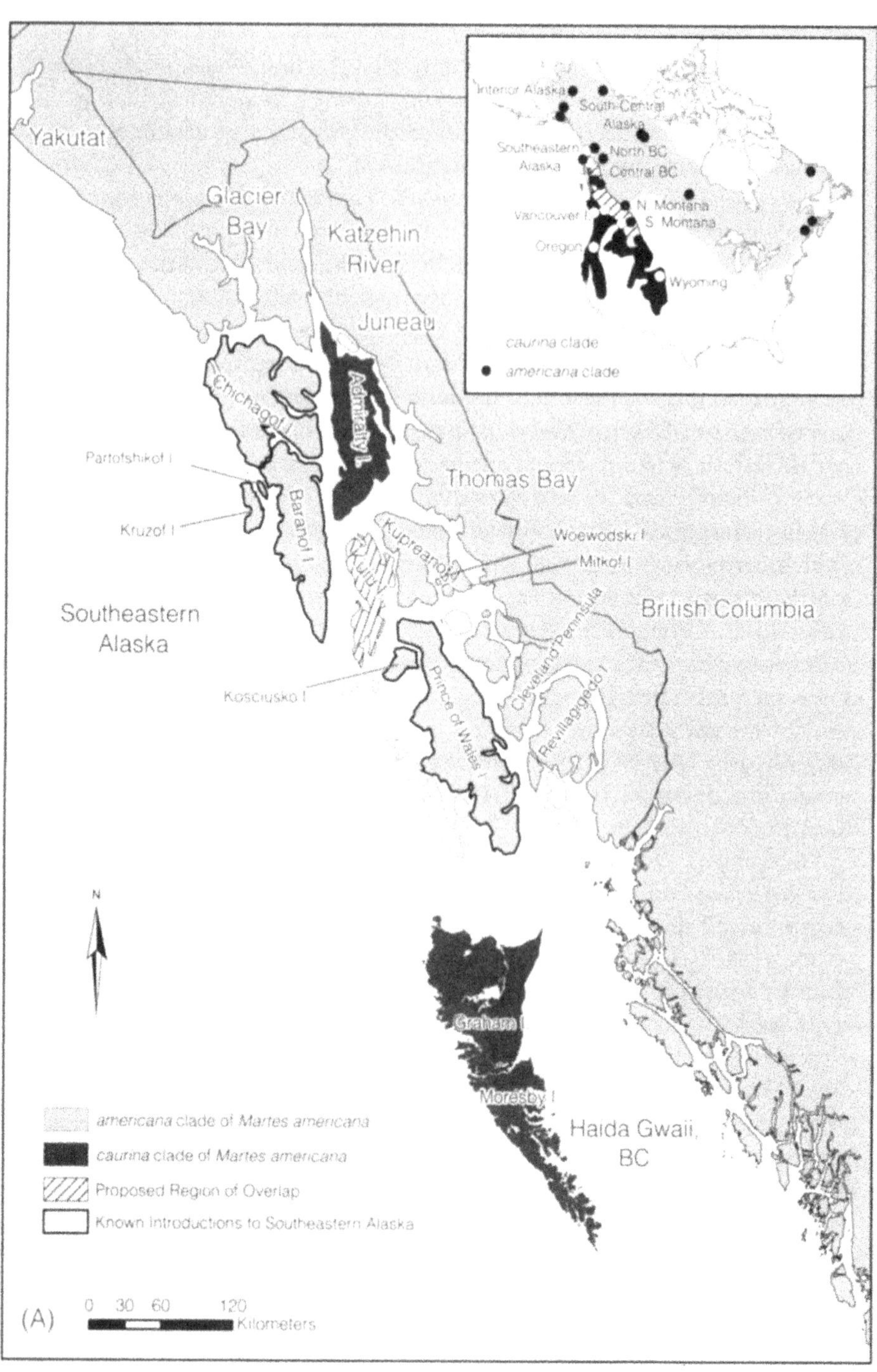
Interior Alaska
South-Central Alaska
Southeastern Alaska
North BC
Central BC
Vancouver I.
N. Montana
S. Montana
Oregon
Wyoming
caurina clade
americana clade
Yakutat
Glacier Bay
Katzehin River
Juneau
Chichagof I.
Admiralty I.
Partofshikof I.
Baranof I.
Kruzof I.
Thomas Bay
Kupreanof
Kuiu
Woewodski I.
Mitkof I.
Southeastern Alaska
Cleveland Peninsula
British Columbia
Kosciusko I.
Prince of Wales I.
Revillagigedo I.
N
Graham I.
Moresby I.
Haida Gwaii, BC
americana clade of Martes americana
caurina clade of Martes americana
Proposed Region of Overlap
Known Introductions to Southeastern Alaska
(A)
0 30 60 120
Kilometers

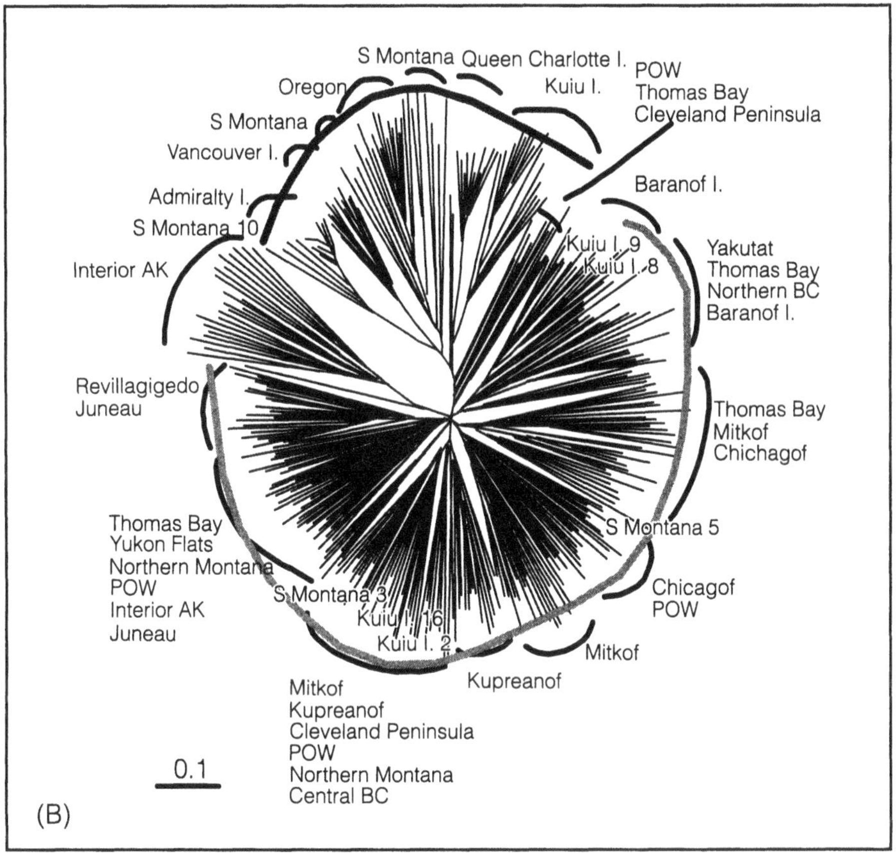

Figure 2.4. Concordance between (A) mitochondrial DNA studies (modified from Stone et al. 2002) and (B) the allele-sharing network based on 14 microsatellite (nuclear DNA) loci (modified from Small et al. 2003), with lines indicating populations of *Martes caurina* (dark gray) and *M. americana* (gray).

M. americana with *M. caurina* appear to have been located within the hybrid zone for these 2 species as defined by molecular markers. Wright (1953) concluded that intergradation between the *caurina* and *americana* groups invalidated their status as distinct species; however, specimens that he assumed were from "pure" populations appear to have been collected within the zone of introgression, such that Wright compared martens only from hybrid populations. In contrast, more geographically extensive sampling in subsequent molecular studies has resulted in a series of markers that are diagnostic for the 2 species (Small et al. 2003; Dawson 2008).

A large number of phylogeographic studies worldwide have identified complex forces that drove sequential range contractions and expansions for many species during the early Holocene; unraveling these histories is key to understanding contemporary diversity (Hewitt 1996; Lessa et al. 2010).

Slauson et al. (2009) investigated phylogeographic diversification within the *caurina* group in coastal California, Oregon, and Washington and concluded that regional genetic differentiation may reflect the isolation of populations in disjunct forest refugia during the Wisconsin glaciation. At the continental scale, Dawson (2008) concluded that both the *caurina* and *americana* clades exhibited signals of recent population expansion, supporting their distinct evolutionary histories as hypothesized by previous researchers (Carr and Hicks 1997; Stone et al. 2002). Although each clade showed signals of expansion from glacial refugia, the refugial localities from which they expanded and the temporal duration of expansion events vary between clades, further reflecting the distinct evolutionary history of each species (Waltari et al. 2007; Dawson 2008).

Relatively few studies have investigated genetic variation among marten populations in eastern North America (e.g., Broquet et al. 2006a; Williams and Scribner 2010). Most studies have focused on the homogeneous structure of marten populations in eastern North America, due either to reintroduction events (Williams and Scribner 2010) or to the hypothesized recent expansion of *M. americana* from a single glacial refugium (Graham and Graham 1994; Kyle and Strobeck 2003). The Newfoundland marten, *M. a. atrata*, is genetically distinct from other *M. americana* populations (McGowan et al. 1999), but it is more closely related to all other *M. americana* than to any *M. caurina* population. Furthermore, Kyle and Strobeck (2003) compared *M. americana* populations across Canada and found little variation among populations.

Of the 14 subspecies of *M. americana* recognized by Hall (1981), 10 are found in western North America (Figure 2.2) where both the *americana* and *caurina* clades occur. The primary zone of contact between these taxa occurs along the central Rocky Mountains of Idaho and Montana and potentially into British Columbia, with an isolated contact zone on Kuiu Island in the Alexander Archipelago in southeast Alaska (Small et al. 2003). Multiple glacial refugia in the west may have contributed to observed geographic variation within the *caurina* clade (Dawson 2008) and is consistent with patterns of genetic variation reported for other species that expanded from glacial refugia during the Holocene (Hewitt 1996; Lessa et al. 2003).

In summary, phylogeographic studies (e.g., Carr and Hicks 1997; Stone et al. 2002; Small et al. 2003; Slauson et al. 2009) largely corroborate the findings of earlier morphometric studies (Hagmeier 1955; Anderson 1970, 1994; Banfield 1974; Clark et al. 1987; Graham and Graham 1994) and reveal independent evolutionary histories for *M. americana* and *M. caurina*. *Martes americana* has recently expanded from glacial refugia in the east, whereas *M. caurina* shows signs of earlier expansion, potentially from multiple refugia in the west (Slauson et al. 2009; N. Dawson, unpublished data).

The historical habitat associations and contemporary population distributions of *M. caurina* largely match Merriam's (1890: 27) description: "The marten inhabiting the dense spruce forests of the heavy rain-fall belt along the northwest coast from northern California to Puget Sound, and doubtless ranging much farther north." Even with contemporary (and likely historical) hybridization between *M. americana* and *M. caurina*, significant differentiation between these species is apparent. Hybridization between species is a relatively common event (Arnold 1997), and other species within *Martes* hybridize with each other (Davison et al. 2001); consequently, lumping these species only because of hybridization is unwarranted. We conclude that there are 2 extant species of martens in North America (Figure 2.2): *Martes americana* (Turton 1806), the American marten, and *Martes caurina* (Merriam 1890), the Pacific marten, and recommend that current taxonomic classifications be modified to restore these previously recognized species to the list of North American mammals.

New Insights for Old Debates

The original taxonomic classification of 14 subspecies among North American martens has been heavily debated. Neither morphometric (e.g., Hagmeier 1955) nor genetic (e.g., Kyle and Strobeck 2003) studies support the recognition of most of the eastern subspecies within *M. americana,* with the exception of *M. a. atrata*, the Newfoundland pine marten (McGowan et al. 1999). In contrast, the disjunct distribution of populations in the western United States may indicate that subspecific differentiation within *M. caurina* more accurately reflects patterns of underlying geographic variation resulting from isolation in disjunct forest refugia during the last glaciation (Slauson et al. 2009).

We analyzed additional genetic data to evaluate the validity of our proposal to formally recognize the occurrence of 2 species of martens in North America. Using nuclear DNA markers described by Koepfli et al. (2008), and additional markers developed at the USGS Alaska Science Center (Table 2.1), we found at least 1 allele per locus among 7 independent loci that distinguished *M. americana* from *M. caurina*. Within the previously identified contact zones (Small et al. 2003), we identified heterozygous individuals for 3 nuclear loci (growth hormone receptor, GHR; feline sarcoma protooncogene, FES; agouti signaling protein, AG), providing additional evidence for introgression between the 2 species in limited areas throughout Montana, southern British Columbia, and southeast Alaska (Carr and Hicks 1997; Stone et al. 2002; Small et al. 2003). We also identified heterozygotic individuals in Idaho. These findings are consistent with independent evolutionary histories for *M. americana* and *M. caurina*, followed by secondary contact as populations expanded in the Holocene. Further study of the extent and dynamics of

Table 2.1. Locus, number of alleles, primer name, and primer sequence for the 7 nuclear loci screened across multiple populations of *Martes americana* and *M. caurina*

Locus	Alleles		Primer	Primer sequence
	M. americana	*M. caurina*		
Agouti Signaling Protein	A	B	AG393(F)	CCT ACT CCT GRC CAC CCT GC
			AG1113	CTG AAA GGA CTA GTT CAT
5-HT1A Serotonin Receptor	A	B	HT1A	TCA GCT ACC AAG TGA TCA CCT
			HT1A (R)	CAG GGA GTT GGA GTA GCC
Retinoid Binding Protein	A,C	B,C	IRBP35F	GGA TGC AGG AAG CCA TCG
			IRBP785	TGG TGT CCC CAC ACA GGG TA
Growth Hormone Receptor	A,C,D	B,C,D	GHR	CCA GTT CCA GTT CCA AAG AT
			GHR (R)	TGA TTC TTC TGG TCA AGG CA
Apolipoprotein B	A	B	APOB	GGC TGG ACA GTG AAA TAT TAT GAA
			APOB (R)	AAT CAG AGA GTT GGT CTG AAA AAT
Nicotinic Alpha Polypeptide	A	B	CHRNA1	GAC CAT GAA GTC AGA CCA GGA G
			CHRNA1R	GGA GTA TGT GGT CCA TCA CCA T
Feline Protooncogene	A,B,C,D,G,H	E,F,G,H	FESi14	GGG GAA CTT TGG CGA AGT GTT
			FESi14R	TCC ATG ACG ATG TAG ATG GG

Sources: Primers were redesigned based on information reported by Kukekova et al. (2004), Yu et al. (2004), Koepfli and Wayne (2003), Våge et al. (2003), Sato et al. (2003), and Eizirik et al. (2003)

Note: Alleles per locus are listed by species to illustrate diagnostic loci, as well as those with alleles shared between species these contact zones would provide additional insights into the evolutionary history of North American martens.

The Role of Taxonomy in Species Conservation

Taxonomic classifications provide a means to characterize and measure biological diversity, and are often used to establish priorities for the conservation of species and populations (McNeely 2002). The degree of conservation concern varies among extant marten populations in North America. American martens are widely distributed, although some populations are at risk of extirpation (e.g., McGowan et al. 1999). The Pacific marten occurs throughout western North America, but there are a number of relatively small and disjunct populations that may exhibit reduced levels of genetic variability (Small et al. 2003). In some cases, isolated populations may be differentiated by a single nuclear allele or mitochondrial haplotype (e.g., Admiralty Island in southeast Alaska; N. Dawson, unpublished data). Higher inbreeding coefficients for Pacific marten populations may signal that reduced population size has resulted in a loss of genetic diversity (N. Dawson, unpublished data). Finally, several populations of this species are believed to have been extirpated, and some remnant populations appear to contain relatively few individuals (Slauson et al. 2009).

A taxonomic classification that accurately reflects geographic variation in genetic characteristics can also be an important tool for designing effective reintroduction strategies for restoring species to their former ranges and protecting the genetic legacy of populations (Schwartz 2007). For example, if translocations are planned to augment remnant populations of Pacific martens along the Pacific coast, it is imperative that translocated individuals be obtained from other Pacific marten populations. Human-mediated introductions on islands in southeast Alaska may have led to the subsequent disappearance of Pacific marten populations from those islands (Hoberg et al., this volume). Maintaining the genetic legacy of species and populations requires collaboration between researchers and management agencies (Isaac et al. 2007). Revising the taxonomic status of North American martens provides a new empirical framework for managing broad- and fine-scale genetic diversity among North American martens, and for gaining a better understanding of the biology of North American *Martes*, including their behavior, ecology, physiology, parasitology, and conservation.

Acknowledgments

Funding for Natalie Dawson's dissertation was provided by the USDI U.S. Fish and Wildlife Service, USDA Forest Service Alaska Region and Forestry Sciences Laboratory, and National Science Foundation (NSF DEB 0196095 and 0415668). Technical assistance was provided by the USDI U.S. Geological Survey Alaska Science Center. We thank Rod Flynn, Anson Koehler, David Nagorsen, Stephen MacDonald, Richard Popko, Tom Schumacher, Karen Stone, Alasdair Veitch, and Jack Whitman for help with various aspects of our work on martens over the past decade. A special thanks to Yadeeh Sawyer for assistance with graphics.

3

Complex Host-Parasite Systems in *Martes*

Implications for Conservation Biology of Endemic Faunas

ERIC P. HOBERG, ANSON V.A. KOEHLER, AND JOSEPH A. COOK

ABSTRACT

Complex assemblages of hosts and parasites reveal insights about biogeography, ecology, and the processes that structure faunal diversity and the biosphere in space and time. Exploring components of parasite diversity in *Martes* and other mustelids can help elucidate the intricate history of the northern biota. Of particular importance in this chapter are issues related to episodic climate variation, biotic expansion, geographic colonization, and host-switching as determinants of the geographic distribution of parasites. We examine these processes for species of *Taenia* (tapeworms), *Trichinella* (nematodes), and, in particular, the large stomach nematode *Soboliphyme baturini*. We address the following questions: (1) Why are parasites important to *Martes* biologists? (2) How do coevolutionary, ecological, and biogeographic processes interact to determine the structure of complex host-parasite faunas in space and time? (3) What signals are revealed by parasites that help elucidate the history, ecology, and biogeography of their hosts? (4) Why are these issues particularly important to conservation biology in a time of accelerating climate change? and (5) Why should mammalogists and parasitologists collaborate to build a comprehensive framework for understanding the biosphere? We argue that, in addition to their role as potential disease agents, parasites provide important information about the origins, distributions, and evolutionary and recent histories of host species. Genetic signatures can be used to demonstrate (1) endemic host populations, (2) persistence of parasite lineages and species in the absence of ancestral hosts (i.e., evidence of broader historical ranges for host species), and (3) introduced populations. Such signatures can also reveal historical interactions among host lineages and species (e.g., ecological relicts, patterns of contact, sympatry, extirpation, extinction). All can have important implications for wildlife management and conservation biology, because genetic signatures of parasites help reveal the history of host species, such as martens, at fine temporal or spatial scales, as well as the interactions of historical and anthropogenic factors that structure the biosphere.

A Parasitological Context

Parasites are ubiquitous and occur in all groups of vertebrates and invertebrates. These relatively cryptic organisms, including both microparasites (prions, viruses, bacteria, protozoans) and macroparasites (flatworms, roundworms, ticks, lice, fleas) exert considerable influence on ecosystems at multiple spatial scales (e.g., Hudson et al. 2006). Although they are a phylogenetically disparate set of organisms, all depend on a host for survival or reproduction or both, and all have the potential to cause disease. Their life cycles may involve a single host (direct), or be considerably more complex (indirect), with adult and developmental stages of the parasite circulating through a series of hosts that may include both invertebrates (usually intermediate hosts) and vertebrates (usually definitive or final hosts where the parasite reproduces). Infection may adversely affect host organisms, including morbidity and mortality, or have less-obvious effects on fecundity, fitness, and behavior. Thus, parasites may contribute to the regulation of host populations and may interact with multiple determinants of disease (e.g., habitat, weather, climate change, large-scale disturbances, contaminants) and other stressors. Although parasites can cause serious diseases in host organisms and adversely affect host populations, a new understanding is emerging that healthy ecosystems are characterized by diverse parasite faunas (Hudson et al. 2006).

Parasites have predictable associations with their hosts, and circulation is often dependent on particular biological pathways; consequently, parasites provide useful information about ecosystem structure, function, and integrity (Hoberg 1997b; Marcogliese 2005). Breakdown of these linkages or pathways from environmental perturbations can have cascading effects on faunal diversity. These pathways often include interactions among one or more host species, the parasite, and its developmental life history stages, as well as the ecosystem occupied by the parasite. For example, many helminth (worm) macroparasites have complex life cycles that reflect intricate food webs. Consequently, parasites are important components of biological structure that also provide links between ecological and evolutionary time scales (Hoberg 1997b; Brooks and Hoberg 2000; Hoberg and Brooks 2008).

Parasites can provide mammalogists and wildlife biologists with novel insights about ecology, evolution, and conservation biology. Fundamental questions include the following: (1) How do parasites influence hosts and host populations (Hudson et al. 2006)? (2) What can parasites reveal about the ecology and evolutionary history of mammals (Brooks and Ferrao 2005; Koehler 2006; Koehler et al. 2007; Galbreath 2009)? (3) What do parasites tell us

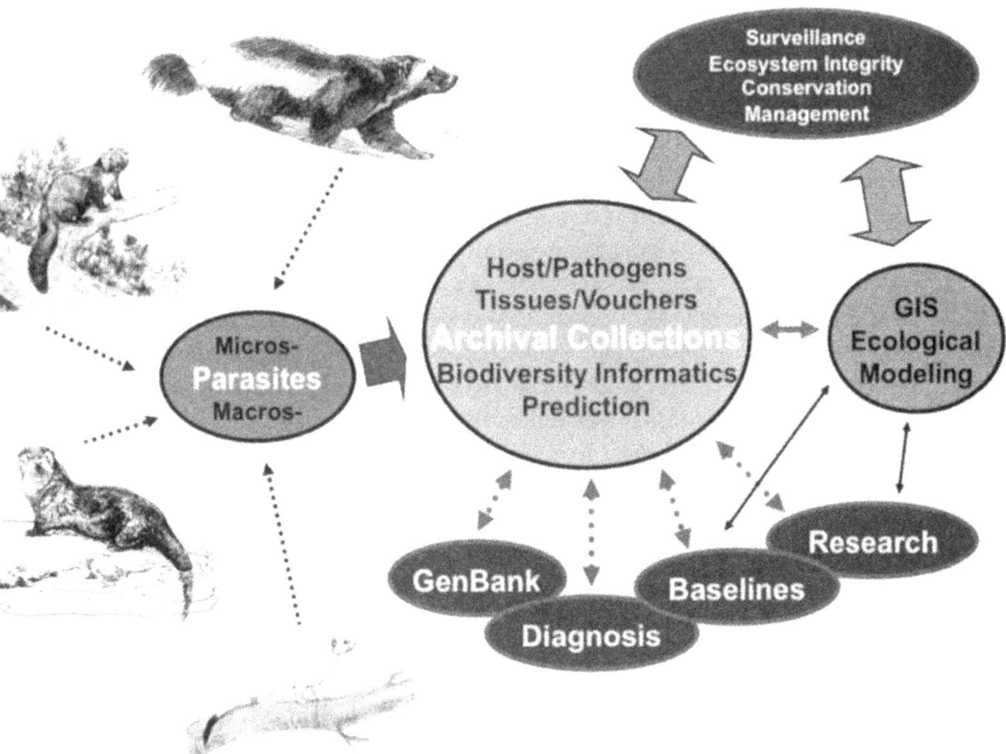

Figure 3.1. Schematic model for ecosystem approaches to document parasite diversity based on integrated surveys and inventories, archival specimen-based collections, and biodiversity informatics resulting from collaborations among state and federal wildlife biologists, commercial fur trappers, mammalogists, parasitologists, and pathogen ecologists. Specimens representing an assemblage of sympatric mustelids (or other mesocarnivores) are collected and preserved as the first step in a rigorous assessment of microparasite and macroparasite diversity. These specimens are then used in an integrated evaluation of parasite communities using both molecular and morphological approaches to characterize diversity (faunal assemblages, species, and populations). Specimens are deposited in archival collections, where georeferenced data are linked to tissues and vouchers for hosts and parasites, along with their definitive identifications. Biodiversity information becomes the focal point for diagnostics, development of temporal and spatial baselines, and diverse research activities, including ecological modeling and prediction in a regime of environmental change. Images of the wolverine and ermine (*Mustela erminea*) were drawn by the late William D. Berry and made available for this figure by Berrystudios (http://www.berrystudios.biz); images of the river otter (*Lontra canadensis*) and marten are by William Hines, and are in the public domain.

about patterns of endemism, contact zones, and gene flow of host species (Haukisalmi et al. 2001; Wickström et al. 2003; Nieberding et al. 2004, 2008; Nieberding and Olivieri 2007; Koehler et al. 2009; Galbreath and Hoberg 2012)? (4) Why is an integrated approach needed for studying and understanding the nature of biodiversity (Brooks and Hoberg 2000; Hoberg et al.

2003; Cook et al. 2005)? and (5) Why should parasites be considered integral to conservation biology and wildlife management, particularly in the context of accelerating environmental changes (Hoberg et al. 2008; Brooks and Hoberg 2006, 2007)?

In this chapter, we present several examples of the ways in which parasites can provide new insights into the ecological and evolutionary processes involved in faunal diversification and persistence. We demonstrate how fundamental principles of biogeography can be elucidated in the context of historical processes linking macroparasite faunas in American marten (*Martes americana*), Pacific marten (*M. caurina*), sable (*M. zibellina*), and related mustelids. We also explore the biological indicators provided by parasites that reveal how evolutionary history and recent wildlife management have affected the distribution of host-parasite systems in martens.

Ultimately, we address the broader implications of an understanding of these issues for wildlife management and the conservation biology of endemic faunas. Our overarching goal is to demonstrate how close collaborations between mammalogists and parasitologists can create a more integrated framework for understanding the biosphere (Figure 3.1), enabling each to gain new insights about the particular systems they study (e.g., Brooks and Hoberg 2000; Hoberg et al. 2003; Cook et al. 2005).

We followed Wozencraft's (2005) mustelid taxonomy, with the exception that we consider *M. caurina* to be a valid species based on both morphological studies (Merriam 1890), and recent molecular evidence (Small et al. 2003; Dawson and Cook, this volume). In addition, because recent research indicates that the wolverine (*Gulo gulo*) is congeneric with *Martes* (Stone and Cook 2002; Koepfli et al. 2008; Hughes, this volume), we also included the wolverine in our assessments, as appropriate.

Parasites in Mustelidae: A Beringian Connection

Mustelids are characterized by a diverse fauna of micro- and macroparasites that has been assembled over evolutionary time (Kontrimavichus 1985; Table 3.1). Survey and inventory data are available for 8 of 9 species of *Martes* and for *G. gulo*; however, the scope and depth of parasitological data representing each species is highly variable (e.g., Kontrimavichus 1985; Hoberg et al. 1990; Foreyt and Lagerquist 1993; Koehler et al. 2007). In addition, both phylogenetic and phylogeographic resolutions for parasite groups and hosts have been patchy (Koehler et al. 2009). Nevertheless, increased knowledge of the historical biogeography and phylogeography (Avise 2000) of certain nematode parasites in martens (*Soboliphyme baturini, Trichinella* spp.) is contributing to a broader understanding of the historical phenomena that link Eurasian and North American faunas (Zarlenga et al. 2006; Koehler

Table 3.1. Macroparasite diversity (species richness) of helminths in Mustelidae and *Martes* spp.

Taxa	Total	Obligate[a]	Facultative[b]	Incidental[c]
Mustelidae	216	106	66	44
Martes spp.	60	33	18	9

Source: Based on compilations by Kontrimavichus (1985)

[a] Obligate: limited in distribution to host species or host groups (genus, family); considered to be host-specific in distribution

[b] Facultative: fauna shared with other mesocarnivores based on dynamics of foraging guilds or exposure through common habitat use; may be considered a component of the typical fauna

[c] Incidental: anomalous records for macroparasites that are characteristic of other vertebrate hosts

et al. 2009; Pozio et al. 2009a). Consequently, we have focused this chapter primarily on the historical biogeography of North America, emphasizing episodes of colonization by parasites of both hosts and geographic regions (e.g., Waltari et al. 2007; Hoberg and Brooks 2008; Galbreath and Hoberg 2012).

Beringia is the region that connected North America to Asia during glacial events resulting from lowered sea level, creating opportunities for biotic interchange between the Palearctic and Nearctic regions (Figure 3.2). These faunal interchanges across Beringia are essential for understanding the history of parasite faunas in *Martes*. A land connection between North America and Asia was present through much of the Tertiary and served as a primary pathway for biotic interchange during times of relatively equitable climate. Tectonics and global cooling in the late Tertiary altered this dynamic. During glacial events in the late Pliocene and Pleistocene, Beringia strongly influenced patterns of species distribution and diversity, alternately serving as a barrier or as a pathway to the expansion of terrestrial biotas and as a center for diversification (Hopkins 1959; Sher 1999). The episodic nature of at least 20 glacial cycles during these epochs (Gibbard and van Kolfschoten 2004), patterns of biotic expansion and isolation at inter- and intracontinental scales, and the formation and dissolution of intermittent ice-free refugia (Figure 3.2) had substantial effects on a complex mosaic of host-parasite systems (e.g., Rausch 1994; Hoberg et al. 2003; Dragoo et al. 2006; Zarlenga et al. 2006; Pozio et al. 2009a; Shafer et al. 2010) from the late Tertiary through the Quaternary (Lister 2004; Haukisalmi et al. 2006; Zarlenga et al. 2006). Biotic expansion has been asymmetrical (Rausch 1994; Hoberg 2005a; Waltari et al. 2007), however, involving primarily eastward dispersal from Eurasia into North America, as evidenced by assemblages of North American mammals, their associated micro- and macroparasites, and other organisms.

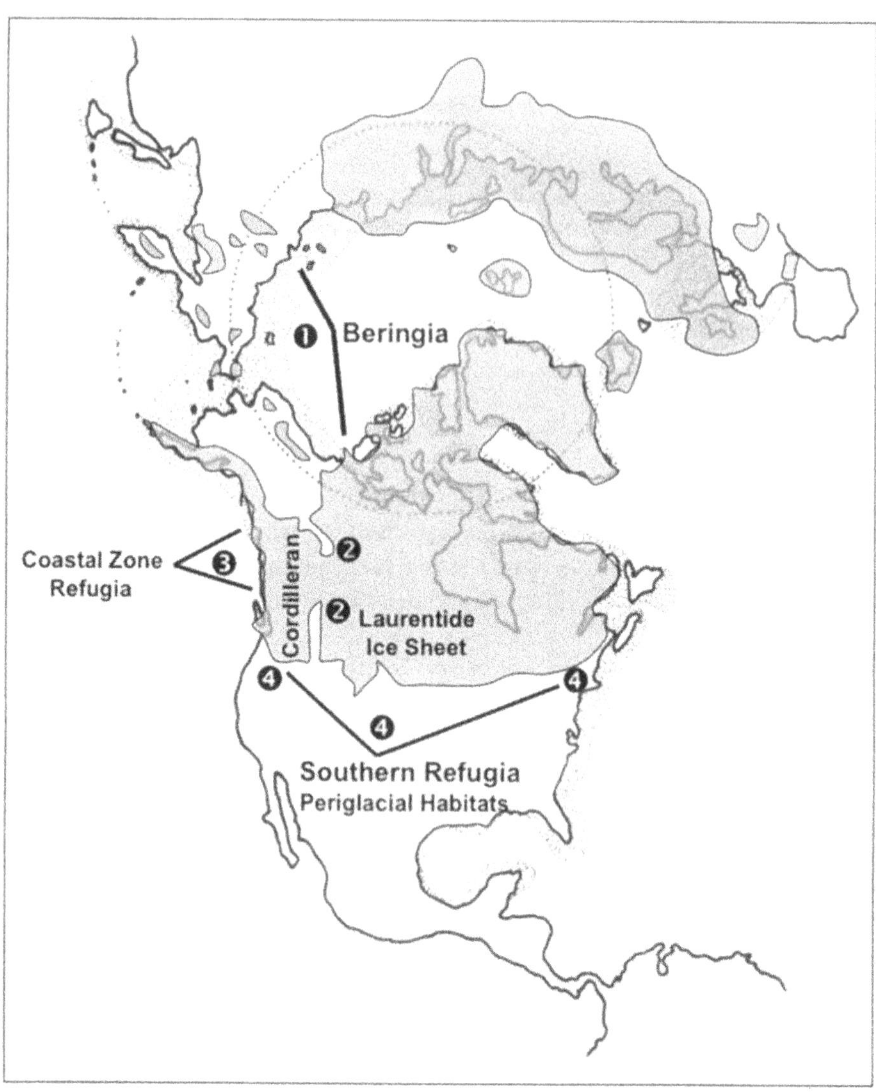

Figure 3.2. Polar projection of the geographic distribution of continental glaciers and glacial refugia in the Holarctic region during the last glacial maximum about 18,000 years ago. The major continental ice masses are shown in gray, and the extent of exposed continental shelf is stippled. The Arctic Circle (66°33'N) is indicated with a dotted line. Substantial refugial zones were present (1) in Beringia and peripheral habitats at high latitudes of North America and eastern Eurasia; (2) in isolated zones between or within the Cordilleran and Laurentide ice sheets, including the ice-free corridor that developed at the end of the Pleistocene; (3) along the western coastal zone, including the Alexander Archipelago and Queen Charlotte Islands; and (4) in periglacial habitats south of the ice sheets. This figure was created from information presented by Pielou (1991), Dyke et al. (2003), Harington (2005), and Shafer et al. (2010).

Diversification of a complex assemblage of parasites has been structured by processes of coevolution (association by descent for host and parasite lineages) but also, more importantly, by patterns of episodic colonization of hosts and geographic regions over varying spatial and temporal scales (e.g., Hoberg 2005b; Hoberg and Brooks 2008, 2010). In addition, some of these host-parasite systems, especially among *M. caurina* and *M. americana* from North America and *M. zibellina* from eastern Siberia (Kontrimavichus 1985; Koehler 2006; Koehler et al. 2009), have been strongly influenced by anthropogenic forces such as translocation and introduction (e.g., Hoberg 2010). Consequently, the parasite faunas of mustelids represent complex spatial and temporal mosaics (Hoberg and Brooks 2010). In general, reticulate host associations are common, there is minimal evidence for co-speciation, and distributions among mustelids have largely been determined by the processes of ecological fitting and colonization (Brooks et al. 2006; Agosta and Klemens 2008; Hoberg and Brooks 2008). These insights have emerged from an integrated series of phylogenetic and phylogeographic studies of both hosts and parasites.

Helminth Faunas in Mustelids

The composition of helminth faunas in mustelids since the Miocene has been strongly influenced by independent waves of biotic expansion of both hosts and parasites (e.g., Koepfli et al. 2008; Koehler et al. 2009). Nematodes, cestodes, and digenean trematodes in mustelids are phylogenetically distinct, reflecting the strong ecological mechanisms involved in faunal assembly over time (Kontrimavichus 1985). Colonization of new hosts has occurred, as well as some level of co-speciation. Host-switching, which has been documented for *Taenia* spp. (tapeworms), *Trichinella* spp. (nematodes), and the large stomach nematode *Soboliphyme baturini*, often involves other mesocarnivores (felids, canids) exploiting common prey resources in the context of foraging guilds (e.g., Hoberg 2006; Zarlenga et al. 2006; Koehler et al. 2007, 2009).

Taenia spp.

Species of *Taenia* are obligate parasites of carnivores that have rodents, lagomorphs, or ungulates as intermediate hosts, and their complex life cycles depend on these predator-prey relationships (Loos-Frank 2000; Hoberg 2006). There are 5–6 known species among mustelids, including 2 in *Martes* (*T. intermedia* from Holarctic regions, *T. martis* from Eurasia) and 1 in *G. gulo* (*T. twitchelli* from Holarctic regions) (Rausch 1977, 2003). *Taenia intermedia* is a typical tapeworm parasite in *M. americana*, *M. caurina*, and the fisher (*M. pennanti*) from North America and in *M. zibellina* from eastern

Siberia, reflecting a history of expansion across Beringia from Eurasia, with subsequent isolation in North America during the Pleistocene. More broadly, their distribution within host species reflects a deep coevolutionary association, whereby sister-group relationships link Procyonidae and Mustelidae with particular species of *Taenia* (e.g., Hoberg 2006; Bininda-Emonds et al. 2007). The designation of *T. mustelae* as the basal species in the genus, however, supports independent histories and some level of host-switching in diversification, providing additional evidence that foraging guild dynamics among an assemblage of small carnivores are important drivers of parasite evolution.

Trichinella spp.

Martens and wolverines are particularly important for the circulation of several species of *Trichinella* nematodes in sylvatic cycles among mustelids and other mesocarnivores, bears, and an array of omnivorous prey species, including rodents and insectivores (Pozio 2005; Pozio et al. 2009a). Carnivory and scavenging represent the primary routes for transmission. For example, the importance of these pathways has been demonstrated by the very high prevalence (>60%) of *Trichinella* in *M. americana* from British Columbia (Schmitt et al. 1976), although these parasites may not be as abundant in other regions (Hoberg et al. 1990).

The contemporary distribution of *Trichinella* in mustelids and other carnivores provides further evidence for the role of pervasive host-switching (driven by prey-sharing) and episodic expansion from Eurasia to North America (Zarlenga et al. 2006; Pozio et al. 2009a). Events that led to the radiation of the North American fauna during glacial epochs were facilitated by the development of land connections and dispersal corridors that modified patterns of geographic isolation. Geographic distributions and patterns of speciation were further determined by refugial effects at inter- and intracontinental scales during the Pleistocene (e.g., relative to the Laurentide [continental] and Cordilleran [Pacific coastal] ice masses (Shafer et al. 2010). These patterns have been demonstrated for an array of phylogenetically and historically disparate host-parasite systems that arose in Eurasia, providing additional evidence of the spatially and temporally congruent forces that have shaped these faunas.

Soboliphyme baturini

This large stomach nematode of mustelids has a geographic distribution centered in Beringia and surrounding areas, and is primarily a parasite of *M. americana* and *M. caurina* in North America, *M. zibellina* in eastern Siberia, and *G. gulo* throughout its Holarctic range (Kontrimavichus 1985; Karmanova 1986; Koehler et al. 2007); it is apparently rare in *M. pennanti*. Infections in other mesocarnivores and small mustelids (*Mustela* spp.) are considered incidental because the parasites do not develop to maturity (Kontrimavichus 1985; Koehler et al. 2007). These robust and conspicuous nematodes attain a length of 2–4 cm and can be very abundant (up to 200 individuals reported in single hosts), with infections persisting for up to 20 months. Pathology associated with *S. baturini* is linked to ulceration of the gastric mucosa and anemia, but the overall impact of infection on the host is poorly understood (Karmanova 1986). More than 50 species of mammals are recognized as hosts, including an assemblage of carnivores (final hosts), rodents, and insectivores (intermediate hosts).

The life cycle of *S. baturini* is complex and involves adults in the stomachs of mustelids (final hosts), eggs deposited in feces, and development of larval nematodes in annelids (segmented worms) residing in the leaf litter (Karmanova 1986). Larvae are ingested by carnivores either by eating infected worms (intermediate hosts), or by preying on rodents and insectivores serving as paratenic hosts (essentially, hosts where larval development does not occur, but which bridge the trophic gap between annelid worms and mustelids); the latter route appears to be most important for maintaining transmission (Karpenko et al. 2007; Koehler et al. 2007).

The *S. baturini* complex is a geographically widespread assemblage involving a single parasite species with multiple hosts (Koehler et al. 2007). Phylogeographic analyses have revealed a protracted evolutionary history with *Martes* in the late Pleistocene, but no evidence that parasite speciation accompanied host diversification. In contrast, among these parasites, substantial population structure has been demonstrated that apparently resulted from geographical and physical events, rather than from an association with host phylogeny (Koehler et al. 2009). An exploration of this history reveals the influences of both abiotic and biotic processes on the distribution of complex host-parasite assemblages in space and time. Additionally, an understanding of the distribution of *S. baturini* in *M. americana* and *M. caurina* provides important insights for the management and conservation of endemic host populations.

Temporal and Geographic Associations of *Martes* spp. and *Soboliphyme baturini*

The biogeographic histories of obligate (host-specific) parasites in mustelids have been strongly influenced by those of the host group. Spatial barriers or connections (e.g., Bering Land Bridge) and temporal events (e.g., glacial advances) are reflected in host histories, but patterns of co-speciation between these hosts and their parasites are not always evident (Koehler et al. 2009). In contrast, the biogeography of facultative (not host-specific) parasites has been driven largely by colonization of both hosts and geographic regions (the dynamic for ecological fitting), as well as by trophic interactions among mustelids and other carnivores (Pozio et al. 2009a; Hoberg and Brooks 2010). Past climatic changes have been the primary abiotic determinants of the distribution and diversification of this fauna. Further, there is a strong ecological signal among both obligate and facultative components of the assemblage. By exploring host and parasite biogeographic histories, we can gain new understandings of these dynamic processes. Our current hypothesis for the biogeographic history of this host-parasite assemblage, which we describe below, is derived from the primary analyses of Koehler et al. (2009) in conjunction with new collections that have revealed intricate patterns of diversity, particularly in former refugial habitats along the coastal zone of Alaska (E. Hoberg, A. Koehler, and J. Cook, unpublished data).

The intricate biogeographic history of *S. baturini* can be understood only within a biogeographical context for species of *Martes* in Beringia and North America. *Martes caurina* and *M. americana* represent the sister group of *M. melampus*, *M. zibellina*, and *M. martes* (Stone and Cook 2002; Stone et al. 2002). Divergence estimates place the split between North American and Eurasian martens about 600–700 kyBP, which coincides with an interglacial preceding the Nebraskan glacial stage (Koepfli et al. 2008). This contrasts with a deeper estimate of about 1 myBP for the origins of *M. caurina* and *M. americana* (Stone et al. 2002). Regardless of this apparent discrepancy, both studies indicate an initial expansion by *Martes* into North America no later than the middle Pleistocene. The critical point is that these dates bracket and set the lower temporal limits for the biogeography and diversification of this host-parasite assemblage.

During the ultimate (Wisconsin) glaciation in North America, martens occurred in ice-free refugia south of the Laurentide ice sheets (Figure 3.2). Paleontological evidence for *M. americana* in the east and *M. nobilis* (an extinct species that may be conspecific with *M. caurina*) in the west and apparently into the Yukon Territory and possibly Alaska, is consistent with the expansion of *Martes* throughout North America by the penultimate (Illinoian) glaciation (Hughes 2009) or earlier (Stone and Cook 2002; Koepfli et al. 2008). *Martes nobilis* was present in Alaska during Wisconsin time (N. Dawson, University of New Mexico, unpublished data). The current distribution of North American martens is consistent with isolation in western (*M. caurina*) and eastern (*M. americana*) refugia during the late Pleistocene, and subsequent expansion during the Holocene (Dawson and Cook, this volume). Molecular clock estimates for a split between these sister species at 600–700 kyBP indicate that their evolutionary histories were influenced by multiple glacial-interglacial cycles (Lister 2004). Although host speciation occurred, similar patterns of diversification in *S. baturini* are not recognized;

thus, the associations of *S. baturini* with *M. caurina* and *M. americana* represent a complex temporal and spatial mosaic.

There is no evidence of multiple colonization events by martens into North America from Eurasia during the Pleistocene, indicating monophyly for extant species of *Martes* and *Soboliphyme*. This contrasts with other North American mustelids that experienced multiple colonization events from Eurasia, including *M. pennanti* and *G. gulo* (Koepfli et al. 2008). These successive colonizations played an important role in establishing the fauna of medium-sized mustelids in North America. Thus, the temporal mosaic represented in the assemblage of this mustelid fauna and its characteristic macroparasites has taken place over many millions of years from the late Miocene through the Pleistocene.

Phylogenetic and phylogeographic structure in *S. baturini* is consistent with geographic colonization from Eurasia across Beringia, and an initial southward expansion with sequential isolation in insular (southeast Alaska and Alexander Archipelago) and coastal refugial zones along the west coast of North America (Figures 3.2, 3.3). Our initial view of this system found that relatively high genetic diversity in coastal and insular populations of *S. baturini* is consistent with an extended period of geographic occupation (since the mid-Pleistocene) and persistence of the assemblage in putative ice-free refugia for hosts and parasites until late-Wisconsin time (Koehler et al. 2009). These conclusions should be further tested with a comprehensive multigene assessment conducted within a coalescent framework (Galbreath et al. 2010). Additional sampling of hosts and parasites in the western contiguous United States would also help evaluate the validity of our biogeographic hypothesis.

During glacial cycles preceding the Holocene, and by Illinoian time, ancestral martens were distributed across temperate latitudes of North America south of the former margin of the Laurentide ice sheet, consistent with divergence estimates by both Stone et al. (2002) and Koepfli et al. (2008). Although martens diverged during isolation in eastern and western refugia, *S. baturini* did not. Persistent populations of *M. caurina* (and *S. baturini*) during the late Pleistocene and Holocene were influenced by waves of expansion eastward to Wyoming and northward into the coastal zone bordering the eastern Pacific (Koehler et al. 2009). Consequently, populations of *S. baturini* in the Alexander Archipelago and adjacent coastal zone are represented by a mosaic of older or paleoendemics with high levels of geographically structured genetic variation (Koehler et al. 2009), and more recently derived colonizers (neoendemics) with low levels of genetic structure (E. Hoberg, A. Koehler, and J. Cook, unpublished data). It appears, however, that *S. baturini* was lost in populations of *M. americana* during isolation in eastern refugia south of the Wisconsin ice sheets (Koehler et al. 2007, 2009).

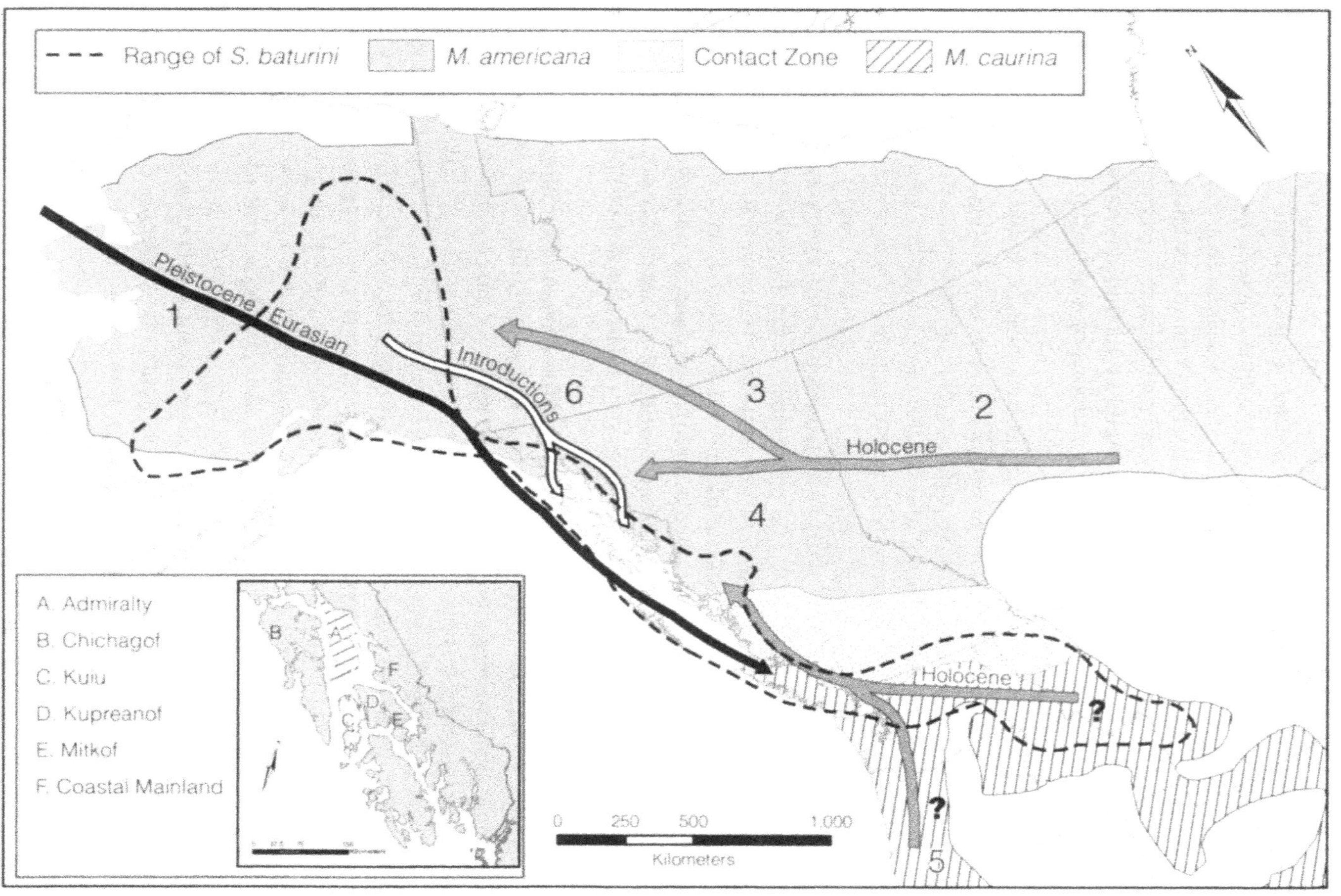
Range of S. baturini
M. americana
Contact Zone
M. caurina
Pleistocene
Eurasian
Introductions
Holocene
Holocene
1
2
3
4
5
6
?
?
A. Admiralty
B. Chichagof
C. Kuiu
D. Kupreanof
E. Mitkof
F. Coastal Mainland
A
B
C
D
E
F
0
250
500
1,000
Kilometers

Figure 3.3.—*cont.*

Biogeographic history of *Martes americana, M. caurina*, and *Soboliphyme baturini* in North America during the Pleistocene and Holocene, based on information reported by Koehler et al. (2007, 2009). (1) *Soboliphyme baturini* colonized North America from Asia with *Martes* during the Pleistocene and expanded its range southward along the Pacific coastal zone and Alexander Archipelago in ice-free refugia. Refugial (relictual) populations of *S. baturini* are linked directly to those of *M. caurina*; (2) *M. americana* expanded from eastern refugia during the Holocene as the continental ice sheet melted; (3) Holocene expansion of *M. americana* northward in the ice-free corridor into eastern Beringia and interior Alaska resulted in the colonization of *M. americana* as a host by *S. baturini*; (4) Holocene expansion of *M. americana* westward as the Pacific coastal ice sheet melted resulted in colonization of the coastal zone and near-shore islands, and contact with *M. caurina* resulted in host-switching by *S. baturini*; and (5) Holocene expansion probably occurred northward by *M. caurina* and resulted in contact with *M. americana* in the north. It remains unclear whether both the Coast Range and southern Rocky Mountains were sources for northward colonization. In addition, (6) natural range expansions and human-mediated introductions of *M. americana* and associated parasites from interior and coastal zones into the Alexander Archipelago resulted in a complex mosaic of endemic and introduced populations of hosts and parasites, and patterns of local extirpation and faunal turnover in contact zones. Map inset shows islands in the Alexander Archipelago that are discussed in the text.

The restricted occurrence of *S. baturini* among populations of *M. americana* (and complete absence from eastern North America) has been attributed to 2 or more independent events during the Holocene involving postglacial colonization of northern regions by *M. americana*. Geographic colonization of interior and southeast Alaska during the Holocene was asynchronous because *M. americana* recolonized northern regions via 2 different pathways: into Beringia through the ice-free corridor that formed as the Cordilleran and Laurentide ice sheets began to melt, and directly into the coastal zone after the ice sheets melted (Figures 3.2, 3.3). These expansions led to secondary contact among *M. americana*, *M. caurina,* other mustelids, and *S. baturini*, with this recent history of sympatry and colonization reflected in patterns of minimal genetic variation among parasites found in different hosts (Koehler et al. 2009). During the last glaciation, offshore islands in southeast Alaska and southwestern British Columbia may have provided ice-free refugia, but coastal and nearshore areas were covered with ice (Carrara et al. 2007). Thus, neoendemic populations of *M. americana* in the coastal zone reflect colonization from the east during the Holocene as deglaciated habitats became suitable. These populations of *M. americana* would have been initially free of parasites, and acquisition of *S. baturini* could have occurred only when they came in contact with populations of *M. caurina* or, less likely, with populations of other resident mustelids.

In the Alexander Archipelago, *M. caurina* currently occurs on Admiralty, Kuiu, Queen Charlotte (now named Haida Gwaii), and Vancouver islands, which provided ice-free refugia during the last glaciation (MacDonald and Cook 2007). All except Kuiu Island (for which no parasites are currently known) have endemic and relictual populations of *S. baturini*. In the Alexander Archipelago and interior Alaska, populations of *S. baturini* that infect

M. americana have switched from their ancestral hosts, *M. caurina*. Recent work on the population genetic structure of *S. baturini* in this region demonstrates the persistence of distinct paleoendemic populations throughout the archipelago in *M. caurina*, which contrasts with the recent geographic colonization of interior Alaska, the coastal zone, and nearshore islands by *M. americana* with host-switching by *S. baturini* during the Holocene (Koehler et al. 2009; E. Hoberg, A. Koehler, and J. Cook, unpublished data).

In the context of our current hypothesis, Chichagof Island provides a good example of the complexity of these host-parasite associations. There is no evidence that martens were present on Chichagof Island historically; however, *M. americana* was introduced to the island between 1949 and 1952 (Stone et al. 2002; MacDonald and Cook 2007). Sources for these translocations included coastal islands (Baranof, Revillagigedo, Wrangell, and Mitkof islands), the coastal mainland (Stikine River), and interior Alaska (near Anchorage). *Martes americana* was also introduced to Baranof and Prince of Wales islands from the coastal mainland (Thomas Bay and Behm Canal) in the 1930s (MacDonald and Cook 2007). Consequently, all nearshore and mainland populations of *M. americana* represent either natural colonizations during the Holocene or human-mediated introductions during the 20th century. Genetic differentiation among populations of *S. baturini* that infect *M. americana* on these islands is minimal, however, providing additional support for a coevolutionary history limited to the Holocene.

In phylogeographic analyses, these nematodes cluster with either interior or coastal clades; thus, they indicate independent geographic sources for introductions and provide important insights about the geographic extent of endemic populations (Koehler et al. 2009; E. Hoberg, A. Koehler, and J. Cook, unpublished data). On Chichagof Island and elsewhere in the archipelago, parasites originated from 3 distinct geographic sources: (1) paleoendemic, ecological relicts (direct evidence of the historical occurrence of *M. caurina* on these islands and of *S. baturini* host-switching to *M. americana* prior to the extirpation of *M. caurina* from Chichagof Island); (2) introductions from interior Alaska (including 3–4 events to Chichagof and Prince of Wales islands); and (3) introductions from the mainland coastal zone to Chichagof Island (multiple events). At a minimum, 4 putative translocation signals are demonstrated by parasites that involve populations on Chichagof and other islands (E. Hoberg, A. Koehler, and J. Cook, unpublished data). These patterns are apparent against a background of endemic populations that, in some cases, have undergone expansion from insular and near-coastal refugial zones during the Holocene. The persistence of these nematodes in *M. caurina*, which currently occurs only on Vancouver, Admiralty, Kuiu, and Queen Charlotte islands, suggests that this assemblage was considerably more widespread historically and emphasizes the urgency of examining and identifying all the insular marten populations in these archipelagos (see also Stone et al. 2002).

The existence of these relictual populations of *S. baturini* suggests that there is considerable genetic diversity remaining to be discovered in this system, including insular ecological relicts and the genetic effects of introductions of *M. americana*. Genetic patterns may be further complicated by local extirpations of hosts, parasites, or both, and whether viable alternative definitive hosts occur in this region. The latter source appears unlikely, given the incidental nature of most infections among other mustelids and mesocarnivores (Koehler et al. 2007).

Genetic patterns among populations of *S. baturini* indicate that *M. nobilis*/*M. caurina* persisted late into the Pleistocene in some areas where it is now extinct (see Hughes 2009). This has significant implications for the conservation of endemic, ancestral populations of *M. caurina* in the archipelago, and ongoing competitive interactions in zones of contact with *M. americana*. Genetic data on relictual populations of *S. baturini* are consistent with the displacement of *M. caurina* by *M. americana*, although Hughes (2009) considered this unlikely because she believed that habitat overlap for these species was minimal historically. Today, at least 2 areas of contact have been documented for these species, with one potentially extending a considerable distance across southwestern Canada and the central Rocky Mountains of the western United States (Dawson and Cook, this volume).

Invasion, Translocation, and Conservation Biology

Martes americana, *M. caurina*, and *M. pennanti* have all experienced substantial range contractions during the past century (e.g., Strickland et al. 1982b,c). Populations on the southern periphery of their historical range often represent isolated, relictual endemics (e.g., fisher in California) or a mixture of ancestral and recently translocated populations. Such translocations may result in genetic mosaics reflecting multiple source populations, including those from geographically distant locations (e.g., MacDonald and Cook 2007). In addition, interactions between the number of individuals involved in translocations of hosts and parasites may introduce sampling bias (i.e., founder effects) that can have strong effects on the prevalence of macroparasites because there are varying degrees of uncertainty associated with being introduced, becoming established, and cycling within host populations (e.g., Hoberg 2010). Sampling bias can affect species diversity and population structure because some species and some components of source populations may not have been included in translocations (e.g., Koehler et al. 2009).

Historical bottlenecks for species of *Martes* in North America will influence the occurrence of macroparasites at local, regional, and continental scales. Where macroparasites are obligate (e.g., *Taenia intermedia*; *S. baturini* may depend on martens for transmission) co-translocation depends on parasite life histories and host densities. In contrast, among facultative parasites that circulate in multiple species of small mustelids (e.g., *T. mustelae*) and mesocarnivores (*Trichinella* spp., *Molineus* spp., *Mesocestoides* spp.), we would expect less-obvious effects from bottlenecks for host species.

Introduction of host species is an invasive process that can result in cotranslocation of parasites, if ecological conditions in source and target localities are comparable (e.g., Hoberg 2010). Colonization of new hosts may accompany introduction through breakdowns in the mechanisms that maintain ecological isolation (Brooks and Hoberg 2006; Hoberg and Brooks 2008), and such ecological perturbations are often linked to the emergence of infectious diseases (Hoberg and Brooks 2010). Naïve hosts can be exposed to an array of new parasite populations, resulting in host-switching among assemblages of mustelids and mesocarnivores at the point of introduction; however, if translocated hosts are too few in number to maintain the transmission cycle, then the parasite will not become established.

Factors that may influence the likelihood of introducing new parasites during mesocarnivore translocations include (1) parasite specificity and longevity (i.e., a long-lived parasite may have enhanced opportunities to become established); (2) the presence of alternative mammalian hosts in the target area; (3) life history of the parasite being translocated (i.e., direct or indirect life cycle); and (4) ecological structure (e.g., role of ecological fitting and breakdown in ecological isolation among translocation events). Reintroductions of hosts are analogous to rapid geographic expansion and may therefore result in loss of associated parasites or loss of genetic diversity in both hosts and parasites (e.g., Torchin et al. 2003). The potential effects of introductions, or rapid naturally driven invasions, may also include infection (or faunal turnover) of translocated hosts by novel parasites from the sharing of prey species among local mesocarnivores (e.g., Hoberg and McGee 1982).

Several questions about host-parasite interactions among translocated populations are of particular interest to *Martes* biologists: (1) How do faunas among introduced (translocated) martens in now disjunct populations compare with those of hosts remaining in the core historical range? (2) Is there ecological release from some array of parasites for those translocated populations? and (3) Have bottlenecks linked to translocations artificially reduced parasite diversity (e.g., Torchin et al. 2003)? We currently lack the data to adequately address these questions because there are few published surveys documenting parasite diversity in *Martes* spp. and *G. gulo* from North America (e.g., Dick and Leonard 1979; Hoberg et al. 1990; Foreyt and Lagerquist 1993; Seville and Addison 1995; Koehler et al. 2007). When conducted in conjunction with the development of archival collections, rigorous field surveys and inventories using both morphological and molecular methods of parasite identification can establish powerful baselines to assess parasite faunal diversity among *Martes* spp. and other mesocarnivores (Figure 3.1) (e.g.,

Cook et al. 2005; Hoberg et al. 2009). They can also provide the empirical basis for direct comparisons of species richness and genetic diversity among parasites. For example, evidence from surveys of *M. caurina* in its core range in the Pacific Northwest suggests that the species diversity of helminth parasites exceeds that observed for *M. americana* in Manitoba, Alaska, and the Northwest Territories (Hoberg et al. 1990). These are not, however, direct comparisons of source and introduced populations; other historical and environmental factors, including latitudinal gradients or local effects, may also influence parasite diversity in martens (e.g., Foreyt and Lagerquist 1993; Seville and Addison 1995). In *M. americana* and *M. caurina*, there was minimal abundance, prevalence, and diversity of helminths other than *Soboliphyme* in coastal zone habitats, and most parasites were characteristic of other carnivores (Koehler et al. 2007; A. Koehler, unpublished data). Thus, faunal diversity in these zones may reflect the legacy of invasion of insular habitats on varying temporal scales.

Based on observations from the Alexander Archipelago, parasites (including those with complex life histories involving intricate trophic pathways and intermediate or paratenic hosts) can become established when translocated with hosts from remote sites (Koehler et al. 2007). In addition, ecological signals and population genetic signatures demonstrate multiple and diverse source populations for these parasites that represent faunal mosaics resulting from translocation events into particular geographic localities. In these instances, genetic profiles for introduced parasites are clearly distinct from those associated with endemic lineages (ecological relicts).

Parasites and the Conservation of Biodiversity

Parasite populations can provide unique information for elucidating the origins, distributions, and histories of host species (Criscione et al. 2005; Nieberding and Olivieri 2007; Nieberding et al. 2008; Galbreath 2009; Galbreath and Hoberg 2012). Genetic signatures can be used to demonstrate the presence of either endemic or introduced host populations. They are also useful for documenting the persistence of parasite lineages or species in the absence of ancestral hosts, which provides evidence of a broader historical range for the host species. Further, such signatures can reveal historical interactions among host lineages and species (e.g., ecological relicts, patterns of contact, sympatry, extirpation, extinction). Thus, these interactions have important implications for wildlife management and conservation, because genetic signatures for parasites can reveal host histories at fine temporal scales and help elucidate the interaction of historical and anthropogenic factors.

The potential and recognized consequences of climate change for biodiversity and ecological structure clearly indicate the need to understand the geographic distributions and histories of complex host-parasite assemblages (e.g., Parmesan and Yohe 2003; Parmesan 2006; Lawler et al. 2009). Environmental shifts linked to climate can eliminate ecological barriers and constraints on the development, transmission, and distribution of parasites (e.g., Brooks and Hoberg 2000, 2006, 2007; Hoberg et al. 2008; Lafferty 2009). Not only will the geographic distributions of many hosts and parasites shift as climate changes, but largely unanticipated ecological effects will also likely result in the emergence of diseases in new locations (Brooks and Ferrao 2005; Brooks and Hoberg 2007). In the Western Hemisphere, some regions may experience near 90% faunal turnover, suggesting that host-parasite systems among *Martes* and other mesocarnivores will be substantially different under global warming (Lawler et al. 2009). Thus, understandings derived from range shifts by both hosts and parasites that occurred during previous glacial or interglacial epochs in Beringia and southeast Alaska represent an important historical analogue for identifying the factors that either limit or facilitate the introduction of potentially invasive species under both natural and anthropogenic control (Cook et al. 2005; Hoberg 2010).

With respect to parasites, potential translocations of *Martes* spp. must be evaluated in an ecological context that includes sympatric mesocarnivores and small mustelids. Ecosystem studies with integrated specimen-based surveys are needed to establish patterns of parasite diversity in both source and target areas (Hoberg 1997a). This will help determine the potential for both introduction and exposure. Museum specimens obtained from wild populations of both hosts and parasites should be linked to comprehensive archival databases with Web-based accessibility (Hoberg et al. 2003; Cook et al. 2005; Hoberg et al. 2009; Figure 3.1). Archives provide a permanent record of faunal structure and represent essential baselines for assessing stability and change. For environments undergoing accelerating changes, these resources serve as indicators of biodiversity losses, introductions, and the origins and structure of new ecological associations. The strong heuristic value of this approach has been demonstrated in southeast Alaska and the Alexander Archipelago and, more broadly, in Beringia (Cook et al. 2005; MacDonald and Cook 2007). Such comprehensive specimen-based collections have enabled researchers to use the genetic characteristics of both hosts and parasites to reveal geographic sources and distributional histories in these assemblages, as well as unexpected patterns of endemism in host species and populations (Galbreath 2009; Koehler et al. 2009).

Parasites can provide unique insights about the structure of the biosphere. Consequently, an understanding of geographic variation and the biogeographic histories of hosts and their parasites form an important cornerstone for conservation. Parasites are elegant indicators of geographic sources and ecological linkages (trophic associations, habitat use) for host species and populations (Hoberg 1997b). Parasites also provide important information about disease conditions (morbidity and mortality), and unique insights about complex biological phenomena that can be critical in future management decisions. Thus, to better inform the conservation of host populations, we urge *Martes* biologists to establish integrated research programs that incorporate an explicit consideration of the nature and history of host-parasite interactions.

Acknowledgments

Funding for integrated field studies of complex host-parasite systems that form the basis for this chapter was provided by the National Science Foundation as the primary sponsor of the Beringian Coevolution Project (BCP) (NSF0196095 and 0415668). Additional field support was provided to the BCP by the USDA Tongass National Forest and Pacific Northwest Research Station, and by the USDI U.S. Fish and Wildlife Service and U.S. Geological Survey. The authors sincerely thank Keith Aubry and our anonymous reviewers for substantial comments that resulted in clarification of a number of concepts and improvements to the chapter.

4

Distribution Changes of American Martens and Fishers in Eastern North America, 1699–2001

WILLIAM B. KROHN

ABSTRACT

Contractions in the geographic distributions of the American marten (*Martes americana*) and fisher (*M. pennanti*) in eastern North America south of the St. Lawrence River between Colonial times (ca. 1650–1800) and the fisher's recent range expansion (ca. 1930–present) are well documented, but causal factors in these range contractions have only partially been studied. Traditional explanations for range contractions by both species are forest clearing and unregulated trapping; little consideration has been given to alternative explanations. It has been hypothesized that deep snow limits the distribution of fishers, and that high fisher populations limit the distribution of martens. I assessed the potential contributions of these factors to observed range contractions for these species by evaluating expected patterns of change in their historical distributions since Colonial times. Using published data on the distribution of martens and fishers in eastern North America, including early and contemporary fur-harvest records ($n = 60{,}702$), I found that broad-scale changes in their geographic distributions in eastern North America were consistent with 3 of those expectations, and partially so with a 4th. I recognize that retrospective analyses cannot establish the relative importance of land clearing, unregulated trapping, and changing climatic conditions on observed range contractions; nevertheless, when historical data from eastern North America are viewed in the context of long-term climate warming and the results of recent ecological studies, they suggest that traditional arguments may only partially explain historical range contractions for both species. This study further suggests that under a warming climate, northern range boundaries for the fisher will expand, and southern range boundaries for the American marten will continue to contract.

Introduction

Documented contractions in the geographic distributions of many species of North American wildlife since the Colonial era (ca. 1650–1800) are usually attributed to habitat loss or overharvest (i.e., trapping, hunting, or a combination of both), with little consideration given to alternative explanations. For example, it is well documented that compared with the late 1700s through 1800s, the ranges of the American marten (*Martes americana*) and fisher (*M. pennanti*) have decreased substantially in eastern North America (e.g., Seton 1909, 1929; Hagmeier 1956; Gibilisco 1994; Proulx et al. 2004). Despite this extensive documentation, formal investigations of the underlying causes have not been conducted, and biologists continue to attribute these range losses solely to unregulated fur harvests and loss of forested habitats. For example, Strickland and Douglas (1987: 532) wrote that the marten's "most suitable habitat has been lost throughout the southern primordial range as a result of land clearance. . . . In some areas where adequate habitat persisted, overtrapping caused local extirpation." Powell (1994a: 11) stated that "American martens and fishers reached their nadir early in this century owing to overexploitation for fur and to habitat loss," and Whitaker and Hamilton (1998: 439) wrote that "in the late 1800s and early 1900s [the fisher] was extirpated over most of its eastern range by over-trapping and loss of habitat." Although timber harvesting (Black 1950; Ahn et al. 2002) and unregulated trapping (Moloney 1931; Ray 1987) were widespread activities historically that likely affected these 2 species at some locations during some time periods, they may not have been the only factors involved.

There is increasing recognition that relatively recent climatic changes have altered the distributions and abundances of numerous plant and animal species (e.g., Schneider and Root 2002). Krohn et al. (1995, 1997) proposed 2 hypotheses relating to the distribution of martens and fishers that involved climate: (1) deep snow limits the distribution of fishers, but not of martens, as a result of much higher energy costs for fishers to travel or hunt in such conditions (a direct climatic effect); and (2) high fisher populations limit the distribution of martens, possibly because of interference competitions by fishers (primarily predation, an indirect climatic effect).

Generally, species distributions are limited at northern range boundaries by abiotic factors, whereas at southern range boundaries, where environmental conditions tend to be less extreme, they are limited by biotic interactions (Brown and Lomolino 1998). The notion that fishers are limited in the north (and at high elevations) by deep snow, whereas martens are limited in the south (and at low elevations) by competitive interactions with fishers, is consistent with this generality. Moruzzi et al. (2003) assessed an unsuccessful reintroduction of martens in the Green Mountains National Forest of southern Vermont and concluded that failure may have been due to competition with fishers. In contrast, there is evidence of a slowly recovering marten population in the highlands of northern New Hampshire (Kelly et al. 2009b), which is an area of considerably deeper and more extensive snowfalls than southern Vermont (Krohn et al. 2004). Carroll (2007) reported correlations across eastern North America in the broad-scale occurrence of American martens with mean annual snowfall >300 cm/year and certain forest types (i.e., mature conifer and mixed). A large genetic study of fishers (n = 769) from southeastern Ontario through northwestern New York concluded that snow depth was an important component of fisher habitat; as snow depth increased, the proportion of immigrants in an area also increased (as occurs in sink habitats), suggesting that deep-snow environments were of lower quality (Carr et al. 2007b). In a recent study in northern Idaho, Albrecht et al. (2009) found that fishers were associated with forestlands with low snowfall, whereas martens were generally associated with deep snowfall areas; in addition, martens were recorded in some low-snowfall forests that lacked fishers, suggesting that fishers may limit martens. Fisher predation was the major cause of natural mortality in studies of >100 radio-collared martens in north-central Maine (Hodgman et al. 1997; Payer 1999). These studies occurred in a region of relatively deep snow and few fishers (Maine Dept. of Inland Fisheries and Wildlife, unpublished harvest data). In northern Wisconsin, predation by fishers and raptors was the major natural source of mortality of adult martens (too few juveniles were captured to estimate mortality in this age class) in a population that failed to expand >30 years after being reintroduced (McCann et al. 2010). Although the preceding studies did not demonstrate that fishers limited marten populations, interspecific predation is common among carnivores (Palomares and Caro 1999), especially among similar species, and predation rates increase with differences in body sizes between predator and prey (Donadio and Buskirk 2006). In the East, the average body mass of fishers is 5.4–6.5 times larger than that of martens (Krohn et al. 2004).

In this chapter, I examine historical changes in the geographic distributions of American martens and fishers, both of which were once abundant and widely distributed throughout the forested regions of eastern North America (i.e., south of the St. Lawrence River and east of the Great Lakes). I give special attention to identifying spatiotemporal patterns in historical fur-harvest records for martens and fishers, and determining whether or not observed patterns are consistent with the expectations that follow from the fisher/snow and fisher/marten hypotheses. If historical patterns are consistent with those expectations, they may provide new insights into reasons the distributions of these species have changed substantially during the last 300 years or so.

Approach, Expectations, and Data

The Little Ice Age (LIA) was a climatically cool period that occurred from ca. 1450 to 1800 (Imbrie and Imbrie 1986). This was also a period of more severe and frequent storms, especially at higher latitudes; in the Northern Hemisphere, the LIA was an exceptionally cold and snowy era (Imbrie and Imbrie 1986; Grove 1988; Mann 2002). In the northeastern United States, the 1600s and 1700s were considerably colder and snowier than they are now (Baron 1992; Zielinski and Keim 2003). Thus, during the LIA, heavier and more persistent snow cover would likely have occurred farther south and at lower elevations in eastern North America than they do currently.

If the fisher/snow and fisher/marten hypotheses are true, then I would expect the historical distributions of martens and fishers to have been much different than they are today, both elevationally and latitudinally. During the Colonial era, which occurred when areas of deep snow presumably reached their maximum extent, I would expect that (1) both species would have occurred farther south than they do now, with fishers occurring farther south than martens; and (2) martens would have had a more continuous distribution than they do now (i.e., less restricted to higher elevations, since deeper snow would have shifted fishers southward). After the LIA ended during the early 1800s, and areas of deep snowfall began contracting in geographic extent, (3) the southern edge of the marten's range would have shifted northward, and martens would have become more restricted to higher elevations. The fisher's northern range boundary would also have shifted northward compared with Colonial times, but fishers would have retained a more southerly distribution than that of martens. Last, given the hypothesized negative effect of dense fisher populations on marten survival, (4) during both time periods, areas supporting high fisher populations would have had low populations of martens, and vice versa.

To obtain data useful for evaluating these expectations, I started with references listed in Seton (1909, 1929) and Hagmeier (1956). Hagmeier's seminal paper on marten and fisher distributions was especially useful because he presents data sources by individual states and provinces, so his references are more spatially explicit than Seton's, which are based on more general sources. Because martens and fishers (along with American beavers [*Castor canadensis*]) were important target species during the early fur trade in North America, 18th- and 19th-century fur records were available for study (Moloney 1931; Ray 1987). Truck (i.e., fur-trade) houses were established in the American colonies during the 17th and 18th centuries, well before the western fur trade developed; although most of these records were lost to fires (especially those archived in Boston), a few have survived. I also examined published reports from natural history surveys conducted during the 19th century in all northeastern states (Merrill 1920), along with 19th- and early 20th-century technical and scientific literature on these 2 species in the East. During these time periods, trapping records were located in sporting (i.e., fishing, hunting, trapping, natural history) books and journals. For example, *Forest and Stream*, a popular outdoor sporting journal during the 1800s and early 1900s, regularly contained articles by prominent scientists and amateur naturalists of the day, as well as some written by trappers.

Previous studies of historical occurrence records have been largely descriptive; in contrast, I gave special attention to locating and evaluating documented fur-harvest records. When studying historical changes in species distributions, written records have 2 major advantages over descriptive accounts: they represent quantitative counts that can be analyzed, and they are less subject to misidentification because the observer had the opportunity to handle and study the animals. Furthermore, because martens and fishers were common in the eastern fur trade, both historically and recently, they are less subject to observational error than are rare or elusive species (see McKelvey et al. 2008). Clearly, however, the reliability of such records is only as good as that of the people who recorded them. Consequently, I used information from nonprofessional observers (e.g., trappers) only when I believed those persons to be reliable sources (see biographical sketches in Krohn and Hoving 2010: 475–506).

There is a strong latitudinal gradient of declining mean annual snowfall from north to south in eastern North America (e.g., Zielinski and Keim 2003: 86; Krohn et al. 2004: 125). To evaluate relative changes in the distributions of these 2 species over time along this spatial gradient, I tabulated fur-harvest records for various time periods to determine whether the ratio of martens to fishers in the harvest was highest in the north (region with greatest snowfall), and lowest in the south (region with little or no snowfall where fishers predominated). Similarly, when assessing changes in marten-to-fisher ratios in a given area over time, I expected a decrease in the ratio of martens to fishers in the fur-harvest as that area warmed near the end of the LIA, and environmental conditions presumably improved for fishers.

By their nature, historical data are incomplete and must be interpreted cautiously. For example, in 1546, the French were trading with native Americans at the mouth of Chesapeake Bay for "as many as a thousand marten skins" (Quinn 1979: 218). While such an early observation is interesting, I did not include it in this analysis. During the initial days of European trade in eastern North America, extensive native trade routes were still intact. For example, skins of the American bison (*Bos bison*) were transported from the Appalachian and Allegheny mountains down major eastern rivers and northward along the coast, where they were traded with natives in eastern Canada. Bison skins were highly prized as flooring in winter quarters. Once Native

Americans formed alliances with the competing European powers, however, trade quickly became more local. For example, with the French and English competing west of the Appalachians, it is highly unlikely that any quantities of northern furs controlled by the French would be traded in Virginia (Crane 1928; Parrish 1972).

Results

Expectation 1: Southern Boundaries for American Martens and Fishers

Ernest Thompson Seton published the earliest detailed range maps for martens (Seton 1909: 905; 1929: 485) and fishers (Seton 1909: 929; 1929: 455), showing that both species occurred much farther south historically than they do today (Figure 4.1). The sources for Seton's range limits are given in the legend of his maps, although it is not clear exactly what occurrence records he used. A later and more thoroughly documented study was that of Hagmeier (1956), who used about 250 references to delineate historical and current ranges for both species; about 80 of these references reported information from eastern North America. Both Hagmeier (1956) and Seton (1909, 1929) reported that, historically, the southern range limit for the marten was north of that for the fisher. Allen (1876), Coues (1877), and Keay (1901) all described martens and fishers as common inhabitants of forests south of New England. The southern range limits for both species were reported to occur farther south during the 1800s than they do now; Audubon and Bachman (1852) recorded fishers as far south as the mountains of North Carolina and Tennessee, and Rhoads (1903) reported that martens occurred as far south as north-central Pennsylvania during the early 1800s.

Fur-harvest records from the British Colonial Office document the number of pelts exported from Virginia to England from 1699 to 1715 (Table 4.1). Both species were regularly shipped in large numbers during this 17-year period, with reported totals of 1432 martens and 3355 fishers (Table 4.1). Interestingly, the wolverine (*Gulo gulo*), which like the marten is associated with deep snow (Aubry et al. 2007), was exported in low numbers in 1699 and 1704 (Table 4.1).

In summary, records from early explorers, fur traders, and naturalists showed that fishers ranged as far south as the Cumberland Plateau from the late 1600s to the early 1800s. Martens occurred north of fishers (Seton 1909, 1929; Hagmeier 1956), and continued to occur as far south as north-central Pennsylvania into the late 1880s (Rhoads 1903). Thus, consistent with Expectation 1, both species were reported to occur farther south in historical times than they do now, and fishers occurred farther south than martens.

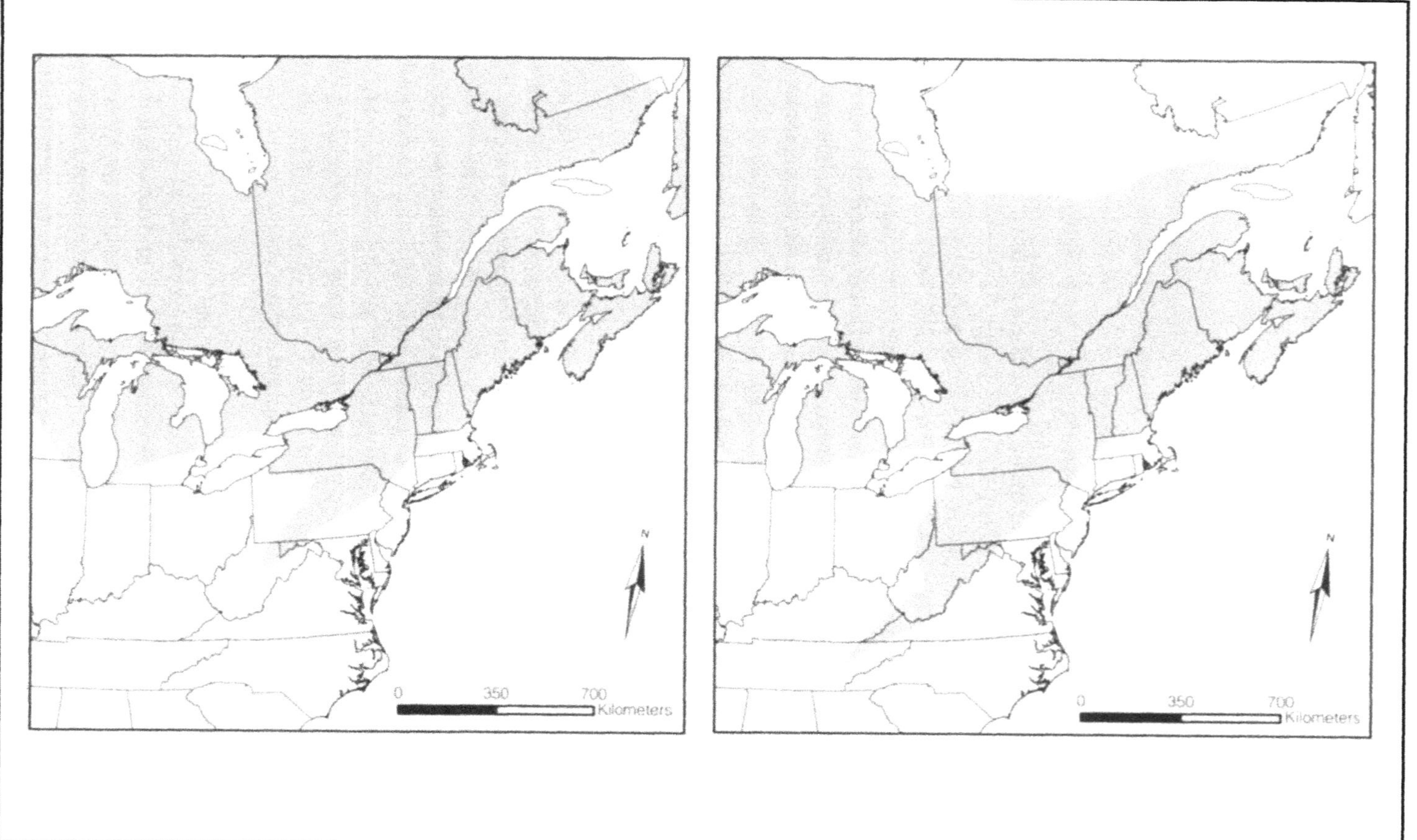

Figure 4.1. Historical distributions of the American marten (*left*) and fisher (*right*) in eastern North America (Seton 1909: 905, 929). Hagmeier's (1956) range maps indicated that the southernmost distribution of fishers included the mountains of western North Carolina and eastern Tennessee, and that American martens occurred along the coast in southern New England.

Table 4.1. Number of American marten, fisher, and wolverine pelts exported from Virginia to England from 1699 to 1715

Year	American marten	Fisher	Wolverine[a]
1699	5	163	7
1700	0	106	0
1701	0	70	0
1702	0	463	0
1703	0	5	0
1704	58	58	4
1705	112	108	0
1706	4	97	0
1707	0	0	0
1708	0	416	0
1709	12	496	0
1710	35	18	0
1711	76	1100	0
1712	1130	151	0
1713	0	0	0
1714	0	90	0
1715	0	14	0
Total	1432	3355	11
Mean	84	197	0.65

Sources: British Colonial Office (CO) Papers (Microfilm: CO class 5, reel 10, frames 133–135); Hagmeier (1956: 162, 164) labeled these data the "Account Showing the Quantity of Skins and Furs Exported Annually . . . from Virginia from . . . 1698 to . . . 1715." His source of this information was "Bailey, J.W. 1946. The Mammals of Virginia . . . Richmond, privately printed, 416 p."

Note: Harvest data were also reported for beaver, deer, moose (*Alces alces*), and other wildlife species (not shown here)

[a] The broad-scale occurrence of the wolverine, like American marten (Carroll 2007) and Canada lynx (Hoving et al. 2005), is strongly associated with deep snows (Aubry et al. 2007). Thus, with historical climate warming in eastern North America, the ranges of all 3 species would be expected to have declined, as has been documented (marten, this study; lynx, Hoving et al. [2003]; wolverine, Krohn and Hoving [2010]).

Expectation 2: Distribution of American Martens—Continuous or Disjunct?

Both species occurred throughout the interior forests of eastern Canada (Hardy 1869; Adams 1873), through New England (Emmons 1840; Thompson 1842; Allen 1876) and New York (DeKay 1842; Merriam 1882), and into north-central Pennsylvania (Rhoads 1903). Keay (1901) reviewed 15 early accounts published between 1524 and 1675, and considered both species to be common and distributed continuously throughout New England during the Colonial era, except along the coast. Allen (1876: 713–714) reported that both martens and fishers "were common inhabitants of not only the whole of New England, but also of the Atlantic States generally as far south as Virginia (excepting possibly a narrow belt along the seaboard)." Fur-harvest records compiled for this study also show that, during historical times, martens and fishers were reported to be more widely distributed than they are today. Thus, consistent with Expectation 2, martens were reported to occur continuously from the interior forests of Pennsylvania north through New York and New England into eastern Canada (e.g., Keay 1901; Rhoads 1903). Although martens did not occur in coastal forests, I found no indication in the historical record that they were limited to interior highlands and mountains.

Expectation 3: Northward Range Shifts after the Little Ice Age

In the 9 states where martens occurred historically, the species was extirpated in all but Maine and New York (Table 4.2); and, in the 15 states where fishers occurred historically, the species was apparently extirpated in all but Maine, Vermont, New Hampshire, and New York (Table 4.2). In New York, both species persisted only in the Adirondack Mountains in the east-central part of the state. In the states that supported both species historically, and where the year of extirpation had been reported, martens tended to be extirpated before fishers (Table 4.2), suggesting that the 2 species were not equally vulnerable to the factors causing extirpation. If deep snow was the main factor reducing contact between martens and fishers, then as the climate moder-

Table 4.2. Approximate years in which American martens and fishers were extirpated in the eastern United States during the 19th and 20th centuries

State	American marten	Fisher
Maine	never extirpated	never extirpated
Vermont	1926	almost extirpated[a]
New Hampshire	1940	almost extirpated[a]
Massachusetts	historically present[b]	~1900
Connecticut	historically present[b]	after 1924
Rhode Island	no record of occurrence	~1938
New York[c]	1930	almost extirpated
Pennsylvania	1900	1903
New Jersey	1853	~1889
Maryland	1880	probably present historically
West Virginia	no record of occurrence	rare in 1911
Virginia	before 1851	1890
North Carolina	probably absent	1854
South Carolina	probably absent	probably present historically
Tennessee	probably absent	1881

Source: Hagmeier (1956) unless otherwise footnoted

[a] Nearly went extinct during the 1920s and 1930s (Hagmeier 1956; Silver 1957)

[b] In the mountains in the western part of the state

[c] Excluding the Adirondack Region, where both species survived at higher elevations (DeKay 1842; Merriam 1882)

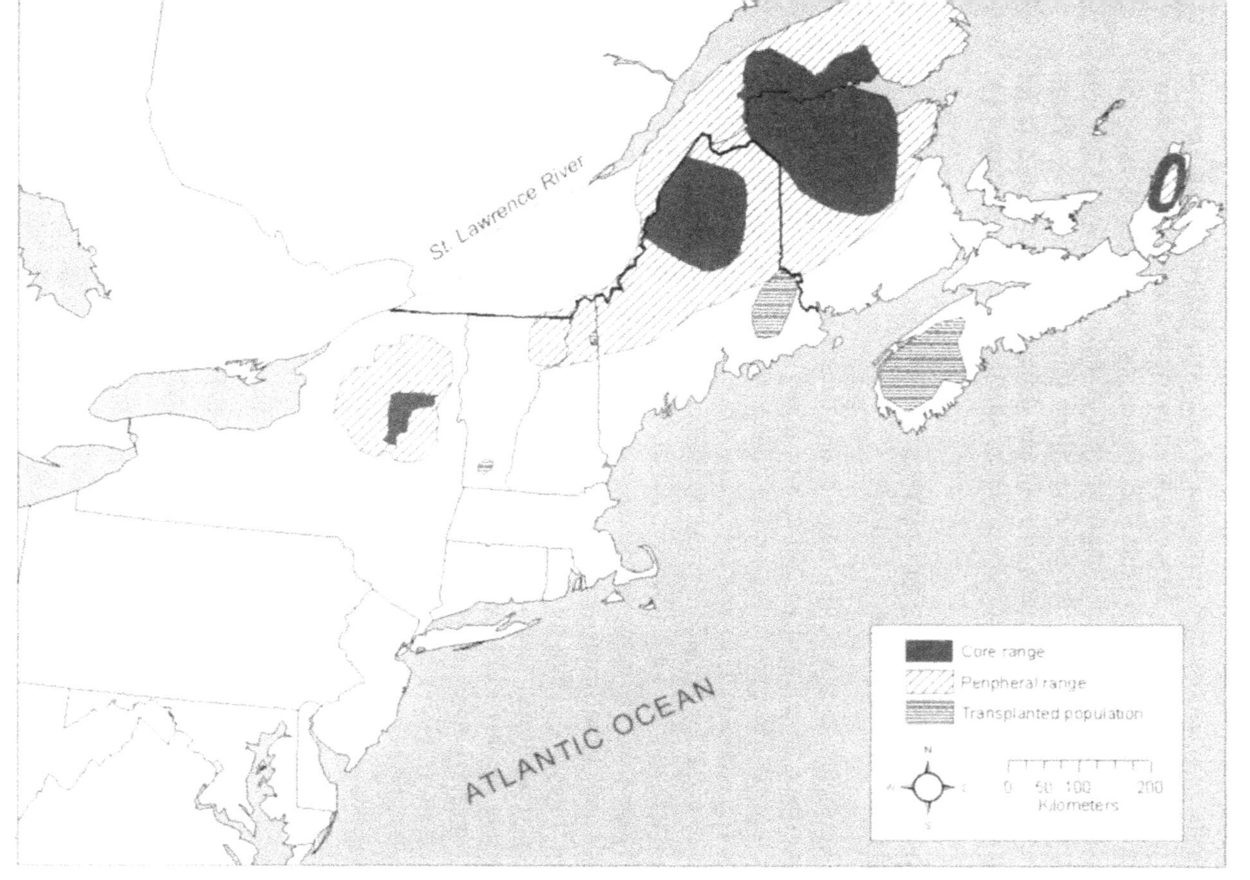

Figure 4.2. Current range of the American marten in eastern North America south of the St. Lawrence River, based on harvest densities (Krohn et al. 1995; Carroll 2005b) and information provided by furbearer biologists in these states and provinces. Note that the highlands of Cape Breton Island and the northeastern Gaspe Peninsula are both high-elevation, deep-snowfall areas, but are not part of the marten's core range, apparently because both are essentially unforested as a result of extreme environmental conditions. The current status of transplanted marten populations is variable, but they do not appear to be robust, high-density populations (e.g., Moruzzi et al. 2003; Kelly et al. 2009b).

ated, marten populations should have contracted northward and to higher elevations, whereas fisher populations should have expanded into areas where snowfall deceased substantially. The current distribution of martens is consistent with this hypothesis; they are no longer distributed continuously in interior forests and have become restricted to relatively high-elevation areas at higher latitudes (Figure 4.2). However, both martens and fishers were extirpated in many eastern states (Table 4.2). Fishers did not persist as expected; rather, their range contracted along with that of martens, although this was not as extensive geographically. Because all my findings were not consistent with Expectation 3, this suggests that climate change was not the only factor causing range contractions in these 2 *Martes* species after the end of the LIA.

Expectation 4: Relative Abundance of American Martens and Fishers

To examine changes in the relative abundance of both species across time and space, I evaluated 12,250 historical and 48,452 contemporary fur-harvest records. The ratio of martens to fishers reported from fur-harvests in Virginia from 1699 to 1715 was 0.43:1 ($n = 4787$) (Table 4.1) compared with 24.8:1 ($n = 6821$) for fur records from Machias, Maine (near the New Brunswick border) and St. John (Pleasant Point), New Brunswick from 1764 to 1778 (Table 4.3). The ratio of martens to fishers documented by trappers in Maine and New Brunswick from 1850 to the early 1900s was 8.58:1 ($n = 642$; Table 4.4). Thus, historical and recent fur-harvest records indicate a steady decrease in the ratios of martens to fishers in Maine and New Brunswick during the last 250 years (Tables 4.4 and 4.5). Although the number of fishers harvested increased relative to martens in both areas, the marten-to-fisher ratio was higher in New Brunswick than in Maine during the latter 2 time periods (1850–early 1900s and 1997–2001; Table 4.5).

Altogether, the ratios of martens to fishers in fur-harvest records was 58 times greater (= 24.8/0.43; $n = 11{,}608$) in the northern part of their range (i.e.,

Table 4.3. Number of American martens and fishers reported harvested from Maine and New Brunswick from 1764 to 1778

Location	Time period	American martens	Fishers	Source
Machias, Maine, USA[a]	1777–1778	367	6	Anonymous (1779)
Portland Point (St. John), New Brunswick, Canada	1764–1775	6190	258	Raymond (1950: 157–158)[b]

[a] Located 42.6 km southwest of the border with New Brunswick, Canada

[b] Numerous other furbearer pelts shipped, including 8 wolverine and 67 Canada lynx

Table 4.4. Documented harvests of 3 furbearers during fur hunts in the mid-1880s to early 1900s in northern Maine, USA and New Brunswick, Canada

		Species harvested			
Year(s)	Location	American marten	Fisher	Lynx[a]	Source
1850	Restigouch and Kedgwick rivers, New Brunswick	108	0	2	Palmer (1949: 3–4)
1858	Northern Maine and upper Tobique and Nepsiquit rivers, New Brunswick	19	1	2	Journal and Letter[b]
1859	Caucogomoc Lake region, northern Maine	~50	4	3	Hardy (1910: 929)
1860	Northwest of Chensuncook Lake, northern Maine	0	2	15	Palmer (1949: 13)
1860–61	Northwest of Chensuncook Lake, northern Maine	100	9	20	Hardy (1903: 263)
1862	Tobique River country, New Brunswick	9	9	10	Palmer (1949: fn 37)
1865	Patten area of north-central Maine	1	1	3	Letter[c]
1868	Patten area of north-central Maine	1	8	3	Letter[d]
1876–77	Upper Magalloway River and Parnachenee Lake, western Maine	49	15	6	Barker and Danforth (1882: 238)
1891–92	North-central New Brunswick	98	0	22	Braithwaite (1892: 6)
1894–95	Rangeley Lakes region of western Maine	30	7	0	Anonymous (1895: 6)
1895–96	Upper Magalloway River and Parmachenee Lake, western Maine	50	4	0	Haywood (1897: 107)
early 1900s	Gulquac River (tributary of the Tobique River), New Brunswick	60	7	0[e]	Shaw (1987: 48)
Totals		575	67	86	

[a] See footnote a in Table 4.1

[b] Journal by Manly Hardy titled "Notes of a Trip to Tobique—1858," and a letter from William H. Staples to Manly Hardy dated February 3, 1859 (both are located in the Manly Hardy Collection [MHC], Special Collections, Raymond H. Fogler Library, University of Maine, Orono), or see Krohn (2005)

[c] Letter from William H. Staples to Manly Hardy dated December 16, 1865 (MHC)

[d] Letter from William H. Staples to Manly Hardy dated April 28, 1868 (MHC)

[e] By 1900, bobcats had moved inland and outnumbered lynx throughout northern New England and eastern Canada (Hoving et al. 2003)

Table 4.5. Temporal changes in the ratio of American martens to fishers in fur harvests from Maine and New Brunswick, 1764–2001

Period (source)	Maine (n)	New Brunswick (n)
1764–1778 (Table 4.3)	Data pooled with New Brunswick[a]	24.8 (6821)
1850–early 1900s (Table 4.4)	5.9 (351)	17.2 (291)
1997–2001 (harvest records)[b]	1.6 (32,000)	3.6 (16,452)

[a] The Maine sample was obtained near the Maine/New Brunswick border; thus, for this time period, samples from Maine (n = 373) and New Brunswick (n = 6448) were pooled

[b] Maine data courtesy of the Maine Department of Inland Fisheries and Wildlife; New Brunswick data courtesy of the New Brunswick Fish and Wildlife Branch

Maine and New Brunswick) than in the southern part (i.e., furs exported from Virginia) from 1699 to 1778. Also, marten-to-fisher ratios were higher in New Brunswick than in Maine. Thus, as climatic conditions warmed and snowpacks decreased, the ratios decreased steadily through time in both areas from 1764 to 2001 (n = 55,915).

Discussion

This was a retrospective assessment, preventing me from evaluating relevant factors independent of potentially confounding conditions. Thus, I cannot rule out overtrapping or forest clearing as important factors contributing to range losses for both *Martes* species in the East. Furthermore, because the fisher failed to expand or even maintain its range when climatic conditions warmed after the LIA, factors other than climate change clearly played a role. The northward pattern of range contraction, however, and the fact that martens apparently responded before fishers, are consistent with climatic warming as a causal factor but inconsistent with historical patterns of European settlement, forest clearing, and agricultural development. Although deforestation occurred throughout the eastern United States (it was much less extensive in eastern Canada) and, in some states, resulted in more cleared than forested land (e.g., Black 1950; Ahn et al. 2002), deforestation did not occur in a south-to-north pattern, nor was it uniform throughout the East (Greeley 1925). Colonial land clearing and settlement occurred westward from the initial European settlements centered in a core area from Boston to Philadelphia (Walker 1872; Greeley 1925), suggesting a pattern that would have split marten and fisher populations into separate northern and southern areas. In 1850, the Appalachian Mountains south of New York State were still covered with virgin forests (Greeley 1925). Even as late as 1920, numerous large areas (>25,000 acres) of virgin forests still remained in the Appalachians south of Pennsylvania (Greeley 1925). Thus, it appears that these forests were still structurally capable of supporting martens and fishers, yet the southern Appalachians experienced the earliest extirpations of both species (Table 4.2).

The fur trade was a major economic activity of the early settlers and occurred throughout eastern North America. Although the fur trade targeted different species in various regions (e.g., white-tailed deer [*Odocoileus virginianus*] in the south and beaver in the north), the geographic expansion of fur harvesting occurred along numerous fronts and was not unidirectional across eastern North America (Crane 1928; Moloney 1931; Parrish 1972; Ray 1987). Thus, the extirpation of martens and fishers by overtrapping would not be expected to proceed from south to north, but such a pattern would occur if climatic warming were a major causal factor in observed range dynamics.

In this study, I have provided evidence that the geographic distributions of the American marten and fisher contracted in eastern North America after the close of the LIA, and that the relative abundance of these 2 species varied in both time and space. If the contraction of southern range boundaries for these 2 species were due simply to forest clearing and unregulated trapping, then both species should have reoccupied their former ranges when these limiting factors were removed. Since Colonial times, eastern forests have recovered (e.g., Black 1950; Ahn et al. 2002), and fur harvests are now regulated by states and provinces (Proulx et al. 2004). The fisher has been expanding its range recently but does not occur as far south as it did historically, whereas marten populations are stable or decreasing (Gibilisco 1994; Proulx et al. 2004). Fishers occur from eastern Canada south through New York State, and populations are increasing throughout southern New England. In addition, reintroduced populations are expanding their distribution throughout Connecticut, southern New York, Pennsylvania, and the highlands of West Virginia (Proulx et al. 2004). In contrast, marten reintroductions have been less successful (e.g., Moruzzi et al. 2003). Not only has the range of the marten failed to increase since the end of the LIA, but populations in northeastern North America also appear to be becoming more insular and disjunct (Figure 4.2). This increased geographic isolation of the marten in the East is occurring despite carefully regulated trapping and extensive forest cover across the region (see satellite map in Google Earth). More recently, however, forest practices in Maine have reduced marten habitat independent of climate change (Simons 2009).

In coastal Maine and southern New England, mean annual temperatures increased by >1 °F (0.6 °C) from 1895 to 2000, whereas in northern Maine they decreased by >1 °F (Zielinski and Keim 2003: 249–250). Northwestern Maine receives greater snowfall than the coastal and southern portions of the state (Zielinski and Keim 2003: 83–88). High snowfall is a consistent feature of the areas in eastern North America that still support core marten populations (i.e., the areas with abundant marten populations based on consistently high harvest records; Carroll 2007). These currently include Cape Breton Island in Nova Scotia, north-central New Brunswick and adjacent forested areas on Quebec's Gaspe Peninsula, northwestern Maine, and the Adirondack Mountains in New York State (Figure 4.2).

In the Hudson Bay region of Canada, Preble (1902: 68–69) reported that fishers were "found sparingly throughout the southern part of the region" whereas "the marten is fairly common throughout the region north to the tree limit." If the range of fishers is strongly tied to low snowfall levels, then I would expect fisher populations to expand their northern range boundary when climatic conditions become warmer. Interestingly, the ratio of martens to fishers in the harvest records of the Hudson Bay Company decreased from 83.9:1 in 1739–1748 ($n = 146{,}086$) to 65.9:1 in 1790–1799 ($n = 197{,}180$), 16.8:1 in 1840–1849 ($n = 900{,}032$), and 18.7:1 in 1890–1899 ($n = 864{,}689$) (Novak et al. 1987: 71–72, 115–116). Harper (1961: 114) noted that fishers occur farther north in western than in eastern Canada (i.e., the Ungava Peninsula), and suggested that the greater mean annual snowfall in the Ungava Peninsula compared with western Canada (330–508 vs. 102–154 cm) might restrict fisher movements and hunting success. Moreover, Brown and Braaten (1998) documented a decrease in snowfall throughout Canada during the past 40–50 years, suggesting the need for a formal assessment of broad-scale changes in marten and fisher populations in relation to snowfall dynamics and other factors (e.g., harvest pressure) in Canada.

Evaluation of Expectation 4 was based on ratios in trapping data, which could be biased if trapping vulnerability varied among areas or changed over time. Although juveniles of both species are more vulnerable to trapping than are adults (Krohn et al. 1994; Hodgman et al. 1997), I found no evidence that suggested the presence of any spatial or temporal biases in the historical data. Although modern trappers commonly use metal body-gripping traps to catch both species, in earlier times deadfalls were widely used for both species (e.g., Adney 1893a,b; Seton 1909: 921). Thus, while the materials used to harvest these 2 species has changed over time (i.e., wood vs. steel), the basic concepts underlying these traps are remarkably similar (i.e., to quickly suffocate the animal). Marten-to-fisher ratios would also be biased if trappers or fur dealers reported captures of 1 species more frequently than the other; however, I had no reason to suspect differential reporting, because both species were highly prized in the fur trade and were commonly trapped, traded, and shipped (Novak et al. 1987; this study).

This study hinges on the assumptions that snowfalls during the Colonial era were heavier and more extensive than they are today, and that snowfall has decreased steadily from south to north since the close of the LIA. Because the period covered by this study predates the widespread measurement and recording of weather data, including snowfall patterns, it is not possible to evaluate the validity of these assumptions. Nonetheless, there is considerable anecdotal evidence suggesting that past climates were colder and snowfall was greater (see Jacobson et al. 2009 for a compilation of studies in northern

New England). In general, the climate in New England has warmed over the last 100 years, whereas the mean annual temperature in northwestern Maine, which still supports viable populations of both American martens and Canada lynx (2 species known to have broad-scale associations with deep-snowfall environments), decreased by 0.4 °F (0.2 °C) from 1895 to 2000 (Zielinski and Keim 2003: 249–250).

Historical data provide a potentially rich and largely untapped source of information for documenting long-term distributional changes in wildlife populations. They may also provide opportunities to evaluate contemporary species-habitat relationship models when such models include a climatic variable as a significant determinant of species occurrence (e.g., snowfall for the marten, lynx, and wolverine). Although historical data have limitations and can be problematic (Edmonds 2001; Sagarin 2002; Krohn and Hoving 2010), this study shows that they can be useful for studying ecological questions related to large-scale distributional changes. Furthermore, my findings suggest that under a warming climate, the fisher will likely expand its range northward, whereas the American marten will continue to retreat from southern regions (see also Lawler et al. 2009, this volume).

Acknowledgments

I thank the furbearer biologists in various provinces and states who provided data for this study and answered my many questions: P. Canac-Marquis (QE), C. Libby (NB), P. Austin-Smith (NS), W. Jakubas (ME), J. Kelly (NH), K. Royar (VT), P. Jensen (NY), and R. Farrar (VA). Randy Farrar, Virginia Department of Game and Inland Fisheries, was especially helpful by suggesting early references related to the southern Appalachians. Shonene Scott did the GIS work to create the marten range map. Initial reviews of this chapter by W. Jakubas, D. Harrison, and R. Powell were greatly appreciated, as were the anonymous reviews of this book. This publication is a contribution of the Maine Cooperative Fish and Wildlife Research Unit (U.S. Geological Survey, Maine Department of Inland Fisheries and Wildlife, University of Maine, and the Wildlife Management Institute, cooperating).

SECTION 2

Biology and Management of *Martes* Populations

5

Population Biology and Matrix Demographic Modeling of American Martens and Fishers

STEVEN W. BUSKIRK, JEFF BOWMAN, AND JONATHAN H. GILBERT

ABSTRACT

In this chapter, we review the population biology of American martens (*Martes americana*) and fishers (*M. pennanti*) and describe important research advances since 1994. These include longitudinal studies of vital rates and population processes at landscape scales. With expanded understanding of vital rates, we constructed matrix demographic models for both species using similar parameterizations. We found that neither species exhibits life histories strongly influenced by fecundity in early life; instead, both species' life histories were most influenced by adult survival. Elasticities for survival of the oldest age classes (≥3 years) were 0.31 and 0.29 for martens and fishers, respectively. The apparent similarity in life histories between the 2 species is surprising considering the difference in body size, which is a frequent covariate of "speed" in species' life histories. The life histories of martens and fishers are consistent, however, with expectations for mammals with prolonged embryonic diapause, and for species of conservation concern.

Introduction

The population biology of *Martes* species has been investigated in various contexts since the 1950s, and was reviewed by Powell (1994b), Powell and Zielinski (1994), and Buskirk and Ruggiero (1994). As do other solitary carnivores, *Martes* species present particular challenges in the study of population biology because of their low densities (Buskirk and Ruggiero 1994), secretive behaviors, and heterogeneous detection rates among species, regions, sex-age cohorts, and detection devices (Zielinski and Kucera 1995; Zielinski and Stauffer 1996). On the other hand, North American *Martes* are relatively tractable subjects of population study because they are easy to lure to detection devices, well suited to external radio transmitters, and readily recovered from trapper harvests (Raphael 1994), which have numbered >10^5/year for American martens (*Martes americana*) and >10^4/year for fishers (*M. pennanti*) in Canada during recent decades (Proulx 2000).

Because North American species have received greater attention at the population level than Eurasian ones, our focus here is on the former. We had 3 primary objectives for this work: (1) to review the state of knowledge of the population biology of American martens and fishers, especially measures of reproduction and estimates of survival, focusing on new knowledge about their vital rates since 1994; (2) to present the results of matrix demographic modeling using parameterizations based on available data, using elasticities of transition rates from those models to draw inferences about important life-history characteristics; and (3) to provide recommendations for improving our understandings of population-related processes, and for using population data to inform conservation planning.

The State of Knowledge

As recently as 30 years ago, attempts to understand *Martes* population biology were based primarily on data obtained from fur harvests. Various efforts were made to understand the characteristics of samples collected by fur trappers (e.g., Bulmer 1974; Strickland and Douglas 1987) and the population-level inferences that could be drawn from such samples. To gather large data sets affordably, biologists sought access to animal carcasses via trappers, conservation officers, and fur buyers. Various jurisdictions have required reporting of fur harvests, or submission of furs for sealing, which provided biologists with opportunities for data collection and analysis (e.g., Statistics Canada 2009). In addition, some researchers still solicit carcasses directly from trappers. Strickland and Douglas (1987) reported sex and age structures for American martens in the Canadian province of Ontario based on large numbers of submitted carcasses. The limitations and biases of these data sources have been discussed previously (Mead 1994; Frost et al. 1999).

In an innovative and prescient study, de Vos (1951a) proposed landscape features that could enhance the sustainability of trapping in a mosaic of trapped and untrapped areas—an early application of source-sink principles to carnivore conservation. Other early emphases related to spatial behavior and population organization: for example, how populations were arranged in terms of home range size, location, and overlap; under what circumstances territories were maintained and abandoned; and how resource availability affected use of space (Powell 1979b, 1994b; Buskirk and McDonald 1989).

Since the syntheses published in 1994 (Powell 1994b; Buskirk and Ruggiero 1994; Powell and Zielinski 1994), several new analyses of temporal dynamics in *Martes* populations have been conducted (Fryxell et al. 1999, 2001;

Haydon and Fryxell 2003; Bowman et al. 2006). Other recent, notable advances in the population biology of martens and fishers have been facilitated by molecular tools, both genetic and isotopic. Tools for extracting, amplifying, genotyping, and sequencing DNA from small samples (e.g., fecal droppings, hairs) have allowed remote and noninvasive determination of identity (Mowat and Paetkau 2002) and sex (Campbell et al. 2010). Identity can be used to construct encounter histories, leading to estimation of detection probabilities, movement rates and distances, abundance, and occupancy rates—all important population metrics. Stable-isotope analyses have been useful for assessing diets (Ben-David et al. 1997), tracking movements of martens across distances larger than home ranges (Pauli et al. 2009a; Pauli 2010), and have the potential to estimate detection probabilities and abundance. Collectively, these tools have rapidly improved our knowledge of various population processes, from short-term behavioral events to systematic and phylogeographic characteristics (Long and MacKay, this volume; Schwartz et al., this volume).

Perhaps the greatest advances in our understanding of *Martes* population ecology since 1994 have been made in studies that consider population processes, such as dispersal, survival, and source-sink dynamics, within real landscapes. A Web of Science search (accession date 18 January 2011) for the term "landscape" combined with "*Martes*" resulted in 96 articles, all published after 1994. The Ontario marten study (e.g., Broquet et al. 2006b; Johnson et al. 2009) is an example of a study with large geographic scale and mechanistic approaches. Carr et al. (2007a,b) used demographic data in combination with a genetic analysis of nearly 800 fishers in Ontario to demonstrate density-dependent dispersal over the scale of hundreds of kilometers. Garroway et al. (2008) subsequently applied a novel graph-theory approach to confirm the findings of Carr et al. (2007b) about fisher dispersal. In Maine, a series of studies (e.g., Chapin et al. 1998; Payer and Harrison 1999) have revealed landscape-scale effects of trapping and timber harvesting on martens. Several other recent examples of studies conducted at the landscape scale (e.g., Hargis et al. 1999; Pauli 2010; Weir and Corbould 2010) have investigated source-sink or metapopulation models of *Martes* ecology.

We reviewed vital-rate estimates for American martens and fishers published since 1994, focusing especially on various measures of reproduction and estimates of survival for American martens and fishers (Tables 5.1–5.2). Authors typically reported a variety of rates for age cohorts, sexes, and other subsets, and we have tried to present the major findings from each study in these tables.

Measures of pregnancy rates and litter sizes have been estimated for many years using counts of corpora lutea obtained from harvested martens and fishers (Douglas and Strickland 1987; Strickland and Douglas 1987). Because there are uncertainties associated with using corpora lutea counts in fecundity estimates, researchers have attempted to quantify reproduction in other

Table 5.1. Demographic attributes and vital rates of American martens reported in the literature since 1994

Demographic attribute	Rate[a]	Location	Reference
Reproduction			
Pregnancy rate	1.5–2.5 yr: 0.86 >2.5 yr: 0.96	Montana	Aune and Schladweiler 1997
	≥1.5 yr with CL: 0.78 1.5 yr: 0.6 ≥2.5 yr: 0.93	Quebec, Canada	Fortin and Cantin 2004
	≥1.5 yr: 0.45 (SE = 0.09) 1.5 yr: 0.13 (SE = 0.03) 2.5 yr: 0.33; 4.5 yr: 0.73	Alaska	Flynn and Schumacher 2009
	>1.5 yr: 0.50 (temperate forest, n = 16) >1.5 yr: 0.33 (boreal forest, n = 27)	Ontario, Canada	Cobb 2000
	1.5 yr: 0.60 (intensive forestry) >2.5 yr: 0.88 (intensive forestry) 1.5 yr: 0.52 (status quo forestry) >2.5 yr: 0.59 (status quo forestry)	New Brunswick, Canada	Pelletier 2005
Females with kits	F with ≥1 kit weaned: 4/13	Oregon	Bull and Heater 2001
Lactation rates	≥2.5 yr: untrapped forest reserve 11/12 ≥2.5 yr: trapped indust. forest 13/16 ≥2.5 yr: untrapped indust. forest 15/20 ≥2.5 yr: all areas 39/48 = 0.81	Maine	Payer and Harrison 1999
Litter size (CL-based)	Mean = 2.6 CL/F with CL	Montana	Aune and Schladweiler 1997
	Mean = 4.1 CL (SE = 0.07)/ovulating F Mean = 3.2 CL (SE = 0.14)/F ≥1.5 yr	Quebec, Canada	Fortin and Cantin 2004
	Mean = 2.86 CL (SE = 0.45)/ovulating F (temperate forest) Mean = 3.11 CL (SE = 0.40)/ovulating F (boreal forest)	Ontario, Canada	Cobb 2000

	1.5 yr: mean = 2.3 CL (SD = 1.0)/ovulating F (intensive forestry) >2.5 yr: mean = 2.7 CL (SD = 1.1)/ ovulating F (intensive forestry) 1.5 yr: mean = 2.3 CL (SD = 1.0)/ovulating F (status quo forestry) >2.5 yr: mean = 2.6 CL (SD = 1.0)/ovulating F (status quo forestry)	New Brunswick, Canada	Pelletier 2005
	≥1.5 yr: mean = 3.2 CL (SE = 0.08)	Alaska	Flynn and Schumacher 2009
Litter size (field)	1.75 kits/F in den	Oregon	Bull and Heater 2001
Survival[b]			
Annual	M: 0.87 (95% CI: 0.75–1.0) F: 0.53 (95% CI: 0.34–0.83)	Maine	Hodgman et al. 1997
1 May to 15 December	>1 yr, M: 0.12 (95% CI: 0.03–0.4) >1 yr, F: 0.39 (95% CI: 0.24–0.66)	Maine	Hodgman et al. 1994
Annual	All ages, F: 0.67 All ages, M: 0.57	Ontario, Canada	Thompson 1994
	Age ≥1 yr, M: 0.85 Age ≥1 yr, F: 0.77 Age ≥1 yr, B: 0.81	Wisconsin	McCann et al. 2010
	Age ≥1 yr, M (forest reserve): 0.95 (95% CI: 0.85–1) Age ≥1 yr, F (forest reserve): 0.62 (95% CI: 0.32–1) Age ≥1 yr, M (trapped industrial forest): 0.51 (95% CI: 0.43–0.8) Age ≥1 yr, F (trapped industrial forest): 0.66 (95% CI: 0.44–0.98) Age ≥1 yr, M (untrapped industrial forest): 0.84 (95% CI: 0.72–0.98)	Maine	Payer and Harrison 1999

Table 5.1—*cont.*

Demographic attribute	Rate[a]	Location	Reference
	Age ≥1 yr, F (untrapped industrial forest): F: 0.82 (95% CI: 0.62–1)		
1 June to 15 December	B: 0.80 (95% CI: 0.69–0.88)	Michigan	Belant 2007
Annual	B: 0.83	Newfoundland, Canada	Hearn 2007
Annualized	Age ≥9 mo, B: 0.63 (95% CI: 0.39–0.87)	Oregon	Bull and Heater 2001
Modelled annualized	≥1.5 yr: 0.67 ≤3 yr: 0.36	Quebec, Canada	Fortin and Cantin 2004
Annual	≤1 yr: 0.61 (SE = 0.15) >1 yr: 0.72 (SE = 0.09)	Quebec, Canada	Potvin and Breton 1997
	B, mean of 8 yearly estimates: 0.68	Alaska	Flynn and Schumacher 2009

Notes: Values are aggregated for trapped and untrapped populations. Field-based rates represent conservative estimates of reproductive attributes.

[a] CL = corpora lutea or data based on counts of corpora lutea; F = females, M = males, B = both sexes

[b] Survival rates are either annual estimates, estimates for a specified interval, or converted to annual rates (annualized)

Table 5.2. Demographic attributes and vital rates of fishers reported in the literature since 1994

Demographic attribute	Rate[a]	Location	Reference
Reproduction			
Pregnancy rate	1.5–2.5 yr: 0.57	Maine	York 1996
	>2.5: 0.79		
	>1.5 yr: 0.86	Ontario, Canada	Kraus 2005
Denning rate[b]	0.63 (annual range 0.33–1)	Maine	Paragi et al. 1994
Litter size (CL-based)	>1.5 yr: mean = 2.41 CL (95% CI: 2.31–2.69)/F	Ontario, Canada	Kraus 2005
	>1.5 yr: mean = 2.86 CL/ovulating F		
	>1.5 yr: mean = 1.83 CL (SE = 0.31)/ovulating F	Vermont	Van Why and Giuliano 2001
Litter size (captive)	2.7	Maine	Frost and Krohn 1997
Litter size (field)	2.8 kits/F (SE = 0.6)	Massachusetts	York 1996
	2.2 kits/F that whelped	Maine	Paragi et al. 1994
Recruitment	0.4-0.7 F kits/F	Maine	Paragi et al. 1994
	Age >2 yr: 1.18 F kits/F (95% CI: 0.63–2.27)	Ontario, Canada	Koen et al. 2007
	Age >2 yr: 2.15 F kits/F (95% CI: 1.16–4.14)		
Survival[c]			
Annual	Age >9 mo, M: 0.33 (95% CI: 0.18–0.60)	Ontario, Canada	Koen et al. 2007
	Age >9 mo, F: 0.63 (95% CI: 0.47–0.86)		
	Age 2 yr, M: 0.45 (95% CI: 0.24–0.83)		
	Age 2 yr, F: 0.81 (95% CI: 0.72–0.91)		
June to December	B: 0.812 (95% CI: 0.374–0.977)	Michigan	Belant 2007
Annual	B: 0.88 (95% CI: 0.50–0.98)	California	Jordan 2007
Trapping season (39 days)	Adult, M: 0.57 (95% CI: 0.42–0.78)	Maine	Krohn et al. 1994
	Adult, F: 0.79 (95% CI: 0.64–0.97)		
	Juvenile, B: 0.38 (95% CI: 0.25–0.57)		
Non-trapping season	Adult, B: 0.89 (95% CI: 0.81–0.99)	Maine	Krohn et al. 1994
	Juvenile, B: 0.72 (95% CI: 0.53–0.99)		
Annual	Adult, M: 0.77 (95% CI: 0.63–0.95)	Massachusetts	York 1996
	Adult, F: 0.90 (95% CI: 0.8–1.0)		
	Juvenile, M: 0.77 (95% CI: 0.46–1.0)		

Table 5.2—*cont.*

Demographic attribute	Rate[a]	Location	Reference
	Juvenile, F: 0.84 (95% CI: 0.65–1.0) Kit, M: 0.56 Kit, F: 0.78 Adult, F: 0.65 (95% CI: 0.50–0.86) Juvenile, B: 0.27 (95% CI: 0.14–0.50)	Maine	Paragi et al. 1994

Note: Values are aggregated for trapped and untrapped populations

[a] CL = corpora lutea, Juvenile = ≤1 year old, F = females, M = males, B = both sexes

[b] Denning rates represent conservative estimates of the number of females giving birth

[c] Survival rates are either annual estimates, or estimates for a specified interval

ways (e.g., proportions of females using reproductive dens, counts of females with kits, field observations of litters), although the use of corpora lutea continues. Tables 5.1–5.2 show the variety of methods used to estimate reproductive performance. Several researchers have documented fisher denning rates (i.e., proportion of females observed using maternal dens) and litter sizes from field studies; these data provide conservative estimates of reproductive rates. Only 2 studies since 1994 have reported on reproductive performance in American martens using data other than corpora lutea counts: Bull and Heater (2001) monitored reproductive output at maternal dens, and Payer and Harrison (1999) captured lactating females to estimate pregnancy rates. Marten pregnancy rates ranged from 45 to 81% for non-corpora lutea estimates, and from 46 to 96% for those using corpora lutea. In the only study that reported litter sizes for American martens based on field observations, Bull and Heater (2001) estimated 1.8 kits per litter. Counts of corpora lutea for American martens ranged from 1.6 to 4.3. Fisher pregnancy rates ranged from 40 to 82%, depending on the age of study animals and the location of the study. Mean litter sizes ranged from 1.8 to 2.8 kits per litter.

For both species, substantially more studies have investigated survivorship than reproduction. As for reproduction, survival estimates varied depending on several factors: whether populations were subjected to trapping, whether timber harvest had occurred in the study area, and the age of study animals. Estimated marten survivorship (annualized) ranged from a low of 0.11 in a heavily trapped area to 0.87 in a preserve with no trapping. Values for fishers ranged from 0.33 to 0.90, again depending on whether the population was trapped and on the age of study animals.

A Matrix Demographic Model

To better understand and compare the key life-stage transitions of American martens and fishers, we developed stage-structured matrix models (see examples in Lefkovitch 1965) for both species. This has not been done in a parallel fashion before, although demographic models have been constructed for American martens (Lacy and Clark 1993). These models require estimates of fecundity and survival by stage or age, but stage-specific estimates of fecundity are particularly rare for untrapped populations of American martens. In constructing our matrices, we used stages rather than ages, because rates could then be generalized across age classes (i.e., we pooled older age classes into a single stage). This is appropriate when there is insufficient information or evidence to derive age-specific fecundity rates (Caswell 2001). We chose to model 4 stages for both species: Stage 1 (age 0–1 year), Stage 2 (age 1–2 years), Stage 3 (age 2–3 years), and Stage 4 (≥3 years).

We used our compiled review of reproductive and survivorship rates (Tables 5.1–5.2) to generate a realistic range of estimated rates for both species.

We found only 2 field-based estimates of fecundity for the American marten, and a few more for the fisher. On the other hand, we found many useful field-based estimates of survival for both species (Tables 5.1–5.2). Because we found no reported measures of reproduction that were common for both species, except counts of corpora lutea (CL), we used CL counts as a basis for estimating fecundity in our models. We used values developed in the Algonquin region of Ontario (Douglas and Strickland 1987; Strickland and Douglas 1987). Thus, our estimates, while likely overestimates of fecundity, are comparable between species and not confounded by region. Corpora lutea counts also have the benefit of being based on large sample sizes for all 4 stages in our analysis. In fact, CL counts for both species were available by year of age, for many more than 4 years (up to age 14.5 for American martens; Douglas and Strickland 1987; Strickland and Douglas 1987). We pooled and averaged rates across years for estimates of fecundity at Stage 4, and weighted averages by cohort size. We estimated fecundity as the product of female survival, pregnancy rate, and litter size (female component) for each stage. Thus, we conducted our analyses on 50% of the population (the "female component"), and assumed a 1:1 sex ratio from conception to death. Tables 5.1 and 5.2 show that, although point estimates for sex-specific survival rates suggest higher survival for females than males, confidence intervals overlap; therefore, we did not find a statistical basis for assuming that survival rates differed between the sexes. Further, the polygynous mating system in *Martes* species strongly suggests that adult sex ratios will not affect pregnancy rates.

Following our literature review, we selected a set of vital rates within the range of published rates and comparable between species as a starting point for model parameterizations, including the fecundity estimates from Ontario (Table 5.3). Initially, we used annual survival rates of 0.25 for Stage 1 and

Table 5.3. Vital rates that were the basis of matrix demographic modeling for the American marten and fisher

		Vital rates		
Species	Stage	Pregnancy rate	Litter size[a]	Survival rate
American marten	1	0	0	0.25
	2	0.78	1.64	0.75
	3	0.92	1.68	0.75
	4	0.96	1.83	0.75
Fisher	1	0	0	0.25
	2	0.97	1.51	0.75
	3	0.96	1.72	0.75
	4	0.99	1.83	0.75

Sources: See Tables 5.1 and 5.2

[a] Litter sizes include only female offspring; values for both sexes would be 2 times these values

0.75 for all other stages, and used the same values for both species. This allowed us to estimate the effect of differential fecundity between species on matrix products. For each species, we estimated lambda (λ), matrix elasticities, the stable age distribution, and the cohort generation time.

Given the range of known vital rates for both species, and the lack of knowledge about some rates (e.g., fecundity), we were interested in varying vital rates for both species and assessing effects on matrix products. We first varied survival rate in probability increments of 0.1 for each stage and reestimated λ and matrix elasticities. We then varied pregnancy rate in 0.1 increments for each species and stage, recalculated fecundity, and reestimated matrix products. Finally, we varied litter size for each stage and species across a plausible range, recalculated fecundity, and reestimated matrix products.

We depicted transition rates, elasticities, and age distributions using a life-cycle diagram. The diagram for the American marten (Figure 5.1a), based on

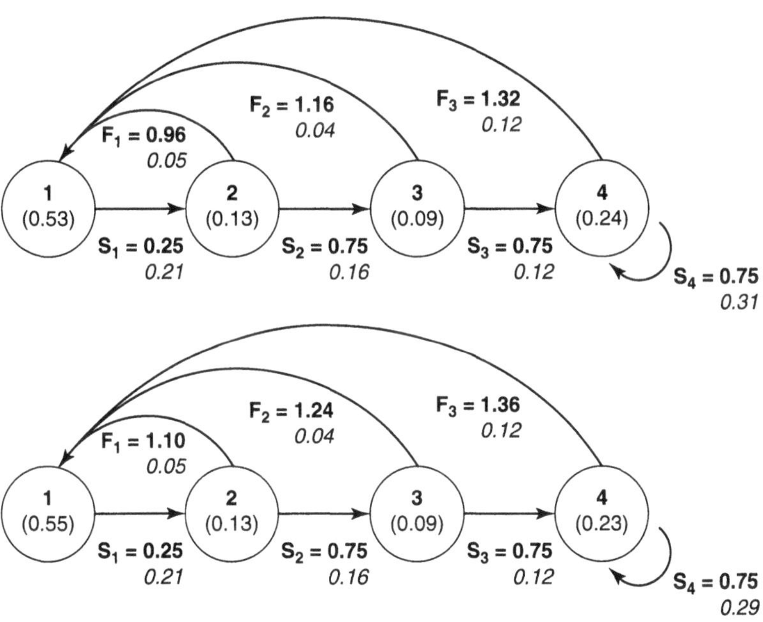

Figure 5.1. Life cycle graphs for the (a) American marten and (b) fisher developed from demographic data obtained from the literature (Tables 5.1–5.2). Circles represent 4 stages in the life history of each species, with the stable age distribution shown in parentheses. The arcs represent stage-specific vital rates, including survival (S_i) and fecundity (F_i), with their associated numerical values. Elasticities (shown in italics) assess the response of λ to small proportional changes in a vital rate. Marten life history strongly responds to changes in Stage 4 (0.31) and Stage 1 (0.21) survival, accounting for 52% of the total elasticity. Fisher life history is very similar, with elasticities for Stage 4 and Stage 1 survival summing to 50%. These demographic measures are typical of species with slow life histories, for which adult survival is typically high and adult fecundity typically modest.

vital rates from Table 5.3, depicts a scenario in which no reproduction occurs during Stage 1 (age 0), survival from Stage 1 to Stage 2 (age 1) is low (0.25), and fecundity for all stages is low, constrained largely by litter sizes that are small, under even the best circumstances. The stable age distribution includes 53% juveniles, 13% yearlings, 9% 2-year-olds, and 24% older animals. Elasticities show that key transition rates are adult (Stage 4) and juvenile (Stage 1) survival (31 and 21% of total elasticity, respectively). Fecundities have low associated elasticities, totaling 21% of total elasticity for all stages (Figure 5.1a). The corresponding diagram for the fisher (Figure 5.1b) is nearly identical, the key difference being the delayed sexual maturity of martens compared with that of fishers (78 vs. 97% pregnancy rates among yearlings, respectively; Table 5.3). This results in stage-specific fecundities for the fisher of 1.10 (F_1), 1.24 (F_2), and 1.36 (F_3), all slightly higher than for the marten. The stable age distribution for the fisher is essentially identical to that for the marten, but elasticities differ slightly, with higher values for survival beyond 3 years, and lower values for fecundity in Stages 2 and 3. The cohort generation time for the marten is 5.1 years, for the fisher, 4.9 years.

Varying marten survival rate at Stage 1 in 0.1 increments, and holding all other rates constant, as shown in Table 5.3, demonstrated that an annual survival rate of 0.25 is required for $\lambda > 1.0$. Higher survival-rate values at Stage 1 resulted in a large increase in λ (Figure 5.2). When the survival rate at Stage 1 was held constant at 0.25, however, no other stage could experience survival <0.75 with resulting $\lambda > 1$ (Figure 5.2). In general, survival rates at Stages 2–4 had less effect on λ than that at Stage 1; among adult stages, the survival rate at Stage 4 had the greatest influence on λ. Thus, varying survival rates provided an outcome consistent with our initial matrix result that Stage 1 and Stage 4 survival rates had the greatest elasticities. Given the similarity in fecundity for martens and fishers, it is not surprising that varying fisher survival rates produced results virtually identical to those for martens.

Varying either marten or fisher fecundity (either through litter size or pregnancy rate) and holding survival rates constant (as in Table 5.3) had very little effect on λ. Among fecundity values for both species, only low fecundity rates at Stage 3 (F_3, the last stage of fecundity) could produce $\lambda < 1$ (Figure 5.3). Provided that all other rates were held constant, even a pregnancy rate as low as 0.05, or a litter size as low as 0.02 in either of the first 2 reproductive stages resulted in $\lambda \geq 1$. This demonstrates the importance of adult survival and F_3 fecundity to marten and fisher life histories.

Model Interpretation

Matrix population models provide a useful framework for simple explorations of important life-history stages of species of interest, the potential for population growth, and the possible effects of management actions. Model

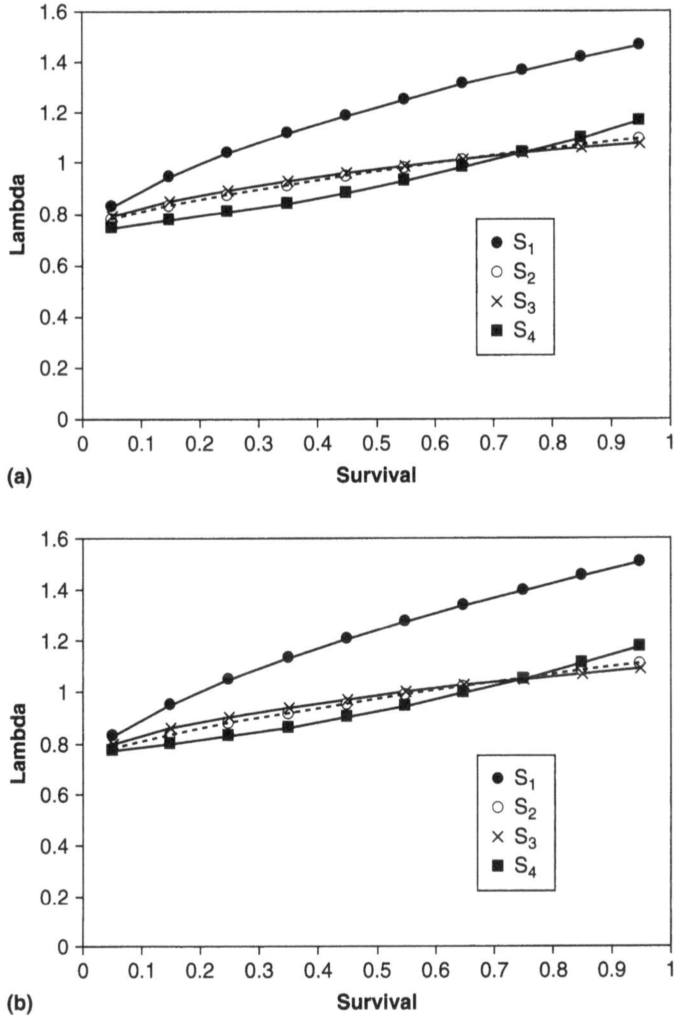

Figure 5.2. Effect of varying survival rate (S) at 4 life stages (1–4) on λ for the (a) American marten and (b) fisher, under the vital-rate assumptions shown in Table 5.3, while holding other vital rates constant. Given adult survival (S_2-S_4) of 0.75, survival at Stage 1 must be >0.2 to achieve $\lambda > 1$. Adult survival rates that fall below 0.7 result in $\lambda < 1$.

explorations can therefore be an important component of species management and conservation planning (Ferrière et al. 1996). Matrix models also allow for comparisons among species, and we have conducted the first such comparison between American martens and fishers.

The tendency for both species to exhibit life cycles that are strongly influenced by adult survival—"slow" life histories (Promislow and Harvey 1990)—is a necessary consequence of the prolonged embryonic diapauses in

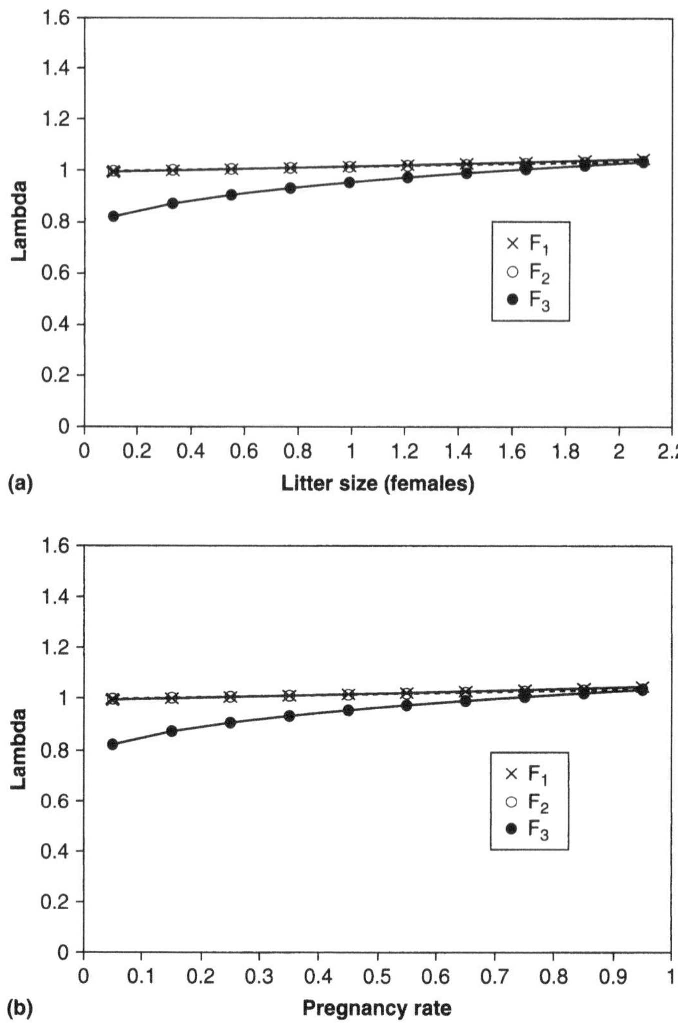

Figure 5.3. Effect of varying (a) litter size and (b) pregnancy rate at 3 life stages (fecundity stages F_1 to F_3) on λ for the American marten, under the vital-rate assumptions shown in Table 5.3, while holding other vital rates constant. For both vital rates, F_1 and F_2 values are not very influential, whereas F_3 values are quite important, with only values near the observed values (Table 5.3) capable of generating $\lambda > 1$.

both species. For both species, the gestational delay is >0.5 year and, for mustelids of similar size in seasonal environments, this translates into a 1-year delay in the age of first parturition, thereby precluding a "fast" life history. How embryonic diapause could have evolved (or been lost) multiple times among lineages of Carnivora has been discussed by Sandell (1990) and Lindenfors et al. (2003), but the salient point here is that American martens and fishers have evolved slow life histories despite the potential costs of delaying reproduction until age 2, even under favorable resource conditions. The mean age of first reproduction for a 1-kg mammal based on simple linear regression without phylogenetic controls is <1 year (Duncan et al. 2007).

In contrast to *Martes*, females in the genus *Mustela* of similar body size produce litters at age 1, resulting in faster life histories. For example, black-footed ferrets (*Mustela nigripes*) in Shirley Basin, Wyoming, had a cohort generation time of 2 years, and 31% of the elasticity of λ was attributed to variation in first-year fecundity (Grenier et al. 2007). In general, species with faster life histories are more sensitive to perturbations affecting early fecundity, whereas species with slower life histories show little variation in early fecundity and are less sensitive to such changes.

The slightly slower life history of the American marten compared with that of the larger-bodied fisher is counterintuitive; body size is an important covariate of population growth rate in mammals (Duncan et al. 2007). The later full sexual maturity in the American marten is the opposite of the pattern expected; the fisher, with an adult body size about 3 times that of the marten, should reach sexual maturity at an older age. The observed pattern may simply reflect species differences in the degree to which we understand vital rates. However, another interpretation is possible: because of their smaller body size, American martens are generally regarded as more arboreal than fishers. Among mammals, arboreality is associated with increased longevity and its life-history correlates (Shattuck and Williams 2010). Thus, the perceived slower life history of American martens compared with fishers could be real, and it may have evolved secondarily to their greater arboreality.

To advance our understandings of the demography and transition rates of *Martes* species, it will be necessary to conduct more intensive and longer-term field studies of survival and fecundity of all species at issue. This need is especially true for populations of American martens not subjected to intensive trapping, and for all fisher populations. To better understand important life-history differences between these 2 North American species, the number of life stages for which vital rates are gathered will need to be increased, likely up to 6 for the fisher. In addition, field-based estimates of fecundity will represent substantive improvements to the laboratory-based estimates we used here, and will likely be more conservative.

Knowing that American martens and fishers have life histories strongly influenced by adult survival has important implications for conservation planning. Rather than expecting that populations will be able to grow quickly in temporarily favorable environments, managers can expect that stable conditions and long time periods will be required for population growth or recovery. This requirement is typical of species of conservation concern. Recommendations for future research on the population biology of these 2 species will depend on the goals of managers and conservationists; however, a perusal of Tables 5.1–5.2 reveals glaring deficits in our knowledge, particularly in survival rates of untrapped populations of American martens, and in field-based fecundity estimates. These are promising areas for the next phases of *Martes* population research.

6

Evaluating Translocations of Martens, Sables, and Fishers

Testing Model Predictions with Field Data

ROGER A. POWELL, JEFFREY C. LEWIS, BRIAN G. SLOUGH, SCOTT M. BRAINERD, NEIL R. JORDAN, ALEXEI V. ABRAMOV, VLADIMIR MONAKHOV, PATRICK A. ZOLLNER, AND TAKAHIRO MURAKAMI

ABSTRACT

Historically, overexploitation, loss and fragmentation of forest habitats, predator control, and climate change caused population decreases and local extirpations of *Martes* populations. Protection led to population recovery in some places but not everywhere. Because these predators are important components of their natural ecological communities and because some can be valuable furbearers, they have been reintroduced to reestablish extirpated populations. In addition, animals have been released to augment small populations and to establish populations at sites never before occupied. Not all reintroductions, augmentations, or introductions have been successful. We modeled reintroductions to predict variables associated with the persistence of populations and tested model predictions using data from actual reintroductions. The model predicted that the more adult females released and the more release sites used, the higher the probability of reestablishing a population. The model also predicted that the number of males released should not affect success, provided that an effective few are released, and that the number of years used to release a given number of animals should not affect success. American martens (*Martes americana*) have been translocated more than 50 times with >50% success. Fishers (*M. pennanti*) have been translocated nearly 40 times with >75% success. European pine martens (*M. martes*), stone martens (*M. foina*), and Japanese martens (*M. melampus*) have been translocated in small numbers with few successes, and these included inadvertent introductions. Nearly 20,000 sables (*M. zibellina*) were translocated in the former Soviet Union, reestablishing many populations. For actual reintroductions, the 2 variables most strongly linked to success were the total number of animals released and the number of release sites; the number of years that animals were released did not affect success. Sex-ratio data for released animals allowed sex-specific analyses for American martens, fishers, and sables. In our analyses of actual reintroductions of these 3 species, the number of females released, the number of release sites, and the number of males released linked strongly to success. Again, the number of years did not affect success. We hypothesize that variation in the breeding ability of released males may explain differences between model results and actual reintroductions. Genetic diversity of the released animals may affect the index of success for a reintroduction, but genetic relatedness of reintroduced and historical populations, protection of released animals, monitoring, and hard versus soft releases appeared not to affect reintroduction success, though most data were plagued by confounding conditions. Tested, functional habitat models are needed to understand the role of habitat quality in reintroduction success.

Introduction

Extensive population decreases of many carnivore species have led to significant range contractions (e.g., wolf [*Canis lupus*], Canada lynx [*Lynx canadensis*], wolverine [*Gulo gulo*], sable [*Martes zibellina*]; Pavlov 1973a; Boitani 2003; Laliberte and Ripple 2004; Aubry et al. 2007). To reestablish locally extirpated populations, wildlife managers have reintroduced animals to unoccupied areas of their historical ranges. While sometimes logistically simple affairs, especially those in the mid-20th century, translocations require considerable planning, can be expensive, are often unsuccessful, and therefore can be controversial (IUCN 1995; Reading and Clark 1996; Miller et al. 1999; Breitenmoser et al. 2001; Lewis and Hayes 2004; Lewis 2006; Callas and Figura 2008). Modern reintroductions of carnivores are also often high-profile events (e.g., Phillips and Smith 1998). When reintroductions fail, support needed for other restoration efforts, especially for imperiled species, can erode. Consequently, understanding the factors associated with successful translocations is critical for evaluating the feasibility and implementation of carnivore reintroductions.

The IUCN (1987, 1995) provided guidelines for translocating animals. Reading and Clark (1996), Miller et al. (1999), and Breitenmoser et al. (2001) provided specific recommendations and cautions for reintroducing carnivores, and Lewis and Hayes (2004), Lewis (2006), and Callas and Figura (2008) gave examples of planning and implementing reintroductions. Nonetheless, for most species, no summary overviews exist for past reintroductions, and specific recommendations for future reintroductions are lacking.

The translocation of animals goes back hundreds to thousands of years in Europe (Alcover 1980; Masseti 1995) and more than a century in North America (reviewed by Bolen and Robinson 2003), casting doubt on recent calls to consider reintroduction biology and restoration ecology new disciplines (Lipsey and Child 2007; Seddon et al. 2007; Armstrong and Seddon

2008). Unfortunately, reintroductions of endangered species in recent decades have experienced frequent failures (Armstrong and Seddon 2008). Many of these failures were due to habitat problems (Griffith et al. 1989; Miller et al. 1999), though Allee effects have not been investigated as thoroughly as perhaps they should be (Deredec and Courchamp 2007). Efforts to counteract failures have led to better planning and to introducing experimental design into reintroductions (e.g., Miller et al. 1990a,b; Lewis and Hayes 2004; Callas and Figura 2008; Biggins et al. 2011). Armstrong and Seddon (2008) have reinforced efforts to put more science into reintroductions by proposing 10 a priori questions at 4 broad ecological scales that reintroduction programs should seek to answer:

POPULATION LEVEL—ESTABLISHMENT

1. How is establishment probability affected by size and composition of the release group?
2. How are post-release survival and dispersal affected by pre- and post-release management?

POPULATION LEVEL—PERSISTENCE

3. What habitat conditions are needed for persistence of the reintroduced population?
4. How will genetic makeup affect persistence of the reintroduced population?

METAPOPULATION LEVEL

5. How heavily should source populations be trapped?
6. What is the optimal allocation of translocated individuals among sites?
7. Should translocation be used to compensate for isolation?

ECOSYSTEM LEVEL

8. Are the target species/taxon and its parasites native to the ecosystem?
9. How will the ecosystem be affected by the target species and its parasites?
10. How does the order of reintroductions affect the ultimate species composition?

Of these 10 questions, detailed evaluations and analyses of historical translocations of *Martes* species can address 5: numbers 1, 2, 4, 6, and 8.

Historically, populations of martens, sables, and fishers have experienced decreases and local extirpations caused by overexploitation (reviewed by

Timofeev and Nadeev 1955; Powell 1993; Slough 1994; Proulx et al. 2004), loss and fragmentation of forest habitats (reviewed by Powell 1993; Slough 1994; Proulx et al. 2004; Broekhuizen 2006; Saeki 2006), predator control (reviewed by Powell 1993; Slough 1994; Proulx et al. 2004), climate change (Krohn, this volume), and, possibly, predation (Lindström et al. 1995; Helldin 2000) in landscapes where habitat generalists, such as red foxes (*Vulpes vulpes*) are favored (Gundersen 1995), though evidence of predation is questionable (Kurki et al. 1998). Protection led to population recovery in some places but not everywhere (reviewed by Powell 1993; Slough 1994; Proulx et al. 2004; Broekhuizen 2006; Saeki 2006). Because these predators are important components of their natural ecological communities and because some can be valuable furbearers, they have been reintroduced to reestablish extirpated populations (reviewed by Monakhov 1978, 1982; Powell 1993; Slough 1994; Proulx et al. 2004). In addition, animals have been released to augment small populations and to establish populations in areas that were unoccupied previously. Slough (1994) reviewed the history of translocations of American martens (*Martes americana*) through the early 1990s, and Lewis et al. (2012) did the same for fishers (*M. pennanti*) through 2011. For sables, Kopylov (1958) reviewed early translocations, Timofeev and Pavlov (1973) compiled a full list of translocations, Monakhov (1978, 1982) assessed the economic results of translocations, and Monakhov (1995, 2006) quantified anatomical changes related to translocations. The purpose of this chapter is to conduct a comprehensive evaluation of translocation efforts for *Martes* species throughout the world. We begin by introducing a model designed to identify important factors affecting reintroduction success, then review the history of translocations of all *Martes* species (hereafter referred to collectively as *martens*; members of individual species are indicated by their full common names), and end by testing model predictions with data from actual reintroductions of martens.

Methods

We define translocation as the transportation and release of animals to establish or to augment a wild population. Translocations include reintroductions, augmentations, and introductions. In a reintroduction, animals are released within a portion of the historical range of the species but where no population presently exists; in an augmentation, animals released are added to, and thereby fortify, an existing population; and, in an introduction, animals are released outside the historical range of the species (IUCN 1987; Nielsen 1988; Armstrong and Seddon 2008).

We considered a reintroduction or introduction to be successful if a population became established and persisted, as documented in scientific and popular literature or in unpublished reports. Failure of a translocation was harder to determine; loss of a population may take years, and absence is difficult to distinguish from an extremely small population (see Vinkey et al. 2006). We considered a translocation to have failed if the responsible agency considered it a failure. We considered an augmentation successful if the literature or unpublished reports documented population growth or if the responsible agency considered it successful. We considered reintroductions of fishers to be independent if the target areas were ≥200 km apart, or if they occurred <200 km apart but the earlier reintroduction failed. We considered release sites <200 km apart to be separate release sites of the same reintroduction. For American martens, Slough (1994) considered sites ≥50 km apart to be independent. For other species, we accepted the definition of independent sites used by the authority that provided the data.

VORTEX Model

We developed a population model using VORTEX (Miller and Lacy 2005) to be similar to that used by Lewis et al. (2012), while general and applicable to all martens. Our goal was a model that predicted the relative probability of success for marten reintroduction programs, given that release sites have adequate appropriate habitat. Wood et al. (2007) used a similar approach to predict the numbers of tree squirrels of 6 species that needed to be released for reintroductions to be successful. The predictions of our model, however, became hypotheses to test with data from actual reintroductions. For hypotheses we accepted, our understandings of marten biology used to build the model were supported. For hypotheses we rejected, some aspect of our knowledge was lacking and will require rethinking and new research. Models are most powerful when they highlight important gaps in our knowledge.

We modeled the reintroduction of marten populations using the parameters shown in Table 6.1. We calculated vital rates from reviews (Powell 1993; Powell et al. 2003; Elmeros et al. 2008; McCann et al. 2010) and from publications that provided information not available in reviews (Raine 1981; Lindström et al. 1995; Helldin 1999; Zalewski and Jędrzejewski 2006; Masuda 2009; Safronov et al. 2009).

We first simulated a baseline population on a landscape that was a mix of public and private land partitioned into a grid of 50 contiguous hexagons, each of which could experience different conditions (Table 6.1). Although we originally conceived the landscape to be forested, and that public and private lands would experience different rates of habitat degradation and loss, we later realized that stone martens (*M. foina*) experience similar impacts of habitat change in agricultural and urban landscapes (Herr 2008). Juveniles dispersing from any hexagon had a 5% probability of reaching an adjacent hexagon and a 3% probability of reaching the next tier of hexagons. Carrying capacity (*K*) was 2000, and the model population was not subject to trapping

Table 6.1. Baseline conditions for population VORTEX analyses

Demographic variable	Value (± SD)	Elasticity	
		−10%	+10%
Starting population size, N_0	1000	4	2
Carrying capacity, K	2000 ± 250	7	8
Mean litter size	2.6 ± 1.0	23	3
Age (yr) at first reproduction	2	—[a]	—[a]
Exponent for density dependence, B	16	0	0
Exponent for Allee Effect, A	0.5	0	0
Mortality rates			
Juveniles (age 0–1)	65 ± 25 %	3	39
Yearlings (age 1–2)	25 ± 20%	2	8
Adults (age ≥2)	12 ± 20%	0	9
Reproduction after logging	50%	4	8
Survival after logging	75%	—[a]	—[a]
Local subpopulations (and timber harvest/100 yrs)			
Total number	50		
Private land	25 (2 harvests/100 yrs)		
U.S. Forest Service land managed for timber	17 (1 harvest/100 yrs)		
U.S. Forest Service land managed as wilderness	8 (0 harvest/100 yrs)		

Notes: Baseline conditions were run 100 times for 100 years using VORTEX with stochastic variation as indicated by standard deviations (SD). Elasticity indexes show the percent change in the probability of extinction when input variables changed by ± 10%.

[a] Elasticity values were not calculated

or hunting mortality; thus, each hexagon could support 40 martens. Because estimates of extinction by VORTEX are only relatively accurate (Caughley 1994; Brook 2000; Holmes et al. 2007), we used 1 minus the probability of extinction (from VORTEX) as an index of population persistence or an index of the likelihood of reintroduction success (i.e., a high index value meant a high probability of reintroduction success). We evaluated the model with 100 runs, each for 100 years. The baseline model had an index of persistence of 0.95. Elasticity analyses suggested that juvenile mortality and litter size had the greatest potential to affect persistence (Table 6.1).

To evaluate the effectiveness of different reintroduction approaches, we simulated the release of martens on an empty landscape of 50 hexagons, then looked for patterns of change in the index of success with the different approaches. The different approaches varied the total numbers of males and females "released," the number of release sites used (hexagons), and the number of years over which we released animals. Values of population variables were the same as for the baseline population except that K = 1000, because a reintroduced population will occupy a smaller area than a source population (that is, each hexagon represented a smaller area than for the base model), and because most extinction occurred in early years, when a reintroduced population still occupied a small area. We released martens most often in groups of

5 per hexagon. Because real reintroductions of martens have involved diverse numbers of animals released over different numbers of years, we modeled 20 martens released in each of 2, 3, 5, or 8 years, releasing animals into new hexagons each year; 20 up to 160 martens all released in 1 year; and 80 martens released into 1, 2, 5, and up to 20 hexagons. We varied the sex ratio from 4:1 (M:F) to 1:4, and varied the number of juveniles released from 0 to 3 in each group of 5 martens released in a hexagon. Our simulations addressed questions 1 and 6 posed by Armstrong and Seddon (2008).

Actual Marten Translocations

We compiled information on translocations of all *Martes* species from the scientific and popular literature; from theses and dissertations; from government agency databases, files, and archives; and through interviews with individuals who had participated in or had knowledge of specific translocations. We combined and updated available information on American martens, fishers, and sables (Timofeev and Pavlov 1973; Monakhov 1978, 1982, 2006; Berg 1982; Roy 1991; Powell 1993; Slough 1994) and expanded the review through 2010 to include all martens except the yellow-throated (*M. flavigula*) and Nilgiri (*M. gwatkinsii*) martens. No data were available for these 2 species, which apparently have never been translocated. The numbers of translocations are necessarily approximate because (1) we counted multiple introductions of American martens or stone martens to islands as 1 translocation, (2) so many sables were translocated in the USSR that small discrepancies exist in the literature, and (3) undocumented translocations of American martens, fishers, and possibly other martens may have occurred.

General and specific characteristics of documented translocations that participants believed may have influenced their success are presented in Table 6.2. We developed geographic range maps based on distribution data published previously, and the information in data tables presented later in this chapter. We plotted all translocation release sites with respect to a species' range before it was known to have been manipulated by humans and with respect to its current range. For fishers and sables, we plotted translocations with respect to the most contracted ranges. These plots illustrate how translocations may have affected species' ranges.

Using Pearson's correlation coefficient, we tested for independence of the variables from actual reintroductions used to test the hypotheses generated by the VORTEX model. Using Akaike's Information Criterion adjusted for small sample sizes (AIC_c; Burnham and Anderson 2002) and data from actual reintroductions, we evaluated the hypotheses generated by the VORTEX model (Table 6.3), allowing us to address questions 1, 2, 4, and 6 posed by Armstrong and Seddon (2008). We calculated likelihood values for AIC_c using Proc GENMOD in SAS (SAS 9.2; http://www.sas.com), and retained all

Table 6.2. General characteristics reported for marten translocations, and factors hypothesized in the literature to influence the success of translocations

General characteristics	Factors that could influence translocation success
Location (country/state/province, specific release site[s])	Feasibility assessment prior to release
	Monitoring post-release
Outcome (success/failure)	Number of martens released
Purpose of translocation	Number of release sites
Source population (country/state/province)	Number of source sites (an index of genetic diversity)
	Number of years martens released
Translocation type (reintroduction/augmentation/introduction)	Ownership of lands where martens released
	Protection from fur-trapping for martens specifically
Year initiated and number of years	Protection from incidental fur-trapping
	Proximity to source population/Genetic relatedness of source and original populations
	Region for American martens and fishers (eastern versus western North America[a])
	Release type (hard versus soft)
	Season of release
	Sex ratio of released martens
	Trapping reestablished following translocation

[a] The 106th meridian (at 106°W) divides Saskatchewan, Canada, into eastern and western halves. We arbitrarily chose this meridian to divide translocations into those that occurred in eastern versus western North America.

hypotheses with $\Delta AIC_c \leq 2.0$. Unfortunately, no age or maturity data existed for reintroduced martens (except for 1 small reintroduction in the Netherlands). Sex-ratio data were available for a subset of reintroductions. Therefore, using AIC_c we first evaluated the following hypotheses: (1) Number of martens released (as a substitute for number of adult females released) affected success; (2) Number of release sites; (3) Number of release years (controlled for total numbers of animals released); and (4–6) the 3 combinations of these 3 variables. For many reintroductions of American martens, fishers, and sables, the number of females (but not their ages) was known. For these data, we next evaluated hypotheses generated by substituting the total numbers of females and males for the total numbers of adult females and adult males. Our hypotheses included: (1) Number of females released; (2) Number of males released; (3) Number of release sites; (4) Number of release years; and (5–10) Number of females or Number of males combined with the other 2 variables.

Although we did not use running means in any of our analyses, we plotted the number of American martens, European pine martens (*M. martes*), fishers, and sables released against the running mean of success to illustrate the effect of the number of animals released on reintroduction success. To calculate the running means, we sorted releases for a given species by the number of martens released. We rated failed releases as success = 0, and successful releases as success = 1. We took the running mean of the success of 7 releases

Table 6.3. Models tested using AIC_c

Models predicted by the VORTEX model	Additional models
First set of models	
Number of animals	Number of animals, Number of release sites, Genetic relatedness
Number of release sites	Number of animals, Number of release sites, Monitoring
Number of years (controlled for number of animals)	Number of females, Number of release sites, Genetic relatedness
Number of animals, Number of release sites	
Second set of models	
Number of females	Number of females, Number of release sites, Monitoring
Number of males	Number of females, Number of release sites, Number of years
Number of release sites	Number of females, Number of release sites, Number of source sites
Number of years	Number of females, Number of release sites, Protection from trapping
Number of females, Number of release sites	Number of females, Number of release sites, Release type
Number of males, Number of release sites	Number of females, Number of release sites, Region
Number of females, Number of years plus interaction	Number of males, Number of release sites, Genetic relatedness
Number of males, Number of years plus interaction	Number of males, Number of release sites, Monitoring
Number of females, Number of years, Number of release sites	Number of males, Number of release sites, Number of years
Number of males, Number of years, Number of release sites	Number of males, Number of release sites, Number of source sites
	Number of males, Number of release sites, Protection from trapping
	Number of males, Number of release sites, Release type
	Number of males, Number of release sites, Region

Notes: The first set of AIC_c models tested the predictions of the VORTEX model for variables affecting success of reintroductions by substituting the total number of animals as an index for the total number of adult females; many datasets of actual reintroductions did not include information on the sexes of animals released, and no datasets had reliable information on age. In the second set of AIC_c models, only datasets in which sex was known were used. Additional models were developed from hypotheses suggested in the literature as possible explanations for the success of failure of reintroductions.

(a target release, plus the 3 releases with next lower and higher numbers of martens released) and plotted the running mean against the number of martens released in the target release.

Finally, we evaluated the variants of the retained hypotheses that included the additional factors (variables) from Table 6.2 for which we had enough data (genetic relatedness, number of source sites as a potential index or surrogate for genetic diversity, monitoring, protection, region, release type) taken one at a time, again using AIC_c to rank the performance of our hypotheses (Table 6.3).

Results

We documented 51 translocations of American martens (39 reintroductions, 1 augmentation, 10 introductions, 1 uncertain), 38 fisher translocations (30 reintroductions, 5 augmentations, 3 introductions), 6 of European pine martens (3 reintroductions, 1 augmentation, 2 introductions), 5 of stone martens (3 augmentations, 2 introductions), 1 introduction of Japanese martens (*M. melampus*) and >304 translocations of sables, for which we had good data from 34 (24 reintroductions, 7 augmentations, 3 introductions).

American Marten

Of the 39 reintroductions of American martens, 20 succeeded, 9 failed, and 10 had uncertain outcomes (Table 6.4, Figure 6.1). For the 29 reintroductions with known outcomes, we obtained nearly or fairly complete datasets for 4 of the 13 variables that could influence translocation success (Table 6.2): number and sex ratio of animals released, number of years that animals were released, and release in eastern vs. western North America. The first 3 are, or relate directly to, variables evaluated with our VORTEX model.

We considered the reintroduction to the Porcupine Mountains in the Upper Peninsula of Michigan to be a success, but Williams et al. (2007) noted the possibility that colonization from other reintroductions in the Upper Peninsula may have been the source (or an augmentation) of the Porcupine Mountain population. We considered 3 other reintroductions in the Upper Peninsula to have unknown outcomes, again because colonization from other sites was possible. Nonetheless, the presence of 3 distinct genetic clusters in geographic proximity to the release sites, as well as the sharing of unique alleles from these clusters with source populations decades after release, suggest that several, if not all, of these releases have succeeded (Williams and Scribner 2007, 2010; but see Swanson et al. 2006). Despite these uncertainties, American martens have clearly been reestablished in the Upper Peninsula of Michigan (Earle et al. 2001; Williams et al. 2007; Williams and Scribner 2010).

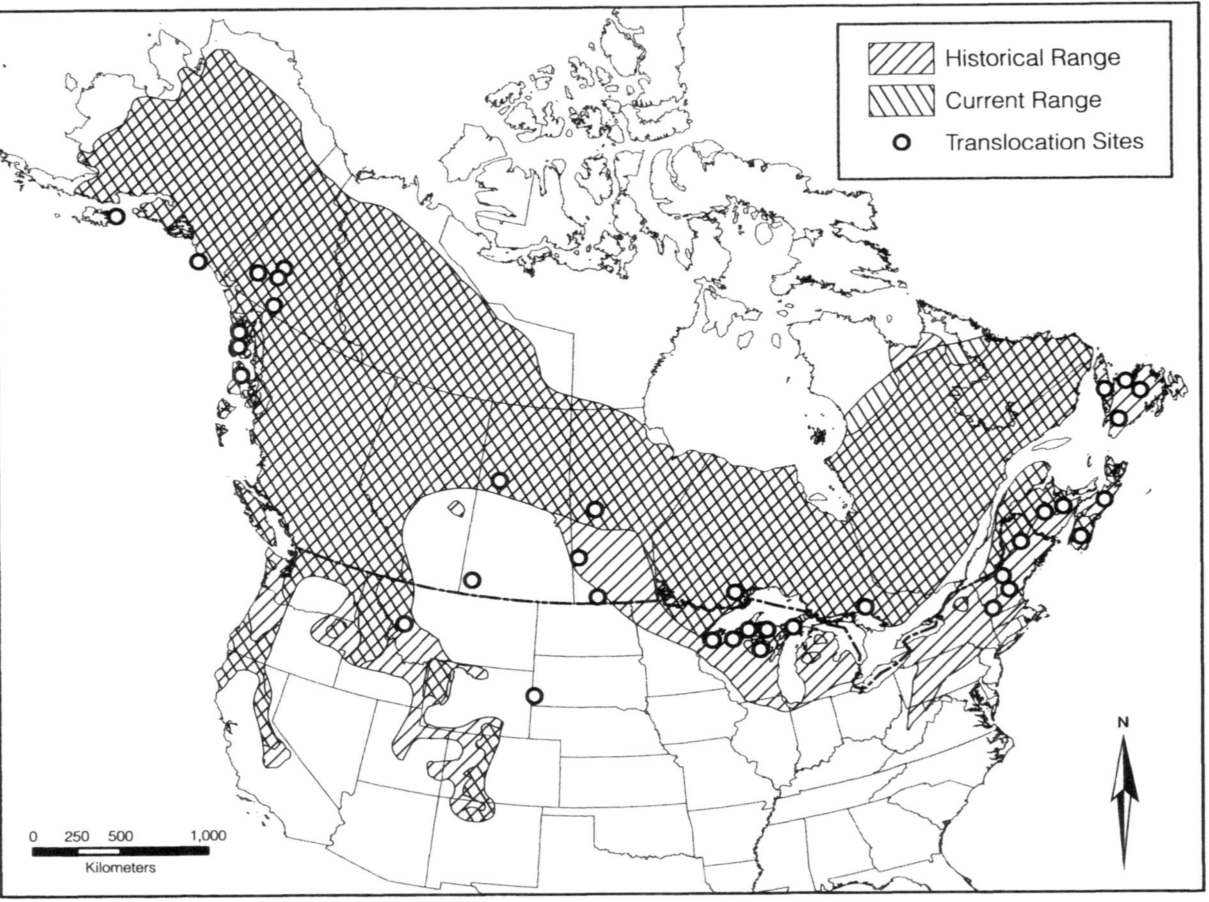

Figure 6.1. Locations of translocation sites in relation to the historical and current ranges of the American marten (updated from Slough 1994).

Table 6.4. American marten translocations, 1934–2012, arranged by release location

Release location (country/state/ province)	Release area (number of sites)	Year(s)	Type[a]	n (♀s)	Source location	Release[b]	Protection[c]	Status[d] (2009)	Source
England	—	early 1900s	I	—	Fur farm	A	Y	G	Kyle et al. 2003
Canada									
Manitoba	Barb Lake	1960–61	R	11 (≥3)	Manitoba	—	S	F	Miller 1961; C. Johnson, Manitoba Dept. of Natural Resources
	Tramping Lake	1961	R	2 (1)	Manitoba	—	S	F	Ibid.
	Minago River/ Williams Lake	1967–68	R	99	British Columbia	—	S	F	C. Johnson and R. Stardom, Manitoba Dept. of Natural Resources
	Duck Mountain	1969	R	42	Ontario	—	S	S	C. Johnson, R. Stardom, and D. Berezanski, Manitoba Dept. of Natural Resources
	Turtle Mountains	1990	I	10	Manitoba, Ontario	H	S	S	G. Armstrong, Manitoba Trappers' Association
	Riding Mountain National Park	1991–93	R	68 (31)	Manitoba	H	P	S	Jensen and Schmidt 2002
New Brunswick	Acadia Forest Research Station	1967–68	R	<10	New Brunswick	—	P	F	D. Cartwright, New Brunswick Dept. of Natural Resources and Energy
	Fundy National Park	1984–91	R	50 (20)	New Brunswick	S	P	S	Quann 1985; Sinclair 1986, 1987; Sullivan 1984; G. Corbett and G. Sinclair, Environment Canada, Parks Canada

Newfoundland	Siviers Island	1976	I	3 (2)	Newfoundland	H	S	F	Bissonette et al. 1988; O. Forsey and L. Mayo, Newfoundland and Labrador Dept. of Environment and Lands
	Notre Dame Bay	1976	—	—	—	—	—	F	Ibid.
	Terra Nova National Park	1982–83, 1998, 1999	R	8, 2 litters totaling 6	Newfoundland	S	P	S	Bateman 1982, 1984; Hoffman 1983; G. Corbett, Environment Canada, Parks Canada
	LaPoile River	1975	R	3 (2)	Newfoundland	H	S	F	Bissonette et al. 1988; Evans 1986; L. Mayo, Newfoundland and Labrador Dept. of Environment and Lands
	Maine River	1976–78	R	11 (4)	Newfoundland	H	S	S	Ibid.
Nova Scotia	Liscomb Game Sanctuary	1956	R	12 (7)	Ontario	—	P	F	Dodds and Martell 1971; B. Sabean, Nova Scotia Dept. of Lands and Forests
	Kejimkujik National Park	1987–94	R	116	New Brunswick	H/S	P	S	Boss et al. 1987; Drysdale and Charlton 1988; Sinclair 1986
Ontario	Sibley Provincial Park	1950–51	R	47 (16)	Ontario	—	P	U	de Vos 1952; M. Novak, Ontario Ministry of Natural Resources
	Parry Sound District	1956–63	R	248 (94)	Ontario	H	N	S	Rettie 1971; M. Strickland, Ontario Ministry of Natural Resources

Table 6.4—*cont.*

Release location (country/state/ province)	Release area (number of sites)	Year(s)	Type[a]	n (♀s)	Source location	Release[b]	Protection[c]	Status[d] (2009)	Source
Saskatchewan	Prince Albert National Park	1954	R	24 (12)	Alberta	H	P	S	P. Galbraith, Saskatchewan Dept. of Parks and Renewable Resources; R. Leonard, Environment Canada, Parks Canada
	Cypress Hills Provincial Park	1986–87	I	33 (19)	Alberta, Yukon	S	P	S	Hobson et al. 1989; W. Runge, Saskatchewan Dept. of Parks and Renewable Resources
Yukon	Takhini Lake/ Wheaton River	1984–86	R	31 (17)	Yukon	H	S	U	B. Slough, Yukon Dept. of Renewable Resources
	Braeburn	1984–86	R	26 (12)	Yukon	H	S	U	Ibid.
	Takhini River	1985–86	R	63 (21)	Yukon	H	S	S/U	Ibid.
	Haines Junction	1984–87	R	51 (17)	Yukon	H	S	S/U	Ibid.
USA									
Alaska	Prince of Wales Island	1934	I	10 (4)	Alaska	H	N	S	Burris and McKnight 1973; Paul 2009
	Baranof Island	1934	I	7 (3)	Alaska	H	N	S	Ibid.
	Kayak Island	1940s	I	—	Alaska	H	N	S	Ibid.
	Patterson Island	1940	I	—	Alaska	H	N	S	Ibid.
	Chichagof Island	1949–52	I	21 (≥4)	Alaska	H	N	S	Ibid.
	Afognak Island	1952	I	20 (12)	Alaska	H	N	S	Paul 2009
Idaho	—	1993–94	R	59	—	—	—	S	W. Melquist, pers. comm. in Proulx et al. 2004
Maine	Northern Maine	1983	R	76	Maine	H	N	S	K. Elowe, Maine Dept. of Inland Fisheries and Wildlife

Michigan	Porcupine Mountain State Park (1)	1955–57	R	29 (11)	Ontario	H	P	S	Harger and Switzenberg 1958; Williams et al. 2007; Earle et al. 2001
	Hiawatha National Forest (1)	1968–70	R	99 (37)	Ontario	H	S	U	Schupbach 1977; Earle et al. 2001
	Huron Mountain State Forest	1979–80	R	78 (31)	Ontario	H/S	S	S	Earle et al. 2001; Churchill et al. 1981
	McCormick Experimental Forest	1980	R	22 (13)	Ontario	H/S	S	S	Ibid.
	Iron County	1980–81	R	48 (27)	Ontario	H/S	S	S	Ibid.
	North Otsego and South Cheboygan Counties (2)	1985	R	49 (24)	Ontario	H	S	S	Earle et al. 2001
	Northeast Lake County (1)	1986	R	36 (16)	Ontario	H	S	S	Ibid.
	Hiawatha National Forest (2)	1989–90	U	20 (9), 27	Michigan	H	S	U	Williams et al. 2007; Michigan Dept. of Natural Resources 1970
	Keweenaw County	1992	U	19 (5)	Michigan	H	—	U	Williams et al. 2007
Montana	Silver Bow County	1944	R	12	Montana	—	—	—	Rognrud 1983
	Lincoln County	1955	R	21	Montana	—	—	—	Ibid.
	Meager County	1956–57	R	9 (5)	Montana	—	—	F	Rognrud 1983; Thompson 1949
New Hampshire	Second College Grant	1953	R	2 (1)	Ontario	—	P	U	Silver 1957
	White Mountain National Forest	1975	R	29 (9)	Maine	—	P	U	Soutiere and Coulter 1975; E. Orff, New Hampshire Dept. of Fish and Game

Table 6.4—*cont.*

Release location (country/state/ province)	Release area (number of sites)	Year(s)	Type[a]	*n* (♀s)	Source location	Release[b]	Protection[c]	Status[d] (2009)	Source
South Dakota	North Black Hills	1980–81	R	42 (17)	Idaho	S	P/S	S	Fecske 2003; Fredrickson 1983
	Central Black Hills	1990–93	R	83 (30)	Idaho, Colorado	S	P/S	S	Fecske 2003; L. Fredrickson, South Dakota Dept. of Game, Fish and Parks
Vermont	Green Mountain National Forest	1989–91	R	115 (27)	New York, Maine	H	S	F	K. Elowe, Maine Dept. of Inland Fisheries and Wildlife; J. DiStefano, Vermont Dept. of Fish and Wildlife
Wisconsin	Stockton Island	1953	R	5	Montana	—	—	F	Davis 1978; Gieck 1986
	Nicolet National Forest	1975–83	R	172 (51)	Ontario, Colorado	H/S	S	S	Davis 1978, 1983; Kohn and Eckstein 1987
	Chequamegon National Forest (3)	1987–90	R	139 (45)	Minnesota	H	S	S/U	Kohn 1991
	Chequamegon National Forest	2008–10	A	58 (36)	Minnesota	S	S	O	Woodford 2009a; J. Woodford, Wisconsin Dept. of Natural Resources

Note: — = not reported
[a] I = introduction, R = reintroduction, A = augmentation, U = unknown
[b] A = accidental and hard, H = hard release, S = soft release
[c] Y = protection for European pine martens, P = protected (no trapping), N = not protected, S = species-specific protection from trapping
[d] G = genes documented, S = successful, F = failure, O = ongoing, U = unknown

Similarly, for reintroductions in the Yukon and New Hampshire, whether reestablished populations resulted from the reintroductions per se, or from colonization by American martens from nearby populations, is unclear (Kelly 2005; B. Slough, unpublished data). For some reintroductions in the Yukon, viability status is unknown because of the lack of well-organized monitoring.

Most failed reintroductions were characterized by small numbers of animals released (i.e., in Montana and Wisconsin in the United States, and in Manitoba, New Brunswick, Newfoundland, and Nova Scotia in Canada). The reintroduction of 115 American martens in Vermont from 1989 to 1991 may have failed as a result of competition with fishers (Moruzzi et al. 2003). In some cases, insufficient monitoring may have failed to detect small numbers of extant individuals.

The reintroduction of American martens to the Nicolet National Forest in northern Wisconsin resulted in an estimated population of 221 (±61) individuals as of 2006 (Woodford 2009a) and is considered successful (Williams et al. 2007). In contrast, the reintroduction of American martens to the Chequamegon National Forest, also in northern Wisconsin, resulted in fewer than 100 individuals as of 2006 (Williams et al. 2007). The latter population was augmented with 90 individuals during autumns 2008 through 2010 (J. Woodford, Wisconsin Department of Natural Resources, personal communication). The ultimate success of this reintroduction and augmentation will be judged in the future.

Although 2 reintroductions in Newfoundland using relatively few animals (11, 14) appear to have succeeded (Gosse et al. 2005; Environment Canada 2009; I. Schmelzer, personal communication reported in Committee on the Status of Endangered Wildlife in Canada 2007), 2 other reintroductions and 1 introduction with few animals (<10) failed.

Six documented introductions to islands in Alaska all succeeded, as did up to 18 or more undocumented introductions to other Alaskan islands. In addition, the population on Admiralty Island may be native or may have been the result of an unreported introduction prior to 1903 (MacDonald and Cook 1996, 2007; Paul 2009).

American martens that escaped from fur farms in England have not established a population, but genes from those American martens have been documented in European pine martens (Kyle et al. 2003; Schwartz et al., this volume).

Fisher

All 5 fisher augmentations and 77% of reintroductions with known outcomes succeeded, but all 3 introductions failed (Table 6.5, Figure 6.2). For reintroductions in eastern North America, 89% (17/19) succeeded, whereas only 43% (3/7) in western North America succeeded. For the 26 reintroductions

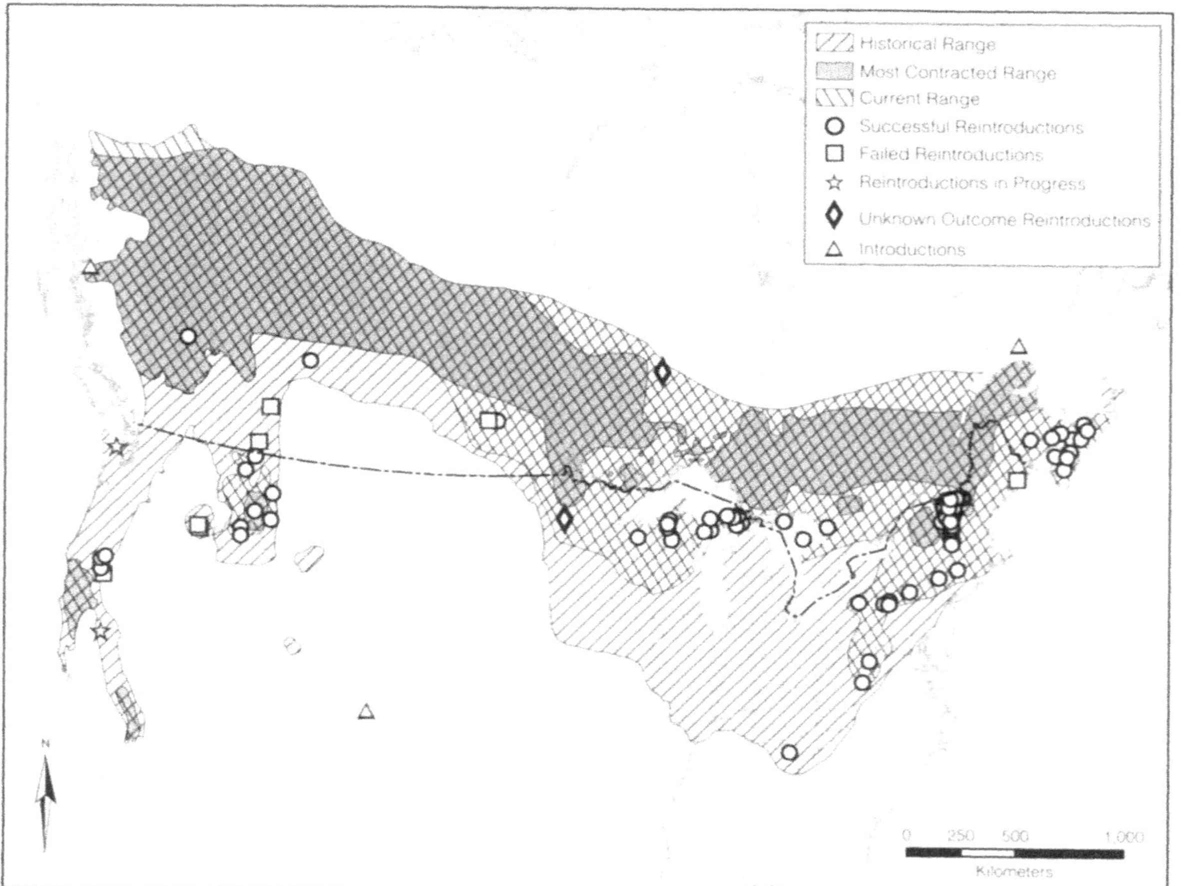

Figure 6.2. Locations of translocation sites in relation to the fisher's historical range, the range at its most contracted state, and the current range. Historical and most-contracted maps were based partly on information in Powell (1993), Gibilisco (1994), and Proulx et al. (2004) (modified from a similar figure by Lewis et al. 2012).

Table 6.5. Fisher translocations, 1896–2012, arranged by release location

Release location (country/state/ province)	Release area (number of sites)	Year(s)	Type[a]	*n* (♀s)	Source location	Release[b]	Protection[c]	Status[d]	Source
Canada									
Alberta	(1)	1990	R	17 (11)	Ontario, Manitoba	S	P	S	Proulx et al. 1994, 2004; Proulx and Genereux 2009
	(1)	1981–83	R	32 (16)	Alberta	H	—	F	Davie 1984; J. Jorgenson, Ministry of Tourism, Parks, and Recreation
British Columbia	(1)	1984–91	I	15 (4)	British Columbia	H	N	F	R. Weir, Artemis Wildlife Consultants; E. Lofroth, Ministry of Forests, Lands, and Natural Resource Operations
	—	1990–92	A	15 (13)	British Columbia	S	N	S	Weir 1995
	—	1996–98	R	60 (36)	British Columbia	H	N	F	Fontana et al. 1999; Weir et al. 2003
Manitoba	Riding Mountain National Park (1)	1972	R	4	Manitoba	H	P	F	Berg 1982; R. Baird, Riding Mountain National Park
	Riding Mountain National Park (1)	1994–95	R	45 (21)	Manitoba	H	P	S	Baird and Frey 2000; R. Baird, Riding Mountain National Park
New Brunswick	Acadia Forest Experiment Station (2)	1966–68	R	25 (15)	New Brunswick	H	S	S	Dilworth 1974; Drew et al. 2003; T. Dilworth, Dept. of Natural Resources

Table 6.5—*cont.*

Release location (country/state/ province)	Release area (number of sites)	Year(s)	Type[a]	*n* (♀s)	Source location	Release[b]	Protection[c]	Status[d]	Source
Nova Scotia	Tobeatic Game Sanctuary (1?)	1947–48	R	12 (6)	Fur ranch	H	S?	S	Benson 1959; Dodds and Martell 1971
	Linscomb and Chignecto Game Sanctuaries; Glenmore, Otter Brook, Stanley and North Mountain (8)	1963–66	R	80 (51)	Maine	—	—	S	Dodds and Martell 1971
	(1)	1993–95	A	14 (6)	Nova Scotia	H	S	S	Potter 2002; M. Boudreau and J. Mills, Nova Scotia Dept. of Fish and Wildlife; D. Potter, Acadia University
	(1)	2000–04	A	28 (21)	Nova Scotia	H	S	S	M. Boudreau and J. Mills, Nova Scotia Dept. of Fish and Wildlife; D. Potter, Acadia University
Ontario	Patricia (1)	1956	R	25	Ontario	H	S	U	Berg 1982; M. Novak, Ontario Ministry of Natural Resources
	Parry Sound (1)	1956–63	R	97 (60)	Ontario	H	S	S	Ibid.
	Manitoulin Is (1?)	1979–81	R	55 (32)	Ontario	H	S	S	Kyle et al. 2001; J. Baker and M. Novak, Ontario Ministry of Natural Resources

	Bruce Peninsula (1?)	1979–82	R	29 (14)	Ontario	H	S	S	Ibid.
Quebec	Anacosti Island (1)	1896–1914	I	2 (presumably 1)	—	—	N	F	Newsom 1937
USA									
California	Butte County (3)	2009–12	R	40 (24)	California	H	P	O	R. Powell, unpublished data
Colorado	(1)	1978 or 1979	I	2 (1)	—	H	S	F	R. Kahn, Colorado Division of Wildlife; Stouffer's Wild America TV show
Connecticut	(1)	1989–90	R	32 (19)	New Hampshire, Vermont	H/S	S	S	Rego 1989, 1990, 1991
Idaho	(3)	1962–63	A	39 (19)	British Columbia	H	S	S	Berg 1982; Luque 1984; Williams 1962b, 1963
Maine	Eastern Maine (1)	1972	R	7 (3)	Maine	—	N	F	Berg 1982; Maine Dept. of Fish and Wildlife, unpublished data
Michigan	Ottawa National Forest (1)	1961–63	R	61 (19)	Minnesota	H	S	S	Brander and Books 1973; Irvine et al. 1964; Powell 1993
	Hiawatha National Forest (9)	1988–92	R	189 (101)	Michigan	—	S	S	R. Earle, Michigan Dept. of Natural Resources
Minnesota	Itasca State Park (1)	1968	R	15	Minnesota	H	P	U	Berg 1982
Montana	(3)	1959–60	A	36 (2)	British Columbia	H	P	S	Heinemeyer 1993; Roy 1991; Vinkey 2003; Vinkey et al. 2006; Weckwerth and Wright 1968

Table 6.5—*cont.*

Release location (country/state/province)	Release area (number of sites)	Year(s)	Type[a]	n (♀s)	Source location	Release[b]	Protection[c]	Status[d]	Source
	(2)	1988–91	R	110 (63)	Minnesota, Wisconsin	—	—	S	Heinemeyer 1993; Roy 1991
New York	Catskill Mountains (1 or 2)	1976–79	R	43 (24)	New York	H	S	S	Wallace and Henry 1985; D. Henry and S. Smith, New York Dept. of Environmental Conservation
Oregon	Wallowa Mountains (4)	1961	R	13 (8)	British Columbia	H	P	F	Aubry and Lewis 2003; Kebbe 1961a,b
	Southern Cascade Range (2)	1961	R	11 (6)	British Columbia	H	P	F	Ibid.
	Southern Cascade Range (5)	1977–81	R	30 (15)	British Columbia, Minnesota	H	S	S	Aubry and Lewis 2003
Pennsylvania	(6)	1994–98	R	190 (97)	New York, New Hampshire	H	S	S	Serfass et al. 2001
Tennessee	(1)	2001–03	R	40 (20)	Wisconsin	H	S	S	Anderson 2002
Vermont	(35 areas throughout Vermont)	1959–67	R	35 (16), 89	Maine	H	S	S	Berg 1982; K. Royar, Vermont Dept. of Fish and Wildlife
Washington	Olympic National Park (9)	2008–10	R	90 (50)	British Columbia	H	P	O	J. Lewis, Washington Dept. of Fish and Wildlife, unpublished data
West Virginia	(2)	1969	R	16 (10), 7	New Hampshire	H	N	S	Pack and Cromer 1981; Wood 1977

Wisconsin	Nicolet National Forest (1)	1956–63	R	60 (24)	New York, Minnesota	H	P	S	Bradle 1957; Irvine et al. 1964; Petersen et al. 1977
	Chequamegon National Forest (1)	1966–67	R	60 (30)	Minnesota	H	P	S	Dodge 1977; Petersen et al. 1977

Note: — = not reported

[a] I = introduction, R = reintroduction, A = augmentation

[b] H = hard release, S = soft release

[c] P = protected (no trapping), N = not protected, S = species-specific protection from trapping

[d] S = successful, F = failure, O = ongoing, U = unknown

with known outcomes, we obtained complete datasets for 5 of the 13 variables thought to influence translocation success (Table 6.2): number and sex ratio of fishers released, number of years that fishers were released, release in eastern vs. western North America, and existence of a post-release monitoring effort. The first 3 are, or relate directly to, variables whose effects we explored with our VORTEX model relative to our index of success for a reintroduction.

Since the mid-1900s, fisher populations have recovered throughout much of their historical range, especially in the United States; some by natural range expansions, aided in many places by protection from trapping, and some from reintroductions (Figure 6.2). By the 1950s, the fisher population in northern Minnesota had begun to expand naturally to the east and south toward Lake Superior. Similarly, populations expanded in Maine and New York. By the 1990s, fishers had recolonized Massachusetts, Rhode Island, and eastern Connecticut. Clearly, however, the many reintroductions of fishers in the 1950s–1990s (Figure 6.2) assisted in their reestablishment in many portions of their historical range after the species' nadir. For example, fishers may have eventually recolonized Wisconsin and the Upper Peninsula of Michigan on their own, but when fishers were first reintroduced to those states, the potential, natural, source population in Minnesota had not reached the western tip of Lake Superior. Thus, these Midwestern states reestablished fishers decades before natural colonization could have occurred. Similarly, the reintroductions in New Brunswick, Nova Scotia, and Ontario in Canada, and in Idaho, Montana, Pennsylvania, and West Virginia in the United States probably contributed significantly to subsequent range expansions. The reestablishment of fisher populations in areas far removed from the margins of the most contracted range highlights the importance and success of reintroductions for the recovery of fisher populations.

Translocations to Idaho and Montana in the early 1960s were thought at the time to be reintroductions where fishers had been extirpated. Recently, however, Vinkey et al. (2006) and Schwartz (2007) identified a native haplotype for fishers trapped in the vicinity of these releases, indicating that the translocations in the early 1960s were actually augmentations to an existing, but extremely small, native population. Without the augmentation, that small native population may not have recovered on its own.

Fishers were introduced outside their historical range 3 times, but none were successful. Whether these introductions were purposeful or considered reintroductions is unknown. Henri Menier released several species native to the nearby mainland onto Anticosti Island, Quebec, at the turn of the 20th century (Newsom 1937), and he may have considered fishers native to the island.

In Ontario in 1956 and in Minnesota in 1968, a nearby fisher population expanded into a reintroduction area before the success of the reintroductions could be evaluated.

Most failures of translocations occurred soon after the last release, many apparently because too few fishers had been released. A few reintroduced populations have remained small for long periods, and some may still be vulnerable to collapse (e.g., southern Cascade Range in Oregon, south-central Alberta, and Tennessee).

European Pine Marten

Paleontological data suggest that European pine martens must have been introduced to the Balearic Islands of Minorca and Majorca by humans in ancient times, where they have now become established (Alcover 1980; Masseti 1995; Table 6.6, Figure 6.3). A small population of European pine martens was augmented in Killarney National Park in Ireland in the 1980s–1990s, and a population was reintroduced to Glengarriff Nature Reserve. Source animals came from within Ireland, and both populations appear now to be reestablished (Lynch 2006; National Parks and Wildlife Service 2007). In 1980 and 1981, European pine martens were reintroduced to Galloway Forest Park in Scotland. Source animals came from within Scotland and that population, too, appears to be reestablished (Shaw and Livingstone 1992). Following a feasibility assessment (Van der Lans et al. 2006), 7 juvenile European pine martens (6 males, 1 female) were reintroduced in Brabant, Netherlands, in a pilot project. As of 2009, only 3 males remained (J. Mulder, Bur Mulder-Natuurlijk, personal communication). In 1962, 5 male and 5 female European pine martens were introduced to the Osh Province of Kirgizia in the former USSR but failed to become established (Pavlov 1973b).

Stone Marten

As with European pine martens, stone martens were introduced to Mediterranean islands by humans in ancient times, but little other information is known (Alcover 1980; Masseti 1995; Table 6.6, Figure 6.3). Stone martens now occur on more than 2 dozen islands, mostly in the eastern Mediterranean. In the western Mediterranean, the species was introduced apparently only to Ibiza, where it became extinct in the early 1970s (Delibes and Amores 1986).

Stone martens were translocated within Denmark and Germany in the 1980s and within Luxemburg in 2006, apparently as experimental translocations (Rasmussen et al. 1986; Skirnisson 1986; Herr et al. 2008). In the early 1970s, stone martens escaped or were released from a commercial fur farm in southeastern Wisconsin. During the subsequent 40 years, they have spread into at least 4 counties, where populations have not grown large but persist (Long 2008). In 1936 in the USSR, 59 stone martens were introduced to Ryazan Province, European Russia (Pavlov 1973c). After 2 years, some of the released animals still survived (mainly near human settlement) but subsequently

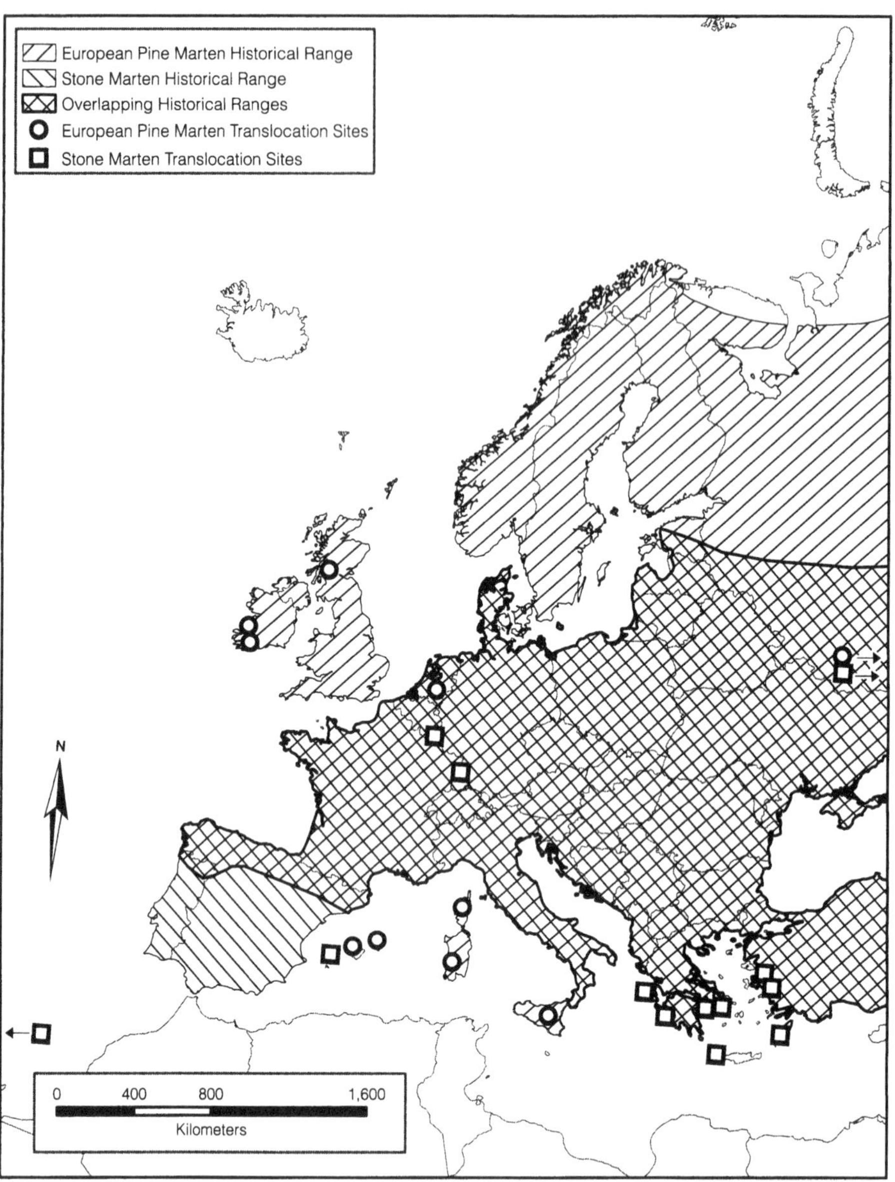

Figure 6.3. Locations of translocation sites of the European pine marten and the stone marten in relation to their historical ranges. The circle and square in Russia accompanied by arrows indicate the failed introductions of European pine and stone martens, which were located further east in the former Soviet Union; the square in the Atlantic Ocean accompanied by an arrow indicates the introduction of stone martens to Wisconsin, USA (see Table 6.6).

Table 6.6. European pine marten, stone marten, and Japanese marten translocations, through 2010, arranged by species and release location

Species	Release location	Release area	Year(s)	Type[a]	n (♀s)	Source location	Release[b]	Protection[c]	Status[d]	Source
European pine marten	Mediterranean Islands	Balearic Islands: Majorca, Minorca	Ancient times	I	—	—	—	—	S	Alcover 1980; Masseti 1995
	Ireland	Killarney National Park	1987–95	A	30	County Mayo and County Clare, Ireland	S	P	S	Lynch 2006; P. O'Leary, National Parks and Wildlife Service
	Ireland	Glengarriff Woods Nature Reserve	2003–04	R	7 (5)	County Kerry and County Clare, Ireland	S	P	S	National Parks and Wildlife Service 2007; C. Heardman, National Parks and Wildlife Service
	Netherlands	Brabant	2008–09	R	7 (1)	—	—	N	F	J. Mulder, Bur Mulder-Natuurlijk
	Scotland, UK	Galloway Forest Park	1980–81	R	12 (6)	Ratagan and Inchnacardock, North Scotland Conservancy	H	P	S	Shaw and Livingstone 1992
	USSR	Sare-Chelek Nature Reserve, Osh Province, Kirgizia	1962	I	5 (5)	Arkhangelsk Province, European Russia	H	—	F	Pavlov 1973b

Table 6.6—*cont.*

Species	Release location	Release area	Year(s)	Type[a]	n (♀s)	Source location	Release[b]	Protection[c]	Status[d]	Source
Stone marten	Mediterranean Islands	>2 dozen islands, East Mediterranean	Ancient times	I	—	—	—	—	S	Alcover 1980; Masseti 1995
	Luxembourg	Bettembourg	2006	T	5 (2)	Same population	H	N	F	Herr et al. 2008
	Denmark	—	mid-1980s	T	53	Same population	—	N	—	Rasmussen et al. 1986
	Germany	—	mid-1980s	T	2	Same population	H	N	F	Skirnisson 1986
	USA	Kettle Moraine State Forest, Wisconsin	early 1970s	I	—	Fur farm	H	N	S	Long 2008
	USSR	Ryazan Province, European Russia	1936	I	59	—	H	—	F	Pavlov 1973c
Japanese marten	Japan	Hokkaido	1940s	I	—	Honshu	A	P	S	Hosoda et al. 1999; Kadosaki 1981; Murakami and Ohtaishi 2000

Note: — = not reported

[a] I = introduction, R = reintroduction, A = augmentation, T = translocation

[b] A = accidental and hard, H = hard release, S = soft release

[c] P = protected (no trapping), N = not protected, S = species-specific protection from trapping

[d] S = successful, F = failure

disappeared. In European Russia, the distribution of stone martens appears to be expanding naturally to the north (Abramov et al. 2006).

Japanese Marten

In the 1940s, Japanese martens were reportedly introduced accidentally to Hokkaido Island (which originally supported only sables) by fur farmers from Honshu Island, where Japanese martens are native (Kadosaki 1981; Table 6.6, Figure 6.4). Japanese martens are now well established on the peninsula forming the southwest sector of Hokkaido (Murakami and Ohtaishi 2000).

Sable

From 1901 through the 1980s, nearly 20,000 sables were translocated in Russia/USSR (Tables 6.7–6.8, Figure 6.5). During the late 1800s and early 1900s, sables were harvested heavily and, by the 1930s, populations were low and had a significantly contracted range (Figure 6.5). In 1901, some sables from Kamchatka were introduced onto Karaginsky Island (near the eastern coast of Kamchatka). More sables were introduced to Karaginsky Island in 1928 and 1930, and small numbers of sables were reintroduced to the Shantar Islands in the Buryatia Republic and in Tyumen Province from 1927 to 1936. A large government program of reintroductions begun in 1939 was curtailed during World War II, but it was ongoing again through the mid-1960s. From 1939 to 1941, a total of 438 sables were released in the Buryatia Republic, Urals, Altai, and the Far East (Timofeev and Pavlov 1973). Thereafter, >17,000 sables were translocated throughout the former Soviet Union (Figure 6.5; Bakeev and Timofeev 1973; Timofeev and Pavlov 1973). Although the government program of sable reintroductions was largely finished by the mid-1960s, some local reintroductions continued until the end of the 1980s (Sinitsyn 2001). Reintroductions were concentrated in the Russian Federation. Apparently, all reintroductions with known outcomes were successful, but the fates of a few remain uncertain.

During the 1940s and 1950s, at least 7 populations were augmented with sables from elsewhere that had more valuable (darker) pelts. Augmentations were generally considered successful because pelt quality did improve in the populations near the release locations (Timofeev and Pavlov 1973; Monakhov and Monakhov 1978; Monakhov et al. 1982; Monakhov 1984, 2006). In 1955, 14 sables from Sverdlovsk Province (middle Urals) were introduced to Chelyabinsk Province (south Urals) and, in the 1960s, a total of 377 sables from Altai in eastern Kazakhstan were introduced to 2 sites in the Alatau Mountains of southern Kazakhstan to extend the species' range toward the southwest. These 3 introductions failed. Translocated sables were captured in

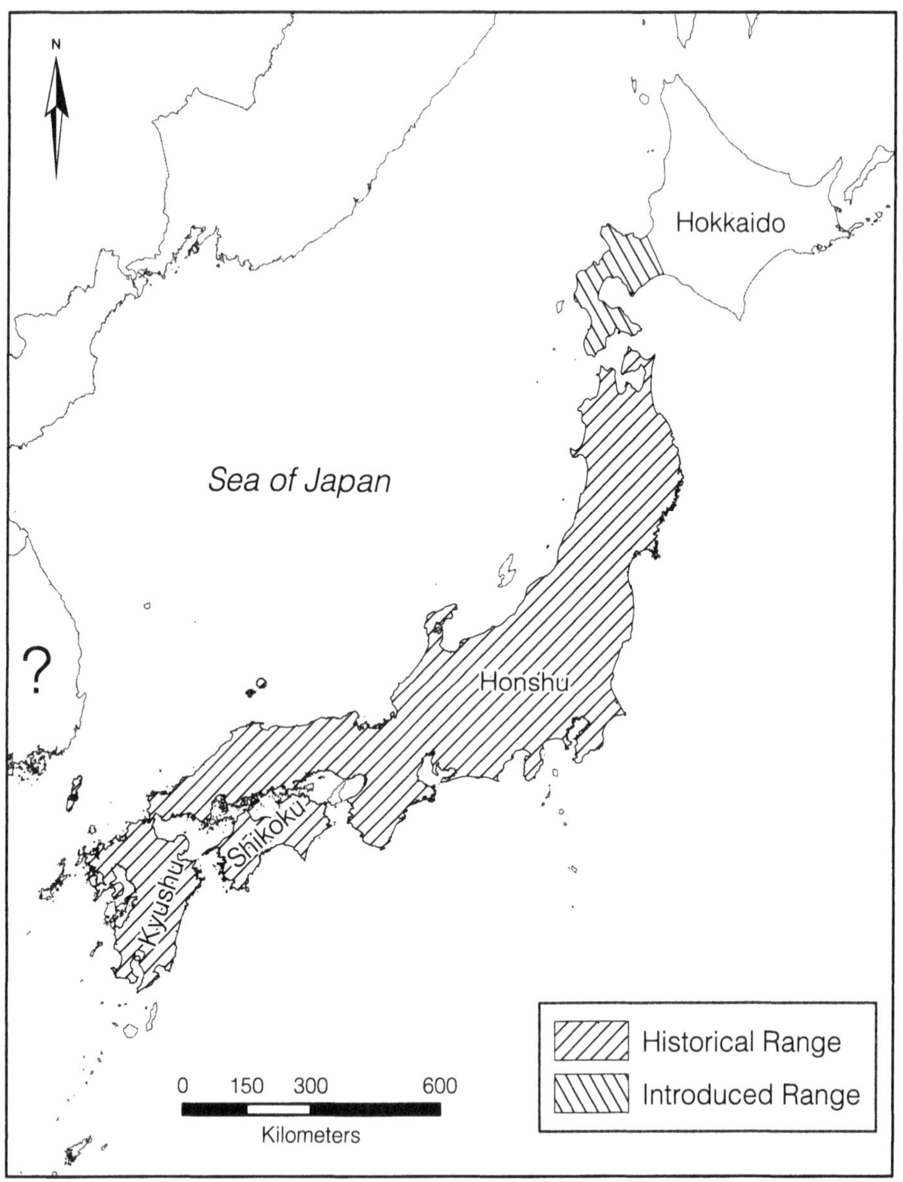

Figure 6.4. Historical range of the Japanese marten and approximate range on Hokkaido, where Japanese martens were introduced.

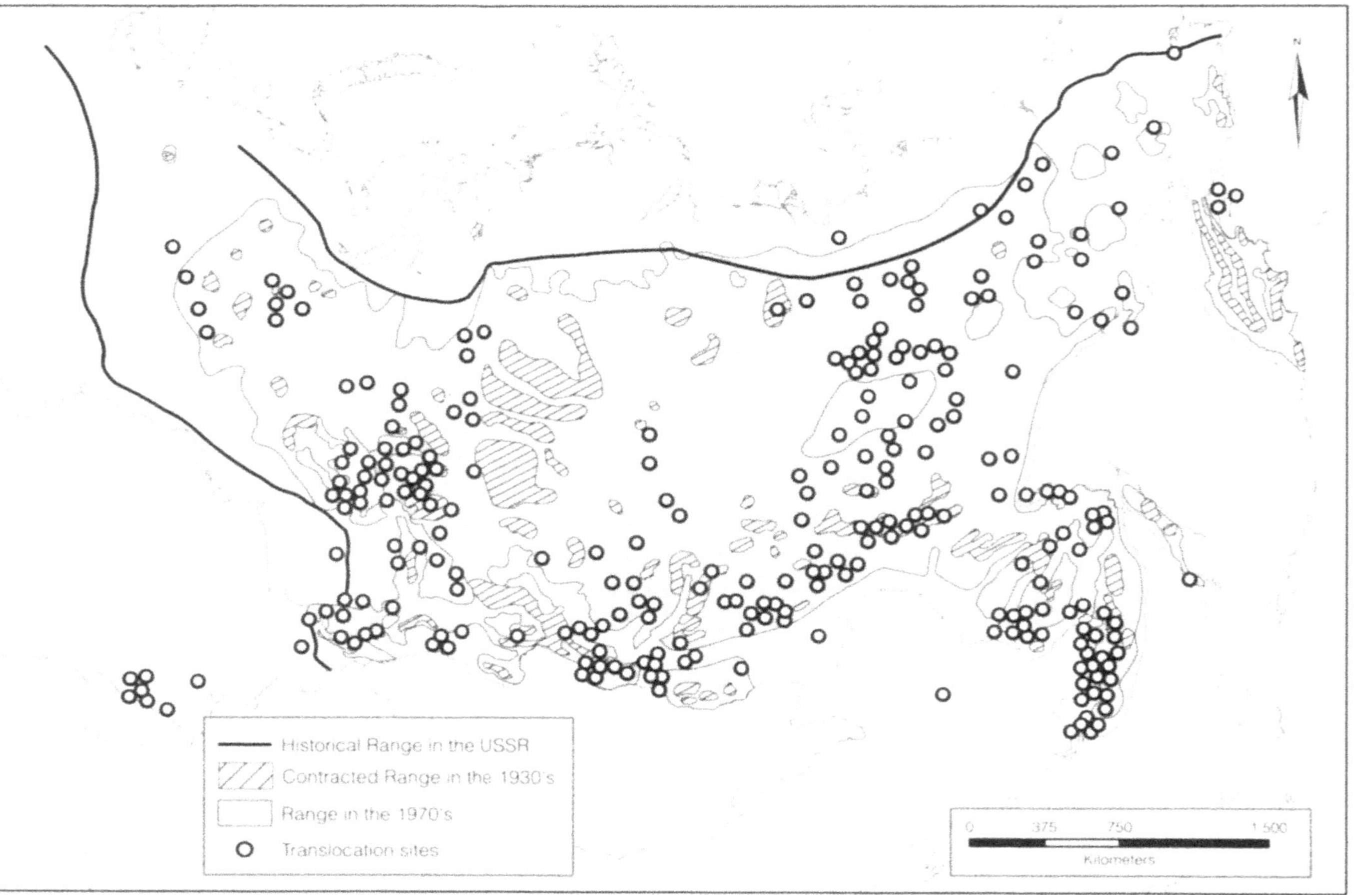

Figure 6.5. Locations of translocation sites in relation to the sable's historical range, the range at its most contracted state in the 1930s, and the expanded range by the 1970s (modified from Bobrov et al. 2008).

Table 6.7. Numbers of sables translocated in Russia and the former USSR, 1901–1970, arranged by release location

Location	'01	'27	'28	'29	'30	'31	'33	'39	'40	'41	'45	'47	'48	'49	'50
Altai Province	—	—	—	—	—	—	—	—	44	49	—	—	—	191	—
Amur Province	—	—	—	—	—	—	—	—	—	—	—	—	—	—	—
Buryatia Republic	—	—	—	—	2	—	—	12	25	45	18	37	51	47	142
Chelyabinsk Province	—	—	—	—	—	—	—	—	—	—	—	—	—	—	—
Chita Province	—	—	—	—	—	—	—	—	—	—	—	—	—	—	58
Irkutsk Province	—	—	—	—	—	—	—	19	50	124	—	37	—	88	297
Yakutia Republic	—	—	—	—	—	—	—	—	—	—	—	—	59	—	73
Kamchatka	10	—	20	—	3	—	—	—	—	—	—	—	—	—	—
Kazakhstan	—	—	—	—	—	—	—	—	—	—	—	—	—	—	—
Kemerovo Province	—	—	—	—	—	—	—	—	—	—	—	79	—	—	—
Khabarovsk Krai	—	12	5	60	—	13	—	—	—	—	—	—	—	40	211
Krasnoyarsk Krai	—	—	—	—	—	—	—	—	—	—	—	—	—	148	88
Magadan Province	—	—	—	—	—	—	—	—	—	—	—	—	—	—	—
Novosibirsk Province	—	—	—	—	—	—	—	—	—	—	—	—	—	—	—
Perm Province	—	—	—	—	—	—	—	—	—	—	—	—	—	—	—
Primorsky Krai	—	—	—	—	—	—	—	—	10	—	—	—	—	—	—
Sakhalin Province	—	—	—	—	—	—	—	—	—	—	—	—	—	—	—
Sverdlovsk Province	—	—	—	—	—	—	—	—	20	—	—	—	—	42	93
Tyumen Province	—	—	—	—	—	—	15	—	—	—	—	—	—	—	—
Tomsk Province	—	—	—	—	—	—	—	—	40	—	—	—	—	—	99
Tuva Republic	—	—	—	—	—	—	—	—	—	—	—	—	—	—	—
Total	10	12	25	60	5	13	15	31	189	218	18	153	110	556	1061

Source: Timofeev and Pavlov 1973

'51	'52	'53	'54	'55	'56	'57	'58	'59	'60	'61	'62	'63	'64	'65	'70	Total
99	—	93	62	—	—	—	—	—	—	—	—	—	—	—	—	538
18	27	73	14	—	115	323	120	—	—	—	—	—	—	—	—	690
63	27	—	97	65	—	9	34	—	—	—	—	—	—	—	—	674
—	—	—	—	14	—	—	—	—	—	—	—	—	—	—	—	14
87		192	239	178	319	281	157	—	—	—	—	—	—	—	21	1532
197	47	74	97	—	—	—	—	—	—	—	—	—	—	—	—	1030
379	378	647	428	1150	693	649	517	104	8	17	—	—	—	—	—	5102
117	—	—	—	—	—	—	—	—	—	—	—	—	—	—	—	150
—	88	93	—	—	—	—	—	—	—	—	54	167	104	52	—	558
182	104	—	—	95	—	—	—	—	—	—	—	—	—	—	—	460
74	281	219	133	—	202	123	108	—	—	—	—	—	—	—	—	1481
77	—	91	114	107	10	61	—	150	—	—	—	—	—	—	—	846
42	87	153	—	255	—	—	279	—	—	—	—	—	—	—	—	816
—	—	34	—	—	—	—	—	—	—	—	—	—	—	—	—	34
—	—	96	—	—	—	—	—	—	—	—	—	—	—	—	—	96
117	95	159	200	306	100	100	98	242	—	45	41	—	—	—	—	1513
—	74	—	—	—	—	1	5	—	—	—	—	—	—	—	—	80
—	—	71	—	—	—	—	—	—	—	—	—	—	—	—	—	226
—	112	—	100	99	112	212	195	216	—	—	—	—	—	—	—	1061
249	317	123	176	303	222	311	159	—	—	—	—	—	—	—	—	1999
—	71	116	100	—	—	—	—	—	—	—	—	—	—	—	—	287
1701	1708	2234	1760	2572	1773	2070	1672	712	8	62	95	167	104	52	21	19,187

Table 6.8. Sable translocations within the former USSR with sufficient data to be used in analyses

Release location	Release area (number of releases)	Year(s)	Type[a]	n (♀s)	Source location	Release[b]	Protection[c]	Status[d]	Source
Altai	Bija and Katun rivers (6)	1940–54	A	389 (218)	Irkutsk Province and Buryatia Republic	H	P	U	Timofeev and Pavlov 1973
	Uba River (2)	1952–53	A	181 (98)	Irkutsk Province	H	P	U	Ibid.
	Charysh and Upper Katun rivers (2)	1949–51	R	149 (66)	Irkutsk Province and Buryatia Republic	H	P	U	Ibid.
Chita Province	Middle Shilka River (5)	1970–82	R	76 (33)	Chita Province (Upper Chikoi River)	H	P	S	Samoilov 2005
Kazakhstan	Zailiysky Alatau Mountains (7)	1962–64	I	325 (152)	East Kazakhstan (Uba River)	H	P	F	Timofeev and Pavlov 1973
	Dzungar Alatau Mountains (1)	1965	I	52 (22)	East Kazakhstan (Uba River)	H	P	F	Ibid.
Kemerovo Province	Abakan River (2)	1951	A	182 (98)	Buryatia Republic	H	P	S	Monakhov 2006
Khabarovsk Krai	Maja River (3)	1949–53	R	147 (79)	Khabarovsk Krai (Uda River)	H	P	S	Timofeev and Pavlov 1973
Khakassia Republic	Abakan River (1)	1949	A	42 (23)	Buryatia Republic	H	P	U	Ibid.
Magadan Province	Kolyma River (4)	1958	R	361 (177)	Khabarovsk Krai (Bureya River)	H	P	S	Ibid.
Middle Siberia	Upper Ket and Middle Chulym rivers (3)	1950–53	A	179 (94)	Buryatia Republic	H	N	F	Ibid.
		1959	A	150 (79)	Irkutsk Province	H	N	F	Ibid.

	Sym and Eloguy rivers (9)	1949	R	106	Krasnoyarsk Krai (Yarcevsky District)	H	P	U	Timofeev and Pavlov 1973
		1954–55	R	282 (158)	Buryatia Republic	H	P	S	Monakhov 2006
Primorsky Krai	Sikhote-Alin Mountains (28)	1951–56	A	135 (70)	Primorsky Krai	H	N	U	Timofeev and Pavlov 1973
		1953–61	A	1008 (494)	Khabarovsk Krai (Bureya River)	H	P	S	Ibid.
		1951–55	A	370 (181)	Irkutsk Province and Buryatia Republic	H	P	U	Ibid.
Tomsk Province	Vassiugan and Iksa rivers (7)	1956	R	105	Irkutsk Province	H	P	S	Monakhov 2006
		1955	R	55 (30)	Tomsk Province	H	P	S	Ibid.
		1940–52	R	476 (257)	Irkutsk Province and Buryatia Republic	H	P	S	Ibid.
	Tym and Ket rivers (20)	1957	R	311	Irkutsk Province	H	P	S	Ibid.
		1954–58	R	596 (310)	Tomsk Province	H	P	S	Timofeev and Pavlov 1973
		1950–54	R	456 (260)	Irkutsk Province and Buryatia Republic	H	P	S	Ibid.
Tuva Republic	Tannu-Ola (4)	1952–53	R	260 (140)	Buryatia Republic	H	P	S	Monakhov 2006
Ural Mountains	Lozva River (4)	1940–53	A	226 (120)	Irkutsk Province and Buryatia Republic	H	P	F	Ibid.
	Vishera River (1)	1953	A	96 (53)	Irkutsk Province	H	P	F	Ibid.
	Chebarkul District (Chelyabinsk Province) (1)	1955	I	14 (5)	Sverdlovsk Province (Lozva River)	H	P	F	Timofeev and Pavlov 1973

Table 6.8—*cont.*

Release location	Release area (number of releases)	Year(s)	Type[a]	*n* (♀s)	Source location	Release[b]	Protection[c]	Status[d]	Source
West Siberia	Agan and Vakh rivers (5)	1952–57	R	535 (284)	Irkutsk Province	H	P	S	Monakhov 2006
	Kazym River (5)	1954–59	R	511 (276)	Irkutsk Province and Buryatia Republic	H	P	S	Ibid.
	Maja River (1)	1957	R	101 (52)	Irkutsk Province	H	P	S	Timofeev and Pavlov 1973
	Jigansk (5)	1948–58	A	393 (193)	Irkutsk Province	H	P	S	Ibid.
Yakutia Republic	Olekma River (4)	1948–55	R	338 (179)	Irkutsk Province	H	P	S	Ibid.
	Kolyma River (6)	1951–58	R	548 (290)	Irkutsk Province, Khabarovsk Krai, Kamchatka	H	P	S	Ibid.
	Yana River (5)	1954–56	R	528 (290)	Irkutsk Province	H	P	S	Ibid.

[a] I = introduction, R = reintroduction, A = augmentation
[b] H = hard release
[c] P = protected (no trapping), N = not protected
[d] S = successful, F = failure, U = unknown

the wild except for a very few that came from fur farms. By the mid-1960s, sables had recovered in much of their historical range (as of 1875) in the former Soviet Union (Monakhov and Bakeev 1981; Figure 6.5).

Model Results and Test of Model Predictions

The index of success increased with the number of release sites and the number of adult females released (Figures 6.6–6.7). Releasing juvenile females and males (adult or juvenile) had little effect on success (as long as 1 adult male was released per group of 5 animals; Figure 6.7). The number of years used to release a constant number of martens had no effect on success (Figure 6.8). In summary, the model predicted that the index of success for a reintroduction would increase only with the number of adult females released and the number of release sites.

The VORTEX population model predicted that the greater the number of adult females released (but not males) and the more release sites used, the higher the index of success for a reintroduction (Figures 6.6–6.7). The model did not predict a clear association between success and the number of years used to release a given total number of martens. Model results did suggest, however, that for releases of ≤40 martens, success decreased with the number of years, whereas the opposite might be true for >60 martens (Figure 6.8).

For actual reintroductions, Number of animals, Number of females, and Number of males released were all highly correlated, as expected ($r > 0.97$ for all combinations). Consequently, we never included these variables together in the same hypothesis evaluated with AIC_c analyses. Number of release sites correlated significantly with Number of animals, Number of females, and Number of males, but the correlation coefficients were all between 0.24 and 0.29 ($P < 0.05$ for all). Because the correlation coefficients were low, we combined Number of release sites with Number of animals, Number of females, or Number of males in hypotheses evaluated with AIC_c. Number of years that animals were released also correlated significantly with Number of animals, Number of females, and Number of males, with correlation coefficients all between 0.36 and 0.41 ($P < 0.05$ for all). When we combined Number of years with Number of animals, Number of females, or Number of males in a hypothesis, we blocked Number of years by the appropriate variable for Number of animals when calculating likelihoods. Number of release sites correlated significantly with Number of years ($r = 0.33$, $P < 0.05$). Because this correlation coefficient was low, we combined Number of release sites and Number of years in our hypotheses. Other variables did not correlate with Number of animals, females, or males.

Because the total number of martens released was known for most reintroductions (Tables 6.4–6.8) but not always the sex ratio, we first evaluated hypotheses generated by the VORTEX model (Table 6.3: first set), substituting

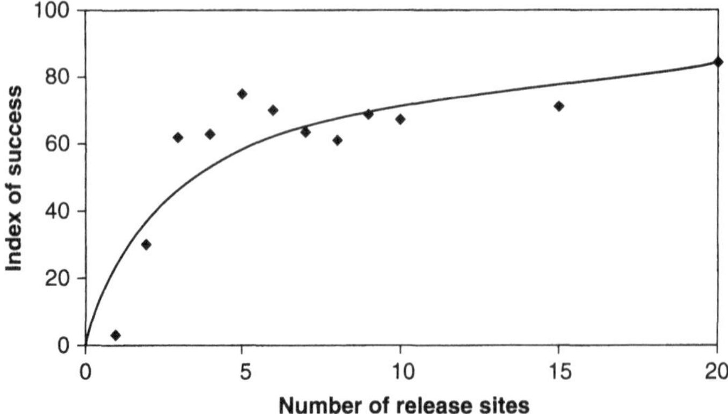

Figure 6.6. Relation between number of release sites and our index of successful reintroductions as predicted by the VORTEX model (modified from a similar figure by Lewis et al. 2012).

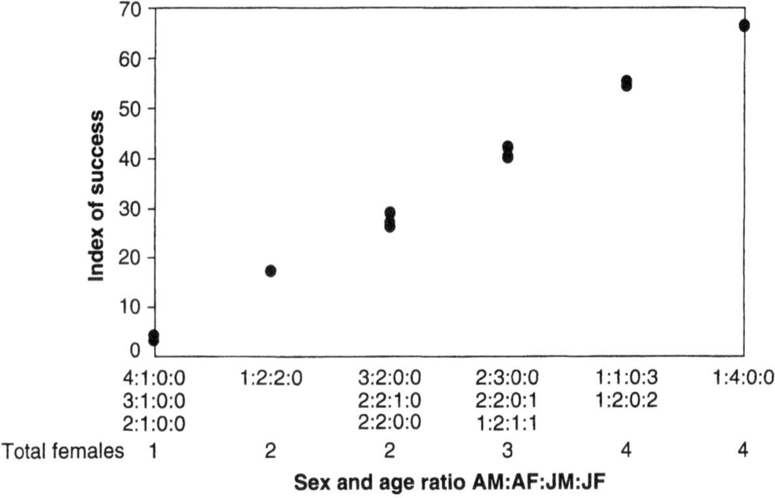

Figure 6.7. Relation between numbers of adult and juvenile female and male martens released and our index of successful reintroductions as predicted by the VORTEX model (modified from a similar figure by Lewis et al. 2012).

total number of animals released for number of adult females, using 5 species (American marten, fisher, European pine marten, stone marten, and sable). We then evaluated the hypotheses using numbers of females and males for American marten, fisher, and sable only.

In our initial evaluation using AIC_c (Table 6.3: first set), we retained 1 hypothesis (Table 6.9), which included 2 variables: Number of animals

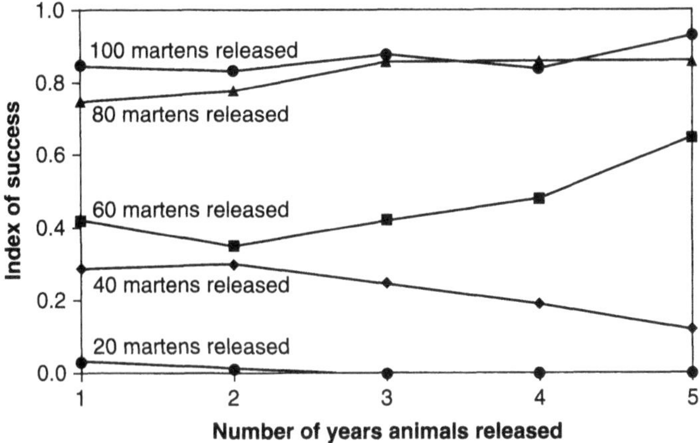

Figure 6.8. Relation between number of release years and our index of successful reintroductions as predicted by the VORTEX model (modified from a similar figure by Lewis et al. 2012).

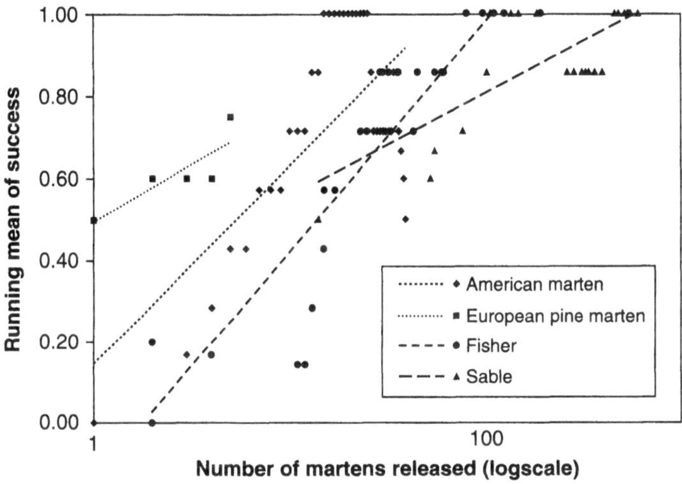

Figure 6.9. Relation between the number of American martens, European pine martens, fishers, and sables released and the running mean of success, showing that reintroduction success increased with numbers of animals released.

(Figure 6.9) and Number of release sites (Figure 6.6). For all other hypotheses, $\Delta AIC_c > 10$. We retained 2 hypotheses (Table 6.10) in our second test (Table 6.3: second set): (1) Number of males combined with Number of release sites, and (2) Number of females combined with Number of release sites. For these 2 hypotheses, AIC_c model weights (w_i) totaled 0.81, suggesting they had identified the important variables. For all other hypotheses, $\Delta AIC_c > 3$.

Table 6.9. Results of tests of predictions from the VORTEX population model, with data from American and European pine martens, fishers, and sables (substituting the total number of animals released for the total number of adult females)

Hypothesis (model)	n	K	log likelihood	AIC_c	ΔAIC_c	Likelihood	w_i
Number of animals and Number of release sites	60	2	–30.05	64.30	0	1	1
All other models	—	—	—	—	>10	0	0

Table 6.10. Results of tests of predictions from the VORTEX population model, with data from American and European pine martens, fishers, and sables (substituting the total number of females released for the total number of adult females)

Hypothesis (model)	n	K	log likelihood	AIC_c	ΔAIC_c	Likelihood	w_i
Number of males and Number of release sites	51	2	–20.09	44.4	0	1	0.45
Number of females and Number of release sites	51	2	–20.31	44.9	0.43	0.81	0.36
Number of males, Number of release sites, and Number of source sites	51	3	–20.50	47.5	3.08	0.21	0.10
Number of females, Number of release sites, and Number of source sites	51	3	–20.60	47.7	3.28	0.19	0.09
All other models	—	—	—	—	>19	0	0

In addition, no hypotheses including genetic relatedness, number of source populations (as an index of genetic diversity), monitoring, protection, or release type (Tables 6.2–6.3) had ΔAIC_c scores <10 when combined with Number of animals and Number of release sites, and none had ΔAIC_c scores <3 when combined with Number of females or males, or Number of release sites. Number of source populations had $\Delta AIC_c = 3.08$ for the hypothesis with Number of females, Number of release sites, and Number of source sites, and $\Delta AIC_c = 3.28$ for Number of males, Number of release sites, and Number of source sites. All other hypotheses had $\Delta AIC_c > 19$.

Discussion

General Characteristics of Translocations

Most reintroductions were designed to reestablish a native species (question 8 of Armstrong and Seddon 2008), though some were also designed for research, some fishers were reintroduced to control North American porcupines (*Erethizon dorsatum*), and most reintroductions and augmentations of sables were designed to improve the quality of fur as well as to reestablish populations. American martens, fishers, and sables have been translocated many times, whereas European pine martens, stone martens, and Japanese martens have been translocated infrequently.

Successful translocations of American martens, fishers, and sables shared several characteristics that were probably critical:

1. Overharvest contributed to population decreases and range contractions (Timofeev and Nadeev 1955; Monakhov and Bakeev 1981; Strickland and Douglas 1981; Powell 1993; Proulx et al. 2004). Consequently, populations could be protected from harvest, providing good conditions for population growth.
2. Even during periods of extreme range contraction and small populations, at least some populations of these species remained robust. Thus, source populations existed that were capable of donating reasonable numbers of martens to be translocated (Timofeev and Pavlov 1973; Powell 1993; Gibilisco 1994; Graham and Graham 1994; Slough 1994; Proulx et al. 2004).
3. Large areas of their former ranges retained suitable habitat capable of supporting reintroduced populations (Gusev 1971; Monakhov and Monakhov 1978; Monakhov and Bakeev 1981; Powell 1993; Slough 1994).
4. Once they were sufficiently reestablished, reintroduced populations could be harvested, providing an economic stimulus for translocating animals.

In contrast, the countries where European pine martens have been reintroduced support high densities of humans, have relatively few large populations of European pine martens suitable as source populations, and have relatively few large areas of suitable habitat capable of supporting reintroduced populations of these martens, compared with North America and Asia. Stone martens have not suffered such large population and range decreases. These martens can maintain populations in forested, agricultural, urban, and suburban areas, and in some places, they have become commensal with humans (Jensen and Jensen 1970; Nasimovich 1973), making reintroductions unnecessary.

To our knowledge, all marten translocations occurred into what was considered good habitat, but only 4 reintroductions chose release sites based on predicted habitat quality (Fontana et al. 1999; Anderson 2002; Weir et al. 2003; Lewis and Hayes 2004; Lewis 2006; Callas and Figura 2008). Releasing animals into suboptimal habitat is a major reason for translocation failure in general (Griffith et al. 1989; Miller et al. 1999; Armstrong and Seddon 2008), highlighting the importance of habitat. Unfortunately, what actually constitutes good habitat for each marten species is not well understood (preventing us from addressing question 3 of Armstrong and Seddon 2008). The literature has many examples of habitat models for martens developed from the literature and researchers' experiences (e.g., Allen 1983, 1984) and using regression techniques with observational data (e.g., Carroll et al. 1999; Payer and Harrison 2003; Davis et al. 2007). Only Allen's model for fisher habitat has been tested with independent data (Thomasma et al. 1991, 1994; Powell 2004), and few characteristics of habitat have been tested to understand their function, with the exception of Andruskiw et al. (2008). No tested, functional models of habitat quality, such as for black bears (*Ursus americanus*; Mitchell et al. 2002; Mitchell and Powell 2003), exist for any marten species. A number of translocations may have failed as a result of inadvertent release of martens into poor habitat. We suspect that a better understanding of habitat may be key to understanding why so few fisher reintroductions failed in eastern North America, whereas so many failed in western North America (Table 6.5). The fisher reintroduction to Montana from 1988 to 1991 succeeded despite high levels of predation on fishers (Roy 1991; Heinemeyer 1993). Good habitat quality may have permitted success, whereas predation could be a proximate cause of failure for reintroductions in lower-quality habitat. Habitat may be key to understanding why successful introductions are common for American martens, European pine martens, stone martens, and Japanese martens, but no introductions of fishers or sables have succeeded (Tables 6.4–6.8).

Augmentations of populations of American martens, European pine martens, and fishers generally succeeded (Tables 6.4–6.6), though the augmentation of American martens in north Wisconsin concluded in 2010 (J. Woodford, personal communication) and, thus, cannot be evaluated. Few augmentations of sables were reported as successful (Table 6.8), but we were forced to define success for these augmentations as it was defined in the literature: an increase in pelt quality.

Test of the VORTEX Model and Other Hypotheses

That we retained no hypotheses with variables other than numbers of males, females, and release sites provides important insights into the success of reintroductions (questions 1 and 6 posed by Armstrong and Seddon 2008). A number of factors have been hypothesized to affect reintroduction success, but none, at present, have much support, except possibly the genetic diversity of released animals (questions 3 and 4 of Armstrong and Seddon 2008). This genetic diversity was represented in our tests by the number of source populations for a reintroduction. The VORTEX model predicted that number of years would have an effect only if releasing martens for more years allowed the release of more martens. Among American martens, fishers, and sables, 58 reintroductions with animals of close proximity had 67% success, whereas 31 with those of distant proximity had 80% success. The number of years that martens were released was 1–8 for American martens, 1–8 for fishers, 1–9 for European pine martens, and 1–14 for sables, yet the number of years did not affect success.

Proximity of the source population to a reintroduction site (presumably also a measure of genetic relatedness of the source population to the original, extirpated population) had no apparent effect on reintroduction success (question 4 of Armstrong and Seddon 2008) although Lewis et al. (2012) found such an effect for fishers alone. Reintroductions and introductions using only 1 source population may introduce a genetic bottleneck, even when the number of animals released is large (Swanson et al. 2006; Swanson and Kyle 2007). Releasing animals from diverse source populations can introduce genetic diversity that counters the bottleneck and could affect the probability of success for the translocation. We were unable to identify a distinct effect of the number of source sites on reintroduction success, but we could identify most source sites only to state or province. The hypothesis that genetic diversity of reintroduced animals affects the probability of reintroduction success deserves further research.

Protection of reintroduced martens appears not to have affected success (question 2 of Armstrong and Seddon 2008). Most reintroduction programs provided some protection for released martens, but 23 did not, suggesting that our negative result was not an effect of low sample size. We find this result counterintuitive and wonder whether an undocumented, confounding factor is involved, such as de facto protection from isolation in areas where no formal protection was provided.

For other variables, small sample sizes and confounding variables prevent conclusions (question 2 of Armstrong and Seddon 2008). For example, soft or gentle releases, whereby martens were allowed to habituate to the release site in cages for weeks or months prior to release (Davis 1983), had 100% success compared with 77% for hard or quick releases, whereby translocated martens were released immediately. Nonetheless, all soft releases involved large numbers of animals, whereas all hard releases involving a few animals failed, and those involving large numbers succeeded. Anecdotal evidence suggests that soft releases have benefitted reintroductions of other mustelids (Biggins et al. 2011).

Monitoring released populations is important because it allows adaptive management and intervention if a reintroduced population appears to be in trouble. Monitoring of released marten populations, however, did not involve adaptive management and, thus, would be predicted not to affect reintroduction success, which is what our analyses found.

Only fishers have been released in all seasons, but data were too few to identify trends. The single summer reintroduction was successful, but winter (65%, $n = 17$), spring (75%, $n = 4$), and autumn (60%, $n = 5$) releases showed mixed success. Two reintroductions of American martens occurred in summer. Proulx et al. (1994) compared releases in summer and winter at the same site; their results suggested that summer reintroductions were more successful. Hobson et al. (1989) compared single females ($n = 5$) and females with kits ($n = 4$ females) released at the same site; the latter remained closer to their release sites. Releases of fishers in late autumn–early winter in the northern Sierra Nevada of California in 2009–2011 led to denning and the birth of kits by released females (A. Facka and R. Powell, North Carolina State University, unpublished data), whereas fishers released in late winter have not been reported to have given birth. These sparse data on American martens and fishers beg further research but suggest that giving animals sufficient time to become familiar with a new landscape, establish home ranges, and find good natal dens before parturition increases the probability of a successful reintroduction.

Few translocations of martens have used captive-reared animals. Kelly (1977) released 2 captive-reared female fishers and both died within a few months of release. Herr et al. (2008) released 4 captive-reared stone martens that survived for several months before their transmitters failed. Anecdotal accounts refer to releases of other captive-reared martens, but insufficient monitoring was conducted to determine the fates of these animals. Other captive-reared mustelids have suffered extremely high mortality rates when released (Biggins et al. 2011).

Our VORTEX model identified the number of adult females released (as long as a few adult males are released) and number of release sites as being the critical variables affecting success of a reintroduction. Our analyses of actual reintroductions identified the numbers of female martens (age or maturity undetermined), the number of release sites, and the number of male martens as critical for success. In fact, the hypothesis ranked highest by AIC_c included the number of males released and the number of release sites. Thus, the number of release sites does appear to be important, and we recommend that, when possible, future translocations should use multiple release sites. Consistent with the predictions of our VORTEX model, we found no effect of the number of years over which a given number of martens is released.

Arguably, our most interesting result is the rejection of the prediction that releasing more than 1 adult male marten per 4 adult females should not affect reintroduction success. The VORTEX model assumed that all adult males were equivalent breeders. In contrast, analyses of actual reintroductions showed that the number of males released affected success rate significantly, possibly because few males released were adults or because adult males may not all be equivalent breeders. A large reproductive skew exists for males of other solitary carnivores exhibiting large sexual dimorphism (Kovach and Powell 2003). All *Martes* species show pronounced sexual dimorphism in body size and greater variance in body size of males than of females (Monakhov and Bakeev 1981; Holmes and Powell 1994; Monakhov 2009), suggesting that a reproductive skew could exist. A large reproductive skew for male martens (all *Martes* species) could explain the contradiction between results of the VORTEX model and data from actual reintroductions. The model could be correct in predicting that few males are needed in reintroductions if those males are effective breeders. Effective breeders, however, may be difficult to identify. Therefore, the data from actual reintroductions may demonstrate that many males need to be released in order to release the few effective breeders needed. We urge researchers to test this hypothesis and the hypothesis that martens have large reproductive skews. If martens do have a large reproductive skew, releasing only males with characteristics associated with reproductive success should increase the probability of a successful translocation (Lewis et al. 2012).

7

Pathogens and Parasites of *Martes* Species

Management and Conservation Implications

MOURAD W. GABRIEL, GRETA M. WENGERT, AND RICHARD N. BROWN

ABSTRACT

The impacts of pathogens and parasites and their associated diseases are integral to understanding potential threats to *Martes* populations. In this chapter, we summarize the known relations of pathogens and parasites to their *Martes* hosts and review the epidemiology and life cycles of 4 selected pathogens that may be particularly important to *Martes* species, including rabies viruses, canine distemper virus, parvoviruses, and *Toxoplasma gondii.* We also address management options for dealing with disease issues and their implications for conservation efforts for *Martes* species. These implications include disease risk in reintroduction programs, handling of potentially diseased individuals, and protocols for disease assessment and prevention. Finally, we suggest future directions and roles of wildlife disease ecology in the research and management of *Martes* species. Our overall goal is to provide information that will be helpful for wildlife biologists, wildlife veterinarians, and others concerned about the biology, management, and conservation of *Martes* species.

Introduction

Despite the extensive and growing body of ecological research conducted on *Martes* species, relatively little is known about their infections by pathogens or infestations by parasites. The threat of disease is integral to conservation programs aimed at protecting members of this genus because of the insular nature of many *Martes* species and concern over the long-term stability of small *Martes* populations (Woodroffe 1999; Broekhuizen 2006; Saeki 2006).

First, we will define some terms frequently used in this chapter but possibly used in a somewhat different sense by other authors. A *pathogen* is any disease-causing organism, and a *parasite* is an organism that lives in or on the host tissue at the expense of its host (Hudson et al. 2004; Collinge and Ray 2006). *Disease* is commonly defined as an impairment that leads to a deleterious change in a host's body condition, which may or may not result in death (Wobeser 2006); this definition is broad but generally refers to some kind of pathology. In many cases, identifying the point when a pathogen or parasite causes disease is difficult, primarily because of the scarcity of studies that have proven a causal relation between the organism and insult through experiment (i.e., Koch's postulates) or of pathological investigations of mortalities. The study of the occurrence of disease in populations is referred to as *disease ecology* or *epidemiology*, whereby a pathogen can either be *enzootic* (occurring at an expected, constant, or low rate) or *epizootic* (occurring at an unexpected rate or pattern; Wobeser 2007).

Potential Effects of Selected Pathogens

Parasites and disease are often lumped into a single category of agents that cause adverse effects on populations. Such generalizations should be avoided, however, because they can create misunderstandings that may lead to poor management decisions. In fact, most parasites cause minimal or only mild signs of disease (Ewald 1995; Lafferty 2008). The remainder commonly cause considerable morbidity or death of individuals, but impacts on host populations can be subtle, and interactions may be complicated; even so, such parasites may influence the evolution of host populations as well as the dynamics of communities and food webs (Collinge et al. 2008; Lafferty 2008). Parasites also have the potential to cause severe diseases and may limit population numbers. These types of effects are usually most profound in small, insular populations, or when diseases act synergistically with other population-limiting factors (e.g., habitat loss or degradation, predation, competition, nutritional stress), or when exotic pathogens are introduced to a population (Daszak et al. 2001; Fenton and Pedersen 2005; Lafferty 2008). Consideration of parasites that can cause severe disease and have the potential to limit populations should be included in conservation planning (Murray et al. 1999; Pedersen et al. 2007; Roemer et al. 2009).

The virulence of parasitic organisms varies among host species and populations, among parasite populations and strains, and through time. Evolutionary pressures may result in a reduction of parasite virulence when, on average, host morbidity or mortality reduces parasite fitness. The severity of disease and pathogen propagation increases, however, as a result of gains in virulence (Ewald 1995; McCarthy et al. 2007; Lafferty 2008). Although it is tempting to encourage managers and conservationists to consider all possible host-parasite interactions, limited resources typically dictate that only the most significant of interactions merit intervention.

Associations of pathogens with *Martes* species have been reported inconsistently in the literature, and most reports are of occurrence rather than of disease. Several pathogens that infect mustelids are known to cause disease in a wide array of carnivore species. Thus, to evaluate the potential impacts of pathogens on *Martes* populations, we must rely to some extent on knowledge extrapolated from related host taxa. The impacts of some pathogens and diseases have been well studied for several mustelid species, including domestic ferrets (*Mustela putorius furo*; Hillyer and Quesenberry 1997; Fox et al. 1998; Langlois 2005), black-footed ferrets (*Mustela nigripes*; Williams and Thorne 1996; Gompper and Williams 1998; Bronson et al. 2007), American minks (*Neovison vison*; Dubey and Hedstrom 1993; Simpson 2001; Cunningham et al. 2009), southern sea otters (*Enhydra lutris neris*; Kreuder et al. 2003; Mayer et al. 2003; Jessup et al. 2007), and northern river otters (*Lontra canadensis*; Hoover et al. 1984; Hoberg et al. 1997; Mos et al. 2003).

Parasites that can infect many host species (generalists) and those associated with exotic species often pose the greatest conservation threats (Begon 2008; Perkins et al. 2008; Kelly et al. 2009a). Generalist parasites are supported by a large assemblage of hosts, and low-density host populations may experience high levels of parasite exposures as a result of interspecific spillover from community reservoirs (Woodroffe 1999; Daszak et al. 2000; Cleaveland et al. 2007). Parasites of exotic hosts, including feral domestics and native hosts undergoing range expansion, are often associated with greater virulence in native species (Daszak et al. 2000, 2001; Begon 2008). Highly virulent parasites can devastate wildlife populations (Woodroffe 1999). The pathogens most often associated with the decline or extirpation of carnivore populations, and therefore of greatest concern for the management and conservation of *Martes* species, are rabies viruses, canine distemper virus, parvoviruses, and *Toxoplasma gondii* (Woodroffe 1999; Cleaveland et al. 2007; Pedersen et al. 2007). We discuss each of these pathogens in the following sections; see Williams and Barker (2001) and Samuel et al. (2001) for more detailed information on these topics.

Rabies

Rabies is caused by a group of widely distributed RNA viruses in the genus *Lyssavirus*, and historical accounts date back several millennia (Rupprecht et al. 2001). Rabies viruses can infect all species of mammals (Rupprecht et al. 2001). Carnivores and bats provide rabies reservoirs throughout most of the world, and viral strains are typically associated with the local reservoir host species (Rupprecht et al. 2001; Krebs et al. 2003; Müller et al. 2004).

Transmission of rabies typically occurs when a susceptible host is bitten or otherwise comes into direct contact with infected saliva (Rupprecht et al. 2001; Langlois 2005). Less common routes of transmission include contact with ocular and nasal exudates or the consumption of infected animals as prey or carrion (Ramsden and Johnston 1975; Charlton 1994).

Rabies viruses display strong tropism for neural tissue, especially in the central nervous system (CNS), where replication occurs at a rapid pace (Rupprecht et al. 2001). Such neural tropism leads to the spread of the virus through various neural pathways to the brainstem, resulting in neurological distress (Rupprecht et al. 2001). As virions pass back to peripheral systems, infection of salivary glands contributes to elevated levels of virions in salivary secretions (Rupprecht et al. 2001; Langlois 2005). Behavioral changes occur after incubation of the virus and are dependent on the strain and proximity of the bite to the CNS (Rupprecht et al. 2001). Both the incubation period and prodromal stage (early signs) can last for days to several months (Rupprecht et al. 2001). Although rabies has been documented in several *Martes* species, clinical signs have been described only in stone martens (*M. foina*; Müller et al. 2004). Individuals were either asymptomatic or exhibited severely abnormal behaviors, including lack of fear, lethargy, and aggression (Müller et al. 2004; Dacheux et al. 2009), as reported in other mammals (Rupprecht et al. 2001). Rabies is generally considered a fatal disease, but survival after infection has been reported (Rupprecht et al. 2001).

Rabies remains a serious conservation concern for many carnivore communities and is especially alarming when species of concern are threatened with local extinctions (Woodroffe 1999; Cleaveland et al. 2007; Knobel et al. 2007). In some cases, domestic dogs (*Canis familiaris*) living in or near areas occupied by wild carnivores have exacerbated the risk of rabies to species in peril (Woodroffe 1999; Cleaveland et al. 2007), and such risks likely extend to *Martes* species.

Canine Distemper Virus

Canine distemper virus (CDV) is a highly labile RNA virus that infects and causes significant disease in many carnivores worldwide, including mustelids (Deem et al. 2000; Williams 2001). Juveniles and immunosuppressed individuals are usually affected in greater proportions than are healthy adults (Van Moll et al. 1995; Deem et al. 2000; Williams 2001). Distemper is believed to be a factor that led to the near extirpation of the black-footed ferret in the wild (Thorne and Williams 1988). The epidemiology of CDV for *Martes* species is not fully understood, but in other carnivores, it has shown cyclical patterns and temporal variance suggestive of density-dependent transmission (Roscoe 1993; Van Moll et al. 1995; Williams 2001).

The transmission of CDV is primarily by direct contact with the oral-respiratory or ocular fluids of infected individuals (Deem et al. 2000; Williams 2001). Although the virus can be transmitted via environmental contamination of feces or urine, infection through these routes is less likely because of the virus' instability in the environment (Deem et al. 2000; Williams 2001). After the virus enters the respiratory system and pervades the respiratory epithelium, it can spread rapidly through the lymphatic and vascular systems and infect other organs (Deem et al. 2000; Williams 2001).

The clinical signs of CDV in *Martes* species have rarely been documented. In one of the few studies available, stone martens exhibited lack of fear, ataxia, and high levels of salivation and ocular discharge (Van Moll et al. 1995). In a suspected CDV epizootic in Newfoundland, Canada, where American martens (*M. americana*) were believed infected, individuals exhibited similar symptoms, including short periods of full-body convulsions; however, attempts to confirm CDV infection by histological, serological, or molecular assays were unsuccessful (Fredrickson 1990). In a recent epizootic in an insular population of fishers (*M. pennanti*) in California, several mortalities were confirmed to be caused by CDV (S. Keller and M. Gabriel, University of California at Davis, unpublished data). Gross clinical signs of infection included hyperkeratosis, skin lesions, and severe emaciation (M. Gabriel, unpublished data; Figure 7.1). In addition, radiotelemetry data provided indirect evidence of lethargy and abnormal movement shortly before mortality (R. Sweitzer, University of California at Berkeley, unpublished data).

In mustelids, CDV tends to affect multiple sites in the central nervous system (Van Moll et al. 1995; Deem et al. 2000; Williams 2001). The strongly immunosuppressive effects of CDV also act synergistically with subclinical or latent infections by other pathogens or parasites to enhance the severity of disease and increase the probability of death (Van Moll et al. 1995; Deem et al. 2000; Williams 2001). For example, clinical signs of a secondary infection by *Hepatozoan* spp. were observed in a stone marten with CDV (Van Moll et al. 1995). Similarly, CDV has been associated with toxoplasmosis in other mustelids (Diters and Nielsen 1978; Van Moll et al. 1995; Frank 2001).

Given the devastating effects of CDV outbreaks on captive and wild populations of carnivores, the conservation implications of infections with CDV could be significant. Likewise, CDV-related mortalities in small or insular *Martes* populations could have a significant effect on their persistence and viability.

Parvoviruses

Parvoviruses are single-stranded DNA viruses, and each strain is generally named for the species in which it was isolated initially (Barker and Parrish

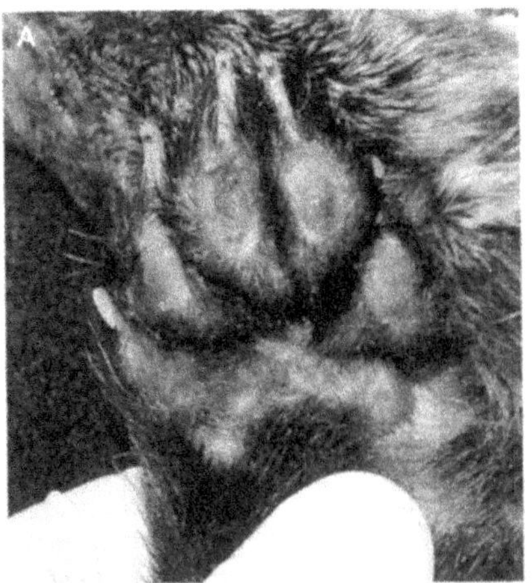

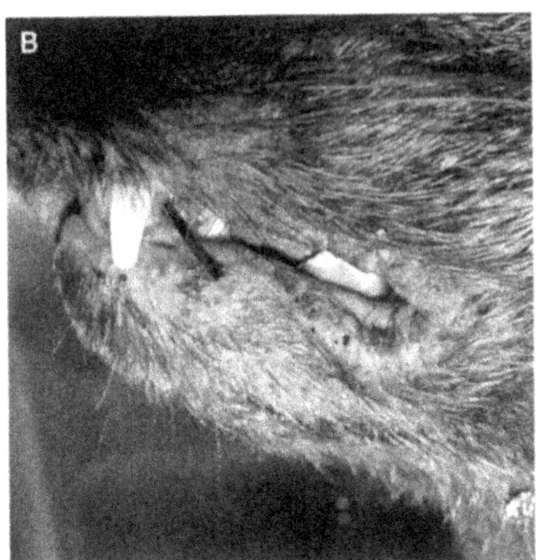

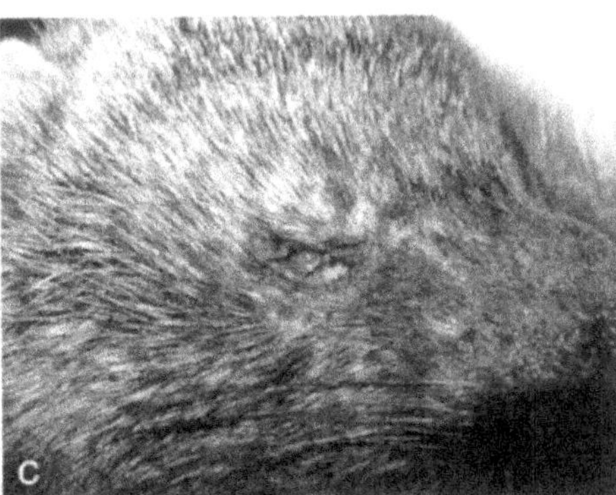

Figure 7.1. Canine distemper virus (CDV) mortality of a fisher displaying gross clinical signs of (A) paw pad hyperkeratosis and ulcerative lesions, (B) ulceration and hyperkeratosis of the skin along the gum line, and (C) eye. (D) Full body of CDV-infected fisher displaying severe emaciation and no subcutaneous fat.

2001; Steinel et al. 2001; Parrish and Kawaoka 2005). The feline parvovirus subgroup (FPV) affects a wide variety of carnivores throughout the world (Steinel et al. 2000, 2001; Barker and Parrish 2001). Several related viruses occur within this subgroup, including feline panleukopenia virus (FPL), mink enteritis virus (MEV), raccoon parvovirus (RPV), blue-fox parvovirus (BFPV), and canine parvovirus types 1 and 2 (CPV-1 and CPV-2, respectively), that have similar genetic and antigen properties (Barker and Parrish 2001; Steinel et al. 2001). Interestingly, CPV-2 emerged rapidly in 1978, such that by the end of that decade, more than 80% of all domestic canids were infected globally (Parrish et al. 1988; Parrish and Kawaoka 2005). Another parvovirus, Aleutian disease virus (ADV), infects various carnivores (including mustelids) but is dissimilar to the FPV subgroup (Barker and Parrish 2001; Steinel et al. 2001). We expect that these viruses have the potential to infect all *Martes* species.

Parvoviruses are highly resistant to environmental degradation and, under suitable conditions, can persist for months and possibly years (Bouillant and Hanson 1965; Gordon and Angrick 1986; Barker and Parrish 2001). Transmission is generally through the fecal-oral route, rather than by direct transmission, and feces deposited at marking sites, latrines, and trap sites are potential sources of exposure for *Martes* species (Barker and Parrish 2001; McCaw and Hoskins 2006). Vertical transmission (the transmission of a pathogen from parent to offspring in utero or during birth) has been shown to occur in pregnant domestic ferrets (Kilham et al. 1967) and might occur in other mustelids.

Parvoviruses display strong tropism toward rapidly dividing cells, facilitating exponential replication of the virus and systemwide infections (Steinel et al. 2001; McCaw and Hoskins 2006). Neonates and juveniles, which have high levels of mitotically active cells, are usually the most susceptible segments of the population (Steinel et al. 2001). Susceptibility of adults is greatest in the rapidly dividing cells of the intestinal epithelium and the lymphatic systems (Steinel et al. 2001). The clinical effects of parvoviruses on *Martes* neonates, juveniles, and adults are unknown. In other mustelids, however, neonates and juveniles have displayed clinical disease and mortality when infected with MEV, FPL, and ADV through natural and experimental infections (Kilham et al. 1967; Duenwald et al. 1971; Steinel et al. 2001). Adults seldom displayed clinical disease with viruses of the FPV subgroup but exhibited clinical signs when infected with ADV (Barker and Parrish 2001; Steinel et al. 2001).

The detection of parvovirus virions in *Martes* feces (Kenyon et al. 1978; Brown et al. 2006; Gabriel et al. 2010) suggests that viral replication is occurring in the crypts and microvilli of the intestinal epithelium, as it does in other carnivores (Steinel et al. 2001; McCaw and Hoskins 2006). The frequency of clinical disease from parvoviruses and its related effects in *Martes* species remains uncertain. In some mustelids, panleukopenia, anemia, gastrointestinal lesions, diarrhea, and dehydration are common symptoms of the disease (Barker and Parrish 2001), and cerebral hyperplasia was reported in fetal domestic ferrets whose mothers were infected experimentally with FPL (Kilham et al. 1967).

Parvoviral infections are unlikely to limit wild carnivore populations, with the possible exception of small or otherwise vulnerable populations (Barker and Parrish 2001). Furthermore, the risk of spillover from nearby infected individuals or fomites (inanimate objects that carry disease organisms) is important because of the environmental resistance of these viruses. Techniques to reduce some of these risks are described in a later section of this chapter.

Toxoplasma gondii

Toxoplasma gondii is an obligate intracellular protozoan parasite with a complex life cycle involving many parasitic stages, and both intermediate and definitive hosts (Dubey et al. 2001). It occurs worldwide and probably has the potential to infect all avian and mammalian species (Tenter et al. 2000). Only felids are capable of shedding the infective stage of *T. gondii* through feces; other species serve only as intermediate hosts (Dubey et al. 2001). Animals can be infected by ingesting either infective oocysts in the environment or the tissues of infected intermediate hosts (Dubey et al. 1998).

Most tissue cysts do not cause obvious harm and remain intact throughout the host's life (Dubey et al. 2001). Although subclinical effects of this parasite are little understood, they are probably consequential (McAllister 2005). Acute toxoplasmosis is thought to be rare in healthy, nonimmunosuppressed individuals (Dubey et al. 2001), but outbreaks have resulted in significant mortality in black-footed ferrets (Burns et al. 2003b) and American minks (Pridham and Belcher 1958) in captive-breeding programs, and in free-ranging southern sea otters (Cole et al. 2000). Some of these cases in mustelids and closely related species were preceded by infection with canine distemper, which causes immunosuppression (Diters and Nielsen 1978; Van Moll et al. 1995; Frank 2001).

Pathogenesis of toxoplasmosis begins when an ingested oocyst releases sporozoites in the intestine that multiply and spread to lymph nodes, where the more effectively spreading (and tissue-damaging) tachyzoites are produced (Jones et al. 1997). In intermediate hosts, the predilection of *T. gondii* for the central nervous system is likely the reason for neurological symptoms and pathology (Webster 2007). The only recorded episode of toxoplasmosis in *Martes* species is a recent case in California in which a fisher died of inflammation of the meninges and the brain, caused by an infection with *T. gondii* (M. Gabriel, unpublished data). Clinical signs observed in other mustelids with toxoplasmosis include head tremors and ataxia, circling, limb lameness, lethargy, blindness, loss of appetite, anorexia, difficulty chewing and swallowing, and abortion (Frank 2001; Burns et al. 2003b; Kreuder et al. 2003; Jones et al. 2006).

The subclinical effects of this pathogen may be important (McAllister 2005). As such, the deleterious effects on behavior from *T. gondii* infection are just beginning to be fully understood (Webster 2007). Norway rats (*Rattus norvegicus*) infected with *T. gondii* were more active and displayed less fear of novel stimuli than uninfected rats (Webster et al. 1994; Berdoy et al. 1995). These effects can cause a greater susceptibility to predation (Webster 2007); for example, in southern sea otters, a strong correlation was found between *T. gondii* encephalitis and fatal shark attacks (Kreuder et al. 2003).

The rare but significant effects of toxoplasmosis epizootics in other mustelids demonstrate the potential importance of this parasite for *Martes* populations worldwide (Dubey et al. 2001). For example, it is possible that a heightened susceptibility to predation from toxoplasmosis could be related to the high predation rates currently experienced by fisher populations in California (G. Wengert, unpublished data). Taken together, this evidence warrants consideration of *T. gondii* as a threat to *Martes* species that should be addressed in current and future research and conservation programs.

Disease Management and Conservation Strategies

Recognition and management of disease risks are becoming common strategies in conservation programs. Managing disease includes efforts to prevent and control the spread of pathogens. In some cases, knowledge of transmission cycles allows avoidance of risks. In other cases, individuals may need to be vaccinated or treated clinically to prevent disease outbreaks. The topics we address below are just a few of the many disease-management options currently available to resource managers, and each could encompass an entire chapter because of differences in *Martes* niche characteristics, sympatric communities, and parasite distributions that may warrant different approaches. Here, we highlight general strategies that can be used to maintain the health of *Martes* populations. We recommend thorough investigations into the appropriateness and feasibility of these methods for all *Martes* populations being managed for conservation purposes.

Vaccinations

One strategy for managing infectious disease in wildlife is the use of vaccines (Haydon et al. 2006). Immunization with vaccines can prevent or reduce the clinical manifestations of many virulent pathogens, thereby reducing transmission to susceptible animals; however, a lack of success in some vaccination programs demonstrates the need for caution (Thorne and Williams

1988; Woodroffe 1999). Vaccines are typically developed for domestic animals and must be assessed before being given to free-ranging wildlife (Woodroffe 1999). In addition, success is often difficult to gauge, since decisions to leave control groups (unvaccinated cohorts) in species or populations of concern may be politically or ethically unjustifiable (Woodroffe 1999).

Several primary concerns can be cited about use of vaccines in wildlife programs. First, a vaccine is usually constructed to protect against a specific pathogen; however, immunized individuals may remain susceptible to infections when wild pathogen strains gain virulence factors and antigenic sites that differ from those used to create the vaccine. Furthermore, protective immunity typically wanes over time, and many vaccines require the immune system to be primed with one or several boosters. Finally, to thoroughly understand a vaccine's efficacy and safety, proactive experimental-challenge studies must be conducted; unfortunately, financial, logistical, and ethical constraints often preclude such studies.

A vaccine should meet several criteria before being used in wildlife species. First, the vaccine should produce no significant disease in the host. In a well-known example of vaccine-induced disease, wildlife managers used a modified live-distemper vaccine in the black-footed ferret conservation program that produced the disease they were intending to prevent (Carpenter et al. 1976). From this and other unfortunate examples, it is now well understood that vaccines developed for domestic or companion animals must be used with caution in wildlife. Second, because of the impracticality of vaccinating a population or many individuals multiple times, a vaccine should produce long-lasting protective immunity. There are several categories of vaccines, and killed or inactivated vaccines often fail to properly stimulate the immune system to produce a long-term response (Thorne and Williams 1988). Third, the vaccine should be able to provide protection from several antigenic variants, thereby affording the host protection from various forms of the pathogen they may encounter. Approved vaccines for *Martes* species are currently limited; however, the American Association of Zoo Veterinarians (2010) has developed a list of vaccines recommended for small carnivores, including mustelids, some of which have been used in *Martes* species.

Distemper

To date, no experimental trials have been conducted to ascertain the effectiveness and safety of current CDV vaccines for use in *Martes* species; however, the currently accepted vaccination of mustelids for distemper virus is a recombinant vaccine marketed as PUREVAX ferret distemper vaccine (Merial Ltd., Duluth, Georgia, USA) (Wimsatt et al. 2006; Lewis and Happe 2008; Jessup et al. 2009). PUREVAX recombinant vaccines were used safely in both black-footed ferrets and sea otters (Wimsatt et al. 2006; Jessup et al.

2009). At present, most of the vaccination programs for *Martes* species have been conducted during translocation efforts that provided limited data on the effects or efficacy of the vaccine (Mitcheltree et al. 1997; Lewis and Happe 2008). Reintroduced fishers vaccinated with the discontinued modified-live Fervac-D (United Vaccines, Madison, Wisconsin, USA) (Mitcheltree et al. 1997) and PUREVAX vaccines (Lewis and Happe 2008) (D. Clifford, California Department of Fish and Game, unpublished data) showed mixed results regarding their performance. Mitcheltree et al. (1997) vaccinated ~45 fishers with Fervac-D prior to release during a reintroduction program in Pennsylvania, with a small subset receiving a secondary booster if they were held in captivity ≥12 days; however, several fishers did not develop antibodies to CDV after either the first or the booster vaccination. Vaccinations with PUREVAX were administered in the Washington and California reintroduction programs (Lewis and Happe 2008; D. Clifford, unpublished data). Available data from these efforts showed elevated titers for IgM and IgG antibodies (M. Gabriel and D. Clifford, unpublished data), but it was unknown whether these responses were sufficient to protect against a natural infection. Indeed, the effectiveness of recombinant vaccines for use in *Martes* species is generally unknown because of lack of data on their effects.

Rabies

Because of the virulence of rabies viruses, killed vaccines are the only ones currently recommended for *Martes* species (American Association of Zoo Veterinarians 2010). Imrab-3 (Merial Ltd., Duluth, Georgia, USA) is a killed rabies vaccine that has been used in various zoo and free-ranging wildlife species, including fishers (Lewis and Happe 2008; D. Clifford, unpublished data). Because this vaccine contains a killed virus, annual vaccination of each individual is recommended (American Association of Zoo Veterinarians 2010). There are numerous accounts of a single dose failing to provide adequate protection, or even measurable seroconversion, within populations (Woodroffe 2001; Haydon et al. 2006). For example, primary vaccinations in Ethiopian wolves (*Canis simensis*) provided protection from infection for <6 months (Randall et al. 2006). It is possible that some American martens and fishers have been exposed to vaccinia-vectored rabies glycoprotein baits distributed as part of proactive measures to control the raccoon rabies-virus variant in eastern North America (Rosatte et al. 2009); however, the efficacy and safety of this vaccine on *Martes* species has not been determined.

The use of vaccines in *Martes* species is a relatively unexplored field, and few data are available to measure the success of previous vaccination programs. Accordingly, managers of *Martes* populations should use only distemper and rabies vaccines recommended by the American Association of Zoo Veterinarians (2010), or those that have been evaluated in experimental trials that included post-vaccination investigation and monitoring. Finally, before a vaccination program is initiated for any *Martes* species, we strongly recommend that those involved consult with a wildlife veterinarian to evaluate the safety of vaccine(s) being considered, and to ensure that the individuals being vaccinated are healthy enough to receive them.

Considerations in Translocation Programs

Disease is rarely the primary cause of a species' extinction, or even the extirpation of a population; however, it can lead to population reductions and make them more susceptible to extirpation under environmental or demographic stochasticity (Woodroffe 1999; Smith et al. 2006). Managers often use augmentations (translocating animals to a portion of their range where population numbers have declined) or reintroductions (translocating animals to a portion of their range from which the species has been extirpated) to offset these threats (IUCN 1998). Because the health of released individuals, and of the wildlife community into which they are released, are essential goals, disease is a critical consideration in the planning of any translocation program,

Many translocations of *Martes* species (Powell et al., this volume) and other wildlife have been conducted; however, the outcomes of most of these programs either have not been monitored or remain unreported (Berg 1982; Proulx et al. 1994; Seddon et al. 2007). Although feasibility studies for the reintroduction of *Martes* species rarely consider disease as a potential contributing factor for their initial extirpation (Earle et al. 2001; Aubry and Lewis 2003; Lewis and Hayes 2004), it should be evaluated carefully in all such feasibility studies (IUCN 1998). Past failures to explicitly address disease implications probably reflect the scarcity of historical health information available to *Martes* researchers upon which to build. Finally, the vaccination of translocated animals, and the selective release of individuals based on disease exposure, remain uncommon in translocation programs (Seddon et al. 2007). We offer the following health considerations for translocating *Martes* species for the purposes of conserving or managing their populations.

Screening Sympatric Carnivores

Before a translocation program is initiated, potential threats (pathogen-related or otherwise) should be evaluated within the augmentation or reintroduction area (IUCN 1998). The pathogens most detrimental to *Martes* species tend to be generalists that infect a wide array of carnivores (Steinel et al. 2001; Williams and Barker 2001). Consequently, the sympatric carnivore community may pose a spillover risk to translocated animals that are susceptible to infection (Munson and Karesh 2002; Mathews et al. 2006). The carnivore community, therefore, should be screened for pathogens before any translocations occur. Sampling can be either intensive, to determine temporal and spatial differences in exposure and infections within the target area, or cursory, using an opportunistic sampling regime. Although both methods provide important information about the pathogens to which reintroduced *Martes* individuals might be exposed, more intensive sampling may enable managers to determine whether a pathogen cycle is enzootic or epizootic. To clarify pathogen cycles, researchers should focus on long-lived carnivore species or sample among age-classes. Spatial and temporal information on pathogen cycles can also aid in determining the optimal timing of translocation, and the target areas that will minimize the risk of pathogenic transmission (Fernández et al. 2006).

In northern California, we recently investigated whether mesocarnivores that were sympatric with a fisher population could act as sentinels for several potentially limiting pathogens (Gabriel et al. 2007). We found no differences in pathogen exposure or active infections between fishers and other carnivores, suggesting that these pathogens cycle naturally in all the mesocarnivores we tested (Gabriel et al. 2007).

Before a reintroduction effort was initiated by the California Department of Fish and Game in the winter of 2010 (California Department of Fish and Game 2010), sympatric mesocarnivores within and around the proposed release site were trapped and screened for pathogens. Most were cosmopolitan species, including the striped (*Mephitis mephitis*) and western spotted (*Spilogale gracilis*) skunks, gray fox (*Urocyon cinereoargenteus*), ringtail (*Bassariscus astutus*), raccoon (*Procyon lotor*), and bobcat (*Lynx rufus*) (M. Gabriel, unpublished data). All captured individuals were tested for the presence of antibodies to canine distemper virus, parvovirus, *T. gondii*, and *Neospora caninum*. Seroprevalence of distemper virus was 33% among all species combined. These findings enabled wildlife veterinarians and biologists from California's natural resource agencies to make informed management decisions, including the use of a prophylactic distemper vaccination for all fishers released into the area (D. Clifford and R. Callas, California Department of Fish and Game, unpublished data). Furthermore, a documented epizootic of rabies in gray foxes in an area where potential founders were to be trapped led managers to avoid trapping in that area to minimize the potential of translocating an infected individual (D. Clifford and R. Callas, unpublished data).

Transportation of Founder Animals

Many translocation programs for *Martes* species employ private trappers to capture founder animals in the source area (Lewis and Happe 2008; Powell et al., this volume). All vehicles used in the transportation of founders must be properly disinfected before and after each use. Traps used to capture and transport individuals must also be properly disinfected between captures. In an unfortunate example in which protocols for disinfecting vehicles were not established prior to transporting river otters in western New York, serum samples for selected viral agents in several otters showed high titers for both canine herpesvirus-1 and canine parvovirus-2 exposure (Kimber et al. 2000). The authors pointed out that these high titers could have been due to several oversights, including exposure to contaminated transport vehicles.

Many traps and transportation enclosures use cubby boxes typically constructed with wood. These boxes can act as fomites for infectious material, requiring diligent disinfection protocols. Such boxes should be lined internally with a nonporous material like fiberglass-reinforced paneling to allow easy disinfection and cleaning of exudates and excrements that may be left behind (Gabriel 2006).

Holding Facilities

Once a translocation program is under way, precautionary measures for animal health still need to be maintained diligently. Adequate disinfection protocols are needed during handling, transportation, and quarantine to minimize disease transmission to founders, and the holding facility should meet guidelines of the International Union for the Conservation of Nature (IUCN 1998). The facility should be large enough to accommodate the expected number of founders and allow easy implementation of disinfection protocols. It should not simultaneously house domestic or companion animals and, if used previously for these purposes, should be properly disinfected. Researchers or technicians working in the holding facility should limit their contact with domestic animals to minimize contamination. For instance, captive programs for black-footed ferrets required the personnel involved to avoid contact with domestic animals for at least 24 h prior to entering the facility, and domestic animals were not allowed near the facility or in transport vehicles (Thorne and Williams 1988). In addition, anyone sick with influenza was not allowed to participate in captures or facility duties until he or she recovered (Thorne and Williams 1988). Finally, the food provided in a holding facility should be inspected carefully, since parasites such as *T. gondii* can be present in uncooked meat (Burns et al. 2003b). Uncooked USDA-inspected rabbit meat infected with *T. gondii* cysts was believed to be the cause of 32% of adult and 20% of juvenile black-footed ferret deaths in a captive facility (Burns et al. 2003b).

The logistic and economic demands for implementing many of these actions are challenging; however, these guidelines will improve the safety of holding facilities, and the likelihood of success of subsequent translocations. Many additional recommendations have been made for preventing disease transmission in holding facilities; for more detailed discussions and recommendations, we suggest consulting the veterinary literature on domestic-animal shelters (e.g., Johnson 2004).

Screening of Founder Animals

Guidelines published by the International Union for the Conservation of Nature (IUCN 1998) recommend that all founder individuals be screened for deleterious pathogens prior to release to avoid transmission to other founders or the target community. Expedited diagnostic testing with validated or sensitive and specific assays can provide managers with the information needed to determine whether a potential founder is suitable for release. If possible, a second sampling or retest should occur just prior to release, enabling researchers to evaluate whether established protocols are effective in minimizing exposure to pathogens. If an outbreak occurs at a holding facility, data on disease exposure can help managers determine whether the facility's protocols need modification.

It should be noted that a rapid diagnostic test developed for domestic or companion-animal use must first be validated for use in wildlife species, and tested to determine its effectiveness for the focal species (Stallknecht 2007). Recent evaluations of rapid diagnostic tests for parvoviruses, and an emerging tick-borne pathogen, determined that these tests currently have poor sensitivity for fishers (Gabriel et al. 2010). The rapid diagnostic tests failed to detect any parvovirus in 66 fisher fecal samples, whereas conventional polymerase chain reaction (PCR) analysis detected it in 13 (20%) of the samples (Gabriel et al. 2010). Based on this study and other evaluations with similar results, we discourage the use of rapid diagnostic tests until they are either validated by experimental studies or run parallel with a conventional diagnostic procedure (Stallknecht 2007; Gabriel et al. 2010).

Serological screening for current and previous exposure to pathogens can be done with many types of assays, some of which use an antibody isotype called IgG, which persists for extended periods of time (Janeway et al. 2007). Alternatively, researchers may choose to use the short-duration antibody isotype IgM, which may reveal very recent or current infection status as a result of its presence in the initial immune response (Janeway et al. 2007). We also recommend screening for active infections. For example, infections with parvoviruses are common problems in animal shelters, where animals are kept in close quarters and may come in contact with contaminated fomites. An isolation protocol was used in the recent California fisher reintroduction program that involved quarantining any animal that shed parvovirus, and repeatedly testing infected and nearby individuals to help evaluate the effectiveness of disinfection and quarantine protocols at the holding facility (D. Clifford and M. Gabriel, unpublished data).

Post-release Monitoring of Diseases and Mortality Causes

Many translocation programs involve the recapture of released individuals to replace monitoring equipment (e.g., VHF or GPS radio-collars). Screening recaptured animals for disease may help determine whether individuals were exposed to pathogens after their initial release. This information can help identify natural pathogen cycles in the translocation area and enable researchers to predict the risk of an epizootic outbreak. If post-release handling does occur, we encourage researchers to collect the necessary biological samples for disease screening, whether or not disease has been identified as a potential problem (Botzler and Armstrong-Buck 1985).

To determine the definitive causes of mortality for translocated individuals, we recommend that a complete and detailed necropsy be performed by a board-certified wildlife pathologist. Many wildlife research projects evaluate the causes of death on the basis of cursory necropsies or visual inspection of carcasses. Having necropsies performed by a trained wildlife pathologist can greatly improve the quality of research by validating or disproving suspected causes of mortality. We recommend the use of pathology as a form of translational science, since it provides context for large amounts of health and ecological data collected during a project, and a reliable basis for determining whether disease is an outcome of exposure to a pathogen or toxicant (Munson 2003).

Necropsies conducted at the University of California at Davis Veterinary Medicine Teaching Hospital on incidental fisher mortalities in California and on those that died during the Washington fisher reintroduction program revealed that fishers that had been predated upon were infected with pathogens that may have predisposed them to predation (M. Gabriel and G. Wengert, unpublished data). Other cases in which fishers died from virulent pathogens would have been overlooked if histology, immunohistochemistry, PCR, and other ancillary tests had not been performed (M. Gabriel, unpublished data).

The pathology laboratories at many universities will reduce necropsy fees to facilitate collaboration on disease research and will archive fixed and frozen tissues for future reference. We recommend that all mortality events for *Martes* species be evaluated by a trained wildlife pathologist whenever it is logistically and financially feasible. However, if carcasses cannot be retained or shipped for examination by a pathology laboratory at the time they are collected, wildlife biologists can be trained to collect and store the tissues needed for later screening for clinical or subclinical diseases (Munson and Karesh 2002). Biologists should always follow established procedures for wild-animal necropsies and for collecting tissue samples (e.g., Munson 2004).

Disinfecting Equipment to Reduce Transmission Risk

Handling and marking animals is often necessary to address demographic and ecological questions about *Martes* populations. Consequently, there is an inherent risk that handling equipment (traps and processing gear) and monitoring equipment (remote cameras and track-plates) may serve as fomites that transfer pathogens to *Martes* or other species. Properly disinfecting equipment reduces these and other zoonotic risks to researchers. Some disinfectants may leave a slight residual odor on the treated surface. Whether such odors can affect trapping success has not been conclusively determined; however, various groups in California working on American martens and fishers have repeatedly captured and processed individuals for demographic studies while thoroughly disinfecting traps after each capture (M. Gabriel, unpublished data).

Many disinfectants are commercially available that can neutralize fungal, bacterial, and viral threats, but their efficiency, toxicity, and field applications vary. For example, some viruses that infect *Martes* species (e.g., parvoviruses) are hydrophilic and non-enveloped, making them highly resistant to many disinfectants (McGavin 1987). In addition, few disinfectants are detergents, so they will not remove residual organic material (exudates and excrement) left by captured animals, and many disinfectants are inactivated by organic material. It is therefore recommended that organic material be removed before disinfecting (Gilman 2004). A list of commonly used disinfectants and their properties and uses is presented in Table 7.1.

Sampling Techniques for Pathogen and Parasite Detection

Collecting Blood

Blood is a very useful and effective tissue for detecting both exposure to pathogens and active infections, so it is important to maximize the amount sampled while considering animal safety. The amount of blood that can be safely drawn from mammals is dependent on a wide variety of physiological and behavioral characteristics. Published guidelines suggest that a volume equal to 1.0% of an individual's body weight can be collected every 14–21 days from individuals in optimal physiological and psychological condition (McGuill and Rowan 1989; Morton et al. 1993; Silverman et al. 2006). Based on this guideline, it is safe to draw 15 ml of blood from a mammal weighing 1.5 kg (1% or 0.01 x 1.5 = 0.015 L or 15 ml). However, biological sampling of wildlife involves uncertainties in both physical and psychological conditions that are not present in the laboratory settings under which most guidelines are established. Thus, the unknown health of individuals, stress of capture, and possible exposure to pathogens or environmental elements should be considered when determining the appropriate volume of blood to draw.

Many serological and molecular techniques require not just whole blood collected with an anticoagulant, but also serum obtained by centrifuging clotted blood. It is important to avoid hemolysis, the breakage of red blood cells, during collection or processing, because hemolysis increases sample degrada-

Table 7.1. List of disinfectants, their proper use, and effectiveness for use in *Martes* research

Disinfectant type	Common names	Recommended dilution (DL) and contact time (CT: minutes)	Effectiveness in hard water (HW) and in organic material (OM)	Effectiveness against resistant viruses	Pros and cons
Sodium hypochlorite	Bleach	DL: 3% (1:32); CT: 10–30	HW: good; OM: poor	Poor	Pros: cheap, readily available; cons: corrosive to metals, irritant
Quaternary ammonium compounds	Roccal-D[a]	DL: variable; CT: 10–30	HW: poor; OM: poor	Poor–fair	Pros: cheap, noncorrosive; cons: toxic to aquatic organisms
Potassium peroxymonosulfate	Virkon-S[b]	DL: 1–2% (1:100–1:50); CT: 10	HW: good; OM: good	Good	Pros: noncorrosive, efficacy color indicators; cons: costly, stock powder is hazardous
Alcohol	Isoproponol Ethanol	DL: 70–95% (7:3–19:1); CT: 1	HW: good; OM: poor	Poor	Pros: noncorrosive, minimum toxicity; cons: costly, fire hazard
Chlorohexidine	Nolvasan[c]	DL: variable; CT: 5–10	HW: good; OM: fair	Poor	Pros: noncorrosive, nontoxic; cons: costly, ineffective as general purpose cleaner
Phenol	Lysol[d]	DL: variable; CT: 10–30	HW: good; OM: good	Poor	Pros: cheap, readily available, deodorizer; cons: high toxicity

[a] Pfizer Inc., New York, NY, USA
[b] Dupont Chemical, Wilmington, DE, USA
[c] Fort Dodge Animal Health, Overland Park, KS, USA
[d] Reckitt Benckiser, Parsippany, NJ, USA

tion (Arzoumanian 2003). Serum samples with a light or dark red tint may be hemolyzed, which can compromise test parameters (Tatsumi et al. 2002; Cornell University Animal Health Diagnostic Center 2010). It is essential to decide beforehand which types of blood containers will be used, because many laboratories and test parameters rely on specific storage methods and containers to ensure uniformity of assays (Tatsumi et al. 2002; Cornell University Animal Health Diagnostic Center 2010). The aliquot needs to be placed in an anticoagulant container (e.g., lavender-top vacutainer) immediately to avoid additional clotting. See Wengert et al. (2012) for additional details on storing and transporting blood samples.

There are numerous risks to study animals whenever they are handled by investigators. Some of these risks include capture stressors, negative pressure from the anesthetizing agent, and the potential for capture myopathy after release; thus, the additional risk involved in drawing recommended amounts of blood is minimal by comparison. Besides the obvious benefits gained from determining the presence of pathogens or parasites, we believe the additional benefits of establishing hematological benchmarks, and providing genetic and archival samples for future use, far outweigh the perceived risks. Accordingly, we recommend that investigators draw blood as a standard practice when handling *Martes* species for any research, management, or conservation program. We direct readers to the many veterinary reference books that describe proper methods for drawing blood, and recommend that researchers reduce the potential for adverse effects by obtaining technical guidance from qualified researchers or veterinarians.

Collecting and Storing Samples for DNA and RNA Analyses

The polymerase chain reaction (PCR) is an effective and commonly used diagnostic tool for detecting the DNA or RNA of pathogens and parasites. DNA and RNA of viruses, bacteria, helminths, and protozoa are commonly extracted from various tissues, exudates, and excrement of captured wildlife. For example, PCR on DNA extracted from fecal samples is the recommended method for detecting the presence of parvoviruses in *Martes* species (Gabriel et al. 2010), whereas canine distemper virus is best detected using RNA extracted from nasal and ocular exudates. Other pathogens that have been identified in *Martes* species, such as *Leptospira interrogans*, are detected by analyzing DNA extracted from urine (Kingscote 1986).

Fecal DNA

Fecal sampling is an effective method for detecting many pathogens that can cause disease in *Martes* species. Fecal samples of parvovirus-infected animals may contain the DNA of the parvovirus (Barker and Parrish 2001; Steinel et al. 2001). Several pathogenic bacteria that infect *Martes* species can be detected by isolation or molecular testing of feces (Toma and Lafleur 1974; Nikolova et al. 2001). Although helminths and protozoa are typically detected in feces with visual screening, PCR can be used to detect their DNA for molecular confirmation or phylogenetic analysis.

A fresh fecal sample can be obtained either from a scat left in a trap or found in the field, or from a swab placed gently inside the rectum. We recommend the use of synthetic-tipped swabs over cotton-tipped swabs, because synthetic fibers are non-porous, allowing sample particles to be removed more easily during the extraction process. See Wengert et al. (2011) for additional details on the safe and effective storage and transport of fecal samples or swabs.

Ocular and Nasal Exudates

Although very difficult to detect in the external exudates of carnivores, CDV is occasionally shed in nasal and ocular exudates (Gillespie 1962; Williams 2001). Other viruses may also be shed in these exudates, and PCR can be used to test for active infections, including canine herpes virus (Carmichael 1970) and influenza virus (Fox et al. 1998). To sample these fluids, synthetic swabs can be used, taking care when swabbing the lining of the eyelids. A separate swab should be used inside the nostrils, gently twisting to obtain nasal exudates. Because CDV is a common virus to screen for in *Martes* species, and this virus is a highly labile RNA virus, extreme care should be taken in preserving these samples. See Wengert et al. (2011) for recommended methods for storing and transporting exudate samples.

DNA/RNA from Blood-borne Pathogens

DNA/RNA from whole blood and sera samples can be extracted and analyzed using PCR to detect certain blood-borne pathogens that infect *Martes* species, including the vector-borne pathogens *Anaplasma phagocytophilum*, West Nile virus, *Rickettsia* spp., *Bartonella hensalae*, and *Yersinia pestis*. Similar to storing blood for serological testing, storing blood to detect DNA or RNA from blood-borne pathogens requires specific methods to ensure the optimal recovery of the pathogen's genetic material.

DNA in Urine

Few pathogens are transmitted in urine, but there are some that pose both wildlife and zoonotic health risks. One that is known to infect *Martes* species is *Leptospira interrogans* (Kingscote 1986). To screen for this pathogen, urine can be collected either opportunistically or directly from the bladder with a hypodermic syringe (Quinn et al. 1999). Leptospires are sensitive to pH and will be undetectable at pH levels <6.8; therefore, urine must first be neutralized with a phosphate-buffered saline solution (Lucchesi et al. 2004). See Wengert et al. (2011) for recommended methods for storing and transporting urine samples.

DNA/RNA Storage

Once DNA or RNA is extracted from a biological sample, it must be stored properly to prevent degradation. Typically, extracted DNA/RNA is stored in Tris EDTA (or a solution provided by an extraction kit) and kept at −80 °C for long-term storage (Santella and Harkinson 2008). When analyzing samples at short intervals (e.g., daily or every few days), however, the sample should be aliquoted with a working DNA aliquot stored at −20 °C to 4 °C, and a working RNA aliquot stored at −20 °C. The remainder should be maintained in the freezer at −80 °C, minimizing the number of freeze-thaw cycles, which will ultimately degrade the sample.

Collecting and Storing Ectoparasites and Endoparasites

Martes species harbor numerous ectoparasites, including fleas, ticks, and mites (Table 7.2). Although ectoparasites themselves often pose no more harm to their hosts than a minor irritation or inflammation, some can serve as vectors for pathogens that can reduce fitness. These include the group of flea species that vector *Yersinia pestis*, the agent of plague, which is known to cause mortality in other mustelids, and is a serious zoonotic risk (Williams et al. 1994). Although many vector-borne pathogens have been associated with carnivores, very little is known about the effects of these pathogens on *Martes* species. Ectoparasite diversity can provide important information about the ecology of host species, since certain ectoparasites are associated with specific habitat types and may emerge only in particular seasons. Furthermore, ectoparasite diversity provides information about disease risk in individuals and in *Martes* populations.

Ticks

Ticks are commonly visible on their hosts, especially when engorged with blood. Often, they are found embedded in the dermis and are unable to escape capture, but study animals should be examined carefully to locate ticks in and around muzzles, ear pinnae, and genital and inguinal areas. Care should be taken when removing ticks to ensure that all body parts are intact, especially the mouthparts (e.g., hypostome, palps), which are needed to identify many species of ticks. Fine-tipped tweezers or a commercial tick-removal tool should be used to grab the tick's mouthparts as close to the skin as possible and gently extract them.

Fleas and Mites

Fleas can be detected and collected using a flea comb, or from the skin using tweezers. Because anesthetizing an animal typically lowers its body temperature, investigators should be aware that fleas can vacate their host quickly.

Thus, we recommend that protective clothing or insect repellent be used to avoid becoming an accidental host. In general, mites are more difficult to detect unless mange is apparent on the animal (Alasaad et al. 2009), in which case, scraping infected areas typically recovers mites that may not be visible on the skin. A razor blade coated with mineral oil and scraped against infected skin generally yields mites that can be detected on a microscope slide for up to 24 h (Alasaad et al. 2009). Investigators who are unfamiliar with this procedure should consult a wildlife veterinarian or experienced researcher prior to performing skin scrapings. See Wengert et al. (2011) for recommended methods for storing and transporting ectoparasites.

Endoparasites

Martes species harbor numerous types of endoparasites (Table 7.2). The most reliable and accurate method for documenting endoparasite faunas is to collect tissue samples from dead hosts. Sampling feces allows for a snapshot of the types of helminths that might be shedding ova at the time, or the occasional passage of protozoans that may be present in the gastrointestinal system. Appropriate procedures for identifying helminth ova and protozoan cysts from fecal samples vary with the types of parasites of interest.

Flotation methods rely on the fact that helminth ova and protozoan cysts are less dense than the medium, so they float to the top, where they can be collected and identified under a microscope (Zajac and Conboy 2006). The choice of flotation solution depends on the expected helminth taxa (e.g., trematode vs. nematode ova), but most parasite ova will float well in solutions with a specific gravity of 1.2–1.3 (Foreyt 2001). Centrifugation procedures prior to flotation will enhance detection probabilities for helminth ova that are present at low abundance (Zajac and Conboy 2006). For organisms that do not float well (typically protozoan cysts), fecal smears or sedimentation techniques can be used. Fecal smears should be interpreted cautiously, because the small amount of feces being scanned lessens the sensitivity of the test, and identifications can be compromised from debris in the sample (Zajac and Conboy 2006). The sedimentation method is based on the premise that, in certain solutions, denser helminth ova will settle to the bottom of the solution (Kauffman 1996). Sedimentation is the recommended method for detecting trematode ova (Foreyt 2001). See Wengert et al. (2011) for recommended methods for storing and transporting fecal samples used for endoparasite assays.

Future Research Considerations

Research on the ecology of *Martes* species has grown tremendously over the past several years (Proulx and Santos-Reis, this volume), although there are significant gaps in our understanding of how disease may impact individuals and populations. Furthermore, the ecology and epidemiology of the most important pathogens affecting *Martes* species are barely understood. Consequently, many health-related questions about *Martes* species have not been investigated. For example, we have few baseline health data for most *Martes* species, especially *M. flavigula* and *M. melampus*, which are greatly underrepresented in the literature. We recommend that all studies on these species that involve capture and handling incorporate sampling for pathogenic exposure or infections. Below, we present several recommendations for *Martes* researchers to consider when investigating disease ecology.

Impacts of Toxicants

The effects of specific toxicants, such as anticoagulant rodenticides, metals (e.g., lead), or insecticides (e.g., cholinesterase-inhibitors), have not been studied for *Martes* species. The impacts of primary or secondary poisoning from these toxicants have been documented in other mustelids (Wickstrom and Eason 1999; Fournier-Chambrillon et al. 2004b). For example, preliminary data from fishers in California and the Washington reintroduction area have shown high exposure rates (>75%) to anticoagulant rodenticides (M. Gabriel, unpublished data). Within these studies, proximate and ultimate mortality factors range widely, but whether these toxicants have sublethal effects is unknown. Studies of baseline parameters of toxicant exposure and possible influences on fitness and survival are warranted for *Martes* species.

Pathogen Impacts on Fecundity

Little is known about the effects of pathogens that cause morbidity but generally do not cause mortality. These infections may decrease fecundity or recruitment in wildlife populations. Since *Martes* researchers often trap females during active pregnancy, opportunities exist to conduct fetal counts and investigate correlations between pregnancy rates or kit survival and pathogen exposure or infection. Clifford et al. (2007) measured pregnancy rates and fetal counts using portable ultrasound equipment to investigate perinatal mortality in Channel Island foxes (*Urocyon littoralis*). Although differences were not significant, foxes that were seropositive for *Toxoplasma gondii* had a 61.5% perinatal mortality rate, whereas seronegative pregnant foxes had a 29.4% rate (Clifford et al. 2007). Studies such as these could help researchers understand how nonlethal infections may ultimately affect vital rates in *Martes* populations.

New Uses for Existing Samples

Many conservation geneticists use the same fecal DNA extraction kits and protocols as those used for fecal-pathogen investigations of *Martes* species, providing a largely untapped source of fecal DNA that could be used in fecal-pathogen or parasite studies. Many of these samples are collected through indirect methods, such as scat-dog surveys, in which feces from other members of the carnivore community are collected. After molecular species-identification procedures, researchers can use these samples to study the community-wide dynamics of certain pathogens.

Vaccine Research

Vaccines may be an invaluable tool for protecting captive-bred and newly reintroduced animals from naturally occurring pathogens that could be population-limiting. Given the frequency of reintroduction programs for *Martes* species, a vaccine that has been tested for safety and efficacy on *Martes* species is instrumental for maximizing the likelihood of a successful translocation or captive-breeding program. Clinical and field trials should be conducted on *Martes* species to determine whether vaccines for distemper viruses, rabies viruses, and other pathogens elicit a response sufficiently strong to protect against natural or experimental infections. Because founder animals in translocation programs for *Martes* species are often held captive for weeks, they may provide important opportunities for conducting vaccination trials to evaluate antibody responses.

Major Histocompatibility Complex

Investigating immunogenetic variation is an emerging field that has begun to shed light on the relation between genetic variation and immunities in wildlife populations. Genetic variation within and between wildlife populations is generally assessed using neutral markers (i.e., not under selective pressure) such as mitochondria, microsatellites, or single nucleotide polymorphisms. On the other hand, immunity can be studied by examining variation in an immunogenetic marker, the major histocompatibility complex (MHC) (Sommer 2005). The role of MHC in resistance to infectious diseases has not been investigated in *Martes* species. Investigating the relations among exposure, active infection, and MHC variation would be valuable for understanding the ability of *Martes* populations to minimize disease impacts. In addition, differences between neutral and MHC markers may reveal additional genetic threats facing insular populations (Aguilar et al. 2004). Once developed, this knowledge might help managers identify suitable founders for reintroduction programs (Acevedo-Whitehouse and Cunningham 2006).

Conclusions

In this chapter, we have highlighted some of the advances, threats, and recommendations for investigating the disease ecology and health of *Martes* species. Many of the accounts described in this chapter are from serological investigations that documented exposure to a pathogen or parasite; however, studies of clinical manifestations in *Martes* species are generally lacking. It is imperative that health and disease be included as an important component of future *Martes* research. An autecological approach to the study of these species should help elucidate current and future threats to *Martes* species. To effectively inform management and conservation efforts for *Martes* species, such studies will require integrated collaboration among a variety of disciplines, including wildlife veterinarians, pathologists, disease ecologists, and wildlife biologists.

Acknowledgments

We thank the associate editor, Gilbert Proulx, 2 anonymous reviewers, and Karen Converse, Sarah Brown, and Brian Luke for providing insightful comments and critiques on earlier drafts of this manuscript.

Table 7.2. Pathogens and parasites associated with *Martes* species

Martes species	Parasitic group	Latin binomial or common name	Parasite or disease	Transmission	Source
M. americana	Arthropoda	*Amalaraeus dissimilis*	Fleas	Fleas occur in leaf litter, dens, burrows, or on prey or conspecifics	Holland 1985
M. americana	Arthropoda	*Ceratophyllus ciliatus protinus*	Fleas	Fleas occur in leaf litter, dens, burrows, or on prey or conspecifics	Haas et al. 1989
M. americana	Arthropoda	*Chaetopsylla floridensis*	Fleas	Fleas occur in leaf litter, dens, burrows, or on prey or conspecifics	Zielinski 1984; Holland 1985; Haas et al. 1989
M. americana	Arthropoda	*Chaetopsylla lotoris*	Fleas	Fleas occur in leaf litter, dens, burrows, or on prey or conspecifics	Benton and Kelly 1975
M. americana	Arthropoda	*Epitedia wenmanni*	Fleas	Fleas occur in leaf litter, dens, burrows, or on prey or conspecifics	Holland 1985
M. americana	Arthropoda	*Euhoplopsyllus lynx*	Fleas	Fleas occur in leaf litter, dens, burrows, or on prey or conspecifics	Holland 1985
M. americana	Arthropoda	*Hystrichopsylla dippiei spinata*	Fleas	Fleas occur in leaf litter, dens, burrows, or on prey or conspecifics	Holland 1985; Haas et al. 1989
M. americana	Arthropoda	*Megabothris atrox*	Fleas	Fleas occur in leaf litter, dens, burrows, or on prey or conspecifics	de Vos 1957; Holland 1985
M. americana	Arthropoda	*Megarthroglossus* spp.	Fleas	Fleas occur in leaf litter, dens, burrows, or on prey or conspecifics	Zielinski 1984
M. americana	Arthropoda	*Monopsyllus ciliatus*	Fleas	Fleas occur in leaf litter, dens, burrows, or on prey or conspecifics	Zielinski 1984

Table 7.2—*cont.*

Martes species	Parasitic group	Latin binomial or common name	Parasite or disease	Transmission	Source
M. americana	Arthropoda	*Monopsyllus eumolpi*	Fleas	Fleas occur in leaf litter, dens, burrows, or on prey or conspecifics	Zielinski 1984
M. americana	Arthropoda	*Monopsyllus protinus*	Fleas	Fleas occur in leaf litter, dens, burrows, or on prey or conspecifics	Holland 1985
M. americana	Arthropoda	*Monopsyllus vison*	Fleas	Fleas occur in leaf litter, dens, burrows, or on prey or conspecifics	de Vos 1957; Holland 1985
M. americana	Arthropoda	*Monopsyllus wagneri*	Fleas	Fleas occur in leaf litter, dens, burrows, or on prey or conspecifics	Zielinski 1984
M. americana	Arthropoda	*Nearctopsylla brooksi*	Fleas	Fleas occur in leaf litter, dens, burrows, or on prey or conspecifics	Holland 1985
M. americana	Arthropoda	*Nearctopsylla genalis*	Fleas	Fleas occur in leaf litter, dens, burrows, or on prey or conspecifics	Holland 1985
M. americana	Arthropoda	*Nearctopsylla grahami*	Fleas	Fleas occur in leaf litter, dens, burrows, or on prey or conspecifics	Holland 1985
M. americana	Arthropoda	*Nearctopsylla hyrtaci*	Fleas	Fleas occur in leaf litter, dens, burrows, or on prey or conspecifics	Holland 1985
M. americana	Arthropoda	*Orchopeas (sexdentatus) agilis*	Fleas	Fleas occur in leaf litter, dens, burrows, or on prey or conspecifics	Hubbard 1947; Holland 1985; Lewis et al. 1988
M. americana	Arthropoda	*Orchopeas caedens*	Fleas	Fleas occur in leaf litter, dens, burrows, or on prey or conspecifics	Hubbard 1947; Holland 1985; Haas et al. 1989

M. americana	Arthropoda	*Orchopeas ciliatus*	Fleas		Fleas occur in leaf litter, dens, burrows, or on prey or conspecifics	Hubbard 1947
M. americana	Arthropoda	*Orchopeas durus*	Fleas		Fleas occur in leaf litter, dens, burrows, or on prey or conspecifics	Holland 1985
M. americana	Arthropoda	*Orchopeas nepos*	Fleas		Fleas occur in leaf litter, dens, burrows, or on prey or conspecifics	Hubbard 1947; Zielinski 1984
M. americana	Arthropoda	*Oropsylla idahoensis*	Fleas		Fleas occur in leaf litter, dens, burrows, or on prey or conspecifics	Zielinski 1984
M. americana	Arthropoda	*Peromyscopsylla longiloba*	Fleas		Fleas occur in leaf litter, dens, burrows, or on prey or conspecifics	Holland 1985
M. americana	Arthropoda	*Peromyscopsylla selenis*	Fleas		Fleas occur in leaf litter, dens, burrows, or on prey or conspecifics	Holland 1985
M. americana	Arthropoda	*Rhadinopsylla alphabetica*	Fleas		Fleas occur in leaf litter, dens, burrows, or on prey or conspecifics	Holland 1985
M. americana	Arthropoda	*Rhadinopsylla difficilis*	Fleas		Fleas occur in leaf litter, dens, burrows, or on prey or conspecifics	Lewis et al. 1988
M. americana	Arthropoda	*Rhadinopsylla fraterna*	Fleas		Fleas occur in leaf litter, dens, burrows, or on prey or conspecifics	Holland 1985
M. americana	Arthropoda	*Tarsopsylla coloradensis*	Fleas		Fleas occur in leaf litter, dens, burrows, or on prey or conspecifics	Holland 1985
M. americana	Arthropoda	*Thrassis spenceri*	Fleas		Fleas occur in leaf litter, dens, burrows, or on prey or conspecifics	Holland 1985

Table 7.2—*cont.*

Martes species	Parasitic group	Latin binomial or common name	Parasite or disease	Transmission	Source
M. americana	Arthropoda	*Stachiella retusa*	Lice	Direct contact with infested animals or fomites (den and rest sites)	Durden 2001
M. americana	Arthropoda	*Listrophorus mustelae*	Mites	Direct contact with infested animals	Cowan 1955
M. americana	Arthropoda	*Ixodes cookei*	Ticks	Ticks quest from vegetation, dens, or in rest sites, or crawl from prey	de Vos 1952; Allan 2001
M. americana	Arthropoda	*Ixodes gregsoni*	Ticks	Ticks quest from vegetation, dens, or in rest sites, or crawl from prey	Lindquist et al. 1999
M. americana	Arthropoda	*Ixodes rugosus*	Ticks	Ticks quest from vegetation, dens, or in rest sites, or crawl from prey	Furman and Loomis 1984
M. americana	Arthropoda	*Ixodes sculptus*	Ticks	Ticks quest from vegetation, dens, or in rest sites, or crawl from prey	Furman and Loomis 1984; Lane 1984
M. americana	Arthropoda	*Ixodes texanus*	Ticks	Ticks quest from vegetation, dens, or in rest sites, or crawl from prey	Furman and Loomis 1984; Allan 2001
M. americana	Bacteria	*Yersinia pestis*	Plague	Flea bite	Zielinski 1984
M. americana	Bacteria	*Yersinia pseudotuberculosis*	Pseudotuberculosis	Ingestion of contaminated material	McDonald and Larivière 2001
M. americana	Cestoda	*Mesocestoides lineatus*	Intestinal tapeworms	Ingestion of larvae in intermediate hosts	Seville and Addison 1995

M. americana	Cestoda	*Mesocestoides* spp.	Intestinal tapeworms	Ingestion of larvae in intermediate hosts	Hoberg et al. 1990
M. americana	Cestoda	*Taenia martis*	Intestinal tapeworms	Ingestion of cysticerci in vole intermediate hosts	Holmes 1963; Poole et al. 1983; Hoberg et al. 1990
M. americana	Cestoda	*Taenia mustelae*	Intestinal tapeworms	Ingestion of cysticerci in rodent intermediate hosts	Holmes 1963; Poole et al. 1983; Hoberg et al. 1990
M. americana	Nematoda	*Pearsonema plica*	Bladder worm	Ingestion of infected earthworm intermediate hosts or paratenic hosts	Seville and Addison 1995
M. americana	Nematoda	*Dioctophyme renale*	Giant kidney worms	Ingestion of second intermediate hosts	Erickson 1946; Seville and Addison 1995
M. americana	Nematoda	*Dracunculus insignis*	Guinea worms	Ingestion of infected copepod host in water	Douglas and Strickland 1987; Seville and Addison 1995
M. americana	Nematoda	*Uncinaria stenocephala*	Hookworms	Direct penetration of larvae	Chowdhury and Aguirre 2001
M. americana	Nematoda	*Baylisascaris devosi*	Intestinal ascarids	Ingestion of infective eggs or of larvae in paratenic hosts	de Vos 1957; Poole et al. 1983; Hoberg et al. 1990
M. americana	Nematoda	*Molineus mustelae*	Intestinal strongylids	Ingestion of, or penetration by, infective larvae	Seville and Addison 1995
M. americana	Nematoda	*Molineus patens*	Intestinal strongylids	Ingestion of, or penetration by, infective larvae	Hoberg et al. 1990
M. americana	Nematoda	*Crenosoma petrowi*	Lungworms	Ingestion of infected slugs or snails	Seville and Addison 1995
M. americana	Nematoda	*Eucoleus aerophilus*	Lungworms	Ingestion of earthworm intermediate host	Seville and Addison 1995

Table 7.2—*cont.*

Martes species	Parasitic group	Latin binomial or common name	Parasite or disease	Transmission	Source
M. americana	Nematoda	*Filaroides martis*	Lungworms	Ingestion of snail intermediate host	Seville and Addison 1995
M. americana	Nematoda	*Sobolevingylus* spp.	Lungworms	Ingestion of infected slugs or snails	Olsen 1952
M. americana	Nematoda	*Capillaria aerophila*	Lungworms	Ingestion of infective eggs	Butterworth and Beverley-Burton 1980
M. americana	Nematoda	*Aonchotheca (Capillaria) putorii*	Stomach worms	Ingestion of infective eggs	Butterworth and Beverley-Burton 1980; Foreyt and Lagerquist 1993; Valentsev 1996
M. americana	Nematoda	*Physaloptera* spp.	Stomach worms	Ingestion of infected intermediate hosts (arthropods) or paratenic hosts (birds, snakes)	Hoberg et al. 1990
M. americana	Nematoda	*Soboliphyme baturini*	Stomach worms	Ingestion of infected intermediate hosts (arthropods) or paratenic hosts (birds)	Hoberg et al. 1990; Zarnke et al. 2004; Koehler et al. 2007
M. americana	Nematoda	*Trichinella* spp.	Trichinosis	Ingestion of infected prey	Seville and Addison 1995
M. americana	Nematoda	*Trichinella spiralis*	Trichinosis	Ingestion of infected prey	Poole et al. 1983; Dick et al. 1986; Hoberg et al. 1990; Rausch et al. 1990; Foreyt and Lagerquist 1993
M. americana	Protista	*Sarcocystis* spp.	Sarcocystosis	Ingestion of infective tissue cysts in prey	McDonald and Larivière 2001

M. americana	Protista	*Toxoplasma gondii*	Toxoplasmosis	Ingestion of infective tissue cysts in prey or infective oocysts in environment	Douglas and Strickland 1987
M. americana	Trematoda	*Alaria mustelae*	Intestinal fluke	Ingestion of mesocercariae in a second intermediate or paratenic host	Hoberg et al. 1990
M. americana	Trematoda	*Alaria taxideae*	Intestinal fluke	Ingestion of mesocercariae in a second intermediate or paratenic host	Holmes 1963; Möhl et al. 2009
M. americana	Trematoda	*Euryhelmis squamula*	Intestinal fluke	Ingestion of frogs with metacercariae	Hoberg et al. 1990
M. americana	Virus	Aleutian disease virus	Parvoviral enteritis	Contact with feces, or fomites contaminated with feces	Douglas and Strickland 1987
M. flavigula	Arthropoda	*Trichodectes* spp.	Lice	Direct contact or via bedding	Emerson and Price 1974
M. flavigula	Arthropoda	*Amblyomma testudinarium*	Ticks	Ticks quest from vegetation, dens, or in rest sites, or crawl from prey	Grassman et al. 2004
M. flavigula	Arthropoda	*Hyalomma hystricis*	Ticks	Ticks quest from vegetation, dens, or in rest sites, or crawl from prey	Grassman et al. 2004
M. flavigula	Arthropoda	*Rhipicephalus haemaphysaloides*	Ticks	Ticks quest from vegetation, dens, or in rest sites, or crawl from prey	Grassman et al. 2004
M. flavigula	Trematoda	*Nanophyetus salmincola*	Intestinal fluke	Ingestion of fish with metacercariae	Schlegel et al. 1968

Table 7.2—*cont.*

Martes species	Parasitic group	Latin binomial or common name	Parasite or disease	Transmission	Source
M. foina	Arthropoda	*Ctenocephalides felis*	Fleas	Contact with fleas in the environment or on prey or conspecifics	Lledo et al. 2010
M. foina	Arthropoda	*Paraceras melis*	Fleas	Fleas occur in leaf litter, dens, burrows, or on prey or conspecifics	Lledo et al. 2010
M. foina	Arthropoda	*Pulex irritans*	Fleas	Fleas occur in leaf litter, dens, burrows, or on prey or conspecifics	Lledo et al. 2010
M. foina	Arthropoda	*Trichodectes melis*	Lice	Direct contact with infested animals	Lledo et al. 2010
M. foina	Arthropoda	*Sarcoptes scabei*	Mites	Direct contact with infested animals	Balestrieri et al. 2006
M. foina	Arthropoda	*Haemaphysalis erinacei*	Ticks	Ticks quest from vegetation, dens, or in rest sites, or crawl from prey	Otranto et al. 2007
M. foina	Arthropoda	*Ixodes hexagonus*	Ticks	Ticks quest from vegetation, dens, or in rest sites, or crawl from prey	Page and Langton 1996; Lledo et al. 2010
M. foina	Arthropoda	*Rhipicephalus pusillus*	Ticks	Ticks quest from vegetation, dens, or in rest sites, or crawl from prey	Perez and Palma 2001
M. foina	Bacteria	*Yersinia pseudotuberculosis*	Pseudotuberculosis	Ingestion	Nikolova et al. 2001
M. foina	Bacteria	*Yersinia enterocolitica*	Yersiniosis	Ingestion	Nikolova et al. 2001

M. foina	Cestoda	*Echinococcus multilocularis*	Intestinal tapeworms	Ingestion of prey with immature tapeworms	Martinek et al. 2001
M. foina	Cestoda	*Insinurotaenia* sp.	Intestinal tapeworms	Ingestion of cysticercoids in insect intermediate hosts	Ribas et al. 2004
M. foina	Cestoda	*Mesocestoides* sp.	Intestinal tapeworms	Ingestion of larvae in intermediate hosts	Cerbo et al. 2008
M. foina	Cestoda	*Mesocestoides* sp.	Intestinal tapeworms	Ingestion of larvae in intermediate hosts	Ribas et al. 2004
M. foina	Cestoda	*Oochoristica* sp.	Intestinal tapeworms	Ingestion of cysticercoids in insect intermediate hosts	Ribas et al. 2004
M. foina	Cestoda	*Taenia crassiceps*	Intestinal tapeworms	Ingestion of cysticerci in intermediate hosts	Jones and Pybus 2001
M. foina	Cestoda	*Taenia hydatigena*	Intestinal tapeworms	Ingestion of cysticerci in intermediate hosts	Jones and Pybus 2001
M. foina	Cestoda	*Taenia martis*	Intestinal tapeworms	Ingestion of cysticerci in vole intermediate hosts	Schoo et al. 1994; Millan and Ferroglio 2001; Ribas et al. 2004
M. foina	Cestoda	*Taenia mustelae*	Intestinal tapeworms	Ingestion of cysticerci in intermediate hosts	Jones and Pybus 2001
M. foina	Cestoda	*Taenia* spp.	Intestinal tapeworms	Ingestion of infective immature tapeworms in intermediate hosts	Cerbo et al. 2008
M. foina	Nematoda	*Angiostongylus* spp.	Arterial worms	Ingestion of snail intermediate host or a variety of paratenic hosts	Millan and Ferroglio 2001

Table 7.2—*cont.*

Martes species	Parasitic group	Latin binomial or common name	Parasite or disease	Transmission	Source
M. foina	Nematoda	*Pearsonema plica*	Bladder worms	Ingestion of infected earthworm intermediate hosts or paratenic hosts	Ribas et al. 2004
M. foina	Nematoda	*Thelazia callipaeda*	Eyeworms	Exposure to arthropod vectors	Otranto et al. 2007
M. foina	Nematoda	*Uncinaria criniformis*	Hookworms	Direct penetration of larvae	Cerbo et al. 2008
M. foina	Nematoda	*Uncinaria stenocephala*	Hookworms	Direct penetration of larvae	Loos-Frank and Zeyhle 1982
M. foina	Nematoda	*Toxocara cati*	Intestinal ascarid	Ingestion of infective eggs	Rodriguez and Carbonell 1998
M. foina	Nematoda	*Strongyloides stercoralis*	Intestinal rhabditoid	Ingestion of infective eggs	Rodriguez and Carbonell 1998
M. foina	Nematoda	*Rictularia proni*	Intestinal Spirurid	Ingestion of insect intermediate host	Miguel et al. 1995
M. foina	Nematoda	*Molineus europaeus*	Intestinal strongylid	Ingestion of, or penetration by, infective larvae	Schoo et al. 1994
M. foina	Nematoda	*Molineus patens*	Intestinal strongylid	Ingestion of, or penetration by, infective larvae	Millan and Ferroglio 2001; Ribas et al. 2004; Cerbo et al. 2008
M. foina	Nematoda	*Capillaria* spp.	Intestinal trichurid	Ingestion of infective eggs	Millan and Ferroglio 2001
M. foina	Nematoda	*Sobolevingylus petrowi*	Lungworms	Ingestion of infected slugs or snails	Ribas et al. 2004; Cerbo et al. 2008
M. foina	Nematoda	*Crenosoma petrowi*	Lungworms - metastrongyloid	Ingestion of infected slugs or snails	Cerbo et al. 2008; Ribas et al. 2004

M. foina	Nematoda	*Eucoleus aerophilus*	Lungworms - metastrongyloid	Ingestion of earthworms intermediate host	Cerbo et al. 2008; Ribas et al. 2004
M. foina	Nematoda	*Skrjabingylus petrowi*	Nasal sinus worms	Ingestion of infected slugs or snails	Anderson 2000
M. foina	Nematoda	*Aonchotheca putorii*	Stomach worms	Ingestion of infective eggs	Schoo et al. 1994; Ribas et al. 2004; Cerbo et al. 2008
M. foina	Nematoda	*Filaria martis*	Subcutaneous filarid	Bite of hematophagous arthropods	Otranto et al. 2007
M. foina	Nematoda	*Trichinella britovi*	Trichinosis	Ingestion of infected prey	Pozio 2000
M. foina	Protista	*Isospora rivolta*	Coccidiosis	Ingestion of infective oocysts	Rodriguez and Carbonell 1998
M. foina	Protista	*Isospora rivolta*	Cryptosporidium	Ingestion of infective oocysts	Rademacher et al. 1999
M. foina	Protista	*Hepatozoon* spp.	Hepatozoonosis	Bite of hematophagous arthropods	Geisel et al. 1979
M. foina	Protista	*Neospora caninum*	Neosporosis	Ingestion of infective tissue cysts	Sobrino et al. 2008
M. foina	Protista	*Toxoplasma gondii*	Toxoplasmosis	Ingestion of infective tissue cysts in prey or infective oocysts in environment	Hejlíček et al. 1997
M. foina	Trematoda	*Brachylaima* sp.	Intestinal fluke	Ingestion of snails with metacercariae	Ribas et al. 2004
M. foina	Trematoda	*Troglotrema acutum*	Sinus fluke	Ingestion of frogs with metacercariae	Koubek et al. 2004
M. foina	Virus	Canine distemper virus	Distemper	Direct contact with saliva, nasal discharge, or feces	Steinhagen and Nebel 1985; Frolich et al. 2000; Philippa et al. 2008

Table 7.2—*cont.*

Martes species	Parasitic group	Latin binomial or common name	Parasite or disease	Transmission	Source
M. foina	Virus	Canine adenovirus	Infectious hepatitis	Direct contact with bodily discharge, urine or feces	Philippa et al. 2008
M. foina	Virus	Influenza virus (H5N1)	Influenza encephalitis	Contact with aerosols or bodily fluids via respiration or from fomites	Klopfleisch et al. 2007
M. foina	Virus	Aleutian disease virus	Parvoviral enteritis	Contact with feces, or fomites contaminated with feces	Fournier-Chambrillon et al. 2004a
M. foina	Virus	Canine parvovirus	Parvoviral enteritis	Contact with feces, or fomites contaminated with feces	Santos et al. 2009
M. foina	Virus	Rabies virus and European bat lyssavirus	Rabies	Bite wounds of infected mammals	Steck and Wandeler 1980; Shimshony 1997; Müller et al. 2004; Dacheux et al. 2009
M. martes	Acanthocephala	*Centrorynchus aluconis*	Spiny-headed worms in small intestine	Ingestion of larvae in intermediate hosts	Segovia et al. 2007
M. martes	Arthropoda	*Ceratophyllus sciurorum*	Fleas	Fleas occur in leaf litter, dens, burrows, or on prey or conspecifics	Lledo et al. 2010
M. martes	Arthropoda	*Sarcoptes scabei*	Mites	Direct contact with infested animals	Coulter 1966; Mörner 1992
M. martes	Arthropoda	*Ixodes hexogonus*	Ticks	Ticks quest from vegetation, dens, or in rest sites, or crawl from prey	Simpson et al. 2005; Lledo et al. 2010

M. martes	Arthropoda	*Ixodes ricinus*	Ticks	Ticks quest from vegetation, dens, or in rest sites, or crawl from prey	Simpson et al. 2005
M. martes	Cestoda	*Echinococcus multilocularis*	Intestinal tapeworms	Ingestion of prey with immature tapeworms	Martinek et al. 2001
M. martes	Cestoda	*Taenia martis*	Intestinal tapeworms	Ingestion of rodent intermediate hosts	Jones and Pybus 2001; Segovia et al. 2007
M. martes	Cestoda	*Taenia mustelae*	Intestinal tapeworms	Ingestion of cysticerci in intermediate hosts	Jones and Pybus 2001
M. martes	Nematoda	*Pearsonema plica*	Bladder worm	Ingestion of infected earthworm intermediate hosts or paratenic hosts	Segovia et al. 2007
M. martes	Nematoda	*Spirocerca lupi*	Esophageal worm	Ingestion of infected beetle or paratenic host	Segovia et al. 2007
M. martes	Nematoda	*Crenosoma petrowi*	Lungworms	Ingestion of infected slugs or snails	Segovia et al. 2007
M. martes	Nematoda	*Eucoleus aerophilus*	Lungworms	Ingestion of earthworm intermediate host	Segovia et al. 2007
M. martes	Nematoda	*Eucoleus trophimenkovi*	Lungworms	Ingestion of earthworm intermediate host	Romashov 2001
M. martes	Nematoda	*Filaroides martis*	Lungworms	Ingestion of snail intermediate host	Segovia et al. 2007
M. martes	Nematoda	*Sobolevingylus petrowi*	Lungworms	Ingestion of infected slugs or snails	Segovia et al. 2007
M. martes	Nematoda	*Skrjabingylus petrowi*	Nasal sinus worms	Ingestion of infected slugs or snails	Anderson 2000
M. martes	Nematoda	*Aonchotheca putorii*	Stomach worms	Ingestion of infective eggs	Segovia et al. 2007

Table 7.2—*cont.*

Martes species	Parasitic group	Latin binomial or common name	Parasite or disease	Transmission	Source
M. martes	Nematoda	*Mastophorus muris*	Stomach worms	Ingestion of beetle intermediate host	Segovia et al. 2007
M. martes	Nematoda	*Physaloptera siberica*	Stomach worms	Ingestion of infected intermediate hosts (arthropods) or paratenic hosts (birds, snakes)	Segovia et al. 2007
M. martes	Nematoda	*Spirura rytipleurites*	Stomach worms	Ingestion of infected cockroach	Segovia et al. 2007
M. martes	Nematoda	*Trichinella britovi*	Trichinosis	Ingestion of infected prey	Pozio et al. 2009b
M. martes	Nematoda	*Trichinella* spp.	Trichinosis	Ingestion of infected prey	Segovia et al. 2007
M. martes	Protista	*Hepatozoon* spp.	Hepatozoonosis	Bite of hematophagous arthropods	Geisel et al. 1979; Simpson et al. 2005
M. martes	Protista	*Neospora caninum*	Neosporosis	Ingestion of infective tissue cysts	Sobrino et al. 2008
M. martes	Protista	*Toxoplasma gondii*	Toxoplasmosis	Ingestion of infective tissue cysts in prey or infective oocysts in environment	Hejlíček et al. 1997
M. martes	Trematoda	*Alaria alata*	Fluke mesocercariae	Ingestion of mesocercariae in an amphibian or mammalian paratenic host	Möhl et al. 2009
M. martes	Trematoda	*Euryhelmis squamula*	Intestinal fluke	Ingestion of frogs with metacercariae	Segovia et al. 2007
M. martes	Trematoda	*Troglotrema acutum*	Sinus fluke	Ingestion of frogs with metacercarae	Koubek et al. 2004

M. martes	Virus	Canine distemper virus	Distemper	Direct contact with saliva, nasal discharge, or feces	Philippa et al. 2008
M. martes	Virus	Aleutian disease virus	Parvoviral enteritis	Contact with feces, or fomites contaminated with feces	Fournier-Chambrillon et al. 2004a
M. melampus	Acanthocephala	*Centrorynchus elongatus*	Spiny-headed worms	Ingestion of larvae in intermediate hosts	Sato et al. 1999b
M. melampus	Bacteria	*Yersinia pseudotuberculosis*	Pseudotuberculosis	Ingestion	Fukushima and Gomyoda 1991
M. melampus	Cestoda	*Mesocestoides paucitesticulus*	Intestinal tapeworms	Ingestion of infective immature tapeworms in intermediate hosts	Sato et al. 1999a
M. melampus	Cestoda	*Spirometra erinaceieuropaei*	Tapeworms	Ingestion of pleurocercoids in intermediate and paratenic hosts	Sato et al. 1999a
M. melampus	Nematoda	*Pearsonema plica*	Bladder worms	Ingestion of infected intermediate hosts (earthworm) or paratenic hosts	Sato et al. 1999a
M. melampus	Nematoda	*Toxocara canis*	Intestinal ascarid	Ingestion of infective eggs	Sato et al. 1999a
M. melampus	Nematoda	*Toxocara tanuki*	Intestinal ascarid	Ingestion of infective eggs	Sato et al. 1999a
M. melampus	Nematoda	*Molineus patens*	Intestinal strongylid	Ingestion of, or penetration by, infective larvae	Sato et al. 1999a
M. melampus	Nematoda	*Molineus* spp.	Intestinal strongylid	Ingestion of, or penetration by, infective larvae	Sato et al. 1999b

Table 7.2—*cont.*

Martes species	Parasitic group	Latin binomial or common name	Parasite or disease	Transmission	Source
M. melampus	Nematoda	*Eucoleus aerophilus*	Lungworms	Ingestion of earthworm intermediate host	Sato et al. 1999b
M. melampus	Nematoda	*Aonchotheca putorii*	Stomach worms	Ingestion of infective eggs	Sato et al. 1999a,b
M. melampus	Nematoda	*Soboliphyme baturini*	Stomach worms	Ingestion of infected intermediate hosts (arthropods) or paratenic hosts (birds)	Sato et al. 1999a,b
M. melampus	Protista	*Hepatozoon* spp.	Hepatozoonosis	Bite of hematophagous arthropods	Yanai et al. 1995
M. melampus	Trematoda	*Brachylaima tokudai*	Intestinal fluke	Ingestion of snails with metacercariae	Sato et al. 1999a
M. melampus	Trematoda	*Echinostoma hortense*	Intestinal fluke	Ingestion of frogs with metacercariae	Sato et al. 1999a,b
M. melampus	Trematoda	*Euryhelmis constaricensis*	Intestinal fluke	Ingestion of frogs with metacercariae	Sato et al. 1999b
M. melampus	Trematoda	*Euryhelmis costaricensis*	Intestinal fluke	Ingestion of frogs with metacercariae	Sato et al. 1999a,b
M. melampus	Trematoda	*Metagonimus miyatai*	Intestinal fluke	Ingestion of fish with metacercariae	Sato et al. 1999a
M. melampus	Trematoda	*Metagonimus takahashii*	Intestinal fluke	Ingestion of fish with metacercariae	Sato et al. 1999a
M. melampus	Trematoda	*Metagonimus yokogawai*	Intestinal fluke	Ingestion of fish with metacercariae	Sato et al. 1999a
M. melampus	Trematoda	*Nanophyetus (Pseudotroglotrema)* sp.	Intestinal fluke	Ingestion of fish with metacercariae	Sato et al. 1999a

M. melampus	Trematoda	*Concinnum ten*	Pancreatic duct / intestinal fluke	Ingestion of insects with metacercariae	Ashizawa et al. 1978; Sato et al. 1999a,b
M. pennanti	Arthropoda	*Oropsylla arctomys*	Fleas	Fleas occur in leaf litter, dens, burrows, or on prey	Hubbard 1947; Holland 1985
M. pennanti	Arthropoda	*Pulex* spp.	Fleas	Fleas occur in leaf litter, dens, burrows, or on prey	R. Brown and M. Gabriel, IERC, unpublished data
M. pennanti	Arthropoda	*Sarcoptes scabei*	Mites	Direct contact	O' Meara et al. 1960; Bornstein et al. 2001
M. pennanti	Arthropoda	*Dermacentor occidentalis*	Ticks	Ticks quest from vegetation, dens, or in rest sites, or crawl from prey	R. Brown and M. Gabriel, IERC, unpublished data
M. pennanti	Arthropoda	*Dermacentor variabilis*	Ticks	Ticks quest from vegetation, dens, or in rest sites, or crawl from prey	R. Brown and M. Gabriel, IERC, unpublished data
M. pennanti	Arthropoda	*Ixodes cookei*	Ticks	Ticks quest from vegetation, dens, or in rest sites, or crawl from prey	Allan 2001
M. pennanti	Arthropoda	*Ixodes gregsoni*	Ticks	Ticks quest from vegetation, dens, or in rest sites, or crawl from prey	Lubelczyk et al. 2007
M. pennanti	Arthropoda	*Ixodes pacificus*	Ticks	Ticks quest from vegetation, dens, or in rest sites, or crawl from prey	R. Brown and M. Gabriel, IERC, unpublished data

Table 7.2—*cont.*

Martes species	Parasitic group	Latin binomial or common name	Parasite or disease	Transmission	Source
M. pennanti	Arthropoda	*Ixodes rugosus*	Ticks	Ticks quest from vegetation, dens, or in rest sites, or crawl from prey	R. Brown and M. Gabriel, IERC, unpublished data
M. pennanti	Arthropoda	*Ixodes texanus*	Ticks	Ticks quest from vegetation, dens, or in rest sites, or crawl from prey	Darsie and Anastos 1957
M. pennanti	Bacteria	*Anaplasma phagocytophilum*	Granulocytic anaplasmosis	Tick bite	Brown et al. 2007
M. pennanti	Bacteria	*Leptospira interogans*	Leptospirosis	Contact with urine or urine contaminated water sources	Douglas and Strickland 1987
M. pennanti	Bacteria	*Borrelia burgdorferi sensu lato*	Lyme borreliosis	Tick bite	Brown et al. 2007
M. pennanti	Bacteria	*Rickettisia rickettsii* or *Rickettsia* spp.	Rocky Mountain spotted fever or related diseases	Tick bite	Brown et al. 2007
M. pennanti	Bacteria	*Francisella tularensis*	Tularemia	Ingestion of muscle tissue of infected prey	Dick and Leonard 1979; Dick et al. 1986
M. pennanti	Cestoda	*Mesocestoides corti*	Intestinal tapeworms	Ingestion of larvae in intermediate hosts	Leiby and Dyer 1971
M. pennanti	Cestoda	*Mesocestoides variabilis*	Intestinal tapeworms	Ingestion of larvae in intermediate hosts	de Vos 1952; Hamilton and Cook 1955
M. pennanti	Cestoda	*Taenia martis*	Intestinal tapeworms	Ingestion of tissue cysts in rodents and insectivores	Dick and Leonard 1979; Jones and Pybus 2001
M. pennanti	Cestoda	*Taenia mustelae*	Intestinal tapeworms	Ingestion of cysticerci in rodent intermediate hosts	Leiby and Dyer 1971

M. pennanti	Nematoda	*Dioctophyme renale*	Giant kidney worms	Ingestion of infected earthworms or paratenic hosts	Erickson 1946; Douglas and Strickland 1987
M. pennanti	Nematoda	*Dracunculus insignis*	Guinea worms	Ingestion of infected copepod host in water	Hamilton and Cook 1955; Douglas and Strickland 1987
M. pennanti	Nematoda	*Uncinaria stenocephala*	Hookworm	Direct penetration of larvae	Erickson 1946; Hamilton and Cook 1955
M. pennanti	Nematoda	*Ascaris mustelarum*	Intestinal ascarid	Ingestion of infective eggs	Dick and Leonard 1979
M. pennanti	Nematoda	*Baylisascaris devosi*	Intestinal ascarid	Ingestion of infective eggs	de Vos 1952; Dick and Leonard 1979
M. pennanti	Nematoda	*Capillaria mustelorum*	Intestinal trichiurid	Ingestion of infective eggs	Hamilton and Cook 1955
M. pennanti	Nematoda	*Aonchotheca (Capillaria) putorii*	Stomach worm	Ingestion of infective eggs	Butterworth and Beverley-Burton 1980; Valentsev 1996
M. pennanti	Nematoda	*Arthrocephalus lotoris*	Intestinal strongylid	Ingestion of, or penetration by, larvae	Hamilton and Cook 1955; Chowdhury and Aguirre 2001
M. pennanti	Nematoda	*Molineus patens*	Intestinal strongylid	Ingestion of, or penetration by, infective larvae	Hamilton and Cook 1955; Dick and Leonard 1979
M. pennanti	Nematoda	*Crenosoma petrowi*	Lungworm	Ingestion of infected mollusks or paratenic hosts (snakes, lizards, and frogs)	Craig and Borecky 1976
M. pennanti	Nematoda	*Sobolevingylus* spp.	Lungworm	Ingestion of infected slugs or snails	Craig and Borecky 1976
M. pennanti	Nematoda	*Trilobostrongylus bioccai*	Lungworm	Ingestion of mollusk intermediate hosts	Craig and Borecky 1976

Table 7.2—*cont.*

Martes species	Parasitic group	Latin binomial or common name	Parasite or disease	Transmission	Source
M. pennanti	Nematoda	*Skrjabingylus petrowi*	Nasal sinus worm	Ingestion of infected slugs or snails	Koubek et al. 2004
M. pennanti	Nematoda	*Physaloptera* spp.	Stomach worm	Ingestion of infected arthropod intermediate hosts or paratenic hosts (birds, snakes)	Dick and Leonard 1979
M. pennanti	Nematoda	*Soboliphyme baturini*	Stomach worm	Ingestion of infected intermediate hosts (arthropods) or paratenic hosts (birds, snakes)	Erickson 1946; Dick and Leonard 1979
M. pennanti	Nematoda	*Trichinella* spp.	Trichinosis	Ingestion of infected prey	Pozio 2000
M. pennanti	Nematoda	*Trichinella spiralis*	Trichinosis	Ingestion of infected prey	Dick and Leonard 1979; Dick et al. 1986
M. pennanti	Nematoda	*Capillaria plica*	Urinary bladder worm	Ingestion of infective eggs	Butterworth and Beverley-Burton 1980
M. pennanti	Protista	*Isospora* spp.	Coccidiosis	Ingestion of infective oocysts	de Vos 1952
M. pennanti	Protista	*Sarcocystis neurona*	Sarcocystosis	Ingestion of infective tissue cysts in prey	Gerhold et al. 2005
M. pennanti	Protista	*Toxoplasma gondii*	Toxoplasmosis	Ingestion of infective tissue cysts in prey or infective oocysts in environment	Tizard et al. 1976; Philippa et al. 2004; Brown et al. 2007
M. pennanti	Trematoda	*Alaria mustelae*	Intestinal fluke and mesocercariae	Ingestion of metacercariae in the lungs of a final intermediate host or mesocercariae of a 2nd intermediate host	Dick and Leonard 1979; Möhl et al. 2009

M. pennanti	Trematoda	*Metorchis conjunctus*	Liver fluke	Ingestion of fish with metacercariae	Dick and Leonard 1979; Dick et al. 1986
M. pennanti	Virus	Canine distemper virus	Distemper	Direct contact with saliva, nasal discharge, or feces	Williams 2001; Brown et al. 2006; Brown et al. 2007
M. pennanti	Virus	Canine herpes virus	Herpes	Direct contact with saliva, nasal discharge, or feces	Brown et al. 2007
M. pennanti	Virus	Canine adenovirus	Infectious hepatitis	Direct contact with bodily discharge, urine or feces	Philippa et al. 2004; Brown et al. 2006; Brown et al. 2007
M. pennanti	Virus	Canine parvovirus	Parvoviral enteritis	Contact with feces, or objects contaminated with feces	Brown et al. 2006; Brown et al. 2007
M. pennanti	Virus	Rabies virus	Rabies	Bite wounds of infected mammals	Rupprecht et al. 2001; Krebs et al. 2003
M. pennanti	Virus	West Nile virus	West Nile viral encephalitis	Mosquito bite, potentially tick bite, or ingesting infected prey	Brown et al. 2007
M. zibellina	Acanthocephala	*Corynosoma strumosum*	Spiny-headed worm in small intestine	Ingestion of larvae in intermediate hosts	Valentsev 1996
M. zibellina	Cestoda	*Diphyllobothrium* spp.	Intestinal tapeworms	Ingestion of prey with immature tapeworms	Valentsev 1996
M. zibellina	Cestoda	*Mesocestoiodes kirbyi*	Intestinal tapeworms	Ingestion of larvae in intermediate hosts	Valentsev 1996
M. zibellina	Cestoda	*Taenia martis*	Intestinal tapeworms	Ingestion of rodent intermediate hosts	Valentsev 1996
M. zibellina	Cestoda	*Taenia mustelae*	Intestinal tapeworms	Ingestion of prey with infective larvae	Jones and Pybus 2001

Table 7.2—*cont.*

Martes species	Parasitic group	Latin binomial or common name	Parasite or disease	Transmission	Source
M. zibellina	Fungi	*Chrysosporium parvum*	Adiaspiromycosis	Direct contact with saprophytic fungi in soil	Burek 2001
M. zibellina	Nematoda	*Anisakidae* spp.	Intestinal ascarid	Ingestion of infective larvae in fish intermediate hosts	Valentsev 1996
M. zibellina	Nematoda	*Anisakis simplex*	Intestinal ascarid	Ingestion of infective larvae in fish intermediate hosts	Valentsev 1996
M. zibellina	Nematoda	*Ascaris columnaris*	Intestinal ascarid	Ingestion of infective eggs	Valentsev 1996
M. zibellina	Nematoda	*Molineus patens*	Intestinal strongylid	Ingestion of, or penetration by, infective larvae	Valentsev 1996
M. zibellina	Nematoda	*Physaloptera siberica*	Intestine	Suspected - ingestion of insect intermediate hosts	Valentsev 1996
M. zibellina	Nematoda	*Crenosoma petrowi*	Lungworm	Ingestion of infected mollusks or paratenic hosts (snakes, lizards, and frogs)	Valentsev 1996
M. zibellina	Nematoda	*Filaroides martis*	Lungworm	Ingestion of snail intermediate host	Valentsev 1996; Monakhov 1999
M. zibellina	Nematoda	*Metathelazia skrjabini*	Lungworm	Life cycle not reported	Valentsev 1996
M. zibellina	Nematoda	*Capillaria (Thominx) aerophilus*	Lungworm	Ingestion of infective eggs	Valentsev 1996
M. zibellina	Nematoda	*Skrjabingylus petrowi*	Nasal sinus worm	Ingestion of infected slugs or snails	Valentsev 1996

M. zibellina	Nematoda	*Aonchotheca (Capillaria) putorii*	Stomach and intestinal worm	Ingestion of infective eggs	Kamiya and Ishigaki 1972; Valentsev 1996; Sato et al. 1999a
M. zibellina	Nematoda	*Soboliphyme baturini*	Stomach worm	Ingestion of infected intermediate hosts (arthropods) or paratenic hosts (birds)	Rausch et al. 1990; Valentsev 1996; Koehler et al. 2007
M. zibellina	Nematoda	*Trichinella nativa or Trichinella T6*	Trichinosis	Ingestion of infected prey	Valentsev 1996; Dick and Pozio 2001
M. zibellina	Protista	*Eimeria sablii*	Coccidiosis	Ingestion of infective oocysts	Nukerbaeva 1981
M. zibellina	Protista	*Eimeria sibirica*	Coccidiosis	Ingestion of infective oocysts	Nukerbaeva 1981
M. zibellina	Protista	*Isospora martessii*	Coccidiosis	Ingestion of infective oocysts	Nukerbaeva 1981
M. zibellina	Trematoda	*Alaria alata*	Fluke mesocercariae	Ingestion of mesocercariae in an amphibian or mammalian paratenic host	Möhl et al. 2009
M. zibellina	Virus	*Pseudorabies virus*	Pseudorabies	Direct contact or via aerosols	Shahan et al. 1947
Martes spp.	Trematoda	*Paragonimus miyazakii*	Lung fluke	Ingestion of freshwater crabs with metacercariae	Sano et al. 1978
Martes spp.	Virus	Canine distemper virus	Distemper	Direct contact with saliva, nasal discharge, or feces	An et al. 2008
Martes spp.	Virus	Rabies virus	Rabies	Bite wounds of infected mammals	Boegel et al. 1977

Note: Although we intend the information to be comprehensive, our literature review was weighted toward readily available literature published in English

8

Ecophysiology of Overwintering in Northern *Martes* Species

ANNE-MARI MUSTONEN AND PETTERI NIEMINEN

ABSTRACT

Winter provides northern *Martes* species with energetic challenges because they do not have exceptionally insulative fur, large body fat reserves, or the ability to utilize a passive wintering strategy with a significantly reduced metabolic rate. These species overwinter successfully, for example, by decreasing their physical activity levels, switching to diurnal foraging, and hunting and resting in the subnivean zone. Moreover, *Martes* species are able to fast for 1–5 days in adverse weather conditions, during which they utilize their limited adipose tissue reserves together with body proteins. The responses to experimental fasting were studied in American martens (*Martes americana*) and sables (*M. zibellina*). They appear to have similar short-term responses to food deprivation as fasting-adapted carnivores, but their low body adiposity limits the extension of phase II of fasting (fat utilization), leading to an earlier initiation of stimulated protein catabolism. Lipids are mobilized efficiently from both subcutaneous and intra-abdominal white adipose tissues, and the mobilization efficiency of fatty acids correlates inversely with their carbon chain length and increases with unsaturation. Hormonal responses to fasting include decreased plasma insulin and elevated cortisol concentrations, which may promote lipolysis and protein catabolism; increased ghrelin levels, which stimulate appetite; and suppressed thyroid activity, associated with a slight metabolic depression leading to energy preservation. In addition to proteolysis, other potentially deleterious health effects caused by fasting include immunosuppression and fatty liver. *Martes* species have no clear differences in fatty acid composition between the inner and outer trunk fat depots. In contrast, the acral body parts (paw pads, tail tip) show an incorporation of polyunsaturated fatty acids with low melting points, creating an unsaturation gradient between the trunk and extremities. This could prevent the solidification of cell membranes in the parts of the body most susceptible to cold air or snow. These ecophysiological characteristics may be essential for the survival of northern *Martes* species during the winter months.

Energetic Challenges during Winter

As inhabitants of northern forests, American martens (*Martes americana*), European pine martens (*M. martes*), fishers (*M. pennanti*), and sables (*M. zibellina*) are adapted to cold environments with wide seasonality in climate (Proulx et al. 2004). Although stone martens (*M. foina*) are not distributed as far north, they also experience subfreezing ambient temperatures (T_a) and snow cover in their habitat. *Martes* species are exposed to energetic stress during the cold season, which could be aggravated by their need to forage almost every day, even during extreme cold, heavy snowfalls, or strong winds (Herman and Fuller 1974; Buskirk et al. 1988). Although snow cover is an excellent thermal insulator, travel and hunting on snow increase locomotor costs (Harlow 1994; Wilbert et al. 2000), which are higher for fishers than for American martens, because the former sink deeper into soft snow due to their higher foot-loading (Buskirk and Powell 1994; Krohn et al. 2004; Renard et al. 2008).

In general, *Martes* species have slim body contours and high ratios of skin surface to body volume (Harlow 1994). The mass-specific basal metabolic rates for these species were assumed to be higher than predicted by allometric relationships; however, many studies of *Martes* species have shown that their metabolic rates can be close to those of other similar-sized mammals, if care is taken to determine the true resting metabolism. Metabolic rate measurements do not always use the same units or similar terminology, making it difficult to obtain a comprehensive overview of the subject, and standardization would be most welcome. Based on data presented by Iversen (1972), Worthen and Kilgore (1981) calculated the basal metabolic rate of European pine martens at 88.18 kcal/day—similar to the resting metabolic rate of 81.27 kcal/day they measured in American martens. According to Iversen (1972), the basal metabolic rate (kcal/day) of European pine martens and other mustelids weighing ≥1 kg was $84.6 \times W^{0.78}$, approximately 20% higher than would be expected from the mammalian standard curve. Heat production by fishers was estimated at 162 kcal/kg $W^{0.75}$ (Davison et al. 1978). The resting metabolic rate of American martens averaged 0.508 ml $O_2/g \times h$ (Buskirk et al. 1988), and the minimal metabolic rate 0.661 ml $O_2/g \times h$ (Worthen and Kilgore 1981). The value reported by Buskirk et al. (1988) was lower than that predicted by body weight (0.63) and those they calculated from the results of Iversen (1972) for mustelids weighing ≥1 kg. When the resting metabolic rate of Worthen and Kilgore (1981) was converted to kcal/day (81.27), it was 6% lower than predicted by calculations based on Iversen's (1972) results, whereas the minimal metabolic rate was 12% higher than based on equations. The field metabolic rate of American martens averaged 204 kcal/day, 40% above the predicted value (Gilbert et al. 2009).

Martes have relatively low thermal efficiencies (Harlow 1994), because their fur is not exceptionally insulative (Scholander et al. 1950) and because of the limited insulation provided by subcutaneous fat (Buskirk 1983). Thermal conductance was 0.0460–0.0527 ml O_2/g × h × °C for American martens, 18–58% higher than predicted for animals with similar body mass (Worthen and Kilgore 1981; Buskirk et al. 1988). During winter, T_a can be well below thermoneutrality for mustelids inhabiting northern coniferous forests (Harlow 1994). Worthen and Kilgore (1981) reported that the T_{lc}, that is, the lowest T_a at which an animal can maintain its body temperature (T_b) without increased metabolism, would be 29 °C for American martens. This is considerably higher than the value of 16 °C reported by Buskirk et al. (1988), which approximates the predicted T_{lc} of 20 °C for similar-sized mammals relatively well. Although T_{lc} could be slightly lower than predicted by body size, martens are morphologically not as well adapted to cold as some other ecologically similar northern mammals. The T_{lc} of fishers, on the other hand, was predicted to be as low as –30 °C for active males and –20 °C for females (Powell 1979a). If correct, this would mean that, unlike American martens, fishers would rarely experience T_a below T_{lc}, and would therefore have lower energetic costs for maintaining their T_b during the cold season.

Martens are considered lean compared with similar-sized mammals (Buskirk 1983; Figure 8.1). According to Harlow (1994), their limited fat reserves can be interpreted as a tradeoff between fasting tolerance and access to subnivean spaces, the latter of which is adaptive when resource predictability is high. Body fat percentages for *Martes* species have been measured previously using several different methods, including Soxhlet extraction, quantitative dissection, and isotopic dilution analysis. In addition, some of the measurements were conducted on skinned carcasses, causing potential loss of subcutaneous fat, and for these reasons, available data may not be directly comparable. Mean body fat percentages for wild American martens were 2.4–5.6% (Buskirk 1983; Buskirk and Harlow 1989; Robitaille and Cobb 2003). Results were similar in farmed animals (2.7%; Nieminen et al. 2006b) but significantly higher when estimated by isotopic dilution analysis for wild individuals (12.3–15.4%; Gilbert et al. 2009). According to Strickland and Douglas (1987), 59–63% of wild American marten males and 40–50% of females had visible fat depots in November–December. Subcutaneous fat was located mostly in the inguinal region, and intra-abdominal fat around the kidneys and in the omentum. Furthermore, there were no indications that the fat reserves of American martens would fluctuate seasonally (Buskirk and Harlow 1989; Gilbert et al. 2009). The average body fat percentage of farmed sables was 8.0% (Mustonen et al. 2006a).

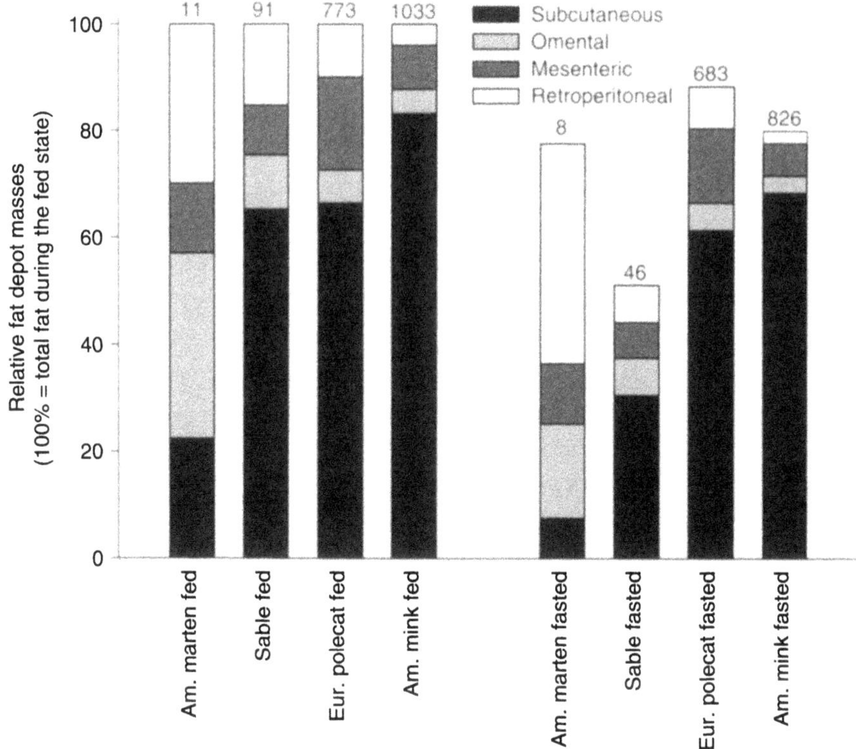

Figure 8.1. Fasting-induced changes in subcutaneous, omental, mesenteric, and retroperitoneal fat depots for the farmed American marten, sable, European polecat, and American mink; because of sexual dimorphism, only data for males are shown. The bars represent relative masses of different fat depots to each other before and after periods of fasting. The values of each species are indexed to make the initial pre-fasting value 100 (total fat during the fed state = 100%), and the figures above the bars represent the total fat mass (g) before and after the fasting experiments. Sources: American marten (fasted for 2 d, *n* = 4–5; Nieminen et al. 2007), sable (4 d, *n* = 8; Mustonen et al. 2006a), European polecat (5 d, *n* = 6; Mustonen et al. 2009), and American mink (5 d, *n* = 10; Mustonen et al. 2005a).

Garant and Crête (1999) suggested that fishers would gather more body fat reserves than American martens, and there are some indications that their fat reserves would increase in autumn (Arthur and Krohn 1991; Harlow 1994). The body fat percentage of fishers averaged 9.0–11.4%, with large variations: 1.5–31.7% for males and 2.6–27.3% for females (Garant and Crête 1999). Slightly lower average values (6.9–8.4%) were obtained by Robitaille and Jensen (2005). Subcutaneous fat accumulated over the rump, on the top of the back, in the inguinal region, and, in the fattest animals, over the fore and hind limbs (Garant and Crête 1999). Intra-abdominal fat was deposited mostly around the kidneys, in the mesentery, and in the lower abdomen. In more obese individuals, the kidneys could be covered with perirenal fat.

Fishers seem to have the physiological capacity to store larger amounts of adipose tissue than American martens, but in the wild, seasonal fluctuations in food availability probably prevent most animals from gaining substantial fat depots. Despite the use of different methods, researchers agree that, in general, *Martes* species have low body fat reserves, with the possible exception of the fisher.

Adaptations to Food Scarcity and Cold Stress during Winter

American martens, European pine martens, and fishers decrease their physical activity levels from summer to winter (Zielinski et al. 1983; Arthur and Krohn 1991; Zalewski 2000). For instance, European pine martens were active for 12 h/day in summer and 2.8 h/day in winter (Zalewski 2000), and the values for American martens were similar, ≤14 h/day in summer and <4 h/day in late winter (Thompson and Colgan 1994). The difference in activity levels between the 2 seasons was caused by the longer resting bouts during the cold season. As the duration of the resting periods increased, the number of activity bouts decreased with the declining T_a (Zalewski 2000). For American martens, T_a but not day length or snow depth influenced activity levels, and a substantial reduction in activity occurred at T_a below −15 °C (Thompson and Colgan 1994). As a result of decreased wintertime activity and field metabolic rate, energy expenditure and food requirements of American martens were estimated to decrease by 28% from fall to winter (Gilbert et al. 2009; J. Gilbert, Great Lakes Indian Fish and Wildlife Commission, personal communication). In sables, T_a ranging from −32 to −40 °C in November–December did not affect normal activity levels of animals inhabiting northeastern Yakutia, but in southern Siberia, physical activity levels decreased when T_a declined below −28 to −30 °C (Safronov and Anikin 2000). Surprisingly, the presence of ≥5 cm of new snow increased the likelihood of wintertime activity in fishers by 85%, but no relations between T_a and diurnal activity were evident (Weir and Corbould 2007).

Martes species can also decrease wintertime energy costs by switching to daytime activity—the period of the greatest insolation. This was documented at least for the American marten (Herman and Fuller 1974; Thompson and Colgan 1994), although contradictory data also exist for this species and for the European pine marten (Zielinski et al. 1983; Zalewski 2000). Hunting could be energetically more efficient in winter, because martens show a greater reliance on larger prey, which provide more energy per capture. Overwintering *Martes* species can also scavenge on ruminant carcasses (Hamilton and Cook 1955; Zalewski 2000) and cache food to be consumed later. Food hoarding was reported for the European pine marten (Nyholm 1961; Oksanen et al. 1985; Siivonen and Sulkava 1994), and there are indications of this type of behavior in the American marten (Murie 1961; Steventon and Major 1982; Zielinski et al. 1983), although it may be rarer in the latter species (Harlow 1994). *Martes* species can also save energy by using snow trails made by other animals and humans (Hamilton and Cook 1955; Buskirk and Powell 1994), and by using rest sites near their foraging areas (Buskirk 1984; Thompson and Colgan 1994). Female fishers and American martens can move on snow more easily than males because of their lower foot-loading (Krohn et al. 2004). Because female fishers give birth in early spring, it can be important for them to be able to move and hunt on snow with less effort. This could translate into selection pressure to reduce foot-loading in the sex with higher reproductive costs.

The subnivean zone is important to *Martes* species because it provides protected microenvironments with prey (Strickland and Douglas 1987; Corn and Raphael 1992; Buskirk and Powell 1994). Subnivean access structures—including live trees, snags, stumps, root masses, logs, log decks, rocks, and holes in snow—provide breaks in the snow that enable martens to access the subnivean zone. It is likely that subnivean spaces are used less frequently by fishers than by American martens (Raine 1987; Strickland and Douglas 1987; Buskirk and Powell 1994). Because snow is a barrier to wind and a good thermal insulator, subnivean air pockets form microenvironments with higher temperatures (T_{me}) compared with supranivean spaces (Buskirk et al. 1989; Taylor and Buskirk 1996). They have more favorable microclimates also as a result of the heat generated by the resting animal. American martens were observed to use subnivean resting sites more frequently when T_a was low and recent snowfall heavy (Buskirk et al. 1989; Wilbert et al. 2000). The preferred sites were often associated with coarse woody debris (Buskirk et al. 1988, 1989; Wilbert et al. 2000) and included, for example, snags, stumps, logs, rock fields, and red squirrel (*Tamiasciurus hudsonicus*) middens (Buskirk 1984; Buskirk et al. 1989). Resting episodes were longer in subnivean sites with coarse woody debris than in rock fields or above the snow (Buskirk et al. 1989). Generally, T_{me} in subnivean sites ranges from –2.5 to –0.5 °C (Buskirk et al. 1988, 1989)—higher than T_a but significantly lower than T_{lc}. Subnivean rest sites are probably not as important to fishers as to American martens (Buskirk and Powell 1994), and fishers were documented to use primarily live trees, snags, stumps, logs, rock crevices and ledges, and fox dens for rest sites (Hamilton and Cook 1955; Purcell et al. 2009). Fishers used rest structures with coarse woody debris, especially when T_a was colder (Weir et al. 2004). They may provide a more favorable microenvironment than branch or cavity structures. European pine martens rested in underground sites in rocky substrate and in arboreal structures (Brainerd et al. 1995). In this species, the use of underground resting sites was inversely correlated with T_a. In Yakutia, Siberia, sables were reported to use terrestrial burrows and spaces under stumps, crags, and root systems of deciduous trees (Safronov and Anikin 2000).

During the most severe weather conditions, *Martes* species rest for prolonged periods and hunt infrequently (Powell 1979a; Strickland and Douglas 1987). Overwintering American martens rested for 10–18 h on average, with a maximum of 55 h (Buskirk et al. 1988; Gilbert et al. 2009). The duration of resting depended on the cooling power of air during the activity period preceding the resting episode (Buskirk et al. 1988), and below −25 °C, martens could rest for >30 h (Thompson and Colgan 1994). The duration of resting periods increased also in overwintering European pine martens that were active for 13 h/day at 25 °C and for 2.5 h/day at −15 °C (Zalewski 2000). Buskirk et al. (1988) reported depressed T_b during the resting bouts of overwintering American martens; the average decrease was 2.9 °C below the active T_b, and the greatest decrease recorded was 5.1 °C. Such decreases in T_b were considered a state of deep sleep during which the thermal gradient between the body surface and the surrounding air decreased. The estimated daily energy saving from hypothermia of this degree was calculated to be 4.0–6.4%, which is relatively small but could be significant if resting bouts were longer than 12 h.

Seasonal Food Scarcity and Adaptations to Fasting

Among the seasonal cycles of food availability and T_a, winter is undoubtedly the most challenging period of the year. Animals employ several adaptive strategies (e.g., migration, food caching, passive wintering) that increase the probability of survival through the unproductive season. Many species prepare for wintering by gathering extensive adipose tissue reserves (e.g., bears [*Ursus* spp.], European and American badgers [*Meles meles* and *Taxidea taxus*, respectively], and raccoon dogs [*Nyctereutes procyonoides*], Mustonen et al. 2007b). Thus, they can rely on these energy stores and remain passive for long periods of time during winter with a reduced metabolic rate and T_b. For *Martes* species, body energy reserves are more limited and animals are unable to use passive wintering strategies or to migrate.

Although *Martes* species do not undergo winter sleep or hibernation, adaptations for fasting may help them cope with winter conditions. Fasting could occur during adverse weather conditions, which can make foraging energetically challenging. The ability to withstand food deprivation, while maintaining the capability for normal physical activity and foraging, may enable them to survive until conditions ameliorate. Periods of fasting for *Martes* species probably last up to a few days, which should be taken into account when examining their adaptations to food deprivation. The response to total fasting has been examined in order to clarify how these lean and active species cope with food deprivation in harsh conditions. The ability of smaller *Martes* species to accumulate fat reserves is less than that of many other carnivores (Buskirk 1983; Mustonen et al. 2006a; Figure 8.1). Thus, they cannot prolong wintertime survival by relying on fat reserves alone, which must influence their response to food deprivation. A limited number of studies are available on the fasting response of *Martes* species. American martens have been studied more intensively in this respect (Harlow and Buskirk 1991, 1996; Nieminen et al. 2006b, 2007), but some data are available for farmed sables as well (Mustonen et al. 2006a; Nieminen and Mustonen 2007). In these studies, periods of food deprivation lasted 1–5 days, similar to what *Martes* would be likely to experience in the wild; however, the study of Harlow and Buskirk (1991, 1996) also included water deprivation and was conducted indoors at 22 °C during February. Other experiments took place at T_a that averaged –9.9 to –2.1 °C during October–November/January, and the animals were given water or ice *ad libitum*. Differences in the experimental settings of these studies could have affected the fasting response of the animals.

The response to fasting can be divided into 3 phases (Le Maho et al. 1981; Castellini and Rea 1992). The short phase I is characterized by the depletion of body carbohydrate stores, especially liver glycogen. During phase II, lipids stored in adipose tissue are the main source of fuel. In phase III, the stimulated use of body proteins for energy eventually leads to death from starvation if food does not become available. During each phase of fasting, all these energy sources are used to some degree, but their relative importance varies. Fasting experiments can be logistically challenging and, without prior knowledge of the duration of the 3 phases, samplings of tissues are hard to pinpoint to the physiological shifts from one phase into another.

Because phase I of fasting is characterized primarily by glycogen degradation, the measurements of body carbohydrate stores and the key enzymes involved in their mobilization can be used to study the duration of this phase. The timeframe of effective glycogen utilization varies between species; for example, liver glycogen stores decline significantly within 4 h in arvicoline voles (*Microtus* spp.; Mustonen et al. 2008) and 1–2 days in domestic dogs (*Canis familiaris*; de Bruijne and de Koster 1983). In farmed sables, liver glycogen stores are depleted from the initial 5 µg/mg (overnight-fasted controls) to 1.5 µg/mg at 4 days of fasting (Mustonen et al. 2006a). It is likely that the depletion occurs earlier, as it was established in fasted American martens that their liver glycogen concentrations decreased from 7–8 to 2–3 µg/mg at 2 days (Nieminen et al. 2007). In both species, muscle glycogen stores decreased significantly. Because the samples in these studies were collected only after overnight fasting and at 2 or 4 days, it is not possible to determine when the depletion of carbohydrate stores became statistically significant. Thus, it is likely that the utilization of carbohydrates could be more rapid than observed after these fasting periods. Consequently, the duration of phase I of fasting in *Martes* species remains uncertain.

One of the main determinants of tolerance to fasting is the mass of fat reserves that can be used during phase II (Goodman et al. 1980). The use of

muscle tissue has a high water content and catabolism of lean body mass thus assists the maintenance of water balance. Such a strategy of energy use could also enable animals to extend the period of fat utilization and to maintain muscle integrity. Robitaille and Cobb (2003) suggested that during phase II of fasting, protein catabolism of martens would be at a higher level than in other similar-sized mammals, leading to the early induction of phase III. Combining the results of Harlow and Buskirk (1991, 1996), Mustonen et al. (2006a), and Nieminen et al. (2007), we hypothesize that after approximately 2 days without food, martens and sables would be in phase II of fasting and display fat mobilization, similar to the patterns found in several other mammals. During days 2–5, protein utilization would gradually become a part of the fasting response. It may be that, despite the early onset of fasting-induced proteolysis, martens can control it by utilizing urea recycling, enabling them to prolong survival (Harlow and Buskirk 1996).

Studying survival during fasting must also encompass the potential detrimental health effects caused by total food deprivation. Among pathological conditions, fasting-induced fatty liver can be observed in several mustelids, including American minks (Mustonen et al. 2005a; Rouvinen-Watt et al. 2010), European polecats (Mustonen et al. 2009; Nieminen et al. 2009), and sables (Mustonen et al. 2006a). Fasting stimulates the mobilization of fat from adipose tissue, and the increased free fatty acid influx into the liver induces lipid accumulation (Rouvinen-Watt et al. 2010). Liver damage can be determined, for example, by histological analyses, from liver fat percentages, and by measuring circulating transaminase activities. Based on these methods, fasted sables developed fatty liver after 4 days (Mustonen et al. 2006a), American minks after 2–7 days (Mustonen et al. 2005a; Rouvinen-Watt et al. 2010), and European polecats after 5 days (Nieminen et al. 2009). In American minks, fasting-induced fatty liver was reversible after 28 days of refeeding (Rouvinen-Watt et al. 2010), suggesting that if it were to occur in nature, it would not necessarily cause permanent damage to the health of an individual. Transaminase activities or other indicators of tissue damage did not increase after 2 days of fasting in American martens (Nieminen et al. 2007). In contrast, lower blood lymphocyte percentages could indicate slight immunosuppression, as was also observed in lymphocyte counts of American minks and European polecats (Mustonen et al. 2005a, 2009). The twice-longer fasting period (4 days) for sables induced increases in the plasma transaminase activities, and also caused lower plasma glucose concentrations and blood lymphocyte percentages (Mustonen et al. 2006a). It must be emphasized, however, that despite significantly decreased plasma glucose concentrations of sables, they and other mustelids do not exhibit plasma glucose values that would be low enough to prevent the animals from foraging when adverse conditions subside. In general, carnivores have relatively high circulating glucose concentrations during fasting and efficient *de novo* glucose synthesis (gluconeogenesis), which may be related to the high protein/carbohydrate ratios in their diets (Sørensen et al. 1995). In this respect, *Martes* species may comply with the common carnivorean pattern.

In summary, phase I of fasting would probably last <48 h in *Martes* species. After 2 days, phase II would clearly be ongoing, at least according to the data on American martens. Phase III would occur somewhere between 2 and 5 days. At present, this cannot be ascertained until the fasting responses in these species are studied at shorter time intervals. In addition to proteolysis, fasting can cause other detrimental health effects, such as fatty liver and immunosuppression.

Physiological Strategies of Energy Saving during Fasting

A significant part of the energy budget in homeothermal animals is used to maintain T_b (Vaughan et al. 2000); thus, reducing thermoregulatory costs can provide significant energy savings during winter. This is a strategy used by true hibernators and, to a lesser degree, by several other northern mammals (Smith et al. 1991; Nieminen and Mustonen 2008). In contrast, farmed European polecats have unchanged T_b during a 5-day wintertime fasting period (Mustonen et al. 2009), and American minks experience increased T_b and physical activity levels during summertime food deprivation, which may be interpreted to reflect an active foraging response to scarcity (Mustonen et al. 2006b). As stated previously, overwintering wild American martens had a lower T_b (≤5.1 °C) during resting periods with potential energy savings (≤6.4%; Buskirk et al. 1988). Although metabolic rates were not measured at the same time, we hypothesize that they would also decrease together with T_b. It is intriguing that the responses of captive mustelids to a relatively long period of food deprivation (5 days) differ significantly from the reduced T_b of American martens resting in the wild—an observation that merits further investigation.

The endocrine system plays a crucial role in the responses to food scarcity. Some endocrine signals participate in thermoregulation, including the thyroid hormones thyroxine and its more active form, triiodothyronine (Despopoulos and Silbernagl 1986). There is extensive evidence that reduced thyroid hormone concentrations are an integral part of the fasting responses for many mammalian species (Mustonen et al. 2009). This has been demonstrated for American martens (Nieminen et al. 2007), sables (Mustonen et al. 2006a), and other mustelids, including American minks and European polecats (Mustonen et al. 2005b, 2009; Figure 8.2). Potential benefits of decreased thyroid activity include energy conservation and limited proteolysis (Goldberg et al. 1980)—both useful strategies for enhancing survival.

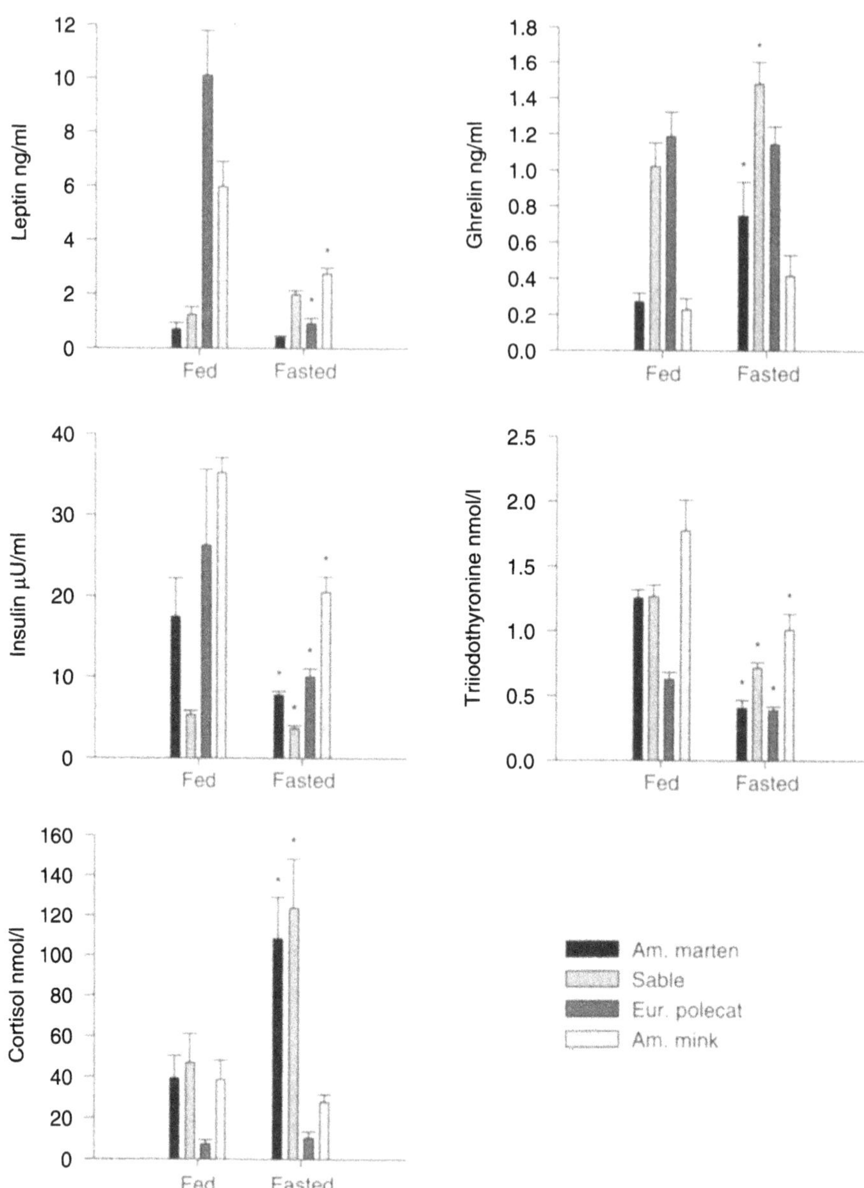

Figure 8.2. Effects of fasting on plasma endocrine variables (mean + standard error) for the farmed American marten, sable, European polecat, and American mink. * = Significant difference between fed and fasted animals ($P < 0.05$). Sources: American marten (fasted for 2 d, n = 7–8; Nieminen et al. 2007), sable (4 d, n = 8; Mustonen et al. 2006a), European polecat (5 d, n = 10; Mustonen et al. 2009), and American mink (5 d, n = 10; Mustonen et al. 2005b, 2006b, unpublished data).

Another well-known endocrine adaptation to fasting and energy restriction is the decrease in circulating insulin concentrations (de Bruijne et al. 1981; Mustonen et al. 2009). Lowered insulin levels lead to glucose sparing, gluconeogenesis, and ketogenesis, and promote fat mobilization (Cahill 1976; Kaloyianni and Freedland 1990). These phenomena contribute to the ability to function during fasting. Cortisol participates in the stress response, but may also enhance lipid mobilization (Divertie et al. 1991). Both sables and American martens display low insulin levels and increased cortisol concentrations during fasting, which could be related to these phenomena and enable adequate levels of proteolysis, producing substrates for gluconeogenesis (Mustonen et al. 2006a; Nieminen et al. 2007; Figure 8.2).

Leptin is an adipocyte-derived hormone (Zhang et al. 1994) interpreted to act as a satiety signal, informing the central nervous system about the amount of energy reserves present in the body. Plasma leptin concentrations in humans and rodents are usually positively correlated with body mass indexes (Maffei et al. 1995). During food deprivation, leptin levels decrease and initiate the neuroendocrine response to fasting, including decreased thyroid hormone and sex steroid concentrations, and increased stress hormone levels (Ahima et al. 1996). Unlike species without clear seasonal cycles in body mass and adiposity, some carnivores in northern latitudes show decoupling of plasma leptin concentrations from body adiposity (Nieminen et al. 2001, 2002). Raccoon dogs, for instance, experience relatively high leptin levels despite the decrease in fat mass that occurs during winter sleep, probably in order to enhance the use of lipid reserves during passivity (Nieminen et al. 2002). In contrast, the plasma leptin levels of American minks and European polecats are reduced during fasting (Mustonen et al. 2005b, 2009), but it is unclear why sables and American martens are in this respect similar to larger carnivores, with no significant changes in their circulating leptin concentrations (Mustonen et al. 2006a; Nieminen et al. 2007; Figure 8.2).

Adiponectin, while also secreted by white adipose tissue similar to leptin (Scherer et al. 1995), displays plasma concentrations inversely correlated to the amount of body fat (Matsubara et al. 2002). Concentrations decrease as a result of obesity and increase after significant weight loss (Swarbrick and Havel 2008), and adiponectin suppresses serum glucose levels by increasing insulin sensitivity (Berg et al. 2001). Adiponectin concentrations would be expected to increase in circulation during food deprivation. In American martens, however, there were no changes in plasma adiponectin levels after 2 days of fasting (Nieminen et al. 2007). This lack of change is similar to some other carnivores (Mustonen et al. 2005b, 2009), and it can be suggested that adiponectin participates mostly in the long-term regulation of body weight, whereas acute changes in nutrition have only minor effects on adiponectin production (Liu et al. 2003).

Ghrelin is a peptide hormone secreted principally by the stomach with increased plasma concentrations during fasting (Tschöp et al. 2000). Unlike leptin, elevated ghrelin concentrations have been shown to increase the appetite of laboratory animals; however, plasma ghrelin levels were unchanged during fasting in farmed American minks and European polecats with large initial fat reserves (Mustonen et al. 2005b, 2009). In contrast, increased plasma ghrelin concentrations were observed in food-deprived *Martes* species (Mustonen et al. 2006a; Nieminen et al. 2007; Figure 8.2). With small initial fat reserves, they would obviously benefit from an increased foraging response during fasting, which may be partly mediated by increased plasma ghrelin levels, which participate in the induction of hunger.

In summary, endocrine responses to fasting in the *Martes* species that have been studied are similar in many respects to those of most mammals. These responses include, for example, decreased plasma insulin and elevated cortisol concentrations that probably promote lipolysis and protein catabolism, increased ghrelin levels that may stimulate appetite, and suppressed thyroid activity that could be associated with a slight metabolic depression that helps to conserve energy.

Fatty Acid Composition of Fat Tissue

Adipose tissue, which was once considered a relatively inert site of energy storage, is actually an active participant in various physiological processes, including endocrine signaling (Zhang et al. 1994; Scherer et al. 1995), selective incorporation of lipids for storage, and their selective hydrolysis (Mustonen et al. 2007a). It also provides substrates for specific fatty acid (FA) derivatives (eicosanoids) involved, for example, in the inflammatory response (Ackman and Cunnane 1992). Lipids are generally stored in body fat depots as triacylglycerols (TAGs) containing a glycerol backbone and 3 FA chains (Mead et al. 1986). Structural membrane lipids contain different phospholipids (PLs) and cholesterol. FAs are classified as saturated FAs (SFAs), monounsaturated, and polyunsaturated FAs (MUFAs and PUFAs) with 0, 1, or ≥2 double bonds in the carbon chains, respectively.

The nomenclature of FAs used in this review is based on carbon chain length and the number and location of double bonds as follows: [number of carbons in the FA chain]:[number of double bonds]n-[position of the first double bond from the methyl end of the FA molecule]. Thus, arachidonic acid is abbreviated 20:4n-6. In addition, the most abundant and biologically important FAs include 16:0 (palmitic acid), 18:0 (stearic acid), 18:1n-9 (oleic acid), 18:2n-6 (linoleic acid), 18:3n-3 (α-linolenic acid), 20:5n-3 (eicosapentaenoic acid), and 22:6n-3 (docosahexaenoic acid). Of these, 18:3n-3 and 18:2n-6 are considered dietarily essential FAs for mammals, because they cannot synthesize these PUFAs *de novo* (Ackman and Cunnane 1992). These two PUFAs act as precursors for longer-chain PUFAs and subsequently for other biological molecules. Certain n-3 and n-6 PUFAs (also called ω-3 and ω-6 PUFAs) can be processed into eicosanoids (prostaglandins, prostacyclins, thromboxanes, and leukotrienes) that mediate, for example, inflammation. Eicosanoids derived from n-6 PUFAs via 20:4n-6 are considered more inflammatory than those from n-3 PUFAs (20:5n-3), and their overrepresentation can also cause increased blood clotting and vasoconstriction (Rouvinen-Watt et al. 2010). For this reason, the n-3/n-6 PUFA ratio in diets and, subsequently, in tissues, is an important indicator of the balance between potentially beneficial and detrimental effects of PUFAs. A high proportion of n-3 PUFAs is surmised to have health benefits, whereas increased percentages of n-6 PUFAs are pro-inflammatory and predispose the individual, for example, to thrombotic diseases (El-Badry et al. 2007).

The acral parts of the body are more easily exposed to cold temperatures than the trunk, and the outer layers of the trunk close to the skin are more easily exposed than the body core. This is emphasized at high latitudes with cold winters and in habitats where animals can be exposed to cold (e.g., during aquatic foraging). Generally, an increase in the number of double bonds lowers the melting point of FAs, and membrane lipids containing FAs with low melting points help to maintain cell membrane fluidity in cold conditions (Nieminen et al. 2006a,b). The double bond index (Mustonen and Nieminen 2006), that is, the mean number of double bonds per FA molecule, can be used as an indicator of the fluidity of fats. In addition, the Δ9-desaturation index, the ratio of the potentially endogenous Δ9-MUFAs to the corresponding SFAs, serves as an indicator of lipid unsaturation in tissues. If lipids solidify because of cold (e.g., in icy water), the functions of cell membranes can be seriously jeopardized, and it is therefore crucial to maintain the structural integrity of membrane fats in the body parts exposed to cold by incorporating unsaturated FAs into cell membranes (Nieminen et al. 2006a,b). The unsaturation of lipids has been shown to increase toward peripheral parts of the body in several mammalian species that inhabit not only arctic and temperate regions, but also tropical areas (Irving et al. 1957).

For *Martes* species, data are available on lipids in energy storage and mobilization, as well as on the anatomical distribution of FAs and their selective enrichment in various body parts. The composition of FAs in, and their mobilization from, anatomically different fat depots were examined in farm-raised American martens (Nieminen et al. 2006b) and sables (Nieminen and Mustonen 2007). Because the general FA profile of an individual animal depends on the composition of FAs in the diet (Bradshaw et al. 2003), the proportions of individual FAs in fat tissues should be compared with the corresponding percentages in the diet, which is feasible in controlled experiments, but less so in the field. In farmed American martens and sables, the FAs with the highest proportions in subcutaneous and intra-abdominal trunk fats were quite similar to those observed in farmed American minks (Mustonen and Nieminen 2006; Nieminen et al. 2006a,b), whereas *Martes* species had higher percentages of SFAs but lower MUFA contents in their trunk fats than farmed European polecats (Nieminen et al. 2009; Figure 8.3). In *Martes* species, SFAs represented approximately 35% of total FAs, and MUFAs (mainly oleic acid), 42–45% of total FAs. The proportions of n-3 PUFAs were higher in American martens (>5%) and sables (3%) than in minks (≤2%; Mustonen and Nieminen 2006; Nieminen et al. 2006b; Mustonen et al. 2007c). In wild stone martens, the situation was different, with 55% of total FAs being SFAs, while the n-3 PUFA sum was about 2.5%, which is fairly similar to those in other studied mustelids (Koussoroplis et al. 2008). Because of dietary differences in these experiments, it is not yet possible to determine whether variations in the FA composition were caused by taxonomic differences, because the FA profiles in diets can also play a significant role.

Sables showed higher percentages of MUFAs in their tissues than in the dietary profile (Mustonen and Nieminen 2006). The same was also found for American minks and European polecats (Mustonen et al. 2007c; Nieminen et al. 2009). This increase in MUFAs was probably caused by the conversion of dietary SFAs into MUFAs. In contrast, the percentages of PUFAs were lower in trunk fats than in the diet. At present, data are not yet available about the incorporation of FAs from the diet into the tissues of American martens; however, if FA signatures of wild *Martes* species are used in future studies to evaluate dietary histories, metabolic modifications of dietary FAs (Iverson et al. 1997; Kirsch et al. 2000; Grahl-Nielsen et al. 2003) will have to be taken into account.

Fasting and Fatty Acid Profiles

During negative energy balance, adipose tissue TAGs are used to meet energetic requirements, as described above. Despite low body fat reserves, *Martes* species can survive food deprivation for moderate periods of time (at least up to 5 days; Harlow and Buskirk 1991, 1996), and they show active use of body lipids during 2–4-day fasting periods (Mustonen et al. 2006a; Nieminen et al. 2007). Generally, FA mobilization is not a random but a selective process in laboratory animals (Raclot and Groscolas 1995; Raclot et al. 1995). FAs are mobilized more effectively when they are short, unsaturated, and their double bonds are close to the terminal methyl group of the FA chain. In addition, the anatomical location of fat depots may also influence FA mobilization, as has been observed in American minks (Nieminen et al. 2006a). The principles of selective FA mobilization were mostly valid also in the studied *Martes* species (Nieminen et al. 2006b; Nieminen and Mustonen 2007). In fasted American martens and sables, the mobilization of SFAs and particular MUFAs was the highest for shorter-chain FAs, similar to European polecats and American

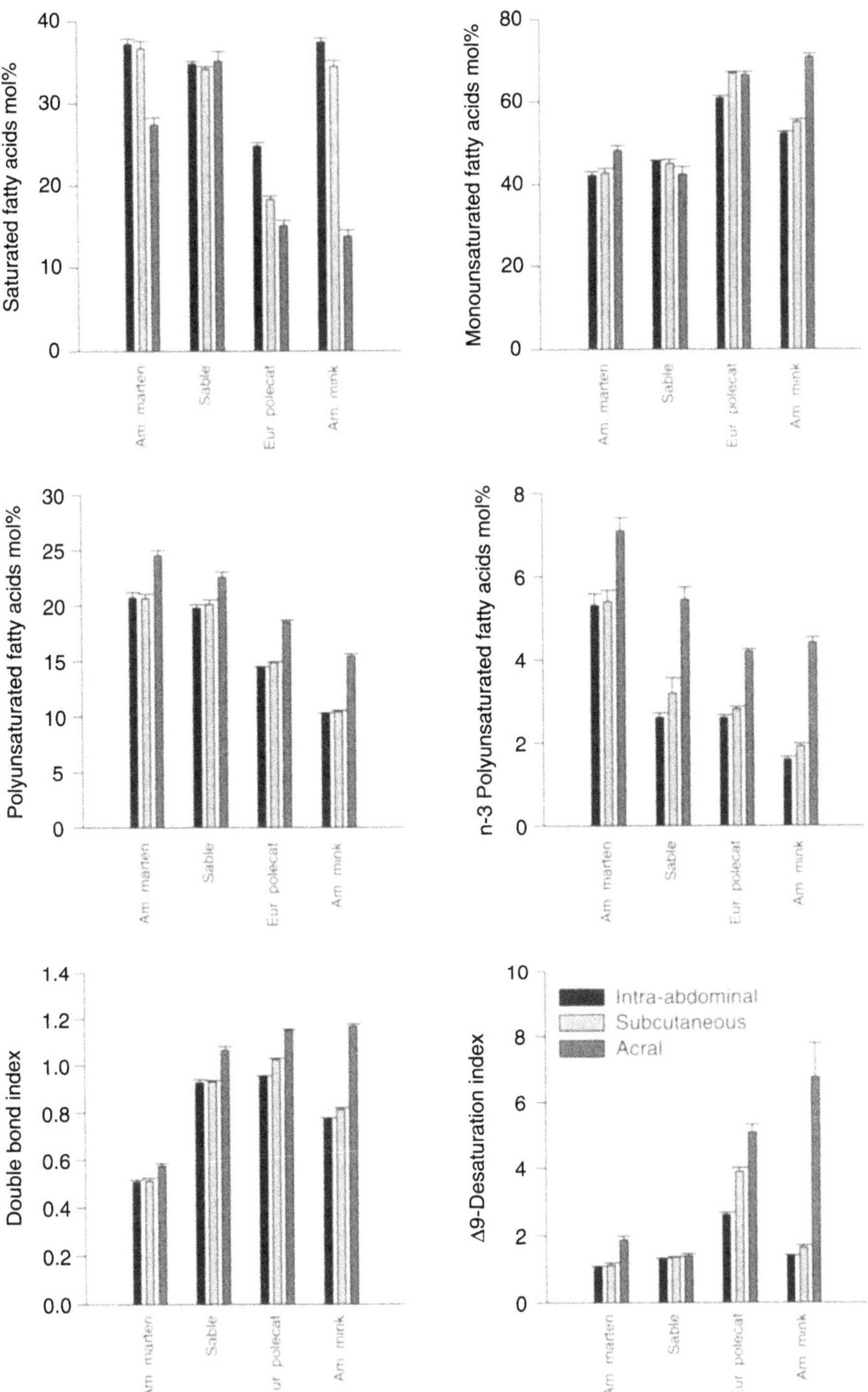

Figure 8.3. Fatty acid sums and unsaturation indexes (mean + standard error) in anatomically different fat tissues (intra-abdominal, subcutaneous, and acral fats including paw pads and tail tip) for the farmed American marten, sable, European polecat, and American mink. Sources: American marten (n = 7; Nieminen et al. 2006b), sable (n = 8; Mustonen and Nieminen 2006), European polecat (n = 10; Nieminen et al. 2009), and American mink (n = 8; Mustonen et al. 2007c).

minks (Nieminen et al. 2006a, 2009). In addition, desaturation of SFAs into the corresponding MUFAs enhanced the FA mobilization in both American martens and sables (Nieminen et al. 2006b; Nieminen and Mustonen 2007). Furthermore, the mobilization of n-3 PUFAs was higher than that of n-6 PUFAs in sables (Nieminen and Mustonen 2007), similar to minks and polecats (Nieminen et al. 2009; Rouvinen-Watt et al. 2010). This led to a decrease in n-3/n-6 PUFA ratios. In contrast, after 2 days of fasting, American martens showed no decreases in n-3/n-6 PUFA ratios, probably because of the shorter fasting period (Nieminen et al. 2006b). The FA composition of appendages was very resistant to fasting.

The greater loss of n-3 PUFAs during fasting could be caused by the principle of competitive inhibition: n-3 substrates are preferred over n-6 ones in desaturation reactions (Ackman and Cunnane 1992). In addition, 18:3n-3 is one of the principal substrates for β-oxidation and it can also be metabolized into longer-chain n-3 PUFAs to be incorporated into membrane PLs or to be modified into eicosanoids, further augmenting the depletion of n-3 PUFAs from adipose tissues (Nieminen et al. 2009). The relative decrease in n-3 PUFAs can have significant effects on the health of animals. In mustelids, fasting-induced decreases in the proportions of hepatic n-3 PUFAs could predispose the animal to the development of fatty liver (Nieminen et al. 2006a, 2009; Rouvinen-Watt et al. 2010), a health hazard commonly associated with both obesity and starvation (Adams et al. 2005). Depletion of n-3 PUFAs can induce fatty liver by favoring FA and TAG synthesis over FA oxidation, and by decreasing lipid export from the liver by suppressing the formation of very-low-density lipoproteins (Videla et al. 2004). It is surprising that similar decreases in hepatic n-3 PUFA percentages were not observed in fasted American martens or sables (Nieminen et al. 2006b; Nieminen and Mustonen 2007); however, after 4 days of fasting, plasma transaminase values and liver TAG concentrations increased in sables, indicating liver damage (Mustonen et al. 2006a). In American martens, there were signs of increased hepatic TAG levels after 2 days, but the difference was not significant (Nieminen et al. 2007). Fasting could cause similar health effects in *Martes* species, but the periods of food deprivation in previous studies may have been too short for such effects to be manifested.

Ecophysiological Implications of Fatty Acid Profiles in Different Body Parts

Although the semiaquatic American mink displays anatomical differences in its FA mobilization—PUFAs are more readily mobilized from the body core than from the subcutaneous fat layer (Nieminen et al. 2006a)—the American marten and sable show no such preference between subcutaneous and inner fats and no clear differences in the FA composition between these 2 fat depots

(Nieminen et al. 2006b; Nieminen and Mustonen 2007). Because *Martes* species are terrestrial, the preservation of fluidity in subcutaneous fats may not be as crucial for them as it is for the more aquatic mink. Thus, the natural history of a species should be considered when interpreting the site-specific FA composition and its modifications caused by food deprivation.

The acral body parts of sables and American martens are more enriched in PUFAs, especially n-3 PUFAs, than the trunk fats (Mustonen and Nieminen 2006; Nieminen et al. 2006b; Figure 8.3). However, the appendages of sables have equal proportions of SFAs as the trunk, whereas American martens have an intermediate position, and American minks display the other extreme, with significantly higher proportions of MUFAs but lower percentages of SFAs in their paws and tail tip compared with the trunk, in addition to highly elevated n-3 PUFA proportions (Mustonen et al. 2007c). Furthermore, the double bond index, while slightly higher in the appendages of *Martes* compared with the trunk, shows more obvious elevation in the acral body parts of minks. Again, the aquatic foraging activity of minks is the most plausible explanation for this pattern. Fats in acral parts of the body have generally little, if any, significance in energy storage, and in many cases, these fats are mostly structural lipids in cell membranes, the integrity of which must be preserved during cold exposure by utilizing FAs capable of maintaining fluidity also in near-freezing water, that is, unsaturated FAs (Nieminen et al. 2006b). However, European polecats also display lower proportions of SFAs and more PUFAs in their acral fats than in the trunk, even though they are mostly terrestrial (Nieminen et al. 2009)—an enigmatic finding that needs to be studied in more detail. A special case in the peripheral FA composition of these species is the nasal lipid profile with high proportions of, for example, 22:6n-3, as has been documented for sables (Mustonen and Nieminen 2006), European polecats (Nieminen et al. 2009), and American minks (Mustonen et al. 2007c). One plausible explanation for this pattern could be related to olfactory requirements (Russell et al. 1989). High proportions of PLs—membrane lipids rich in PUFAs—are crucial for the sense of smell, which is critical for hunting.

Previously, Irving et al. (1957) hypothesized that the unsaturation of FAs toward the periphery (as found in *Martes* species, and even more clearly in American minks) would be genetically determined and not an adaptation to cold, because similar gradients were also observed in mammals from temperate and tropical regions. The differences in the unsaturation gradient (for *Martes* species, polecats, and minks) and the anatomical differences in FA mobilization (for minks), however, suggest that these phenomena could also have adaptive value. Such adaptations would not necessarily be confined to arctic and boreal regions, because nighttime T_a can be quite low also in tropical regions, T_a may decrease during the rainy season, and animals can be exposed to water significantly lower in temperature than the air.

Implications of Ecophysiology for the Conservation of *Martes* Species

Many studies indicate that *Martes* species can have higher metabolic rates than other mammals of similar size (Iversen 1972; Gilbert et al. 2009) and that their body fat reserves are limited, but can be mobilized effectively during negative energy balance (Buskirk 1983; Mustonen et al. 2006a; Nieminen et al. 2006b). From a conservation perspective, however, we need to know whether the physiology of *Martes* could make them more susceptible to environmental changes, including habitat destruction, xenobiotic exposure, and climate change. Because the metabolic rates of *Martes* species can be considered high, especially during hunting and cold T_a, habitat destruction that forces animals to allocate more resources to locomotion, foraging, and/or thermoregulation can have potentially adverse effects on body condition and, eventually, also on populations.

Structurally complex forests containing stands of diverse ages can provide martens with the best availability of subnivean travel routes and rest sites (Hargis and McCullough 1984), which would mean a reduced risk of extensive energy expenditure. Such forests may also provide *Martes* species with high-quality hunting sites and high prey diversity, making food deprivation less likely. Logging leads to enormous changes in forest habitats for *Martes* (Strickland and Douglas 1987) and can indirectly cause energetic challenges. American martens benefit from red squirrel middens, which are often located in old-growth stands of white spruce on south-facing slopes (Buskirk 1984). The warmer microclimates and additional insolation on south-facing slopes in late winter probably make these shelters attractive to martens. Consequently, reductions in the availability of resting sites could affect their energy balance in a deleterious manner, which should be taken into account when planning future economic use of these forest types.

In conclusion, winter provides northern *Martes* species with energetic challenges because they do not have exceptionally insulative fur, large body fat reserves, or the ability to utilize a passive wintering strategy with a significantly reduced metabolic rate. These species survive winter conditions, for example, by decreasing their physical activity, switching to diurnal foraging, hunting and resting in the subnivean zone, and fasting for 1–5 days during adverse weather conditions, during which they effectively utilize their limited adipose tissue reserves along with body proteins. These adaptations are essential for the survival of American martens (and possibly other *Martes* species) during the winter months.

Acknowledgments

The authors thank Dr. Gilbert Proulx and 2 anonymous referees for valuable comments on the manuscript.

SECTION 3

Ecology and Management of Habitat for *Martes* Species

SECTION 3

Ecology and Management of Habitat for *Martes* Species

9

Improved Insights into Use of Habitat by American Martens

IAN D. THOMPSON, JOHN FRYXELL, AND DANIEL J. HARRISON

ABSTRACT

We reviewed habitat selection by American martens (*Martes americana*) across North America based on information published from 1993 to 2010. Habitat use by the species is variable across the continent, with populations occupying a range of forest ages and types, making generalities regarding habitat preferences difficult to infer at the stand level. Clear differences exist between habitat selection in boreal and montane forests and selection in midcontinent and eastern transitional mixed forests. Most models indicate that habitat selection occurs at 3 spatial scales: landscapes (i.e., home range scale), stands within home ranges (i.e., stand scale), and individual structures at the scale of specific sites within stands (within-stand or site scale). Overall, the strongest patterns of selection occur at the landscape scale, suggesting a strong connection between home range composition and individual fitness. Although American martens are not restricted to mature and old mixed-wood and conifer forests, most studies have suggested that these forest types receive the highest relative use. Use of managed landscapes by American martens is common where sufficient cover and structures important to fitness are present and where the species can exhibit landscape-scale selection to maintain 70% or more of home ranges in suitable habitat conditions. Habitat selection may be a function of availability but may also be influenced by density dependence, mortality from commercial trapping and predation, the capacity for dispersal, and the availability of untrapped reserves as source populations. Few studies have tested multiple hypotheses of the mechanisms affecting habitat choice. Further, since most dispersing animals are juveniles, observed patterns of habitat selection by unmarked American martens could represent use of poor habitats by subordinate or inexperienced individuals. Hence, we argue that individual fitness, rather than simply the presence of the species, should be used to evaluate habitat selection and quality.

Introduction

American martens (*Martes americana*) live primarily in boreal forests, but their distribution extends south into some temperate and montane forest areas, including as far south as central California and northern New Mexico. Like all predominantly boreal forest species, American martens must have become adapted to the unpredictability resulting from various disturbance regimes, because, over ecological time, fire, blowdown, and insect outbreaks alter forest structure and age composition across the landscapes they occupy (e.g., Fisher and Wilkinson 2005; Drapeau et al. 2009). At any point in time, the boreal forest landscape presents animals with a range of habitat types and forest ages as a matrix at large spatial scales. Selection from among these habitats is ultimately related to the fitness of individual animals as mediated through a variety of choices made and factors acting at those various scales (Morris 1987; Levin 1992). Choice of habitats affects the capacity of individuals to acquire resources while minimizing energy expended and avoiding predation to gain those resources (Charnov 1976; Krebs 1978). Habitat selection represents a suite of choices at several scales, each of which may influence individual fitness (e.g., Johnson 1980; Morris 2002). Understanding the scale at which key habitat choices are made, and the mechanisms behind the choices, improves our capacity to predict how animals may react to management alternatives (e.g., Buskirk and Ruggiero 1994; Ciarniello et al. 2007; Oatway and Morris 2007).

Habitat selection has several key components: first, animals might be expected to seek out high-quality patches, and, second, among these patches, variance in the availability of key resources (food, cover) should be reduced (Mayor et al. 2007); both will affect fitness, and selection at the larger scale may constrain selection at smaller scales. Furthermore, habitat selection could be influenced by the density of conspecifics (Fretwell and Lucas 1969; Morris 1987; Oatway and Morris 2007). At low densities, animals may readily find high-quality patches, but as density rises, fewer such patches are available and mean fitness declines. Greene and Stamps (2001) have argued that Allee effects may be observed at low densities such that, up to an inflection point, fitness may increase as density rises before becoming negatively density dependent.

Here we review information on habitat selection by American martens from studies across North America from 1992 to early 2010, and following from summaries by Buskirk and Ruggiero (1994), Buskirk and Powell (1994), and Thompson and Harestad (1994). Since these reviews were published, the focus of marten research has changed in 4 key respects. First, most of the work after 1992 has focused on habitat selection across a landscape matrix of forest types and ages, rather than on comparisons between old forests and younger regenerating forests. Such heterogeneous landscapes are common in most jurisdictions, where a matrix of managed and unmanaged forest stands of various ages and types extends over large spatial scales. Second, researchers have increasingly recognized that American martens make habitat choices at multiple spatial scales. This understanding began with work by John Bissonette and Daniel Harrison and their respective students (e.g., Bissonette et al. 1989; Bissonette et al. 1997; Chapin et al. 1997a, 1998). Although much of the earlier research on habitat use by martens (e.g., Soutiere 1979; Bateman 1986; Snyder and Bissonette 1987; Thompson and Colgan 1987) did not necessarily ignore the landscape scale, the main focus was clearly on habitat selection at the stand scale and, specifically, on the relative selection of recent clear-cuts versus residual forest stands within the home range. Third, improved statistical models became widely available after about 1990, at about the same time as an understanding that pseudoreplication from telemetry data could influence conclusions about habitat use. Hence, studies after about 1990 often employed superior statistical modeling and reduced sampling error. Finally, many recent studies have focused specifically on selection of individual within-stand habitat features, such as den sites, resting sites, and cover attributes. We have structured our review around 3 of Johnson's (1980) 4 orders of habitat selection, although we prefer to use the term *scale*. There are strong linkages across these scales (or orders); thus, we review landscape (second order) and stand-scale (third order) selection together, followed by within-stand or site-level selection (fourth order).

We reviewed 11 published papers that provided sufficient information to derive a use/availability index (w_i). We also report on >30 published articles that either examined marten habitat selection or provided specific information about habitat choices at the stand or site scale (e.g., den sites). We have attempted to report results only for resident adult martens, given the possibility that juveniles may be observed often in low-quality habitats (Thompson and Colgan 1987; Ruggiero et al. 1988; Buskirk and Ruggiero 1994). When necessary, we have referenced older studies as published support for various aspects of marten ecology.

Buskirk and Powell (1994) and Thompson and Harestad (1994) proposed several hypotheses for mechanisms that may influence marten habitat selection, including predator avoidance, presence of key site-scale habitat features (structural complexity: coarse woody debris, cover) that increase the quality of the habitat, and abundance of preferred prey. Here, we examine support from studies conducted during the past 18 years for the proposed hypotheses for habitat selection, and argue for the need to improve our understanding of the role that density dependence might play in observed patterns.

Some Forestry Definitions

We refer to forest ages as *young regenerating*, which, for any forest type, is the shrubby and young-tree stage that follows a major disturbance; *regenerating*, which we consider as forests where the tree component (not shrubs) dominates, often at high stem density, and where rapid growth in height is occurring; *mature* refers to forests where the dominant tree cohort has attained close to maximum height and stem density is often lower than in regenerating stands; and *old* or *old-growth* as a stage where some of the dominant tree cohort have died, snags and canopy gaps are common, and young trees have begun filling in gaps. Old forests are often referred to as *overmature* in forest planning. The age of the trees at each stage depends on the particular forest ecosystem. We defined *mixed woods* as forest stands with >25 and <75% conifer (or deciduous) trees, otherwise they are *conifer* or *deciduous*. In Great Lakes-St. Lawrence forests and Acadian forests, a transition zone between the boreal and southern deciduous zones, managed stands may be maintained in the mature or old age classes through selection harvesting (see below).

Forests are *managed* if trees are harvested commercially or pre-commercially. *Unmanaged* forests are those originating as a result of natural disturbances (fire, wind, insects, ice storms). *Clear-cut harvesting* refers to the even-aged silvicultural method of removing most or all trees of the dominant cohort of trees at the same time; this method is common in boreal and some Acadian forest types. *Selection harvesting* refers to an uneven-aged silvicultural technique whereby certain trees are selected for harvesting from the same stands over long periods of time. The latter technique is a common silvicultural practice in most Great Lakes-St. Lawrence forests. *Patch-cutting* refers to clear-cut harvesting where the size of the stands removed is restricted to small patches, for example <50 ha, with buffer areas of mature forest between and surrounding the patches. *Pre-commercial thinning* refers to the removal of stems to a specified basal area during regeneration or at an early mature stage, to improve stem growth of commercial tree species.

Habitat Selection at Landscape and Stand Scales

Forest landscapes can take various forms, but 2 broad types are reported in the literature: managed/unmanaged and multi-aged managed mosaic. In most areas of the boreal forest, progressive clear-cutting leaves large areas, often several thousand km^2, with a range of forest ages but containing little mature or old-growth forest, except in small, remnant patches (e.g., along rivers). Larger areas of older forest are usually found toward the ends of the roads (i.e., at the edges of the managed landscape) where harvesting has yet to occur, or in protected areas (e.g., Thompson and Colgan 1994; Gosse et al. 2005; Johnson et al. 2009). However, in many managed temperate and montane forests, and in some boreal forests where selection harvesting or patch-cutting has occurred, forest landscapes contain a matrix of forest types and ages (including old-growth) (e.g., Hargis et al. 1999; Payer and Harrison 2000; Potvin et al. 2000; Fuller and Harrison 2005). These 2 types of landscapes present different habitat-selection problems for dispersing martens. In the progressive clear-cut landscapes, martens must often make a choice between living in a regenerating forest, or dispersing to try to find a mature or old forest landscape. In the mosaic landscapes, martens have the opportunity to occupy a home range that contains some mature or old forest and some regenerating stands.

Results presented by Thompson and Harestad (1994: 361), based on studies conducted prior to 1992, were largely confirmed by more recent research (Table 9.1, Figure 9.1) that showed an increase in relative use with forest age. Nevertheless, there are many important exceptions, including studies in Maine (Payer and Harrison 2003), British Columbia (Poole et al. 2004), Alaska (Paragi et al. 1996), Quebec (Potvin et al. 2000), and Newfoundland (Hearn et al. 2010). In several of these newer studies, marten density was similar in pole-sized or older regenerating, mature, and old forests. For example, Hearn et al. (2010) found no apparent difference in selection between regenerating conifer and old-conifer habitats with a closed canopy, although martens did select for old insect-killed stands and open old-conifer forests. Overall, mature and old forests were selected preferentially by martens in 9 of 11 studies reviewed.

We also summarized results from 4 studies of marten preference among 3 broad forest types: conifer, mixed woods, and deciduous. This required recalibration of 1 study (Potvin et al. 2000) that had used a more liberal definition (<50% conifer) than we used for deciduous forest. Martens generally preferred mixed-wood forests, and conifer forests to a lesser extent, whereas use of deciduous forests was variable (Figure 9.1). Preference for mixed-wood forests is not surprising, given the generally higher primary productivity of mixed-wood forests compared with most conifer forests, which is related to favorable site conditions (e.g., Sims et al. 1989; Longpré et al. 1994), and to the greater amounts of structure in mixed-wood forests compared with more monotypic stands (e.g., Fuller et al. 2004; Payer and Harrison 2004; Fuller and Harrison 2005). Higher primary productivity can be directly related to prey populations, which have been shown to have a direct effect on marten abundance (Thompson and Colgan 1987; Fryxell et al. 1999). Mixed-wood forests often maintain higher small mammal populations than most coniferous forests (e.g., Orrlock et al. 2000; Fuller et al. 2004). Further, at least in Maine, partial harvesting did not reduce small mammal prey densities (Fuller et al. 2004). In abundant seed years, however, populations of small mammals in conifer forests may also be high (e.g., Fryxell et al. 1998; Elias et al. 2006), especially in older forests (e.g., Thompson and Colgan 1987; Rosenburg et al. 1994).

In western montane populations of American martens, the species appears to prefer mature and old conifer forests (e.g., Buskirk and Ruggiero 1994; Bull et al. 2005; Mowat 2006; Slauson et al. 2007), whereas in eastern North America, especially toward the southern limit of the species' distribution, a

Table 9.1. Studies that reported habitat selection by American martens based on comparisons of use vs. availability (w_i) in 3 forest types and 4 forest developmental stages (see text) following disturbances

Author	Location	Scale[a]	Season	Marten density	Marten status[b]	n[c]	Technique	Forest types preferred	Shrub/ sapling	Immature/ pole	Mature	Old
Potvin et al. 2000[d]	Quebec	Landscape	Winter	Unk[e]	Ad and J with HR[f]	20	Telemetry	Mixed woods	0.6	(1.3)	0.9	—
Potvin et al. 2000	Quebec	Stand	Winter	Unk	Ad M with HR	20	Telemetry	Mixed woods	0.9	1.2	(1.7)	—
Poole et al. 2004	British Columbia	Stand	All	Low	Ad and J with HR	31	Telemetry	Mature conifer	0.9	1.2	(1.7)	—
Gosse et al. 2005	Newfound- land	Stand	All	Low	Ad with HR	23	Telemetry	Mature conifer	0.6	0.3	(1.8)	—
Fuller and Harrison 2005	Maine	Stand	All	High	Ad with HR	18	Telemetry	None, avoided young stands	0.25	—	(1.0)	—
Bull et al. 2005	Oregon	Landscape	All	Unk	Ad	20	Telemetry	Mature and old conifer	0.08	0.4	1.7	(1.9)
Proulx 2006b	Alberta	Stand	Winter	Unk	Unk	44 tracks	Track transects	Mature conifer and mixed wood	0.0	1.0	(1.7)	—
Mowat 2006	British Columbia	Landscape	Winter	Unk	Unk	177 hairs	Hair snags	No preference	0.9	0.8	1.1	(1.3)
Dumyahn et al. 2007	Wisconsin	Landscape	Winter	Low	Unk age with HR	13	Telemetry	Deciduous	0.5	0.5	(1.3)	—
Slauson et al. 2007[d]	California	Stand	Summer/ fall	Low	Unk	26 tracks	Track plate	Old conifer[g]	0.0	0.0	0.5	(3.2)
Godbout and Ouellet 2008	Quebec	Landscape	All	Mod.	Unk age with HR	15	Telemetry	Mature conifer	0.5	0.5	(1.7)	—

Table 9.1—*cont.*

Author	Location	Scalea	Season	Marten density	Marten statusb	n^c	Technique	Forest types preferred	Shrub/ sapling	Immature/ pole	Mature	Old
Hearn et al. 2010	Newfound- land	Landscape	All	Low	Ad and J with HR	58	Telemetry	Open old conifer	1.2[h]	1.9[i]	0.8	1.1[j]
Hearn et al. 2010	Newfound- land	Stand	All	Low	Ad and J with HR	58	Telemetry	Insect-killed conifer with understory	0.75	1.0[k]	0.9	1.0[j]

Note: Values in parentheses are the maximum w_i for each study

[a] Scale at which use and availability data were provided (stand = within home range scale)

[b] Ad = adult, J = juvenile; M = male only; with home ranges (HR) only if specified

[c] n = sample size of martens with home ranges, unless noted, for all forest types

[d] Estimated w_i only

[e] Unk = unknown

[f] HR = home range

[g] Data from "non-serpentine stands"; sample from "serpentine stands" was very low

[h] Inferred from nonsignificant selection for recent cuts ($P = 0.7$) based on 19 martens whose use exceeded availability and 16 martens whose use was less than availability for this type

[i] Inferred from positive selection ($P = 0.006$) for regenerating cuts with dense conifer regeneration >3.5 m in height as indicated by 55 martens with positive selection for this type and 29 martens with negative selection for this type

[j] Mean for tall open conifer (1.3) and tall closed conifer (0.9), for landscape scale; and 1.0 and 1.1, respectively, for stand scale; highest stand selection index was 1.5 for insect-killed old conifer with a regenerating understory

[k] Combined pole-sized and young regenerating forest; pole-sized was used and <3.5 m was avoided

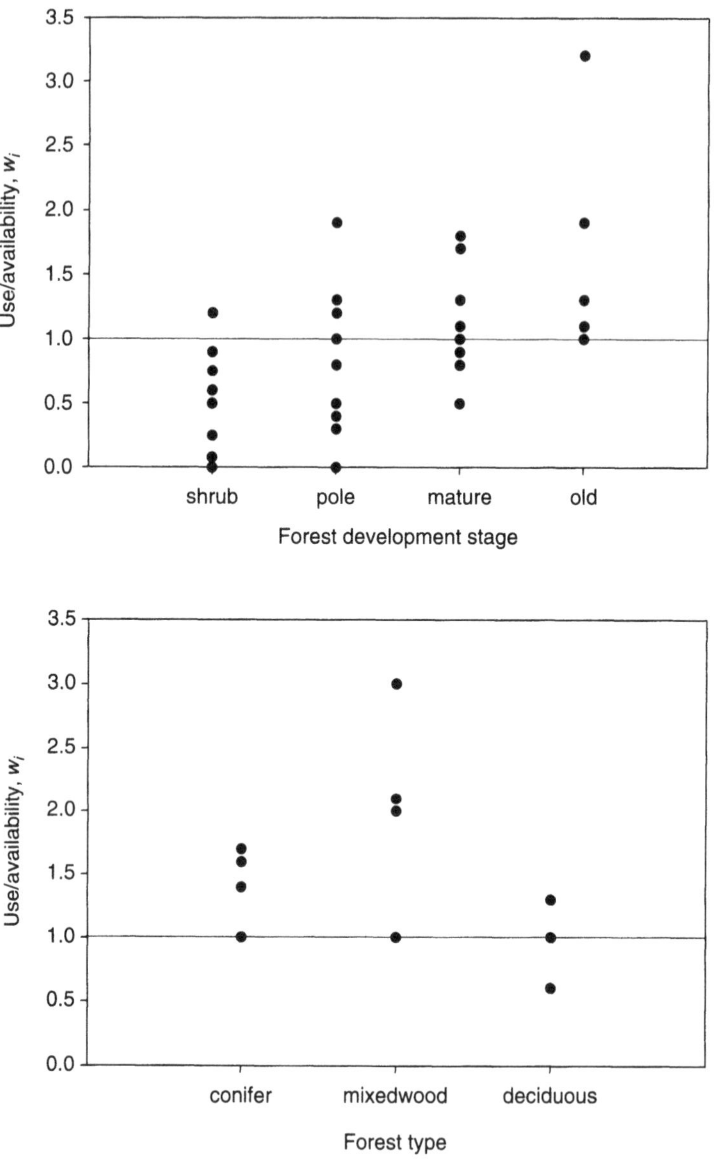

Figure 9.1. Habitat selection by American martens for broad forest stage and type classes (w_i), from studies reported in Table 9.1 that also reported selection of forest types.

broader selection of forest conditions has been observed (e.g., Francis and Stephenson 1972; Chapin et al. 1997a; Dumyahn et al. 2007; Hearn et al. 2010). Boreal forests constitute >90% of the marten's geographic distribution, and in that zone, mixed woods and conifer forests are preferred, whereas deciduous forests are avoided (e.g., Thompson and Harestad 1994; Potvin et al.

2000; Proulx 2006b; Godbout and Ouellet 2008). In Newfoundland, where almost all forests are coniferous, martens used old open and insect-killed forests and selected for dense regenerating stands >3.5 m tall (Hearn et al. 2010). In 2 studies where deciduous forests were used extensively by martens (e.g., Chapin et al. 1997a; Dumyahn et al. 2007), selection was observed for mature and old forests and the species composition of deciduous trees was diverse. Poole et al. (2004) also observed use of deciduous stands where a single genus (*Populus* spp.) of deciduous trees predominated; however, they also reported selection by American martens for the limited amount of mature conifer forest found on their landscape, along with use of the regenerating deciduous stands composing most of that landscape. The overall implication is that martens may exhibit substantial use of a diverse array of forest types, if the required habitat structures are present and if sufficient food is available. Studies of habitat selection generally assume that occupancy of a particular habitat reflects individual fitness (Buskirk and Powell 1994); nevertheless, as discussed by Van Horne (1983), Buskirk and Powell (1994), and Powell (2004), in the absence of data on fitness, we must interpret patterns of habitat occupancy cautiously. This is particularly true in cases where indirect methods, such as hair snags or tracks, were used as an index of use, in circumstances where juvenile animals or animals of unknown age and residency status could confound interpretations, and in extensively managed landscapes (e.g., boreal) where martens occur at low densities and there is a high probability of Allee effects (e.g., Morris 1992, 2002).

Where habitat selection was examined at both landscape and stand scales, results have generally indicated that models tended to predict occurrence better at the landscape scale than at the stand scale (Bissonette et al. 1997; Potvin et al. 2000; Fuller and Harrison 2005; Mowat 2006; Slauson et al. 2007; Baldwin and Bender 2008; Godbout and Ouellet 2008). This finding may be partly explained by the fact that territorial animals must regularly visit all of their territory to exclude other adults of the same sex and may traverse a variety of forest types, including those that are not preferred. Further, American martens have large spatial requirements and may exhibit strong selection when positioning their home ranges on heterogeneous landscapes to maximize fitness. The species responds to habitat loss through habitat selection at the landscape scale by maintaining high densities in high-quality, unmanaged habitats (e.g., Chapin et al. 1997a), at least until a habitat-loss threshold is reached (Thompson and Harestad 1994; Bissonette et al. 1997). Occupancy by American martens across a forested landscape declines when 25–30% of the landscape is regenerating following timber harvesting (Chapin et al. 1998; Hargis et al. 1999; Potvin et al. 2000; Fuller and Harrison 2005) (Table 9.2). Further, a link has been found between landscape pattern, habitat quality, and home range size. For example, Potvin and Breton (1997) reported that increased area in openings resulted in larger home ranges, Thompson (1994) found larger ranges in regenerating than in old forests, and Fuller and

Table 9.2. Forest variables or metrics measured at landscape or home range (HR) scales that suggested which forest type or condition was important to resident American martens in managed forests

Study	Location	Methods	n[a]	Significant variables (values if given)	Fitness
Chapin et al. 1998	Maine	Telemetry / GIS grid cells	33	Median area of used forest patches 27 ha; at 250 ha grid, 59% of entire landscape was residual forest; male and female residents averaged <22% of regenerating forest in HRs	Low[b]
Hargis et al. 1999	Utah	Livetrap / 18 9-km^2 plots / GIS Fragstats	46	Area of clear-cut >25% resulted in reduced use	N/A
Potvin et al. 2000	Quebec	Telemetry / GIS Fragstats	20	HR size proportional to area of mature forest (mixed and conifer); HRs had >40% uncut and <30% regenerating forest	N/A
Poole et al. 2004	British Columbia	Telemetry	29	Stands with >25% conifer preferred	Population stable[c]
Fuller and Harrison 2005	Maine	Telemetry	18	Maximum proportion of partial harvest in HR in winter 34% (increases to 73% in summer)	N/A
Bull et al. 2005	Oregon	Telemetry	20	Stands with no harvesting preferred; especially old subalpine fir/Engelmann spruce types	N/A
Dumyahn et al. 2007	Wisconsin	Telemetry / GIS grid cells	13	At least 70% suitable cover types; high use of mature upland deciduous forest	N/A
Hearn et al. 2010	Newfoundland	Telemetry / GIS	58	Open old conifer preferred at landscape scale; insect-killed old forest with regenerating understory preferred at stand and landscape scales. High use of regenerating forests but avoidance of recent cuts.	N/A

[a] n = number of adult marten with home ranges, except for Hargis et al. (1999) in which n = number of adults live-trapped
[b] High human-caused mortality with annual replacement; no measure of lambda
[c] No measure of lambda

Harrison (2005) reported that, in fall and winter, home ranges containing partially harvested forests were twice the size of those composed primarily of mature forest. In addition, Fuller and Harrison (2005) reported that martens exhibited landscape-scale selection during winter for habitats with higher overstory canopy cover by expanding their home ranges, and suggested that this may be a predator-avoidance behavior. Chapin et al. (1997a) and Phillips et al. (1998) reported no habitat shifts in winter at the scale of the home range, and Payer et al. (2004) observed no differences in seasonal or annual home-range fidelity among martens inhabiting managed versus unmanaged landscapes. The resident adult martens studied by Payer et al. (2004) selected for mature forest within heterogeneous landscapes and exhibited high fidelity despite large variation in landscape composition. Collectively, these findings argue strongly in favor of landscape-scale habitat selection as a primary influence on the individual fitness of martens.

Density-dependent effects have rarely been explicitly examined for martens as a hypothesis to explain habitat selection (Table 9.3) by testing predictions among competing models. However, several studies (Hodgman et al. 1994; Thompson 1994; Johnson et al. 2009) have reported density-related behaviors, such as dispersal to habitats with a low density of resident martens, and Payer et al. (2004) reported that the degree of territorial overlap was density

Table 9.3. Support (yes/no) for 4 habitat-selection hypotheses from studies of American martens published since 1993 that considered stand- or landscape-scale habitat selection

Study	Prey abundance	Habitat features	Predator avoidance	Density dependence
Thompson and Curran 1995[a]	Yes	Yes	N/A	N/A
Paragi et al. 1996	Yes	No	N/A	Yes[b]
Chapin et al. 1997a	N/A	Yes	N/A	N/A
Hargis et al. 1999	No	Yes	N/A	N/A
Potvin et al. 2000	N/A	Yes	N/A	N/A
Payer and Harrison 2003	N/A	Yes	Yes[c]	N/A
Poole et al. 2004	No	Yes	No	(Stable density)
Gosse et al. 2005	N/A	Yes	N/A	N/A
Fuller and Harrison 2005	Yes	Yes	Yes[c]	N/A
Bull et al. 2005	N/A	Yes	N/A	N/A
Mowat 2006	N/A	Yes	N/A	N/A
Dumyahn et al. 2007	N/A	Yes	N/A	N/A
Slauson et al. 2007	N/A	Yes	N/A	N/A
Coffin et al. 1997	Yes	Yes	N/A	N/A
Godbout and Ouellet 2008	N/A	Yes	N/A	N/A
Hearn et al. 2010	Yes[c]	Yes	No[c]	N/A

Note: N/A = factor not tested

[a] Marten density inferred from earlier work

[b] Selection influenced by density although lambda was not reported

[c] Inferred but not measured directly

dependent. Fryxell et al. (1999) related marten density directly to prey levels, and Hearn et al. (2010) suggested that marten habitat choice was related to snowshoe hare (*Lepus americanus*) abundance (as prey) in Newfoundland. In Ontario, however, high hare abundance in regenerating forests did not entice marten to use these habitats, even during years of high densities of martens or hares (Thompson and Colgan 1987), suggesting that factors other than food alone were strongly influencing habitat selection in that area. Similarly, Fuller and Harrison (2005) reported high predation on hares by American martens during winter and an order of magnitude higher hare densities in regenerating stands compared with mature forest stands. Despite apparent foraging advantages in those regenerating habitats, American martens strongly avoided regenerating forests at the stand scale, which the authors postulated was a result of inadequate escape cover for the species in regenerating stands. Thus, recent work has demonstrated that population densities of American martens and their prey are related to many aspects of habitat use, but uncertainty remains regarding possible density-dependent mechanisms of habitat selection and their effects on the individual fitness of American martens.

Habitat Selection and Dispersal at the Landscape Scale

It is adaptive for animals to select habitats that will maximize their individual fitness (Stephens and Krebs 1986; Powell 2004). During periods of high population density, however, juvenile American martens born in superior habitat may be forced to disperse to inferior habitats (Johnson et al. 2009). Thus, managing marten habitat at very large spatial scales may be particularly important if selection at the landscape scale affects home range- and site-scale habitat selection in a density-dependent manner; however, the process of habitat selection may be limited in scale (Morris 2002). For example, such a threshold could be determined by the capacity of a species to disperse. Johnson et al. (2009) found that 80% of dispersing juvenile martens settled <20 km from where they were born to where they established a home range, although in some cases the animals moved much farther than 20 km before returning (Broquet et al. 2006a). The dispersal process is likely dependent to some extent on the availability of suitable habitat, but it provides some guidance about the appropriate scale for studying habitat selection.

Johnson et al. (2009) showed that dispersal distance and success (survival and settling) were highly dependent on the quality of the landscape through which the martens were moving. Marten dispersal success and distance, and hence habitat selection, were probably related primarily to the population density of adult animals, and hence territorial vacancies, with lower population density on a regenerating managed landscape than on an unmanaged mature–old forest landscape. If surrounding preferred habitat is saturated, animals might be compelled to move farther than in a landscape where space is available, including moving through or into poor habitat to find a vacant home range. In very large progressively logged landscapes, most martens do not move far enough to leave the landscape in which they were born and settle in an unmanaged landscape (Johnson et al. 2009). Thus, it may be that use of managed forests reflects a long-term (e.g., decadal-scale) accumulation of dispersal events. In years when American martens are abundant, some animals must leave the unmanaged forests and settle in managed landscapes, suggesting either a source-sink or an ideal despotic (Fretwell and Lucas 1969) density model. Certainly, some American martens are capable of long-distance dispersal (Thompson and Colgan 1987; Broquet et al. 2006a; Johnson et al. 2009), although survival and fitness are generally lower in managed than in unmanaged forests as a result of commercial trapping and increased natural (mostly predation-caused) mortality (Hodgman et al. 1994; Thompson 1994; Johnson et al. 2009; but see Hodgman et al. 1997). A population in poor habitat (sink) could persist with periodic emigration from preferred habitats (source), as was suggested by Paragi et al. (1996) for young regenerating burned forests adjacent to unmanaged old forests in Alaska, and by Hodgman et al. (1994) for managed forests next to a preserve in Maine. Other population models, such as ideal despotic for example, might apply for marten populations living in a more heterogeneous matrix of forest ages and types.

Habitat Selection Within Stands

Stand-scale habitat selection has been observed in most studies, and forest structures are important features that influence marten selection of individual forest stands, often regardless of stand age or type (Chapin et al. 1997b; Payer and Harrison 2003). Linkages have rarely been made between stand-scale variables and the individual fitness of American martens. Various studies have described the stand conditions martens prefer, however, with some (e.g., Payer and Harrison 2000, 2003) providing considerable detail on complex forest structures (Table 9.4). For example, Payer and Harrison (2003) indicated that, in Maine managed forests, martens used stands with a basal area >18 m^2/ha, mean tree (dbh ≥7.6 cm) height ≥9 m, and snag (dbh ≥7.6 cm and >2 m tall) volume ≥10 m^3/ha. These variables are believed to have functional importance to martens by providing cover from predators, facilitating foraging for prey, enabling resting or denning, and providing other benefits. Coffin et al. (1997) and Bowman et al. (2000) noted a relationship between southern red-backed voles (*Myodes gapperi*), a main prey of martens in many regions, and forest structures. Similarly, Andruskiw et al. (2008) illustrated the functional importance of coarse woody debris to hunting success. A nutritional link to breeding success has previously been well established for American

Table 9.4. Stand- or site-scale variables that were important in models or univariate tests of habitat use by American martens within home ranges

Study	Location / method	Forest stage	n[a]	Significant variables (and values if given)
Thompson 1994[b]	Ontario / telemetry	Old and regenerating	37	Mean tree density (>10 cm) 680/ha; conifer trees 80%; canopy 60%; mean tree (>10 cm dbh) height 18 m
Bowman and Robitaille 1997[c]	Ontario / winter tracks	Regenerating	151 used sites in regenerating forest	Conifer trees 76%; mean tree height 13 m; canopy 80%; logs/100-m² quadrat 3.5
Chapin et al. 1997a	Maine / telemetry	Mature and insect-killed	38	No selection among stand types
Smith and Schaefer 2002	Labrador/ telemetry	Old and mature	26 (includes some juveniles)	Canopy >20%
Payer and Harrison 2000, 2003	Maine / telemetry	Mature and regenerating	24	Mean tree basal area >18 m²/ha; tree height (>8 cm dbh) >9 m; snag volume >10 m³/ha; snag and cull tree BA >18 m²/ha; CWD 55 m²/ha; mean diameter CWD >22 cm; winter canopy >30%[d]
Poole et al. 2004	British Columbia / telemetry	Mature and regenerating	29	Shrub cover >15%; snag density 4.5/ha; conifer >25%
Payer and Harrison 2004	Maine / telemetry	Mature	57	Volume of exposed root masses 25.2 m³/ha
Bull et al. 2005	Oregon / telemetry	Old and regenerating	20	Canopy >50%; large snags (25 cm) 22/ha; trees (>25 cm dbh)100/ha
Fuller and Harrison 2005	Maine / telemetry	Partial-harvest	18	Tree basal area >18 m²/ha; winter canopy >30%
Slauson et al. 2007[c]	California / track plates	Old	26 detections	Shrub cover; % conifer

[a] n = number of adult home ranges, unless specified differently
[b] Study did not explicitly test used vs. random or unused sites
[c] Home ranges were not determined
[d] BA = basal area, CWD = coarse woody debris

martens (e.g., Thompson and Colgan 1987). Substantial variation occurs among published values for some key variables, however, such as canopy closure, percent conifer in a stand, woody debris volume, and tree and large-snag density (Table 9.4). These differences likely reflect variation among types of forest ecosystems (e.g., montane vs. boreal vs. transitional) resulting from differences among endemic tree species and the ecology of the systems, as well as to regional differences in prey use by American martens.

Given that habitat selection by martens has been documented to occur at the within-stand (site) scale, analyses conducted at broader scales may not capture some important density-dependent variation. Few models have been developed for marten habitat selection at site scales. Models described in Chapin et al. (1997b), Ruggiero et al. (1998), and Porter et al. (2005) found predictability for resting- and den-site selection. These models suggested that, within home ranges, martens may make important choices about fine-scaled features such as large snags, overhead cover, and conifer-tree density.

Natal Dens, Maternal Dens, and Resting Structures

American martens appear to be more selective of habitat conditions at den sites than at resting sites. Ruggiero et al. (1998) noted the importance of structures associated with late-successional forests to den selection in Wyoming; both the characteristics of denning structures and the features of the stands where the structures occurred seemed to influence den-site selection. Ruggiero et al. (1998) cautioned, however, that selection among natal dens, maternal dens, and resting structures needed to be evaluated separately, given the possibility of greater selectivity for natal dens than for other structures or sites. The most common feature reported in studies of natal and maternal dens was a large dead or live tree that provided a hollow space that could contain a mother and her kits (Table 9.5). Such hollow spaces could result from a split in the trunk, a woodpecker cavity, or interior rot accessible from the base of the tree. Other common den structures included rock crevices and stumps. Resting structures were most commonly associated with live trees, especially those where various fungi had caused excessive growth on branches (Bull and Heater 2000), tip-ups, and large coarse woody debris (Table 9.5).

None of the studies on fine-scale habitat selection (Table 9.5) provided any indication that den structures were limiting at the population level. Several researchers have found little difference in the abundance of important structures, such as coarse woody debris, between used and unused habitats (e.g., Thompson and Colgan 1987; Payer and Harrison 2004). This suggests that, at least in some ecosystems, certain fine-scale habitat elements were either not limiting or were not quantified at a sufficiently refined level (e.g., volume or number of coarse woody debris >25 cm in class 2 as opposed to all coarse

Table 9.5. Selection of resting and denning structures by American martens

Study	Location / method	n[a]	Structures used (%)	Mean characteristics (if values given)
Gilbert et al. 1997	Wisconsin / telemetry	32 rw 19 rs 7 d	rw: tip-up (44), CWD[b] (28) rs: live tree (68) d: live tree (71)	d: >50 cm tree
Chapin et al. 1997b	Maine / telemetry	73 rw 67 rs	rw: subnivean (40), tip-up (18), stump (18) rs: live tree (53), snag (15)	—
Raphael and Jones 1997	Oregon / telemetry	95 rw 163 rs 31 d	rw: subnivean (41), CWD (31) rs: slash (60), CWD (16) d: CWD (32), slash (29)	—
Raphael and Jones 1997	Washington / telemetry	140 rw 240 rs 26 d	rw: live tree (53), snag (20) rs: live tree (46), snag (22) d: live tree (54), snag (31)	r: live tree 88 cm d: live tree or snag 98 cm
Ruggiero et al. 1998	Wyoming / telemetry	105 d	d: rock crevice (28), snag (25), squirrel midden (19)	d: snag 55 cm
Bull and Heater 2000	Oregon / telemetry	1184 r w and s 30 d	r: live tree (58), log pile or hollow log (w 23, s 9), snag (17) d: cavity (40), hollow log (11)	r and d: live tree 52 cm, hollow log (CWD) 66 cm, snag 79 cm
Wilbert et al. 2000	Wyoming / telemetry	190 rw	rw: subnivean CWD (40), live tree (35), snag (15)	rw: snags and live trees 52 cm; CWD 39 cm
Porter et al. 2005	British Columbia / backtracking	52 rw	snag (unk)	rw: snags Class 3+

[a] n = number of resting structures (r) in summer (s) or winter (w), and natal or maternal den structures (d); only the 2–3 most abundant structures are provided from each study

[b] CWD = coarse woody debris

woody debris). However, high-quality natal den structures could be a limiting feature to a marten population, especially in managed forests with small-diameter trees, few snags, and little woody debris (Ruggiero et al. 1998). Future work should be directed toward determining which fine-scale features may limit the population persistence of American martens in various forest types. For example, Raphael and Jones (1997) observed that, where natural structures were scarce in managed areas, American martens used slash piles for denning, but the implications for relative fitness were not explored.

Energetic Considerations

Martens are believed to have high energy demands because of their body form and exposure to winter weather (Buskirk and Harlow 1989; Harlow 1994). Thompson and Colgan (1994) showed a linear decline in marten activity at ambient temperatures below –5 °C, and Wilbert et al. (2000) found that accumulating snow reduced activity levels. An important component of habitat for American martens is access to structures that enable them to conserve body heat, especially in winter. For example, resting sites in winter are often subnivean (Table 9.5), and Buskirk and Powell (1994) and Taylor and Buskirk (1994) found that the microenvironments of resting sites enabled behavioral thermoregulation in all seasons. Use of subnivean sites for resting in Wyoming was more likely when snow cover was deep, when snowfall the previous day had been heavy, and when low temperatures prevailed (Wilbert et al. 2000). Although snow as well as features enabling subnivean access (Corn and Raphael 1992) and those providing thermal cover (Chapin et al. 1997b) likely influence resting-site selection by martens, several authors (Corn and Raphael 1992; Thompson and Colgan 1994; Chapin et al. 1997b) have concluded that these features are not necessarily limiting to populations. Again, additional work is needed on the relationships between the availability of denning and resting structures (including those that are subnivean) and the fitness of American martens. Finally, ambient temperature and snow are key environmental influences on microhabitat selection by American martens, but unless they are studied at fine spatial scales, subtle behavioral and fitness implications may not be apparent.

Effects of Forest Management

Over the long term, an important question in our understanding of the effects of forest management is whether or not second-growth forests will converge with primary forests in terms of ecosystem processes, patterns, and structures. Findings from several areas (Maine, Quebec, Ontario, and British Columbia) suggest that, in some forest types and with management guidance, martens may be able to persist in managed forests, although long-term data are lacking. Further, densities in managed landscapes may be similar to, or lower than, densities in unmanaged forests, depending on forest type and prevailing forest management practices (Thompson 1994; Poole et al. 2004; Fuller and Harrison 2005). Mechanisms that are important to the relationships among marten density, forest management, and available forest types require consideration of alternative explanations tested against multiple hypotheses over time to fully elucidate these complex relationships.

Effects of Clear-cut Harvesting

As noted above, where clear-cut harvesting is practiced, marten population declines can be expected for ≥40 years if >25–30% of the forest is composed of regenerating stands (Thompson and Harestad 1994; Chapin et al. 1998; Hargis et al. 1999; Potvin et al. 2000; Poole et al. 2004). An exception was reported from Newfoundland, where martens fully exploited landscapes in which much of the forest comprised regenerating harvested areas or insect-killed stands, but where most predators and competitors of American martens were uncommon or absent (Hearn et al. 2010). Few data are available to determine whether American martens will persist in regenerated clear-cut forests once they reach maturity, because none of these stands have reached old age. During a 4-year study in British Columbia, Poole et al. (2004) reported adult resident martens living at moderate densities in second-growth forests about 40 years after clear-cutting and previous agricultural use. In Maine (Chapin et al. 1998; Payer and Harrison 2003; Fuller and Harrison 2005) and Quebec (Potvin et al. 2000), martens were studied in matrix landscapes comprising a mixture of stands regenerating after modified clear-cut and patch harvesting, interspersed with large stands of unharvested mature forest. These residual stands influenced home range placement and movements across the landscape by American martens, but the animals made some use of regenerating stands and avoided the youngest stands. Median size of used residual stands in Maine was 150 ha, whereas in Quebec, the size of home ranges was inversely proportional to the amount of residual uncut forest they encompassed, which on average was 60–70% of a home range. Landscapes with progressive and large-scale clear-cutting with few remaining old or mature stands, however, supported substantially fewer animals during the regenerating period than matrix landscapes (e.g., Thompson 1994; Potvin et al. 2000; Payer and Harrison 2003; Godbout and Ouellet 2008).

The length of time in which regenerating habitats are unsuitable to American martens seems to depend strongly on landscape context, forest type, and silvicultural practices. Certain forest types lack suitable structures for long periods of time, whereas in others, the structures needed by the species are in more continuous supply (e.g., Chapin et al. 1997a; Payer and Harrison 2004). Although understory conditions may not be limiting in some areas, required overstory conditions (e.g., overstory canopy closure and the basal area of residual trees) may be limiting for longer periods in managed landscapes (Payer and Harrison 2003). For example, low residency of American martens has been observed in extensively clear-cut boreal forests up to 50 years after harvest (Thompson 1994; Godbout and Ouellet 2008; Thompson et al. 2008). In contrast, Newfoundland martens used stands that were considerably less than 50 years old (including dense regenerating stands with open overstories and pre-commercially thinned stands) but where ecological release from predation may have influenced habitat selection (Hearn et al. 2010). In many forest types, even young regenerating stands may be used to obtain berries that are seasonally abundant in these areas, but these areas are usually avoided at all other times (Steventon and Major 1982; Thompson and Colgan 1987; Poole et al. 2004). Paragi et al. (1996) found American martens occupying 30-year-old regenerating burned habitats surrounded by older forests, but suggested that the animals were mostly juveniles and could not persist. Ecosystem type and regeneration following disturbances are related to structural features and prey densities, which may influence occupancy in both time and space (e.g., Chapin et al. 1997a; Payer and Harrison 2003; Poole et al. 2004; Johnson et al. 2009).

Effects of Partial Harvesting

Consistent with the concept that the American marten can persist in managed landscapes if certain thresholds are not surpassed, Fuller and Harrison (2005) found that the species tolerated partial harvesting to a basal area of 18 m^2/ha with ≥30% canopy closure in transitional forests in Maine; however, martens expanded their home ranges in winter to include more residual forest. In Quebec, Godbout and Ouellet (2008) found that mature conifer forest was the main forest type (49%) in American marten home ranges and that the animals tolerated partial harvesting that left at least 40% canopy closure, while avoiding stands that were pre-commercially thinned. Their results were similar to the findings of Potvin et al. (2000) in an area of mixed modified and clear-cut harvesting in mixed-wood forest. American martens sometimes tolerate more partial harvesting than clear-cut harvesting within the forest matrix; however, they appear very sensitive to the residual basal area of mature overstory trees and the extent of winter canopy cover remaining after partial harvests (Fuller and Harrison 2005). This suggests that with careful consideration and additional studies, some opportunities may exist to manage for timber production and still maintain viable populations of American martens. For any forest system in which selection harvesting is conducted, threshold values for canopy closure and species composition need to be developed if the species is to be conserved. It should also be recognized that partial harvesting may reduce the total available uncut forest by requiring a larger harvesting footprint to meet timber-harvest goals.

Testing Hypotheses for Habitat Selection

Few explicit tests of the hypotheses proposed by Buskirk and Powell (1994) and Thompson and Harestad (1994) have been conducted for habitat selection by American martens at the stand or landscape scale. Among the 16 studies reviewed (Table 9.3), only 6 attempted to measure variables other than vegetative characteristics to explain habitat use by American martens. Support for the site-scale habitat features hypothesis was suggested in 15 studies, whereas 4 of 6 studies provided evidence that prey abundance was important (Table 9.3). Density-dependent habitat selection was suggested by Paragi et al. (1996) in burned and unburned forests in Alaska. The possibilities of predator avoidance and food limitation were invoked in several studies but not assessed with data on predator abundance or mortality rates. Although many habitat-selection studies have been conducted during the past 18 years, few have controlled for potentially mechanistic variables, including relative food abundance, predators, or density dependence.

Conclusions

American martens generally prefer mature or old forests over regenerating stands in areas where both are available in the landscape. This finding is especially true in conifer and mixed-wood forests, but less so in deciduous forests, particularly those dominated by only 1 or 2 species. Martens seem to have greater affinity for mature deciduous stands where transitional northern hardwood forest stands intersperse with mixed-wood and coniferous stands across the landscape. Where they still occur, the importance of mature and old forests to martens is clearly evident. When the availability of these habitats in a landscape is reduced to <70% from wildfire or forest management, the population density of American martens declines. In matrix landscapes that include a mixture of high- and low-quality habitats, American martens appear to select home ranges dominated by high-quality habitat types, but also include some recent clear-cuts and regenerating forests (e.g., Chapin et al. 1997a; Potvin et al. 2000; Fuller and Harrison 2005). In progressively clear-cut boreal habitats, declines in the population density of American martens result from reduced habitat quality, and there is limited evidence of recovery to pre-harvest levels in regenerating forests after 40–50 years (e.g., Thompson 1994; Godbout and Ouellet 2008), which represent the oldest second-growth boreal forests currently available. A notable exception occurs in Newfoundland, where martens use regenerating forests and pre-commercially thinned stands, but where there are few predators and competitors of martens (Hearn et al. 2010). In contrast to findings from most boreal and montane forests, American martens in the transitional forests of the northeastern United States and southeastern Canada can persist in managed forest landscapes where residual forests remain the dominant background matrix within occupied home ranges (Taylor and Abrey 1982; Chapin et al. 1998; Fuller and Harrison 2005; Dumyahn et al. 2007). Although there is a preference for mature stands in these forests, American martens persist in stands with basal areas >18 m^2/ha and trees >9 m tall, but avoid young regenerating forests (Chapin et al. 1997a; Payer and Harrison 2004; Fuller and Harrison 2005).

The apparent disparity among studies suggesting a requirement by American martens for old forest stages and those finding the species persisting either in mixtures of young and old forest on a landscape or in regenerating forests appears to be linked to key aspects of marten ecology in different forest types. Other factors include differences in prey abundances and diet preferences, variation in the local predator community, or perhaps differing climatic conditions. Martens need certain structures to improve hunting success (Andruskiw et al. 2008) and for denning (Ruggiero et al. 1998). Studies have shown that in some forest types, these structures are available in stands well before the mature or old stages, such as in insect-killed stands (Payer and Harrison 2000; Hearn et al. 2010), or they can potentially be maintained during some forms of partial harvesting if sufficient residual tree basal area and overstory canopy closure in winter are maintained (e.g., Fuller and Harrison 2005). In most forest types, however, the structures required by American martens are absent for >50 years after clear-cutting (e.g., Godbout and Ouellet 2008; Thompson et al. 2008). Further, martens seem to require overhead cover, possibly in multiple layers, presumably to avoid or escape predators (e.g., Buskirk and Powell 1994; Thompson and Harestad 1994). Such cover can occur in regenerating conifer forests at very early stages, especially where stem densities are high (Hearn et al. 2010), whereas in deciduous-dominated forests (especially during winter), these cover requirements may be met only in mature deciduous stands with high species richness (Chapin et al. 1997a). Where adequate overhead cover is not maintained during winter, American martens may seasonally expand their home ranges to include more cover (e.g., Fuller and Harrison 2005). Poole et al. (2004) also found that the species used open, deciduous-dominated forests that maintained high structure and food levels, but they did not provide data on potential predators of American martens. Hearn et al. (2010) suggested that, under reduced risk of predation, American martens may make greater use of open stands, especially where there are higher prey densities than in older stands. Marten habitat use varies among regions; consequently, if maintaining marten populations is a management objective, assumptions about stand-scale habitat relationships should be ecosystem-specific and not based on general understandings.

Since the 1994 reviews, our understanding of use and selection of habitats by martens has improved considerably. This is especially true regarding the importance of spatial scale from sites to landscapes. At the same time, our understanding of the mechanisms affecting habitat selection has not advanced to the same degree. Most authors have implicitly assumed that habitat choices made by martens maximize or improve individual fitness. In general, this is a reasonable assumption for American martens (Buskirk and Powell 1994; Buskirk and Ruggiero 1994), although it is widely understood that density and fitness may not be correlated (Fretwell 1972; Van Horne 1983; Hobbs and Hanley 1990). For American martens and other *Martes* species, this disconnect may be related to nonoverlapping home ranges and, possibly, an ideal despotic distribution. There are limitations to habitat-selection models, especially in the absence of fitness correlates to improve our understanding of the ways in which selection might change under varying conditions. For example, habitat-selection theory suggests that individual fitness declines as a function of population density (e.g., Fretwell and Lucas 1969). For American martens, however, we might not expect such a relationship if females always occupy and protect the best habitats to buffer any negative effect (despotic habitat selection). At a minimum, we believe that habitat choices can be understood only in the light of some measure of both population density and survivorship. For martens, few data are available on the relationships among demography, dispersal, and habitat selection. Predicting the likelihood of long-term population persistence is not possible without this information. Consequently, the challenge for future research is to provide a more detailed demographic approach to understanding the ways in which martens will respond to altered forest conditions and climate change (Carroll 2007) by employing testable predictions based on the knowledge we have gained over the past 18 years.

Acknowledgments

This chapter benefited immensely from reviews by an anonymous reviewer and by Steve Buskirk, and from editing by Keith Aubry.

10

Habitat Ecology of Fishers in Western North America

A New Synthesis

CATHERINE M. RALEY, ERIC C. LOFROTH, RICHARD L. TRUEX, J. SCOTT YAEGER, AND J. MARK HIGLEY

ABSTRACT

In this chapter, we present a synthesis of the habitat associations of fishers (*Martes pennanti*) in western North America based on information produced since 1994. Contrary to limited results from previous studies, evidence from contemporary research indicates that fishers in western North America are not dependent on old-growth conifer forests for survival. Rather, fishers were associated with complex vertical (e.g., large trees and snags) and horizontal (e.g., large logs and dense canopy) structure characteristic of late-seral forests. Fisher distribution (first-order selection) was associated consistently with expanses of low- to mid-elevation mixed-conifer or conifer-hardwood forests with relatively dense canopies. Fisher home ranges (second-order selection) were characterized by a mosaic of available forest types and seral stages, including relatively high proportions of mid- to late-seral conditions, but low proportions of open or nonforested environments. Patterns of habitat use or selection by fishers were strongest at finer spatial scales (third- and fourth-order selection), and demonstrated that the fisher is a structure-dependent species in western North America. Female fishers are obligate cavity users for reproduction; tree cavities appeared to provide secure environments for kits by regulating temperature extremes and limiting access by predators. Compared with availability, fishers consistently selected large live trees, snags, and logs for resting that resulted from long-term forest growth and decay processes. Thermoregulation is more important to fishers than was recognized previously, and appeared to influence selection of rest structures and sites. Tree pathogens (e.g., heart-rot fungi, mistletoe) are essential for creating the microstructures used for reproduction (cavities) and resting (e.g., cavities, branch platforms), and represent important components of fisher habitat throughout the species' range in western North America. Our understanding of fisher habitat ecology has improved substantially since 1994. Nevertheless, focused investigations of the mechanisms that may influence habitat selection by fishers at multiple spatial scales (especially at the home range scale) and correlating use or selection of environments to measures of individual fitness, are needed to better understand fisher habitat quality and improve the conservation and management of fisher populations in western North America.

Introduction

The geographic distribution of the fisher (*Martes pennanti*) in western North America (the Rocky Mountains west to the Pacific Ocean) has contracted substantially since European settlement, primarily as a result of overtrapping, predator control, and habitat loss through timber harvesting and other anthropogenic changes to forest landscapes (Powell and Zielinski 1994; Zielinski et al. 1995; Aubry and Lewis 2003; Lofroth et al. 2010). Although the commercial harvest of fishers has been closed for 6–20 years in southern British Columbia and >60 years in most other jurisdictions (Idaho, Washington, Oregon, California), and harvest limits have been reduced in Montana, fisher populations have not recovered in these portions of their historical range (U.S. Fish and Wildlife Service 2004, 2010; Lofroth et al. 2010).

Understanding the habitat relations of fishers and the influence of various environmental conditions on survival, reproduction, and other life needs is critical for the successful conservation of this species in western North America. The most recent synthesis of fisher habitat ecology was published almost 2 decades ago. In this comprehensive review, Buskirk and Powell (1994) hypothesized that the fisher was a habitat specialist and, in western North America, may require old-growth conifer forests for survival. Generally, the habitat associations of fishers were assumed to reflect those of their preferred prey, such as the snowshoe hare (*Lepus americanus*), which in the Pacific Northwest states coincided with the historical distribution of late-seral Douglas-fir forests (*Pseudotsuga menziesii*; Buskirk and Powell 1994). Evidence also suggested that fishers in western North America favored riparian forests. Overall, however, the structural characteristics of forest stands appeared more important to fishers than tree species composition (Buskirk and Powell 1994). These conclusions, also reflected in Powell and Zielinski's (1994) review of fisher ecology in the western United States, were based primarily on insights gained from fisher habitat studies conducted in eastern North America. Only 3 studies had been completed in the West prior to 1994: 1 telemetry study in northern California (Buck 1982; Buck et al. 1983, 1994), 1 detection study in northern California (Raphael 1984; Rosenberg and Raphael 1986), and 1 telemetry study in north-central Idaho (Jones 1991; Jones and Garton 1994). Although Powell and Zielinski (1994) included details in their review from 2 ongoing telemetry studies in northern California, research on fisher populations in the West was still in its infancy. Since 1994, numerous field studies on fisher habitat ecology in British Columbia, Montana, Idaho, Oregon, and California have been conducted or are currently in progress (Lofroth et al. 2010).

Our objectives in this chapter are (1) to identify new advances in our understanding of fisher habitat ecology since 1994, (2) to synthesize key findings and compare them with previous understandings of fisher habitat relations, and (3) to identify key information gaps and new avenues for research that would benefit conservation efforts and habitat management for this species in western North America. In this chapter, we relied extensively on the detailed literature review and data summaries of fisher habitat associations we presented in Lofroth et al. (2010) and do not repeat those details here. Rather, this chapter is a synthesis of the overarching patterns that emerged during that review; patterns that represent key components of fisher habitat, fisher life needs, and other factors that influence the habitat choices fishers make. We have presented these patterns in the order of their perceived importance to fishers based on our evaluations, beginning with the strongest and most consistent patterns, which also reflects a gradient of increasing spatial scale. We then summarize our key findings within the context of hierarchical habitat use or selection by fishers in western North America.

Fisher Habitat Studies

We reviewed 18 published papers and 24 unpublished reports, dissertations, or theses from 23 geographic areas in western North America that presented results on fisher habitat use or selection at various spatial scales (Table 10.1). This body of literature represents new information on fisher habitat ecology generated from 1994 through 2010. We also included 1 dataset (Simpson Resource Company, unpublished data) that augmented information presented in Thompson et al. (2007). For ongoing studies that produced progress reports (e.g., Thompson et al. 2010), we included relevant data on the characteristics of forest environments used by fishers, but not results from interim analyses of resource selection. Information on fisher habitat ecology was available for all geographic areas in western North America in which extant fisher populations occur (Figure 10.1), but more studies have been conducted in northern California and the southern Sierra Nevada than elsewhere. For detailed summaries of the individual field studies that produced information on fisher habitat ecology, including objectives, duration of field sampling, and methods, we refer readers to Lofroth et al. (2011). In the Rocky Mountain region, 2 telemetry studies, 1 in north-central Idaho (J. Sauder, Idaho Department of Fish and Game, personal communication) and 1 in west-central Montana

Table 10.1. Information on fisher habitat ecology in western North America produced from 1994 through 2010

Geographic area	Published	Unpublished	Method and order of selection
Pacific coastal provinces and states			
Regional		Buskirk et al. 2010	T, 3
Central British Columbia			
Williston Reservoir	Weir and Corbould 2010		T, 2
	Weir et al. 2004		T, 4
		Weir and Corbould 2008	T, 2–4
McGregor	Proulx 2006a		S, 3
Chilcotin		Calabrese and Davis 2010	n/a[a], 4
		Davis 2003	S, 3
		Davis 2009	T, 2–4
Beaver Valley	Weir and Harestad 1997		T, 3
	Weir and Harestad 2003		T, 3–4
	Weir et al. 2004		T, 4
Rocky Mountain provinces and states			
Regional	Carroll et al. 2001		O, 1
Southern Oregon			
Cascade Range		Aubry and Raley 2006	T, 3–4
Siskiyou National Forest		Slauson and Zielinski 2001	D, 3
Northern California			
Regional	Carroll et al. 1999		D, 1, 3
	Carroll et al. 2010		D, 1
	Davis et al. 2007		D, 1
	Zielinski et al. 2010		D, 1
		Carroll 2005a	D, 1
Coastal region	Thompson et al. 2007		T, 4
		Beyer and Golightly 1996	D, 1, 3
		Klug 1997	D, 1, 3
Redwood National and State Parks		Slauson and Zielinski 2003	D, 1, 3
Hoopa Valley Indian Reservation		Higley and Matthews 2009	T, 2–4
		Yaeger 2005	T, 3–4

Pilot Creek, Six Rivers National Forest	Zielinski et al. 2004a		T, 3–4
	Zielinski et al. 2004b		T, 2
		Truex et al. 1998	T, 3–4
Sacramento Canyon		Reno et al. 2008	T, 4
		Self and Kerns 2001	T, 2–4
Shasta Trinity National Forest		Dark 1997	T, 3; D, 1, 3
		Seglund 1995	T, 3–4
		Yaeger 2005	T, 3–4
Hayfork Summit		Reno et al. 2008	T, 4
Mendocino National Forest		Slauson and Zielinski 2007	D, 3
Sierra Nevada, California			
Regional	Carroll et al. 2010		D, 1
	Davis et al. 2007		D, 1
	Spencer et al. 2011		D, 1
		Campbell 2004	D, 1, 3
Yosemite National Park	Chow 2009		O, 1
Kings River, Sierra National Forest	Purcell et al. 2009		T, 3–4
	Thompson et al. 2011		T, 2
	Zielinski et al. 2006c		T, 3
		Mazzoni 2002	T, 2–4
		Thompson et al. 2010	T, 4
Blodgett Forest and Sequoia National Park		Truex and Zielinski 2005	T, 3; D, 3
Sequoia and Kings Canyon National Parks		Green 2007	D, 1
Tule River, Sequoia National Forest	Zielinski et al. 2004a		T, 3–4
	Zielinski et al. 2004b		T, 2
	Zielinski et al. 2006c		T, 3
		Truex et al. 1998	T, 3–4

Note: To determine fisher presence or use, investigators of fisher habitat ecology used telemetry (T), detection methods using track-plates or remotely triggered cameras (D), snow tracking (S), or occurrence data from museum specimens, trapping records, and sighting reports (O). Many studies investigated fisher habitat ecology at >1 order of selection: (1) first order, geographic distribution; (2) second order, home range; (3) third order, environments within a home range; and (4) fourth order, specific resources within an environment. Geographic areas are presented by latitude from north to south and, for California, from the coast inland; some sources apply to >1 geographic area.

[a] Study sampled potential reproductive den structures but did not sample fisher presence or use

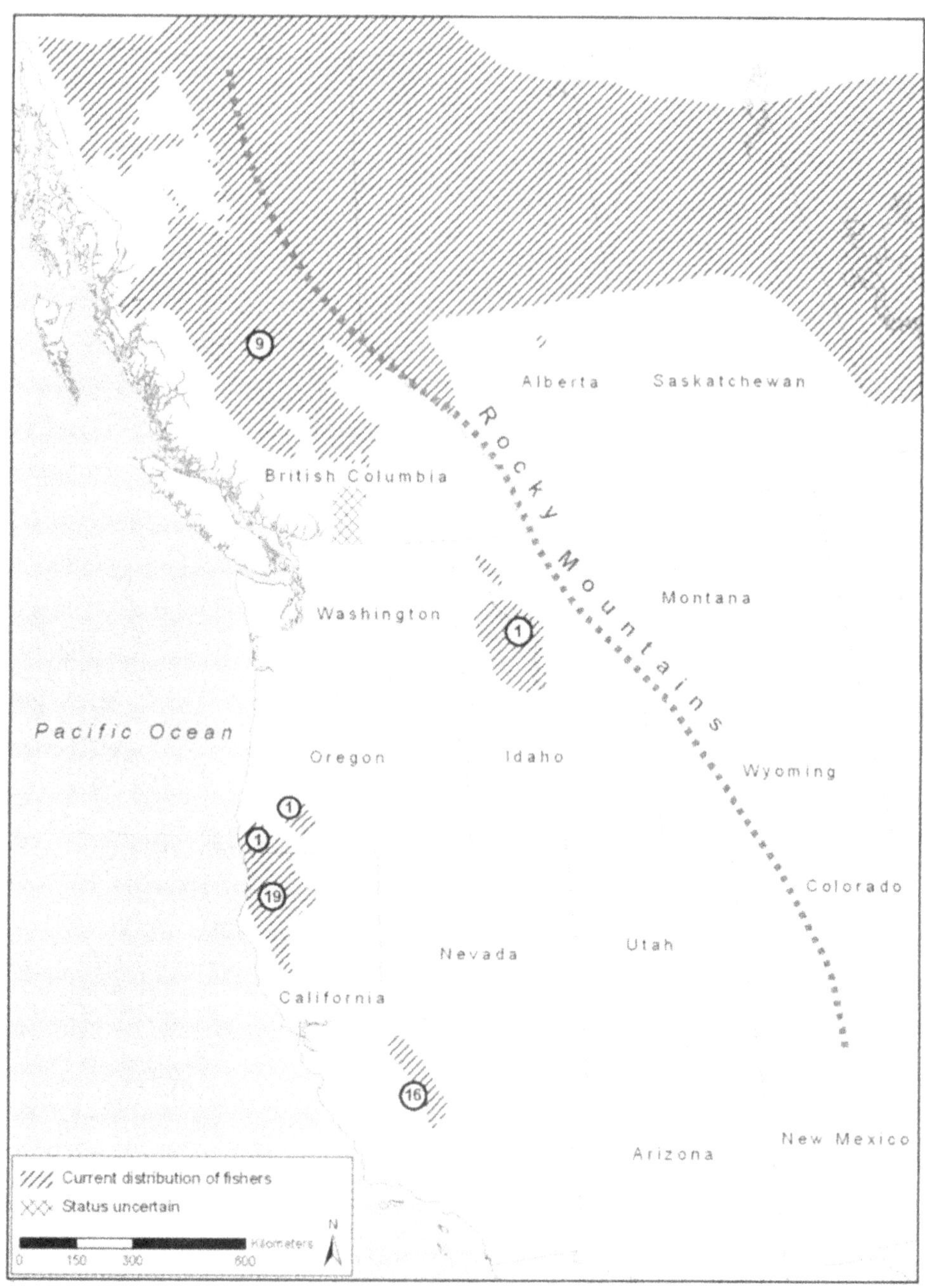

Figure 10.1. Number of published and unpublished sources of information on fisher habitat ecology in western North America produced from 1994 to 2010. Some sources generated information for >1 region. Map does not include 1 regional source that applies to British Columbia, Oregon, and California (Buskirk et al. 2010). Current distribution of fishers is based on Lofroth et al. (2010).

(M. Schwartz, USDA Forest Service, personal communication), were completed, but the results were not available at this time.

For this synthesis, we have focused on the habitat ecology of established fisher populations in western North America, including both indigenous and reintroduced populations. Although we recognize the value of investigating the habitat associations of recently translocated fishers in British Columbia (e.g., Fontana et al. 1999), Washington (Lewis et al. 2011), and California (Callas and Figura 2008), we have not included information from such studies here. Translocated animals are naïve to their new environment and may not yet have learned which habitat conditions are available or provide the best suite of resources to meet their life needs, such as escape from predators, foraging, resting, and reproduction.

Fishers select habitat at multiple spatial scales for different activities or behaviors (Powell and Zielinski 1994; Weir and Harestad 1997). First-order selection determines a species' geographic distribution; second-order, selection or composition of individual home ranges within that distribution; third-order, selection or use of different environments within a home range (e.g., stand types, rest sites); fourth-order, selection or use of specific resources within those environments (e.g., rest structures, food items; see Johnson 1980). We were mindful of these concepts in our evaluations, and the conclusions and hypotheses we present are based on interpreting available evidence at appropriate and comparable spatial scales. Although many fisher studies investigated habitat use or selection at multiple spatial scales, third- and fourth-order studies were most prevalent (Table 10.1).

Investigators of fisher habitat ecology use various terminologies to describe the characteristics of environments or resources used by their study animals. In this synthesis, we use the term *habitat* to indicate the suite of resources or environmental conditions that meet the life needs of fishers. The terms *selection* and *preference* are often used interchangeably in the literature, but selection refers to the process of choosing resources, whereas preference is the likelihood an animal will choose a resource, given the equal availability of other resources (Johnson 1980). Hence, we use the term *selection* when use of a resource was demonstrated to be greater or less than its availability. Investigators have measured forest canopy in different ways, and it was not always possible to determine which attributes of the canopy had been measured (e.g., canopy cover by trees vs. all overhead vegetation). Thus, we use the term *canopy cover* to represent all such measurements, and then provide additional clarification as needed. The term *forest structure* refers to numerous components that contribute to the vertical and horizontal complexity of forests, including distributions of tree sizes and ages, standing and down dead wood, vertical foliage distribution (understory and overstory vegetation layers), and horizontal canopy distribution (e.g., Spies 1998). The term *structure* refers to a live tree, snag (dead tree), or log (including all coarse down wood) used by a fisher, whereas the term *microstructure* refers to various features within the structures, such as a mistletoe broom in a live tree, cavity in a snag, or hollow in a log. Descriptions of the *sites* used by fishers for resting or denning include the rest or den structure and the forest conditions measured in the immediate vicinity of that structure (typically <0.5 ha). Finally, to provide meaningful insights on denning ecology, we categorized dens used by reproductive females according to the developmental stage of the kits (see Lofroth et al. 2010): (1) *natal dens* are the tree cavities in which parturition occurs; (2) *pre-weaning dens* are any subsequent structure used after the natal den but before the kits are weaned; (3) *post-weaning dens* are any structure used after the kits are weaned; and (4) *reproductive dens* refer to all dens used by an adult female during a single reproductive effort, or when the developmental stage of the kits was unknown.

Importance of Forest Structure at Small Spatial Scales

Den Structures

An extensive body of literature demonstrates that the fisher is a structure-dependent species in western North America. Reproductive females are obligate cavity users; they use cavities in large-diameter live trees and snags exclusively for birthing and rearing kits until weaning. Fisher kits are born during the late winter and early spring when weather conditions are cold and wet. Because kits are altricial at birth, and their ear canals and eyes remain closed until about 8 weeks of age (Powell 1993), we conclude that tree cavities are essential components of fisher habitat that provide thermoregulatory advantages to kits and protect them from predators.

With the exception of 2 reproductive dens in California and 1 in Idaho (Powell and Zielinski 1994), little was known about fisher denning habitat in western North America prior to 1994. New information from >330 reproductive dens in 12 geographic areas, however, has greatly improved our understanding of fisher denning ecology. In most cases, the cavities that reproductive female fishers used for natal and pre-weaning dens were created by heartwood decay through the action of heart-rot fungi (e.g., Aubry and Raley 2006; Reno et al. 2008; Weir and Corbould 2008). Most post-weaning dens were also in tree cavities, although females with older kits occasionally used other types of structures, including hollow logs (Aubry and Raley 2006). Reproductive den trees were always among the largest trees available; 1.7–2.8 times larger in diameter on average than other trees in the vicinity of the den (e.g., Reno et al. 2008; Weir and Corbould 2008; Davis 2009). Trees used for denning were old; the average estimated age of reproductive den trees in British Columbia was 372 years for Douglas-fir, 177 years for lodgepole pine (*Pinus contorta*), and 96 years for trembling aspen (*Populus tremuloides*;

Davis 2009). Typically older (thus, larger diameter) trees have a higher incidence of heartwood decay than younger trees (Manion 1991). In some regions, conifers were used for denning more than hardwoods (e.g., Aubry and Raley 2006; Davis 2009; Thompson et al. 2010), whereas, in other regions, hardwoods were used more frequently even if they were a minor component of the forest (e.g., Weir and Harestad 2003; Weir and Corbould 2008; Higley and Matthews 2009). However, available evidence indicates that the incidence of heartwood decay and cavity development is more important to fishers for denning than is the tree species.

Other characteristics, such as the size and height of the cavity opening and the interior dimensions of the cavity, may also influence females' choice of natal and pre-weaning den structures. The cavity must be large enough to accommodate an adult female and 1–4 growing kits, and have a relatively small opening (just large enough for a female to fit through) high off the ground (15–26 m on average; e.g., Aubry and Raley 2006; Weir and Corbould 2008; Thompson et al. 2010). These characteristics may be important for excluding potential predators and aggressive male fishers. During the breeding season, adult males locate a natal den and wait for an opportunity to mate with the female when she exits the cavity (Lofroth et al. 2010); the relatively small cavity opening may prevent males from entering the cavity and injuring or killing the kits. Consequently, the large diameter of structures used for denning may reflect the age of the tree, the presence of heartwood decay, and the need for a large and structurally sound interior chamber in the mid- to upper bole, where most natal and pre-weaning dens are located. Presumably, the cavity must also have adequate thermal properties to protect kits from weather extremes. Compared with ambient temperatures, tree cavities provide stable microclimates with narrow temperature fluctuations (Sedgeley 2001; Weir and Corbould 2008; Coombs et al. 2010). Most (75%) of the dens used by reproductive female fishers were in live trees. Cavities in relatively large live trees appear to have more stable temperatures during the day, and stay warmer at night, than those in relatively small snags (Wiebe 2001; Coombs et al. 2010). Other factors, such as the orientation of the cavity and exposure to sunlight (i.e., amount of canopy cover), may also influence the thermal properties of cavities, but quantitative evidence is lacking.

Rest Structures and Sites

Fisher resting habitat in western North America is also strongly tied to forest structure. Fishers typically rest in large deformed or deteriorating live trees, snags, and logs, and forest conditions around the rest structures (i.e., the rest site) frequently include structural elements characteristic of late-seral forests. Although this generally agrees with previous descriptions of fisher resting habitat (Buskirk and Powell 1994; Powell and Zielinski 1994), we now have evidence from >2260 rest structures from 12 different geographic areas (compared with <200 rest structures from 2 areas prior to 1994) that the characteristics of structures used by fishers for resting are overwhelmingly consistent throughout western North America.

Fishers rested primarily in deformed or deteriorating live trees (54–83% of all rest structures identified in individual studies), and secondarily in snags (6–26%) and logs (3–20%; e.g., Weir and Harestad 2003; Zielinski et al. 2004b; Aubry and Raley 2006; Purcell et al. 2009). The species of trees and logs used for resting appeared to be less important than the presence of cavities, platforms, and other microstructures. In live trees, fishers rested primarily in rust brooms in more northern study areas (Weir and Harestad 2003; Weir and Corbould 2008; Davis 2009) and mistletoe brooms or other platforms elsewhere (e.g., Self and Kerns 2001; Yaeger 2005; Aubry and Raley 2006). In contrast, fishers primarily used cavities when resting in snags (e.g., Self and Kerns 2001; Zielinski et al. 2004b; Purcell et al. 2009). Fishers used hollow portions of logs or subnivean spaces beneath logs more frequently in regions with cold winters (e.g., Weir and Harestad 2003; Aubry and Raley 2006; Davis 2009) than those with milder winters (e.g., Yaeger 2005; Purcell et al. 2009; Thompson et al. 2010). These results suggest that fishers use structures associated with subnivean spaces to minimize heat loss during cold weather (Weir et al. 2004; Weir and Corbould 2008). Many of the microstructures fishers need are created through the actions of particular organisms (e.g., rust fungi, heart-rot fungi) or ecological conditions (e.g., tree species in decline) and take decades to develop. Because fishers frequently rest in cavities in large trees or snags (similar to structures used for denning), a suitable microstructure may require >100 years to develop. Live trees, snags, and logs used for resting were, on average, 1.4–3.4 times larger in diameter than available structures (e.g., Weir and Harestad 2003; Zielinski et al. 2004b; Purcell et al. 2009). The large size of these structures is likely related to tree age and the long time periods required for various microstructures to develop.

Forest conditions around rest structures also influence a fisher's choice of rest sites. Compared with available sites, those used by fishers for resting had ≥1 of these structural components: greater average tree diameter or greater abundance of large-diameter trees, presence of large-diameter snags or greater abundance of snags, greater average size of logs or greater abundance of logs (primarily British Columbia and Oregon studies), and greater average diameter or greater abundance of hardwood trees (primarily California studies) (e.g., Zielinski et al. 2004b; Davis 2009; Purcell et al. 2009). A meta-analysis that included data from 8 studies in British Columbia, Oregon, and California confirmed this general pattern and demonstrated that, throughout the West, fishers consistently select sites for resting that have larger diameter conifer and hardwood trees, larger diameter snags, more abundant large trees and snags, and more abundant logs than random sites (Buskirk et al. 2010).

Several hypotheses have been proposed to explain these patterns. Because deteriorating trees, snags, and logs tend to be patchy in distribution (Bull et al. 1997), fishers may expend less energy finding suitable rest structures by selecting sites with a greater abundance of potential structures from which to choose (e.g., Seglund 1995; Weir and Corbould 2008; Davis 2009). A hardwood component in rest-site selection models for the Sierra Nevada was believed to represent the importance of black oaks (*Quercus kelloggii*) to fishers for rest structures (Zielinski et al. 2004b). An abundance of nearby large trees may also create favorable microclimates for resting fishers, especially in hotter and drier regions (Zielinski et al. 2004b; Purcell et al. 2009). Others have speculated that greater abundance or size of structures, or prevalence of hardwoods (especially mast-producing species), may reflect sites with more abundant prey (e.g., Yaeger 2005; Davis 2009; Purcell et al. 2009), thus providing energetic benefits for foraging. Rest sites with a relatively high density of large structures may also provide fishers with escape cover and protection from predators. We believe these are all plausible hypotheses and propose that, although rest sites may not be as clearly linked to fitness as reproductive den sites, they provide fishers with multiple advantages that improve individual fitness (e.g., thermal, security, and proximity to prey).

Forest Structure Associated with Active Fishers

When engaged in active behaviors (e.g., foraging, traveling), fishers in western North America were frequently associated with complex forest structure. In general, active fishers were associated with the presence, abundance, or a greater size of ≥1 of the following characteristics: logs, snags, live hardwood trees, and shrubs (e.g., Carroll et al. 1999; Slauson and Zielinski 2003; Weir and Harestad 2003; Campbell 2004). Although these results indicate that complex vertical and horizontal structure is important to active fishers, we did not find strong overarching patterns of use or selection. Some have suggested that the environments used by active study animals reflect foraging habitat (e.g., Buskirk and Powell 1994), but this idea has not been evaluated critically. Most methods used to study fisher habitat ecology (e.g., telemetry, noninvasive techniques) do not allow the investigator to distinguish among active behaviors such as foraging, traveling, seeking mates, or dispersal. Thus, the lack of strong habitat-association patterns for active fishers may reflect the sampling of multiple behaviors, each of which could be linked to different forest conditions. Alternatively, if the behavior sampled is primarily foraging, the lack of consistent patterns could reflect the diverse diets of fishers (e.g., Zielinski et al. 1999; Weir et al. 2005; Aubry and Raley 2006) and varying habitat associations among prey species, or the forest conditions in which fishers are most effective at capturing prey (see Buskirk and Powell 1994). Regardless, more-focused investigations are needed to understand the habitat associations of active fishers and the relations between fishers and their prey. Improved sampling methods that enable investigators to better distinguish among active behaviors would benefit such efforts.

Importance of Tree Pathogens for Creating Den and Rest Structures

Fisher habitat in western North America is intricately linked to a complex web of ecological processes that include natural disturbances (e.g., wind, fire), tree pathogens, and other organisms (e.g., primary excavators) that create and influence the distribution and abundance of microstructures (e.g., cavities, mistletoe brooms) in live trees, snags, and logs. Because female fishers rely exclusively on tree cavities for reproduction, we conclude that heartwood decay by heart-rot fungi, the process by which most reproductive den cavities are created, is an essential component of fisher denning habitat in western North America. This ecological process is also important for creating the microstructures that fishers use for resting (cavities in live trees and snags, and hollows in logs).

Heartwood decay by heart-rot fungi (specialized fungi that can tolerate the chemical defenses of live trees) is a complex process that can be initiated only in living trees (Manion 1991; Bull et al. 1997). Consequently, for live trees, snags, or logs to contain the types of cavities or hollows needed by fishers for denning and resting, heart-rot fungi must infect a live tree and persist long enough to decay a sufficiently large core of heartwood before the tree dies. Older, suppressed, or unhealthy trees are less capable of responding to injuries and defending themselves against infection by heart-rot fungi than are younger or healthier trees (Wagener and Davidson 1954; Manion 1991). Thus, advanced tree age and environmental stressors are important factors contributing to the susceptibility of live trees to infection by heart-rot fungi (Wagener and Davidson 1954; Manion 1991) and, consequently, the development of suitable cavities for fishers. Compared with conifers, hardwoods typically have thinner bark and their open growth form make them more susceptible to breakage (Gumtow-Farrior 1991; Bunnell et al. 2002). Hence, relatively young hardwood trees may have higher rates of infection by heart-rot fungi than similarly aged conifers. Nevertheless, for any tree to provide a cavity large enough for a fisher to use, decades are required for the tree to attain a large diameter, for damage or other stress factors to weaken its vigor, and for heartwood decay to reach an advanced stage and develop a suitable cavity.

Several investigators have reported fishers commonly using pileated woodpecker (*Dryocopus pileatus*) cavities as natal or pre-weaning dens (Aubry and Raley 2006; Reno et al. 2008; Davis 2009; Higley and Matthews 2009). This primary excavator creates a relatively large nest cavity in live trees or snags that have been softened by heartwood decay and, for roosting, excavates openings into portions of trees that have been hollowed out by advanced decay (Bull et al. 1992; McClelland and McClelland 1999; Aubry and Raley 2002a). In addition, through both cavity and foraging excavations, woodpeckers may facilitate the inoculation of live trees with heart-rot fungi (e.g., Aubry and Raley 2002b). Thus, in some regions, pileated woodpeckers may benefit fishers by initiating heartwood decay in live trees that will eventually create cavities that fishers can use for denning, or by providing fishers with access to naturally formed tree cavities that did not previously have an opening.

In each locality, forest management likely influences which tree species may become infected by heart-rot fungi and develop potential den and rest structures for fishers. Tree species managed for timber production (primarily conifers in western North America) are generally harvested before they reach older age-classes and become more susceptible to infection by heart-rot fungi. In contrast, species not managed for timber production (primarily hardwood species, but also conifer species without a high market value) are more likely to reach older ages and be exposed to various stressors that contribute to infection by heart-rot fungi. Some management practices retain older or deteriorating trees in leave patches or buffer zones (e.g., stream or riparian buffers). In this context, both conifer and hardwood species may have a higher incidence of decay by heart-rot fungi than do the same species in stands managed for timber production. Furthermore, if leave patches or buffer zones are permanent features on the landscape, the younger trees retained in these areas provide for future recruitment of older trees with heartwood decay.

Brooming or platforms in live trees caused by parasitic plants such as dwarf mistletoes (*Arceuthobium* spp.) or rust fungi (*Chrysomyxa* spp. or *Melampsorella* spp.), and the development of branch platforms in older trees, are additional ecological processes that are important for creating and maintaining fisher resting habitat in western North America. Because dwarf mistletoes and rust fungi can have a negative impact on timber production (i.e., tree deformities, stunted growth, and mortality [especially dwarf mistletoes]; Scharpf 1993; Allen et al. 1996; Parks and Bull 1997), infected trees are frequently removed to control or eradicate these pathogens from a stand. Consequently, some forest-management practices can interrupt these ecological processes and decrease the availability of microstructures that provide suitable resting habitat for fishers.

A better understanding of how these key processes function would provide valuable insights about regional variation in the availability of denning and resting habitat, and the ways in which fisher habitat may shift over time in response to disturbance events or changes in forest-management strategies. Also, few researchers have estimated the age of trees used for denning or resting (Davis 2009). Determining the ages of den and rest trees would improve our understanding of the time periods needed for these ecological processes to create suitable microstructures and, consequently, how forest managers can best maintain or promote them to benefit fishers.

Importance of Rest Structures and Sites for Thermoregulation

The need to minimize heat loss during cold weather is more important to fishers than was recognized previously. Buskirk and Powell (1994) speculated that, because of their larger size, thermal losses while resting are not as important for fishers as they are for American martens (*M. americana*), and access to subnivean spaces did not appear to influence their choice of resting habitat during winter. Although the metabolic demands and energetic constraints experienced by these 2 species during winter may differ, new evidence indicates that fishers in western North America select rest structures during cold weather that provide them with thermal benefits.

The strongest evidence that thermoregulatory constraints influence habitat selection by fishers comes from studies conducted in the northern part of their range in western North America. In British Columbia, fishers primarily used arboreal structures (branch platforms and cavities in trees) for resting throughout the year, but switched to logs or other ground structures (e.g., middens, burrows) when temperatures dropped and the snow pack was deep enough to provide subnivean spaces (Weir and Corbould 2008; Davis 2009). Fishers were most likely to select subnivean spaces associated with logs rather than arboreal structures for resting when ambient temperatures fell below –11 °C (Weir and Corbould 2008), and used subnivean spaces exclusively when temperatures were colder than –15 °C (Weir et al. 2004; Weir and Corbould 2008). Although other studies have not measured the temperature or precipitation regimes associated with the use of different rest structures by fishers, there is a general pattern of greater use of ground structures in regions that have colder winters (19–39% of all rest structures in British Columbia and the Cascade Range of Oregon; e.g., Weir and Harestad 2003; Aubry and Raley 2006; Davis 2009) than in those with milder winters (2–10% of rest structures in California; e.g., Zielinski et al. 2004b; Yaeger 2005; Purcell et al. 2009).

Available evidence also indicates that selection of rest sites by fishers in western North America may be influenced by their need for thermal relief during hot weather. Several investigators have hypothesized that, in the hotter and drier portions of their range (e.g., southern Sierra Nevada in California), selection by resting fishers for steep slopes, dense canopy cover, and proximity to water represents selection for cool sites and favorable microclimates (Zielinski et al. 2004b; Purcell et al. 2009). Although no data were available for warmer regions, there is evidence that the types of microstructures used by fishers for resting during hot weather (e.g., branch platforms, cavities in trees or snags) may also be influenced by thermoregulatory needs. We hypothesize that selection of rest structures and sites by fishers is influenced, in part, by their need to maintain thermoneutrality during both cold and hot weather to minimize their energetic costs. Focused investigations are needed to determine whether these patterns are consistent throughout western North America, or if they are pronounced only in regions that experience more extreme temperature or precipitation conditions. To elucidate the importance of thermoregulatory constraints on fishers, we recommend that future studies include focused sampling during periods of temperature and precipitation extremes, rather than simply comparing differences among seasons, which could mask important influences on resting habitat selection by fishers.

Importance of Canopy Cover at Multiple Spatial Scales

In western North America, a moderate to dense forest canopy is one of the strongest and most consistent predictors of fisher distribution and habitat use or selection at all spatial scales. The association of fishers with high amounts of canopy cover is further demonstrated by their avoidance of open environments. Early studies in the West indicated that canopy cover was important to fishers, but avoidance of areas with no tree or shrub cover was a more consistent pattern (Buskirk and Powell 1994; Powell and Zielinski 1994). Nevertheless, specific information on these associations at different spatial scales was lacking, and the biological significance of canopy cover to fishers was unclear. Based on the wealth of information now available, we conclude that moderate-to-dense canopy cover is a critical component of fisher habitat throughout western North America that is linked to multiple aspects of the fisher's life needs.

At regional and landscape scales, an increasing amount of forest canopy was the most consistent predictor of fisher occurrence in California (Carroll et al. 1999; Carroll 2005a; Davis et al. 2007; Zielinski et al. 2010). Similarly, fisher occurrence in the Rocky Mountain region was positively correlated with canopy cover up to an apparent threshold of 60% (Carroll et al. 2001). Fisher home ranges included primarily forests with moderate-to-high canopy-cover values for the region being studied. In British Columbia, fishers selected home ranges with ≥30% canopy cover (Weir and Corbould 2010). In the Sierra Nevada, one study showed mean canopy cover of 63% within female home ranges (Thompson et al. 2011), whereas another reported that 66% of the area within all fisher home ranges was composed of the densest (60–100%) canopy-cover class (Zielinski et al. 2004a). Fishers avoided open areas when selecting home ranges in British Columbia (Weir and Corbould 2010) and, in California, the average proportion of fisher home ranges that included open environments was ≤5% (Self and Kerns 2001; Zielinski et al. 2004a;

Higley and Matthews 2009). At smaller spatial scales, fishers selected sites for resting that had denser canopies than random sites (e.g., Zielinski et al. 2004b; Davis 2009; Purcell et al. 2009). The generality of this association throughout the West was confirmed by a meta-analysis of rest-site selection that included data from 8 study areas in British Columbia, Oregon, and California (Buskirk et al. 2010).

Previously, it was thought that the positive association of fishers with canopy cover (and their avoidance of open areas) was related to predator avoidance, but Buskirk and Powell (1994) noted that little evidence existed of predation on fishers to support that hypothesis. However, recent studies have demonstrated that predation on fishers in western North America is relatively common; documented predators included bobcat (*Lynx rufus*), cougar (*Puma concolor*), Canada lynx (*Lynx canadensis*), coyote (*Canis latrans*), and wolverine (*Gulo gulo*) (Truex et al. 1998; Weir and Corbould 2008; Higley and Matthews 2009); predation by raptors appears to be uncommon (e.g., Truex et al. 1998). Thus, it seems likely that selection for relatively dense canopies by fishers is explained, at least in part, by the vertical escape cover (i.e., tree boles) they provide from terrestrial predators (e.g., Weir and Corbould 2010). Forests with greater amounts of canopy cover are also likely to provide more favorable microclimatic conditions for fishers. The amount and structure of forest canopies have a profound influence on microclimates, including absorption of solar radiation (Spies 1998) and interception of snowfall, which affects snow-accumulation patterns (Storck et al. 2002). Recent studies have demonstrated that thermoregulation is important to resting fishers in cold (and potentially hot) weather; thus, denser canopies may be correlated with physiological optima for fishers. Data on foot-loading in *Martes* species indicate that traveling in deep, soft snow is energetically demanding for fishers (Krohn et al. 2004). Thus, interception of snow by the forest canopy probably creates more favorable traveling and foraging conditions for fishers. Canopy cover may also be linked to other habitat conditions important to fishers but not measured directly, such as abundant prey and the presence of trees that provide many of the critical habitat features on which fishers depend for reproduction and resting.

It is clear that canopy cover is an important component of fisher habitat in western North America, and that results from previous studies of habitat ecology will continue to be widely used by resource managers. However, because investigators used different field and analytical methods to estimate canopy cover (Table 10.2), we were not able to make direct comparisons among studies or evaluate critical thresholds of canopy cover for fishers at different spatial scales. Thus, we encourage investigators to carefully consider issues of both terminology and measurement for canopy cover when presenting their findings and designing new studies. The term *canopy cover* is commonly used to describe various vegetation measures, regardless of the angle of

Table 10.2. Selected examples of different analytical and field methods used by investigators to estimate the amount of forest canopy associated with fisher distribution or occurrence (first-order habitat use or selection), fisher home ranges (second-order), or sites used by fishers (third-order) in western North America

Source	Attribute[a]	Method
First-order		
Carroll et al. 2001	Canopy cover	GIS data derived from classified satellite imagery[b] (30-m resolution), moving-average within a 30-km^2 window
Davis et al. 2007	Canopy cover	GIS data derived from classified satellite imagery[b] (1-ha resolution), values calculated within a 10-km^2 sample unit
	Canopy closure	Ground estimates of canopy closure based on visual assessments
Zielinski et al. 2010	Canopy cover	GIS data derived from classified satellite imagery[b] (1-ha resolution), values calculated within a 5-km^2 moving window
Second-order		
Thompson et al. 2011	Canopy cover	GIS data derived from a combination of aerial photo interpretation and stand exam data
Weir and Corbould 2010	Canopy cover	GIS data derived from map-based ecosystem data
Zielinski et al. 2004a	Canopy cover	GIS data derived from a combination of several classified satellite images and aerial photography
Third-order		
Purcell et al. 2009	Canopy cover	Ground measurements using a moosehorn
Weir and Harestad 2003	Canopy cover	Ground estimates of vegetation cover based on visual assessments
Zielinski et al. 2004b	Canopy closure	Ground measurements using a densiometer

[a] Canopy cover variables derived from GIS data were typically categorical, whereas canopy cover or closure variables derived from ground measurements (e.g., moosehorn, densiometer) were typically continuous

[b] Methods used to classify satellite images, or to develop new aggregations of cover types from existing classified images, were different among first-order studies

view (Fiala et al. 2006). Specifically, however, canopy cover is the "proportion of the forest floor covered by the vertical projection of tree crowns," whereas *canopy closure* is the "proportion of the sky hemisphere obscured by vegetation when viewed from a single point" (Jennings et al. 1999: 62). Using data from field plots on the Sierra Nevada National Forest, at which canopy measurements were collected using different methods (Landram 2002), Purcell et al. (2009) created regression equations for converting canopy estimates at fisher rest sites among 3 different measurement methods. Although the applicability of these regression equations to other forest types and regions has not been evaluated critically, Purcell et al. (2009) demonstrated that the estimates often varied in size by 20–30% between 2 commonly used ground methods to measure canopy cover (moosehorns) and closure (densiometers). Different ground methods used to estimate the same canopy attribute (e.g., canopy closure) also produced variable results (e.g., Fiala et al. 2006). Similarly, canopy-cover estimates derived from satellite imagery or other GIS data may not be directly comparable among studies because of differences in classification methods, resolution, and scaling of data (Table 10.2).

There is no ideal method for estimating canopy cover or closure; rather, the method should match the research objectives and the spatial scale of interest. For some objectives (e.g., identifying potential canopy-cover thresholds for fishers in the West), we recommend the development of a coordinated approach that can be used in all ecosystems where fishers occur that includes standardized methods for estimating canopy cover at different spatial scales. However, we also believe that additional insights about the importance of canopy to fishers are still needed, and we urge investigators to design studies that test specific hypotheses and identify other attributes to measure that are potentially linked to the amount of canopy, but may be of greater importance to fishers (e.g., microclimate, prey abundance).

Importance of Forest Composition and Age

Buskirk and Powell (1994) hypothesized that fishers in western North America may require old-growth conifer forests for survival, especially Douglas-fir forests. Habitat studies conducted during the past 2 decades, however, demonstrate that fishers are not dependent on old-growth forests, per se, nor do they appear to be associated with any particular forest type. Rather, fishers occur in a variety of low- and mid-elevation forest types and use a diversity of plant communities. Perhaps the most consistent attribute of fisher home ranges is that they comprise a mosaic of forest plant communities and seral stages, but often include relatively high proportions of mid- to late-seral forests. Thus, it is clear that habitat conditions other than those found in old-growth forests are capable of supporting fishers in western North America if adequate canopy cover, large structures for reproduction and resting, vertical and horizontal escape cover, and prey can be found.

Fisher distribution in the West has been associated consistently with low- to mid-elevation forests (e.g., Carroll et al. 2001; Zielinski et al. 2010; Spencer et al. 2011) and the proportion of the landscape that contains mid- to late-seral mixed-conifer or mixed conifer-hardwood forests (i.e., increasing amounts of medium and large-sized trees and complex structure; Carroll et al. 1999; Davis et al. 2007; Zielinski et al. 2010). Studies in eastern North America indicated that fishers are inefficient energetically when traveling and hunting in terrain covered by soft, deep snow (Raine 1983; Krohn et al. 2004). Thus, higher elevations in western North America that receive substantial snowfall appear to be less suitable for fisher occupancy because of increases in snowpack (Aubry and Houston 1992; Powell and Zielinski 1994; Krohn et al. 1997). We hypothesize, however, that other factors associated with increasing elevation, including lower primary productivity and changes in forest structure (Franklin and Dyrness 1988; Meidinger and Pojar 1991), may also limit fisher distribution or abundance through their influence on the abundance of large structures needed for denning and resting and for providing an abundance of prey.

Fisher home ranges typically had moderate-to-high proportions of mid- and late-seral forests (on average, 42–72% of fisher home ranges; e.g., Zielinski et al. 2004a; Davis 2009), but we found no overarching patterns of selection for particular seral conditions or species composition at this scale (e.g., Higley and Matthews 2009; Weir and Corbould 2010). Results from analyses at other spatial scales were also variable (e.g., stands used within home ranges; Weir and Corbould 2008; Davis 2009; Higley and Matthews 2009); however, fishers rarely used the earliest seral conditions (e.g., herbaceous stage) or nonforested vegetation types (e.g., Weir and Harestad 2003; Aubry and Raley 2006; Purcell et al. 2009). We hypothesize that, when they establish their home ranges, it benefits fishers to include a diversity of available forest conditions, thereby increasing their access to a greater diversity and abundance of prey species while still providing habitat features important for reproduction and thermoregulation. The lack of any overarching patterns of selection by fishers for particular forest types or seral stages may also be due, in part, to differences in management history among locales and subsequent influences on forest structure. In northern California, young pole (10–29 year-old) and conifer-hardwood (≥30 year-old) stands selected for denning by reproductive females appeared to provide habitat similar to that typically found in older forests (≥80 years) because of the relatively high amounts of residual structure left in those stands when they were harvested (Higley and Matthews 2009). In the southern Sierra Nevada, forests have been altered substantially by almost a century of selective timber harvesting (McKelvey and Johnston 1992) and decades of fire suppression (Purcell et al. 2009). In portions of this region, fisher resting habitat is characterized by large legacy trees and snags surrounded by dense stands of smaller trees (Purcell et al. 2009).

Previously, it was thought that fishers in western North America may favor riparian forests (Buskirk and Powell 1994; Powell and Zielinski 1994); however, results from recent studies do not support this hypothesis. Although riparian forests were important to fishers in some locales (e.g., black cottonwood [*Populus balsamifera trichocarpa*] forests provided denning habitat in British Columbia; Weir and Corbould 2008), consistent use or selection for riparian forests has not been demonstrated. Several studies found that fisher rest sites were located closer to a stream or a body of water than random sites (e.g., Zielinski et al. 2004b; Yaeger 2005; Purcell et al. 2009). We caution against the assumption that all such results represent selection of riparian forests, however, because differences in vegetation composition between used and random sites were not evaluated, and the width of riparian zones in mountainous terrain can be quite narrow (Brinson and Verhoeven 1999). Also, some studies used maps to identify the nearest stream to a fisher rest site, but they did not always determine the stream type (e.g., perennial, ephemeral) nor verify the presence of water at the time the sites were used by fishers (e.g., Seglund 1995; Purcell et al. 2009). Use of areas near water, in conjunction with other site conditions (e.g., low topographic position, steep slopes, high canopy cover), more likely reflects selection by fishers for cooler microclimates and, perhaps, for more productive sites (e.g., Zielinski et al. 2004b; Yaeger 2005; Purcell et al. 2009).

Several investigators have hypothesized that forests with a hardwood component, especially mast-producing species, may be more productive than others and provide fishers with diverse and abundant prey, and more den and rest structures (e.g., Carroll et al. 1999; Yaeger 2005; Purcell et al. 2009). Zielinski et al. (2004a) hypothesized that the greater abundance of forest types with black oak in the southern Sierra Nevada compared with coastal California enabled female fishers to occur at higher densities and meet their life needs within smaller home ranges. At finer spatial scales, some hardwood species clearly provide important structures for denning and resting (e.g., Reno et al. 2008; Weir and Corbould 2008; Higley and Matthews 2009). Selection by fishers for mixed conifer-hardwood forests has not been demonstrated, however, nor do we know whether such forests provide more resources for fishers. A better understanding of the importance of hardwood tree species for fishers, and of potential variation in their role in fisher habitat ecology among regions, represent important research needs.

Hierarchical Habitat Selection

Prior to 1994, few investigators incorporated spatial scale into their study designs explicitly. However, the overarching patterns we describe in this chapter indicate clearly that fishers use or select habitat in a hierarchical fashion. Hierarchical habitat selection is based on the premise that large-scale ecological processes occurring at relatively slow rates constrain those occurring at finer spatial scales at faster rates (e.g., Johnson 1980; King 1997). Hence, the fisher's geographic distribution (first-order selection) establishes the ecological niche that fishers can occupy successfully, which is further refined at the home range scale (second-order selection). The environments (third order) and resources (fourth order) selected by fishers within their home ranges are influenced and ultimately constrained by landscape-scale conditions and processes. Overall, environments and resources used or selected at all spatial scales by fishers are linked and collectively provide the annual and daily life needs of persistent populations.

First-order habitat selection by fishers has been researched extensively in California (e.g., Carroll et al. 1999; Davis et al. 2007; Spencer et al. 2011), and to a lesser extent in the Rocky Mountains (Carroll et al. 2001). These studies have provided important insights about the landscape conditions that currently support fisher populations. Fisher occurrence was associated consistently with expanses of dense, structurally complex, and productive forests (based on tree biomass, greenness, wetness, and other indices of primary productivity). Landscape-scale abiotic factors, such as annual precipitation, topographic relief (i.e., landscape ruggedness), and elevation were also important in predicting fisher occurrence. The distribution of fishers was correlated with mid-elevation areas (Carroll et al. 2001; Zielinski et al. 2010) that had moderate levels of annual precipitation (Davis et al. 2007; Spencer et al. 2011), supporting the hypothesis that deep, persistent snow may limit fisher distribution (e.g., Krohn et al. 1997). Metrics of landscape ruggedness may actually represent forests with limited anthropogenic alterations, rather than the particular physiographic conditions that are selected by fishers. Although rigorous studies of first-order habitat selection have been conducted only in California, descriptive information from studies conducted elsewhere in western North America support the conclusion that fishers occur in a variety of low- to mid-elevation forests with relatively dense canopies (see Lofroth et al. 2010).

Within their distributional niche, fishers established home ranges comprising a mosaic of available forest types and seral stages, but contained relatively high proportions of mid- to late-seral forest conditions, moderate-to-dense canopy cover, and few open areas (e.g., Zielinski et al. 2004a; Higley and Matthews 2009; Weir and Corbould 2010). Our review did not reveal particularly strong selection at this scale, but that may reflect the limited number of studies that have been conducted at this scale (Table 10.1). Also, other factors that may influence the establishment of home ranges by fishers (e.g., landscape fragmentation, heterogeneity, edge ecotones) have not been well studied (but see Thompson et al. 2011).

Habitat use or selection by fishers was strongest and most consistent at finer spatial scales (third- and fourth-order selection). Fishers were associated with complex vertical and horizontal structure (e.g., large live trees, snags, and logs, and moderate-to-dense canopy cover), and they used or selected large structures characteristic of late-seral forests for reproduction and resting. What appeared most important to fishers at finer spatial scales were the ecological processes that create large structures with suitable microstructures for reproduction (cavities) and resting (e.g., cavities, mistletoe and rust brooms), and complex forest structure that provides security cover, favorable microclimates for thermoregulation, and, potentially, abundant prey.

Information Needed to Improve Management and Conservation

Research findings produced during the last 2 decades have now provided a foundational understanding of fisher habitat ecology in western North America, but we have identified various information needs throughout our review, as appropriate. Here, we present other significant information gaps that, if addressed adequately, would advance our understanding of the habitat-related factors that may influence fisher abundance and distribution and, ultimately, improve management and conservation efforts. Although our focus is habitat, we recognize that other factors not covered in this chapter, such as predation, competition, disease (Gabriel et al., this volume), and anthropogenic impacts (e.g., vehicle-related fisher mortalities), may also influence or limit the distribution and abundance of fishers.

Currently, there is limited information on the factors that may influence the selection of home ranges by fishers (second-order selection). Understanding whether home range establishment is mediated by forest type, canopy cover, abundance of structures, or prey (or a combination of factors) will be essential for maintaining or expanding fisher populations in western North America. This information is also needed to gain a better understanding of the amount and spatial configuration of forest conditions that fishers may need to meet their life needs, and the potential thresholds at which forest fragmentation (natural and anthropogenic) may preclude home range establishment and, thus, landscape occupancy.

Because fishers depend on structural elements to meet their reproductive and thermoregulatory needs, their habitat does not necessarily parallel that of their primary prey, as suggested by Buskirk and Powell (1994). Our understanding of fisher food habits has improved considerably; we know that the fisher's diet is diverse, varies regionally (Zielinski et al. 1999; Weir et al. 2005; Aubry and Raley 2006; Golightly et al. 2006), and is not as strongly linked to snowshoe hares as was suggested previously (Buskirk and Powell 1994). However, fisher foraging habitat remains undescribed and, although differences in prey diversity, abundance, and catchability among forest conditions may influence home range establishment by fishers (second-order selection) and habitat use at finer spatial scales (third- and fourth-order selection), no studies have tested these hypotheses.

Recent studies provide some evidence that female fishers are more selective than males for forest types at the home range scale (second order; Zielinski et al. 2004a) and for resting habitat at finer spatial scales (third and fourth order; Zielinski et al. 2004b; Yaeger 2005; Aubry and Raley 2006). Because fishers have a low reproductive capacity, and males do not contribute to the raising of young (Lofroth et al. 2010), the habitat needs of females are critical for population growth. More-focused investigations, or meta-analyses of existing data, are needed to determine whether these apparent patterns of gender-based selection are consistent throughout western North America.

Few studies have investigated the abundance or spatial distribution of potentially limiting resources for fishers (Aubry and Raley 2006; Calabrese and Davis 2010). Given strong third- and fourth-order habitat use and selection by fishers (e.g., females are obligate users of tree cavities for reproduction), we recommend developing a common research approach for determining whether reproductive den structures are a limiting resource. Such an effort should be based on a sampling design that is applicable in all ecosystems where fishers occur in western North America, and focused on quantitative comparisons of the availability and spatial distribution of suitable structures (especially large live trees and snags with cavities) within female home ranges to correlates of fitness, such as reproductive success or kit recruitment. A similar approach could be used to test hypotheses regarding the potential scarcity of other resources (e.g., rest structures).

Finally, none of the studies we reviewed related correlates of fitness (e.g., reproduction, survival) to different forest conditions used or selected by fishers. Buskirk and Powell (1994) recognized the difficulty of collecting such measurements, but cautioned against the implicit assumption that resource selection confers greater fitness to individuals making those choices or that environments being selected are of higher quality than others. We believe that continuing advances in monitoring and analytical methods (e.g., radiotelemetry, noninvasive survey methods, occupancy analysis, and genetic and stable-isotope analyses; see Long and MacKay, this volume; Schwartz et al., this volume; Shirk et al., this volume; Slauson et al., this volume; C.M. Thompson et al., this volume) offer important new opportunities to researchers for relating fitness to habitat use or selection at multiple spatial scales and, ultimately, for quantifying fisher habitat quality in ways that will enable the development of more-comprehensive predictive models.

The overarching patterns of fisher habitat use or selection we have described in this chapter demonstrate strong support for some of the hypotheses proposed by Buskirk and Powell (1994; e.g., physical structure of stands is important, fishers avoid open environments) but limited support for others (e.g., importance of riparian forests). We found no support for their speculations that fishers in western North America depend on old-growth conifer forests or are primarily associated with Douglas-fir forests, or that the thermoregulatory needs of fishers do not influence habitat selection. However, many of the suggestions put forth by Buskirk and Powell (1994) to fill important knowledge gaps have not been implemented (e.g., investigate the effects of habitat fragmentation, relate prey catchability to forest structure and prey abundance, compare the structural conditions and prey associations selected by fishers in mixed broad-leaved forests in the East with those in mixed-conifer forests in the West). Thus, we encourage fisher researchers to design new kinds of studies that will effectively address these remaining knowledge gaps. Additionally, during the past 2 decades, an unprecedented number of high-quality datasets have been collected on fisher habitat ecology in western North America. These datasets provide potentially important opportunities for conducting quantitative meta-analyses (e.g., Buskirk et al. 2010) that would further elucidate the overarching patterns of habitat selection by fishers in western North America that we have described in this chapter.

Acknowledgments

This chapter was improved with helpful reviews by Fraser Corbould, Kurt Jenkins, and Craig Thompson. We also benefited from insightful discussions and suggestions from Keith Aubry and Bill Zielinski.

11

Habitat Ecology of *Martes* Species in Europe

A Review of the Evidence

EMILIO VIRGÓS, ANDRZEJ ZALEWSKI, LUIS M. ROSALINO, AND MARINA MERGEY

ABSTRACT

We reviewed published information about the habitat ecology of the European pine marten (*Martes martes*) and stone marten (*M. foina*) in Europe. The European pine marten has long been considered a forest specialist; however, recent studies indicate a greater habitat flexibility for this species. European pine martens can persist in fragmented (fine-grained) landscapes and can exploit unforested areas on Mediterranean islands. Patterns of habitat selection can be explained as trade-offs among available food sources, predation risk, and access to resting sites that provide thermal benefits, yet most studies lack adequate quantification of these factors. The stone marten is synanthropic in most of its geographic range in northern, eastern, and central Europe. It inhabits more-natural landscapes in Mediterranean areas, and it can inhabit more fragmented and unforested areas than the European pine marten. The accepted explanation for this geographic difference in the stone marten's association with human-impacted habitats is greater competition with European pine martens in northern areas, but recent studies have shown that the 2 species coexist in some parts of central and southern Europe. Thus, more studies are needed to determine how the current and historical availability of forested and urban areas may explain geographic differences in the synanthropic behavior of the stone marten. Future research needs include multiscale studies, studies designed to test specific hypotheses, and better quantification of the different factors hypothesized to be important for both species.

Introduction

Habitat is a key element in the successful completion of an individual's life history, and the selection of a particular habitat can greatly affect fitness

(Morris 1987). *Habitat* has been defined as the physical elements used by individuals during their daily activities (Morrison and Hall 2002). In contrast, a species' *ecological niche* is the entire set of environmental factors that determine the distribution of individuals, populations, and species along a series of *n*-dimensional axes (Chase and Leibold 2003). Confusion surrounding these definitions has had consequences for the study of species-environment relations. Here, we use the concept of habitat in the narrower sense; that is, as those landscape units or cover types that are selected (i.e., used disproportionately to their availability; Manly et al. 2002) by individuals in a particular region.

Habitat selection has been studied using various methods and at various spatial scales (including both grain and extent; Farina 1998). In our review, we evaluated the impact of variations in approach and scale on the conclusions drawn. Selection of habitat types or structural elements can be constrained by the scale of analysis (Johnson 1980; Wiens et al. 1993). For example, the preference for pine forests over urban settlements by stone martens (*Martes foina*) in central Spain was detected only at spatial scales large enough to include all available habitat types (e.g., Virgós and Casanovas 1998); however, with more-detailed studies at smaller spatial scales, it is possible to detect the use of small human settlements within forested areas for resting or denning. In other words, habitat selection varies with the spatial scale of the analysis. A robust explanation of the underlying mechanisms for habitat preferences needs to address how individuals and populations select habitat elements available at many spatial scales: from the use of particular areas within home ranges, to changes in abundance among habitat types in a region, to the entire distribution of the species (Johnson 1980; Ims 1995).

For all *Martes* species, forested habitats are key features (Buskirk and Powell 1994; Bissonette et al. 1997; Zalewski and Jędrzejewski 2006). In Europe, the genus *Martes* is represented by the European pine marten (*M. martes*) and stone marten (Mitchell-Jones et al. 1999; Proulx et al. 2004). The European pine marten is found in boreal and Eurosiberian regions, with some populations also occurring in the Mediterranean area (Mitchell-Jones et al. 1999; Proulx et al. 2004), but it is absent in large areas of Portugal and Spain. In contrast, the stone marten is distributed from the Mediterranean region to central and eastern Europe but not found in Scandinavia or the northern part of European Russia (Mitchell-Jones et al. 1999; Proulx et al. 2004). Thus, although the European pine marten occurs farther north, both species co-occur throughout much of Europe. In this region, both species have adapted to use a variety of habitat types, from coniferous forest in northern Europe, to deciduous and Mediterranean forests in southern Europe (Grakov 1981; Marchesi 1989; Clevenger 1994b; Zalewski 1997a). Both species may occur in human-impacted habitats, including largely fragmented forests and urban and suburban areas of villages and towns (e.g., Broekhuizen and Müskens

2000; Goszczyński et al. 2007; Herr et al. 2009). Although many studies have been conducted on both species in a variety of habitats, most were limited to only 1 spatial scale (but see Santos and Santos-Reis 2010), and no review has been conducted of multiscale habitat selection.

In this chapter, we report results from a comprehensive review of published literature on the habitat ecology of both species. We searched for scientific literature in the journals included in the Science Citation Index system (SCI, Thomson Reuters, USA), and also evaluated findings published in national and regional journals, including some information published as abstracts or proceedings in national and international congresses. The geographic locations of the studies we considered are shown in Figure 11.1.

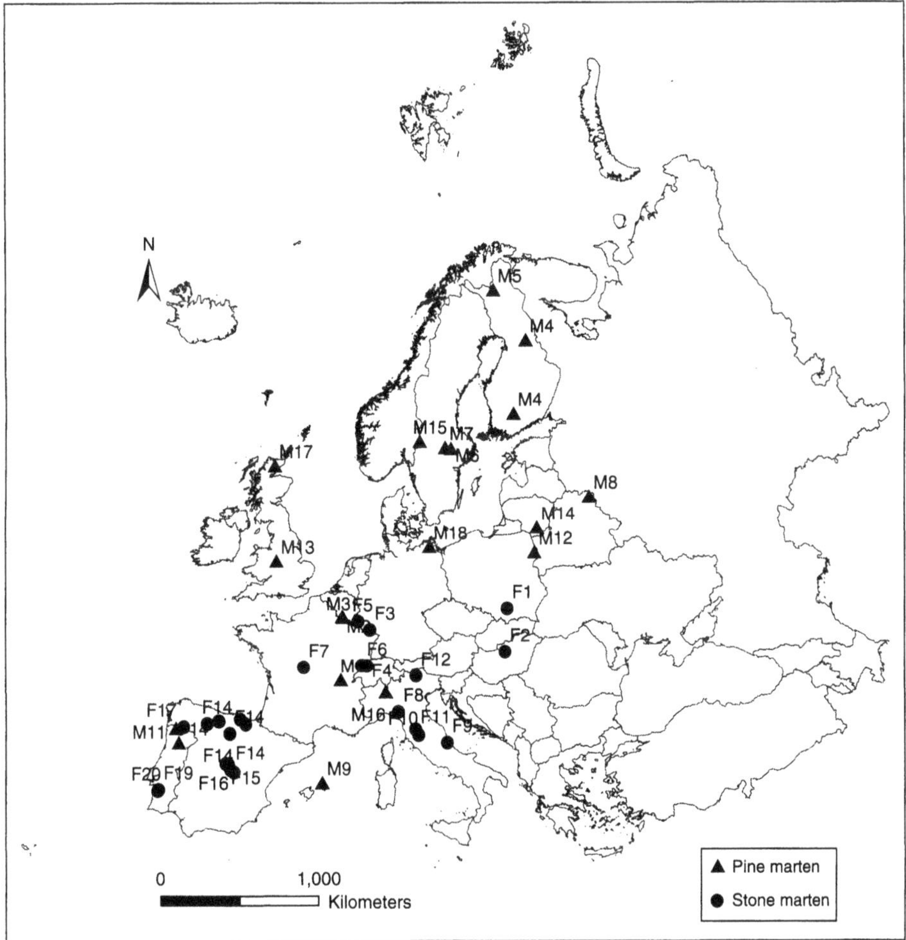

Figure 11.1. Map of Europe showing the locations of the studies used in this review. Studies of the European pine marten are indicated with "M," and studies of the stone marten are indicated with "F." Additional details on each study are presented in Tables 11.1 and 11.2.

European Pine Marten

Most studies of European pine martens were conducted in study areas of <200 km^2, which we consider a small spatial scale (Johnson 1980). Although one-third of these studies used radiotelemetry (Table 11.1), some used sign surveys or camera trapping. Studies conducted in larger areas (>1000 km^2) were relatively common but generally based on questionnaires, with some use of sign surveys.

We observed a striking shift in perceived habitat selection by this species over time. The earliest studies on this topic reported the pine marten to be an old-forest specialist (Grakov 1981; Pulliainen 1981c; Wabakken 1985; Lindström 1989; Storch et al. 1990) associated with vast, undisturbed stands of older, mixed forests dominated by spruce (*Picea* spp.). However, all these studies were conducted in the boreal region (i.e., the conifer-dominated forest belt below the Fennoscandian mountains, characterized by a cool and mainly continental climate), and these findings were not consistent with those of more recent studies conducted farther south. In the last decade, the pine marten has been shown capable of living in small wooded areas, although it is still considered to require forests (Storch 1988; Schröpfer et al. 1989). Kurki et al. (1998) were among the first to show that European pine martens can reach high abundances in fragmented landscapes, although a correlation between fragmentation and population abundance has not yet been demonstrated. Population trends indicate an increase in numbers since the 1980s, even though agricultural lands have expanded in extent during the same period. In the British Isles, the species survived extensive deforestation (Birks et al. 2005). Moreover, Ruette et al. (2005), Pereboom et al. (2008), Mergey et al. (2011) in France, and Balestrieri et al. (2010) in Italy have documented the persistence of European pine martens in primarily agricultural landscapes, where the size of forest patches ranges from hedgerows (<1 ha) to woodlots (>100 ha). Hedgerows have been shown to be important for resting sites or as corridors between forest patches. Thus, although forest cover is still necessary for their persistence, pine martens can tolerate a broader range of landscape conditions than was previously believed, including highly fragmented landscapes (Zalewski and Jędrzejewski 2006; Balestrieri et al. 2010; Mergey et al. 2011). For example, Brainerd et al. (1994) showed that they can live in young forest stands, and that a preference for old forests is not a general rule. Similarly, Brainerd and Rolstad (2002) reported that forest structure (e.g., tree height, developed understory) was a more important determinant of habitat selection by pine martens than forest composition and age. Studies conducted in Lithuania (Baltrūnaitė 2006b) and Belarus (Sidorovich et al. 2005) corroborated these findings. In the temperate regions of Europe, pine martens inhabit forests characterized by a diversity of tree species. Indeed, forests can be deciduous (Castién and Mendiola 1985; Álvares and Brito 2006; Matos

Table 11.1. Habitat-selection studies of the European pine marten considered in this review, including some that also studied the stone marten

ID[a]	Location	Spatial extent	Method	Primary habitat used	Source
M1	France	260 km^2	Radiotelemetry	Woods and ecotones	Ruette et al. 2005
M2	France	300 km^2	Radiotelemetry	Woods and ecotones	Pereboom et al. 2008
M3	France	900 km^2	Radiotelemetry	All forest patches	Mergey et al. 2011
M4	Finland	3000–4000 km^2	Sign surveys	Young forest stands	
M5	Finland	200 km^2	Sign surveys	Coniferous and mixed forests	Pulliainen 1981c
M6	Sweden	30 km^2	Hunting records	Old-growth forests	Lindström et al. 1995
M7	Sweden	140 km^2	Sign surveys	Old-growth forests	Storch et al. 1990
M8	Sweden and Norway	75 km^2, 140 km^2	Radiotelemetry	Spruce-dominated forests with large trees	Brainerd and Rolstad 2002
M9	Belarus	300 km^2, 250 km^2	Radiotelemetry	Forests and ecotones	Sidorovich et al. 2005
M10	Spain (Balearic Islands)	30 km^2	Sign surveys	Forests, coastal scrublands, and open areas equally	Clevenger 1994b
M11	Portugal	45,000 km^2	Atlas dataset	Mature broad-leaved forests	Matos and Santos-Reis 2006
M12	Portugal	1974 km^2	Questionnaires	Broad-leaved forests	Álvares and Brito 2006
M13	Poland	47 km^2	Radiotelemetry	Mature forests with warm temperatures	Zalewski 1997a
M14	United Kingdom	National	Atlas dataset	Broad-leaved forests	Velander 1983
M15	Lithuania	150 km^2	Sign surveys	Forests	Baltrūnaité 2006b
M16	Italy	17 km^2	Sign surveys	Riparian vegetation	Balestrieri et al. 2010
M17	Scotland	60 km^2	Radiotelemetry	Mature coniferous forests and rough grassland	Caryl 2008
M18	Germany	17 km^2	Radiotelemetry	Deciduous forests with a minor coniferous component	Stier 2000

[a] IDs correspond to the study locations shown in Figure 11.1

and Santos-Reis 2006), mixed (Labrid 1986; Marchesi 1989; Zalewski 1997a; Barja 2005), or coniferous (Velander 1983; Ruiz-Olmo et al. 1988; Ruiz-Olmo and López-Martín 1995).

As the number of studies increased, reported patterns of habitat selection by pine martens became more complex. More recent findings are strengthened by improved accuracy in both geographic location and species identification (e.g., DNA-based identification), which has provided a clearer picture of habitat selection. From 2000 to 2010, the number of studies based on radiotelemetry increased, and results from those studies have raised concerns about the accuracy of identifications based on sign in areas of sympatry (e.g., sign surveys, questionnaires, Atlas dataset; Birks et al. 2005).

Habitat-selection studies should address the potential mechanisms explaining spatial and temporal patterns of habitat use. Individuals select habitat as a consequence of various habitat factors, including food quality and availability, predation risk (e.g., Brown 1988; Lima and Dill 1990), and intra- and interspecific competition (Fretwell 1972; Rosenzweig 1981; Morris 1989). For the European pine marten, 3 main factors have been proposed to explain these patterns: food availability, predation risk, and thermoregulation during resting times (Pulliainen 1984; Buskirk and Powell 1994; Thompson and Harestad 1994; Zalewski 1997a). Choices by individual martens might reflect trade-offs among these factors.

In most boreal forests, old forests dominated by spruce are selected by pine martens because they represent the best combination of abundant food, access to well-insulated resting and denning sites, and low risk of predation (Grakov 1981; Pulliainen 1981c, Wabakken 1985; Lindström 1989; Storch et al. 1990; Kurki et al. 1998; Brainerd and Rolstad 2002; but see Brainerd et al. 1994); nonetheless, in some boreal landscapes, habitat fragmentation has produced new costs and benefits in available habitat types. In fragmented landscapes, young stands support higher concentrations of prey than large, continuous forests (Brainerd et al. 1994; Brainerd and Rolstad 2002) while maintaining favorable conditions for thermoregulation and risk of predation. The same pattern has been observed in temperate regions, where food availability is higher in heterogeneous landscapes, and where the co-occurrence of forests, hedgerows, riparian areas, agriculture patches, and grasslands promotes higher biodiversity (Rosalino et al. 2009) than in continuous forests. Accordingly, the selection of hedgerows and riparian vegetation for establishing a home range may be due to higher abundances of small mammals in these habitat types, as was shown by Sidorovich et al. (2005). Decaying snags and tree cavities that are frequently available in these forest patches also provide well-insulated resting sites and protection from predators. Habitat-selection patterns described by Clevenger (1994b) in the Balearic Islands provide additional support for the importance of these 3 factors. In that region, the species preferred scrublands and open areas to continuous pine stands, again because of greater food availability and the lack of predators, which preclude the use of these habitat types in other parts of their distribution. In the British Isles, Birks et al. (2005) also showed that pine martens modified their habitat-selection patterns, using open habitats and rocky areas under conditions of extensive deforestation.

Finally, the presence of European pine martens in apparently unsuitable environments, such as agricultural lands, suggests significant behavioral flexibility. Flexibility in feeding behavior (Marchesi 1989; Jędrzejewski et al. 1993) is probably the primary trait that allows the species to adapt to diverse environmental conditions. Thus, their strong aversion to using open habitats for foraging, resting, or traveling could be due to predation risks and thermoregulatory constraints. Storch et al. (1990) suggested that the absence of the species in suitable habitats could be explained by intraguild predation by red foxes (*Vulpes vulpes*) or wildcats (*Felis silvestris*), or, as Clevenger (1993b) reported, by intraguild competition with stone martens. Contrary to this hypothesis, however, Lindström et al. (1995) could not demonstrate a higher use of those habitats when red fox population numbers decreased after an outbreak of sarcoptic mange. They proposed that the risk-aversion hypothesis may also explain underutilization of some food-rich habitats in boreal landscapes. Unfortunately, predator abundance in each habitat was not measured in previous studies, so this hypothesis still needs field testing.

No findings are available from Europe providing strong inferences on the role of thermoregulation in habitat selection, especially for resting sites; however, because the European pine marten has the northernmost distribution of European martens, which includes many areas with very cold winters, we believe that both thermoregulatory needs and prey availability may influence habitat selection by pine martens. We recommend that future studies address these questions by measuring food abundance and thermal conditions simultaneously and consider potential interactions between these factors.

Stone Marten

Most studies of stone martens have been conducted at small spatial scales (<100 km^2); only a few have covered larger areas (>500 km^2) encompassing a variety of biogeographic regions (e.g., Ruiz-González et al. 2008) or habitat types (Table 11.2). Almost half the studies of stone martens (9 of 20) involved the use of radiotelemetry (Table 11.2); the remainder used mostly sign surveys. However, the relative importance of methods used has shifted during the last 2 decades, with indirect methods (sign surveys) becoming more common compared with radiotelemetry studies. This pattern differs from that for the pine marten, for which most recent studies have used radiotelemetry, and may have led to questionable results for some stone marten studies.

Table 11.2. Habitat-selection studies of the stone marten considered in this review, including some that also studied the European pine marten

ID[a]	Location	Spatial extent	Method	Primary habitat used	Source
F1	Poland	<100 km^2	Radiotelemetry	Urban	Eskreys-Wójcik et al. 2008
F2	Hungary	525 km^2	Sign surveys and interviews	Urban areas	Tóth et al. 2009
F3	Germany	30–59 km^2	Radiotelemetry	Woodland patches	Herrmann 1994
F4	Switzerland	30 km^2	Radiotelemetry	Forests and pastures	Lachat-Feller 1993a
F5	Luxembourg	1.6–5 km^2	Radiotelemetry	Urban areas	Herr et al. 2009
F6	France	100 km^2	Radiotelemetry	Forests and pastures	Michelat et al. 2001
F7	France	Small but undefined	Radiotelemetry	Forests and pastures	Lodé 1991b
F8	Italy	109 km^2	Sign surveys	Shrubs	Sacchi and Meriggi 1995
F9	Italy	36 km^2	Radiotelemetry	Forests and shrubs	Rondinini and Boitani 2002
F10	Italy	800–1000 km^2	Sign surveys and camera trapping	Forests with abundant fruit	Mortelliti and Boitani 2008
F11	Italy	107 km^2	Radiotelemetry	Forests and shrubs	Genovesi et al. 1997
F12	Italy	24 km^2	Sign surveys	Urban areas and mixed woods	Prigioni et al. 2008
F13	Spain	2236 km^2	Visual censuses and sign surveys	Meadows and open habitats	Zabala et al. 2009
F14	Spain	>5000 km^2	Sign surveys	Forests and riparian woodlands	Virgós and García 2002
F15	Spain	500–600 km^2	Sign surveys	Forests	Virgós and Casanovas 1998
F16	Spain	150–200 km^2	Sign surveys	Forests interspersed with rocky areas	Virgós et al. 2000
F17	Spain	67 km^2	Sign surveys and camera trapping	Forests	Rosellini et al. 2008
F18	Spain	>5000 km^2	Sign surveys	Forests	Ruiz-González et al. 2008
F19	Portugal	20 km^2	Radiotelemetry	Forests	Santos and Santos-Reis 2010
F20	Portugal	35 km^2	Radiotelemetry	Riparian vegetation and cultivated fields	Santos-Reis et al. 2004

[a] IDs correspond to the study locations shown in Figure 11.1

Stone martens show greater habitat generalization than other *Martes* (Delibes 1983; Libois and Waechter 1991; Proulx et al. 2004), inhabiting steppes, rocky areas, forests, and urbanized landscapes. Human associations with stone martens are well studied (Waechter 1975; Skirnisson 1986; Lodé 1991b; Herrmann 1994; Broekhuizen and Müskens 2000; Herr et al. 2009) and occur primarily in central and northeastern Europe. In the southern part of their distribution, they prefer more natural habitats (Delibes 1983; Libois and Waechter 1991). It has been suggested that this geographic variation in habitat preferences is associated with the co-occurrence of European pine martens in the northern part of their range, where they outcompete stone martens. Thus, the stone marten's association with human features may have arisen to reduce interspecific competition, and is believed to be a more likely explanation than thermal or energetic constraints (Delibes 1983). In fact, the stone marten originated in the Middle East and is therefore well adapted to dry, unforested landscapes (Anderson 1994). The species prefers to den and rest in refuges with efficient thermal insulation (e.g., hollow trees; Santos-Reis et al. 2004). This pattern may also explain its preference for rocky habitats in most of its geographic range (Delibes 1983; Libois and Waechter 1991; Virgós and Casanovas 1998), because rocky areas can provide both insulated and well-protected refugia. It has been suggested that stone martens use human buildings instead of forest or rocky refuges in landscapes where European pine martens co-occur (e.g., Eskreys-Wójcik et al. 2008). In areas where pine and stone martens coexist, however, the relative abundance of stone martens is strongly affected by such factors as landscape composition, human disturbance, and prey availability (Ruiz-Olmo and López-Martín 1995; Vadillo et al. 1997; Pilot et al. 2007; Rosellini et al. 2008; Ruiz-González et al. 2008; Balestrieri et al. 2010). Thus, these species can coexist, and the preference by stone martens for human settlements in some parts of Europe may also have resulted from other factors, such as human population density and its historical dynamics. For example, in southwestern Portugal, where the European pine marten is absent, the stone marten uses abandoned barns and cork piles as resting sites in villages (Santos and Santos-Reis 2010). In the same area, another study showed that the core area of a male stone marten monitored with radiotelemetry was located within a rural village, and almost 90% of its locations were in human-made structures near or in the village (Santos-Reis et al. 2004). To test this apparent preference for anthropogenic structures, future studies should investigate habitat selection by stone martens in areas with lower current and historical forest availability associated with long-term human occupation and urbanization. Such situations could force this flexible species to expand its habitat niche from forest into urban areas. Where human population density is lower and forest constitutes a larger proportion of the landscape, both martens can coexist because subtle differences in competitive abilities are mediated by landscape composition (Virgós and Casanovas 1998).

In areas where forests are abundant, the preference of stone martens for forested areas is clear; for example, stone martens are associated with forests in the mountains of central Spain (Virgós and Casanovas 1998), although a more-detailed study indicated their preference for a mosaic of forest and rocky areas (Virgós et al. 2000). This latter preference likely reflected a trade-off between sites for thermal protection in rocky areas, and higher food availability and lower predation risk in forests. The preference for mosaics with moderate levels of forest fragmentation was shown in several studies (Sacchi and Meriggi 1995; Michelat et al. 2001; Rondinini and Boitani 2002). All these studies indicated a preference for areas with some human presence (agricultural areas, hedgerows, and woodlots), where food and protected sites are interspersed. Stone martens are capable of establishing home ranges in fine-grained, fragmented landscapes, if some forest cover is available (Genovesi et al. 1997). Similar conclusions were reported from other studies conducted in fragmented landscapes that demonstrated the importance of woodland-patch size (Rondinini and Boitani 2002; Virgós and García 2002; Mortelliti and Boitani 2008).

All these studies showed that relatively large forest patches are needed to maintain viable populations of stone martens in fragmented landscapes. Although stone martens are capable of persisting in other habitat types, these findings underscore the importance of forests for this species. The presence of stone martens is also linked to riparian areas, especially in cultivated landscapes, where riparian vegetation provides the only overhead cover available (Virgós 2001; Rondinini and Boitani 2002; Matos et al. 2009). Thus, both forests and riparian areas provide vertical cover within the landscape that may also function as travel corridors and play important roles in dispersal, foraging, and scent-marking movements (Virgós 2001; Matos et al. 2009). Vertical cover may be especially important for species such as the stone marten that are better adapted to arboreal than cursorial movements (Nowak 2005).

The presence of stone martens in urban environments is further evidence of their ecological flexibility; for example, stone martens occur throughout the city of Krakow, Poland, where they rest in attics or roof spaces (Eskreys-Wójcik et al. 2008). In Budapest, Hungary, stone martens use older houses with courtyards, small gardens, and circular galleries that provide both food (household garbage, birds, small mammals) and refuges (attics, roof spaces, suspended ceilings, church towers) (Tóth et al. 2009). The same behavior has been noted in Bettembourg and Dudelange, Luxembourg, in western Europe, where stone martens occupy urban home ranges, rest and den mostly in buildings, and have adapted their activity patterns to avoid humans and vehicular traffic (Herr 2008). Furthermore, in that area, animals often climb into car-engine compartments in search of shelter and warmth, where they may gnaw on insulation and rubber or plastic components (Herr et al. 2009). Stone martens also kill poultry, raid gardens, or damage house ceilings with urine and feces. In some urban areas, these conflicts have resulted in human persecution of the species.

The stone marten is the most frugivorous carnivore in Europe (Rosalino and Santos-Reis 2009), and fruit availability is an important determinant of its distribution and abundance. Mortelliti and Boitani (2008) showed that the availability of fruits and other food resources may help explain the distribution of the species in fragmented landscapes. The potential inclusion of food-availability estimates in habitat models has not been examined extensively, but a recent study in central Spain showed how the abundance of fruiting tree species shaped local abundances of the stone marten (Virgós et al. 2010).

For the European pine marten, the abundance of red foxes and other predators or competitors is believed to be one of the main factors determining abundance at various spatial scales (Storch et al. 1990). Similarly, European pine martens and genets (*Genetta genetta*) may affect the distribution of stone martens (Barrientos and Virgós 2006). Several studies have indicated that stone martens and genets can coexist (Santos-Reis et al. 2004; Zabala et al. 2009; Santos and Santos-Reis 2010), but others showed that coexistence required habitat partitioning (Mangas et al. 2007). Domestic cats, red foxes (Lachat-Feller 1993a), and eagle owls (*Bubo bubo*) (Virgós et al. 2000) can prey on stone martens and may therefore limit their occurrence. Unfortunately, available data on this topic are largely anecdotal, although a recent study at the regional scale in central Spain indicated a clear negative association between the abundance of stone martens and the number of reproductive pairs of eagle owls (Baniandrés and Virgós 2009). No study of habitat selection by stone martens has investigated the importance of behavioral thermoregulation in relation to the location of resting sites, which could be an important factor during winter, especially in northern portions of their range.

In summary, as for the European pine marten, we lack comprehensive studies about the importance of thermal refuges, predation risk, competition, and food abundance for explaining patterns of habitat selection by the stone marten. We also lack published studies on habitat selection by the stone marten in northern Europe: 13 of the 20 stone marten studies we reviewed were conducted in the Mediterranean region, which could bias our understanding of their habitat niche.

Research Needs

Habitat selection is a hierarchical process that must be understood at various spatial scales. At each scale, different mechanisms or processes may influence habitat selection. At large scales (e.g., geographic range), habitat edges can be influenced by climatic factors (Caughley et al. 1988; Lawton 1993; Gaston 2003). For example, based on the distribution of both species (Proulx et al.

2004), it could be hypothesized that the range of the European pine marten is limited in the south by dry conditions and high temperatures (but see Clevenger 1994b), and that the range of the stone marten is limited in the north by extreme winter conditions. We might also expect that, under conditions of global climate change, European pine martens will be extirpated from southern areas in Mediterranean peninsulas and islands, whereas stone martens will expand their range northward. These hypotheses have not yet been evaluated with large-scale niche modeling for either species.

Small-scale studies, especially those based on radiotelemetry, predominate in the literature. In future studies, a greater emphasis on testing explicit hypotheses is needed to understand the relative importance of various factors (e.g., food availability, abundance of predators and competitors) in determining local distributions and habitat-selection patterns. Moreover, studies at local scales should not be focused only on habitat selection within home ranges, but also on the effects of changes in environmental factors on abundance at this scale. Finally, more studies are needed at larger spatial scales, especially those designed to investigate selection of home ranges within landscapes and regional variation in habitat selection. Such studies will provide important information for the conservation of these species and build on the findings of previous studies conducted at other spatial scales.

SECTION 4

Advances in Research Techniques for *Martes* Species

12

Scale Dependency of American Marten (*Martes americana*) Habitat Relations

ANDREW J. SHIRK, TZEIDLE N. WASSERMAN, SAMUEL A. CUSHMAN, AND MARTIN G. RAPHAEL

ABSTRACT

Animals select habitat resources at multiple spatial scales; therefore, explicit attention to scale-dependency when modeling habitat relations is critical to understanding how organisms select habitat in complex landscapes. Models that evaluate habitat variables calculated at a single spatial scale (e.g., patch, home range) fail to account for the effects of other scales on the probability of species occurrence. Although single-scale approaches are common, we hypothesize that such models will be less predictive and incur greater risk of false inferences than multiscale models that account for the effects of habitat variables that are individually scaled to reflect their optimum relation with species occurrence. This follows from the knowledge that an animal's location is not well defined by the effects of habitat conditions occurring at any 1 spatial scale; rather, animal locations represent the cumulative influence of habitat selection across a broad range of spatial scales, from the landscape to the microsite. In this chapter, we describe an approach that uses bivariate scaling, logistic regression, and information theory to derive a multiscale model combining habitat variables at the scales that optimally predict species occurrence. We demonstrate the utility of this approach by investigating habitat relations of the American marten (*Martes americana*) in northern Idaho using 2 modeling approaches. By comparing how a multiscale model differs in terms of fit, performance, variable use, and spatial prediction from a single-scale model constrained to the scale of habitat patches (i.e., within 90 m of the sampling location), we reveal the importance of considering all relevant spatial scales in habitat models. Our results show that both the strength and the nature of apparent habitat relations are highly sensitive to the scale of predictor variables, and that models that are naïve to scale may easily misconstrue the nature of wildlife-habitat relations. We discuss this issue in the context of other marten habitat studies and conclude by describing the implications of our findings for managing marten habitat.

Introduction

Previous studies of habitat selection by American martens (*Martes americana*) were often focused on the relation between marten occurrence and field data collected within a relatively small area centered on a sampling point. This approach has and continues to yield valuable insights about the factors determining marten habitat selection at relatively small spatial scales (e.g., Spencer et al. 1983; Bowman and Robitaille 1997). Recent advances in geographic information systems (GIS) and spatial statistics and the availability of remotely sensed spatial data have made it practical to model marten habitat relations at larger spatial scales. The importance of understanding ecological processes that occur across a range of relevant scales is a central tenet of landscape ecology (Urban 1987; Turner 1989) and has long been considered in studies of wildlife-habitat relations (Wiens 1976). Species respond to their environment across a hierarchy of spatial scales (Johnson 1980; Schaefer and Messier 1995; Rettie and Messier 2000). Johnson (1980) identified 4 levels within this hierarchy: (1) first-order selection is the species' geographic range; (2) second-order selection is the choice of home ranges within that range; (3) third-order selection is the choice of patches within a home range; and (4) fourth-order selection is the choice of habitat elements within a patch. Although this conceptual framework has been modified to account for additional scales of selection, including the location of populations within a geographic range and metapopulation dynamics (Baguette and Mennechez 2004; Meyer and Thuiller 2006), habitat-association models that have accounted for hierarchical selection have provided important insights about the ecological relations of a broad array of species (e.g., Bergin 1992; McIntyre 1997; Harvey and Weatherhead 2006).

The importance of scale and hierarchy in habitat selection has also been reported in studies of the American marten and other *Martes* species (Bissonette et al. 1997). For example, at the home-range scale, marten habitat selection has been related to canopy cover and habitat fragmentation (Chapin et al. 1998; Hargis et al. 1999; Potvin et al. 2000; Payer and Harrison 2003). At the stand scale, however, marten habitat selection has been related to forest structural characteristics such as age class, tree species, and understory composition (Chapin et al. 1997a). At smaller spatial scales, martens have been reported to select habitat in accordance with particular behaviors such as foraging, resting, denning, and predator avoidance. Fine-scale habitat selection related to these behaviors includes selection of structures based on the availability of coarse woody debris, vertical escape routes, tree cavities, and subnivean access sites (Buskirk et al. 1989; Corn and Raphael 1992; Taylor and Buskirk 1994).

The complexity of habitat selection at different spatial scales for martens and other wildlife species underscores the need to consider a multiscale approach for inferring habitat relations (Bissonette and Broekhuizen 1995; Bissonette et al. 1997; Cushman and McGarigal 2004). Thus far, the predominant approach to account for the influence of spatial scale on marten habitat selection has been first to identify 1 or more scales of interest (e.g., patch, stand, home range), and then to find relations between marten occurrence and habitat variables at each scale. Habitat variables sampled in the field are often collected and calculated at the patch scale. At broader spatial scales, habitat variables are frequently obtained from raster GIS models of landscape attributes, then calculated by taking the average value of that attribute within a given radius from locations where the study organism was either present or absent (Boyce 2006). Once habitat variables are calculated at the scale of interest, a resource selection function may be developed (Boyce and McDonald 1999; Manly et al. 2002; Johnson et al. 2006) to relate those variables to probability of occurrence.

Although this approach has yielded important insights about habitat selection for many species (including martens) across a hierarchy of spatial scales, it does not consider the cumulative influence of all scales of habitat selection in a single model or resource selection function; instead, it constrains all the variables to act at a single scale. Just as early habitat-association studies that focused on the patch scale did not account for potential influences at larger spatial scales, more recent studies constrained to analyses at the largest spatial scales do not account for influences occurring at finer scales. Developing several models at different spatial scales using the same data does not fully address this issue. Rather than referring to a series of scale-constrained models (e.g., individual models scaled to the patch, stand, and home range) as representing multiscale habitat selection, as is commonly reported in the literature, we refer to each as a *single-scale* model. We reserve the term *multiscale* for a single model that accounts for the cumulative influences of habitat associations occurring across a range of relevant scales.

We hypothesize that a predictive model of species occurrence based on the influence of habitat variables acting at all relevant spatial scales would generally offer greater specificity, sensitivity, classification accuracy, and predictive power than a model based on variables constrained to a single scale. We also predict that such multiscale models may provide new insights into species habitat associations that would not be apparent when variables were constrained to a single scale. Indeed, recent studies based on a multiscale approach have yielded highly predictive models of bird and moth rarity (Grand et al. 2004) and bald eagle (*Haliaeetus leucocephalus*) habitat selection (Thompson and McGarigal 2002). Among marten habitat studies, Slauson et al. (2007) found that a model combining stand- and home-range scale habitat selection was more predictive of marten occurrence in coastal

California (USA) than either a stand- or home-range scale model alone. These examples suggest the power of a multiscale approach.

Although previous studies and expert knowledge can suggest scales that might be important to consider for evaluating the influence of habitat variables, it is not possible to know a priori the spatial scale at which a species perceives a particular habitat variable. Constraining these variables to 1 or a few potential spatial scales and ignoring the cumulative effects of habitat selection at all possible scales is likely to reduce the predictive power of the model and may lead either to invalid inferences or to obscure valid ones. Instead, we propose an approach that does not make unsupportable assumptions regarding the scale at which potential habitat predictor variables may be acting.

Our objectives in this chapter are (1) to assess whether the multiscale modeling approach reveals hierarchical habitat relations and (2) to compare this multiscale approach with a single-scale approach and evaluate differences between model performance, interpretation, predictive power, and probability maps of species occurrence. We use bivariate scaling, logistic regression, and information theory to examine marten occurrence in northern Idaho with respect to relevant habitat variables across multiple spatial scales. We then repeat the analysis using the same occurrence data and habitat variables, but restrict the analysis to a single spatial scale. We then discuss the differences between the multiscale and single-scale models in the context of previous marten habitat models and conclude with a discussion of the implications of our findings for managing marten habitat.

Methods

Study Area

The study area is a 3000 km^2 section of the Selkirk, Purcell, and Cabinet mountains, encompassing the Bonners Ferry and Priest River ranger districts of the Idaho Panhandle National Forest (Figure 12.1). Elevation ranges from about 700 to 2400 m above sea level. The climate is characterized by cold, wet winters and mild summers. The area is heavily forested, with subalpine fir (*Abies lasiocarpa*) and Engelmann spruce (*Picea engelmannii*) codominant above 1300 m, and a diverse, mixed forest of Douglas-fir (*Pseudotsuga menziesii*), lodgepole pine (*Pinus contorta*), ponderosa pine (*Pinus ponderosa*), western white pine (*Pinus monticola*), grand fir (*Abies grandis*), western hemlock (*Tsuga heterophylla*), western redcedar (*Thuja plicata*), western larch (*Larix occidentalis*), paper birch (*Betula papyrifera*), trembling aspen (*Populus tremuloides*), and black cottonwood (*Populus trichocarpa*) dominating below 1300 m.

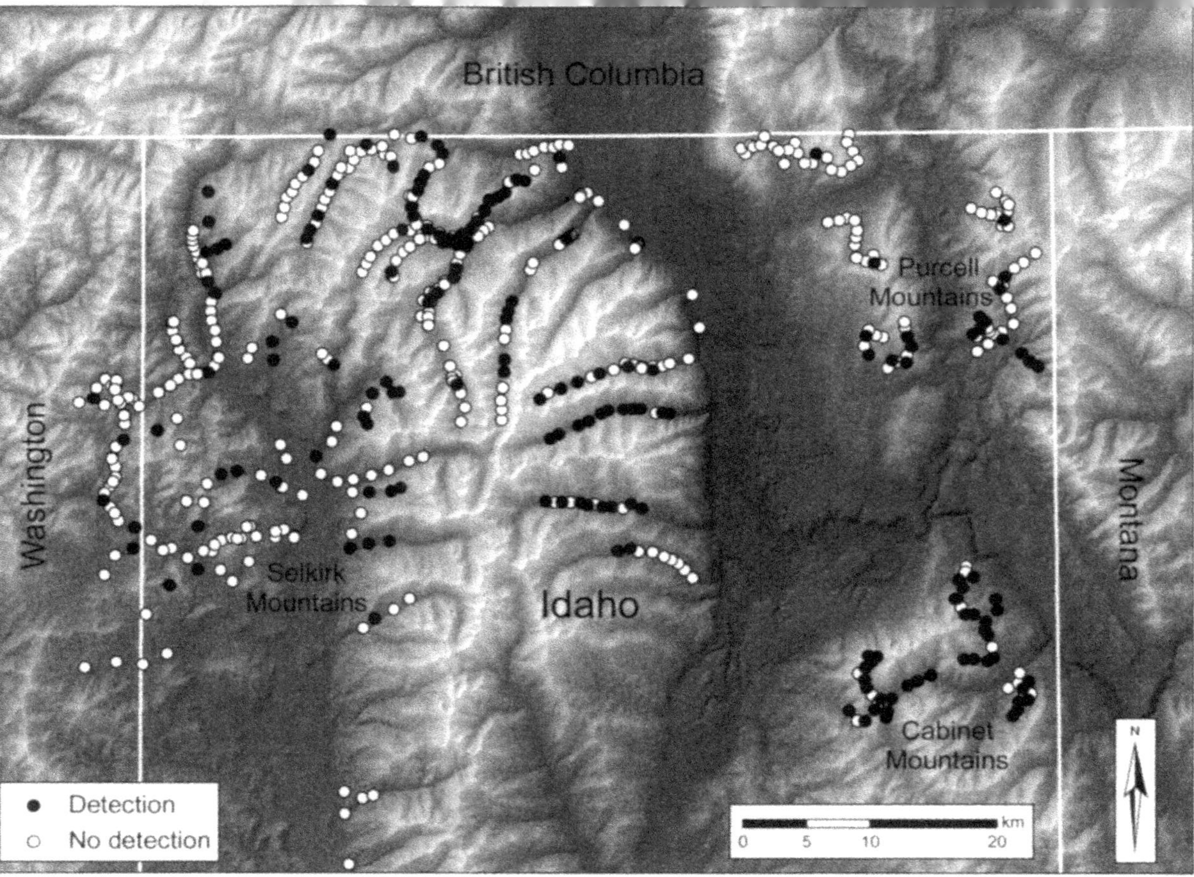

Figure 12.1. Study area extent and digital elevation model in the panhandle of northern Idaho where marten surveys were conducted. Lighter shades indicate higher elevations and darker shades indicate lower elevations. The Kootenai River valley is oriented north to south and is bordered by the Selkirk Mountains on the west, the Purcell Mountains to the east, and the Cabinet Mountains to the southeast.

American Marten Occurrence Data

To document marten occurrence, we set hair snares to collect genetic samples along transects in various portions of the study area from January through March 2005–2007. We selected sampling transects to provide near-complete coverage in all areas of federal or state ownership that are accessible via snowmobile. We set snares for 2 weeks, after which we revisited each station to collect hair samples and rebaited each for another 2-week cycle. Altogether, we surveyed for martens at 361 different locations.

Species identifications from hair-snare samples were based on genetic analyses conducted at the Rocky Mountain Research Station (RMRS) Wildlife Genetics Lab in Missoula, MT using the method of Riddle et al. (2003). Briefly, DNA was extracted with the DNeasy tissue kit (Qiagen) from hair samples taken from each detection station and amplified by polymerase chain reaction (PCR) using primers specific to the mitochondrial gene cytochrome *b*. After PCR products were digested with the restriction enzymes *Hinf* I, *Hae*III, and *Mbo*I, restriction fragments were separated by agarose gel electrophoresis and compared with those that are diagnostic for the American marten.

Variable Selection

We selected a set of variables a priori that we believed would be strongly related to American marten occurrence based on previous research (e.g., Buskirk and Ruggiero 1994; Hargis et al. 1999; Mowat 2006). These variables include elevation, landscape metrics based on seral stage, percent canopy cover, road density, and predicted habitat suitability for tree species associated with riparian habitats (western redcedar), warm and xeric habitats (Douglas-fir), and cool and mesic habitats (subalpine fir) in the study area

Table 12.1. Candidate habitat variables

Habitat variable	Description
ELEVxSDy	Gaussian function of elevation (x = optimum, y = SD)
CWED	Contrast-weighted edge density
PLAND_LST	Percentage of focal area in large sawtimber
PROX_AM_LST	Area-weighted mean proximity index for large sawtimber
AREA_AM_LST	Area-weighted mean patch size for large sawtimber
PD_NST	Patch density of nonstocked timber
PLAND_NST	Percentage of focal area in nonstocked timber
CANCOV	Percentage canopy cover
ROADS	Road density
HAB_DF	Predicted habitat suitability for Douglas-fir
HAB_WRC	Predicted habitat suitability for western redcedar
HAB_SF	Predicted habitat suitability for subalpine fir

(Table 12.1). We co-rectified all input variables to a UTM projection with a 30-m cell size.

We obtained elevation source data from the Shuttle Topographic Radar Mission downloaded from the USGS national map (available at http://nationalmap.gov). We modeled the influence of elevation on the probability of marten occurrence as a Gaussian function based on the expectation that martens should show a unimodal optimum in habitat quality with respect to elevation. Specifically, we generated a factorial combination of elevation variables that differed in the optimum elevation (ranging from 1200 to 2000 m in 100-m increments) and the standard deviation of the optimum elevation (ranging from 300 to 1000 m in 100-m increments). We named these elevation variables ELEVxSDy, where *x* represents the optimum elevation and *y* represents the SD of the optimum elevation.

We created a seral-stage GIS layer by merging the Idaho Panhandle National Forest forest stands map (A. Zack, USDA Forest Service, unpublished data) with the Idaho Department of Lands timber type map (Idaho Department of Lands 2006). Seral stage was defined based on tree diameter at breast height (dbh) and included 6 categories: nonforest, nonstocked forest land (<10% stocked with trees after logging), seedling/sapling (<1.4 m high, <8 cm dbh), pole timber (8–21 cm dbh), small sawtimber (21–41 cm dbh), and large sawtimber (>41 cm dbh).

We used FRAGSTATS 3.3 (McGarigal et al. 2002) to derive 2 landscape metrics based on the seral-stage map. These included patch density (PD) and contrast-weighted edge density (CWED). We selected these variables based on previous work showing that martens avoided fragmented landscapes with a large amount of high-contrast edges (Hargis et al. 1999). We also used FRAGSTATS 3.3 (McGarigal et al. 2002) to calculate 4 stand-scale metrics for seral stages believed to influence marten habitat associations (Buskirk and Ruggiero 1994). Specifically, we calculated the percentage of the focal landscape (PLAND_LST), area-weighted mean patch size (AREA_AM_LST), and area-weighted mean proximity index (PROX_AM_LST) for large sawtimber, because previous work has shown strong effects of area, patch size, and connectivity of late-seral forests on the probability of marten occurrence (Hargis et al. 1999). Likewise, we calculated the percentage of the landscape (PLAND_NST) and patch density (PD_NST) for nonstocked timber land consisting of unregenerated timber-harvest areas, because previous work has shown that martens avoid landscapes with even moderate amounts of recent clear-cuts (Hargis et al. 1999). We calculated all landscape metrics at multiple, nested spatial scales using FRAGSTATS 3.3 (McGarigal et al. 2002) to define 12 circular analysis windows with radii ranging from 90 to 1080 m in 90-m increments around each marten detection station.

We derived a canopy-cover GIS layer from the National Landcover Database (2001). We calculated percent canopy cover (CANCOV) within focal landscapes centered on each marten detection station. To model the effect of roads on marten habitat selection, we calculated the kernel density of all road classes following procedures used by Cushman et al. (2006). Roads within the study area include both maintained and unmaintained roads (including abandoned roads in various stages of recolonization by vegetation). We identified unmaintained roads from previous road layers; these represent roads that have been abandoned but may still affect landscape processes, such as habitat selection by wildlife. We calculated the kernel density of roads within focal landscapes centered on each marten detection station.

We created variables representing the habitat suitability of 3 dominant tree species, including Douglas-fir (HAB_DF), western redcedar (HAB_WRC), and subalpine fir (HAB_SF), using a random forest-ensemble modeling approach based on climatic, topographic, and spectral predictor variables (Evans and Cushman 2009). We determined the dominant tree species within focal landscapes centered on each marten detection station.

We used a presence-absence statistical design (Fielding and Bell 1997) whereby we compared habitat variables at stations where martens were detected with those at stations where they were not detected. For each candidate habitat variable listed in Table 12.1, we identified the scale that showed the greatest difference in the means between occupied and unoccupied sites using a nonparametric difference in means test (PROC NPAR1WAY Wilcoxon; SAS Institute 1999–2000) following procedures used by Grand et al. (2004). Next, for each candidate variable, we selected the scale that produced the smallest P value <0.05, and excluded all other scales from further analysis. If all scales of a variable produced a $P < 0.05$, that variable was eliminated from further analysis. We refer to this process as "bivariate scaling." After the scale-selection process was completed, we identified pairs of correlated variables (Pearson's $r > 0.5$) and eliminated the variable with the higher P value in each pair.

Logistic Regression Analysis

We used R version 2.8 (Venables and Smith 2008) to conduct 2 logistic regression analyses. Both analyses were based on the presence or absence of martens at detection stations. The first analysis, a multiscale model, included logistic regressions for the full factorial combination of all habitat variables remaining after the bivariate-scaling analysis described above. The second, a single-scale model, included logistic regressions for the full factorial combination of all habitat variables calculated only at the patch scale (90-m radius).

For both the multiscale and single-scale analyses, we selected the model with the smallest AIC_c for further analysis. For both models, we calculated several measures of model performance, including sensitivity, specificity, Kappa statistic (Cohen 1968), area under the receiver operator curve (AUC), and the percentage of observations that were classified correctly. We also calculated the sign and coefficent of each habitat variable in the model as well as odds ratios representing the effect size of each variable's influence on the probability of marten occurrence. We calculated the odds ratio as the percent change in probability of occurrence, given a change from the 25th to the 75th percentile for each habitat variable. We then used ArcGIS 9.1 (ESRI) to generate a spatial map of predicted marten occurrence for both the multiscale and single-scale models, based on the model intercept and coefficients. We then calculated the difference between the 2 prediction maps to evaluate spatial differences in model predictions.

Results

We detected martens at 159 of the 361 (44% detection rate) hair-snare stations located throughout the study area (Figure 12.1). There were large differences in the scale at which habitat variables optimally predicted marten occurrence in the multiscale model compared with the single-scale model derived from these presence-absence observations (Table 12.2). Although the 90-m radius (patch scale) was the optimum radius for ELEV, HAB_WRC, and PLAND_LST in the multiscale model, the other variables, including CANCOV, ROADS, PD_NST, AREA_AM_LST, and PLAND_NST, were most related to marten habitat selection at much larger spatial scales approximating the size of marten home ranges (990- and 1080-m radii; Table 12.2).

In addition to differences in the optimum scale for habitat variables, 2 variables (CANCOV and ROADS) changed signs between the multiscale and single-scale models, and there were large differences in the optimum elevation

Table 12.2. Multiscale and single-scale model variables, scale at which the variable was calculated, variable coefficients, and *P* values for evaluating statistical significance ($P \leq 0.05$)

	Multiscale model			Single-scale model		
Habitat variable	Scale (m)	Coefficient	*P* value	Scale (m)	Coefficient	*P* value
(Intercept)		−0.320	0.0066		−0.153	0.138
CANCOV	990	0.398	0.0085	90	−0.046	0.716
ROADS	1080	−0.218	0.0798	90	0.002	0.106
ELEV1400SD400	90	0.381	0.0058			
ELEV2000SD900				90	0.181	0.015
HAB_WRC	90	0.198	0.0822	90	0.319	0.009
PD_NST	990	−0.269	0.0478	90	−0.001	0.991
AREA_AM_LST	990	−0.335	0.0330	90	−0.326	0.833
PLAND_NST	990	−0.394	0.0948	90	−0.911	0.986
PLAND_LST	90	0.284	0.1069	90	0.013	0.750

Table 12.3. Differences in odds ratios between the multiscale and single-scale models

Habitat variable	Odds ratio	
	Multiscale	Single-scale
CANCOV	0.73	−0.94
ROADS	−0.78	1.40
ELEV1400SD400	1.66	—
ELEV2000SD900	—	1.65
HAB_WRC	1.27	1.19
PD_NST	−0.69	−1.00
AREA_AM_LST	−0.54	−0.46
PLAND_NST	−0.02	−0.00
PLAND_LST	1.64	3.20

and standard deviation of elevation (Table 12.2). Moreover, only 2 of 8 variables in the single-scale model were significantly different from 0 ($P < 0.05$), whereas in the multiscale model, 4 of 8 variables were significantly different from 0 (Table 12.2).

Based on odds ratios, we detected a large apparent difference in variable importance between the multiscale and single-scale models (Table 12.3). This change in apparent effect size was particularly large for PLAND_LST, which decreased from 3.2 to 1.64 from the single-scale to the multiscale model, and for ROADS and CANCOV, which reversed sign and changed from 1.4 to −0.78 and from −0.94 to 0.73, respectively.

We found a substantial difference in model fit and performance between the multiscale and single-scale models. The multiscale model had a significantly lower AIC_c value and higher sensitivity, specificity, Kappa statistics,

Table 12.4. Comparison of model parameters between the multiscale and single-scale models

Model parameter	Multiscale model	Single-scale model
AIC_c	468.70	484.34
ΔAIC_c	0	15.64
Threshold	0.50	0.50
PCC[a]	0.665	0.576
Sensitivity	0.579	0.421
Specificity	0.733	0.698
Kappa statistic	0.314	0.122
AUC[b]	0.70	0.64

[a] Percentage classified correctly

[b] Area under the receiver operator curve

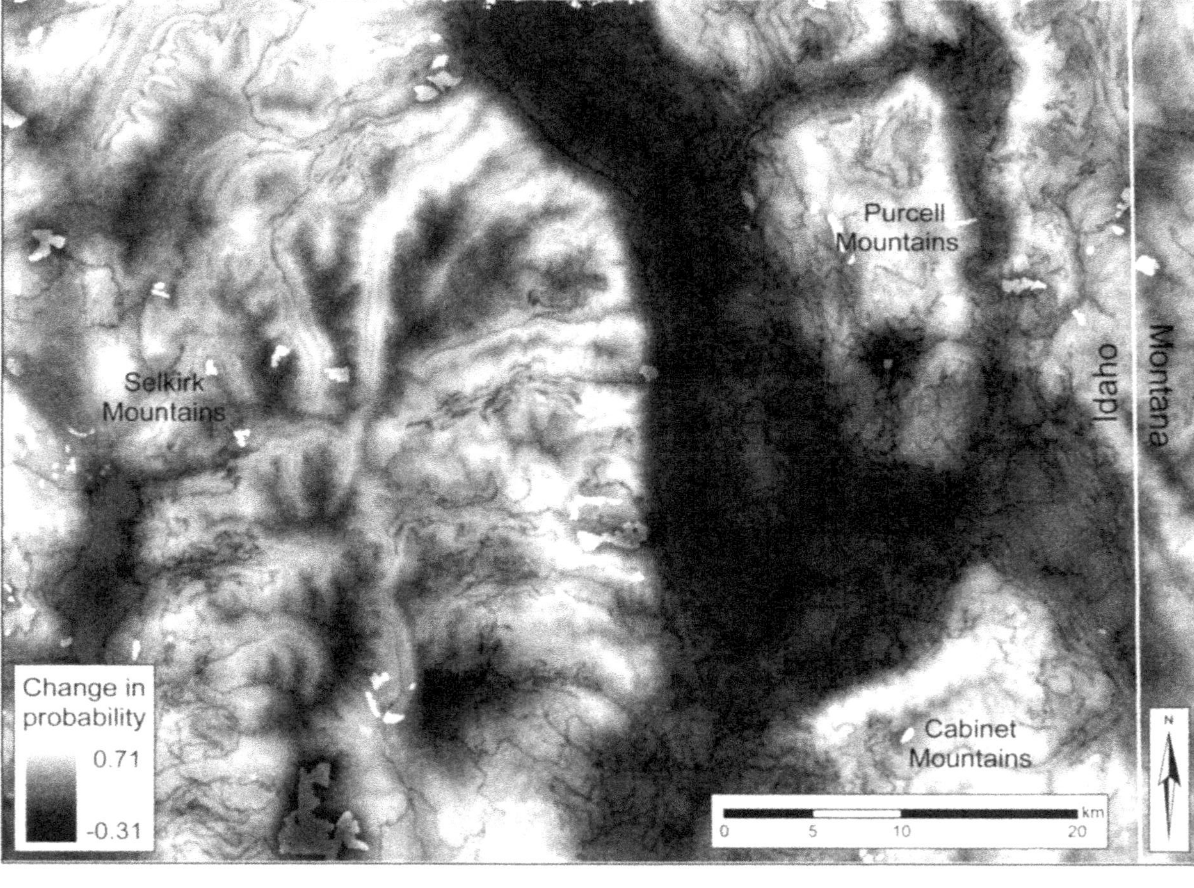

Figure 12.2. Difference in probability of occurrence between multiscale and single-scale models. Darker shades indicate areas where the multiscale model predicts higher probability of occurrence than the single-scale model. The single-scale model would make substantial errors of omission in those areas. Lighter shades indicate areas where the multiscale model predicts lower probability of occurrence than the single-scale model. The single-scale model would make substantial errors of commission in those areas.

AUC values, and percent classified correctly (Table 12.4). In addition, differences in the spatial prediction of marten occurrence between the multiscale and single-scale models ranged from 0.712 to −0.317, respectively (Figure 12.2).

Discussion

Hierarchical Habitat Relations of the American Marten

As in previous studies that identified hierarchical habitat relations for the American marten, we found strong evidence of habitat preferences at both the patch and home-range scales. Rather than impose these scales on the analysis, the bivariate scaling approach allowed us to infer the appropriate scale based on the optimum relation with marten occurrence. By evaluating the full factorial of possible models across a range of relevant spatial scales, we were able to construct a single model that accounted for the hierarchical effects of multiple predictor variables acting at the scales at which martens perceive habitat conditions in the study area.

Based on the multiscale analysis, habitat variables were most related to marten occurrence at 2 hierarchical levels. At a scale approximating the size of marten home ranges (1080-m radius), we observed a negative relation with ROADS (Table 12.2). Avoidance of areas where low-use roads occur at high densities has been reported for other *Martes* species (Goszczyński et al. 2007), but previous results have been equivocal for the American marten (e.g., Mowat 2006). We expected martens to avoid heavily roaded areas because of increased risk of predation or mortality, and perhaps because of behavioral avoidance of road-related disturbances (Forman and Alexander 1998; Coffin 2007). We also found a positive relation with CANCOV at the home-range scale (990-m radius). Dense canopy cover and avoidance of clear-cuts have been positively related to marten occurrence in North America (Bissonette et al. 1997; Potvin et al. 2000) and linked to greater prey availability, subnivean access, resting sites, den sites, and vertical escape routes for predator avoidance (Corn and Raphael 1992; Buskirk and Ruggiero 1994; Bissonette et al. 1997).

We found negative habitat relations at the home-range scale for PD_NST, AREA_AM_LST, and AREA_PROX_LST. In accordance with other home range-scale studies of habitat selection by martens (Chapin et al. 1998; Potvin et al. 2000; Payer and Harrison 2003), these relations indicated that, at large spatial scales, martens avoided areas where stands of large-diameter trees were fragmented into small or less-proximal patches, and where regenerating clear-cuts represented a large proportion of the landscape.

We found positive selection occurring at the patch scale (90-m radius) for ELEV1400SD400, HAB_WRC, and PLAND_LST. Together, these results indicate selection for riparian habitats in montane forests (where western red-cedar predominate) dominated by mature closed-canopy forests. Such habitats are likely to provide for finer-scale winter habitat requirements related to prey availability, subnivean access, and cavities or coarse woody debris suitable for resting, denning, and predator avoidance (Buskirk et al. 1989; Bowman and Robitaille 1997; Mowat 2006).

Variables analyzed at the stand scale (180–360-m radii) were absent from the multiscale model. This may reflect the absence of significant stand-level heterogeneity within the sampled portions of the study area. This is likely because all detection stations were within national forest land consisting of relatively homogeneous montane forest managed using similar forest practices. The lack of a stand-scale effect may also simply reflect the lack of a limiting habitat factor at this scale (Rettie and Messier 2000). Interestingly, Minta et al. (1999) found weak stand-scale habitat selection to be a general phenomenon among marten studies (see I.D. Thompson et al., this volume).

Habitat Relations Based on Multiscale vs. Single-Scale Models

We hypothesized that constraining a habitat-selection model for martens to a single spatial scale of analysis could result in false inferences or obscure valid habitat relations. This hypothesis was based on the knowledge that martens select habitat across a range of spatial scales (Bissonette et al. 1997; Nams and Bourgeois 2004), and our expectation that a model incorporating all scales of influence would more accurately predict marten occurrence. The differences between the multiscale and single-scale models in this analysis provide support for this hypothesis. For example, the single-scale analysis found a positive relation between marten occurrence and ROADS, and a negative relation with CANCOV (Table 12.2). These inferences contrast with previous work showing strong habitat preferences for roadless areas with high canopy cover at small spatial scales (Spencer et al. 1983; Robitaille and Aubry 2000). As expected, however, the multiscale model identified a negative relation with road density and a positive relation with canopy cover. Similarly, the multiscale analysis indicated a significantly lower optimum elevation with a smaller standard deviation than the single-scale analysis.

There were also differences in the relative importance of habitat variables in the multiscale and single-scale models. Based on odds ratios, the single-scale model was most influenced by PLAND_LST (Table 12.3), yet the coefficient of this variable was not significantly different from 0, indicating uncertainty about its influence on marten habitat selection. Notably, only 2 of 8 variables in the single-scale model were significantly different from 0 (ELEV2000SD900 and HAB_WRC), and both were optimally related to marten occurrence in the multiscaled model within the 90-m radius. In contrast, the multiscale model comprised 4 variables with coefficients significantly different from 0. Together, these differences highlight the greater emphasis of the single-scale model on 1 significant variable (PLAND_LST) at 1 spatial scale, whereas the multiscale model accounts for the influences of an array of habitat variables acting across multiple spatial scales. In other habitat-selection studies that have compared multiscale and single-scale models, results were similar in that the multiscale models were generally better predictors of species occurrence (Poizat and Pont 1996; Graf et al. 2005; Slauson et al. 2007).

We observed significant differences in spatial predictions between the multiscale and single-scale models. Given that the multiscale model is more sensitive, specific, and predictive, large differences in spatial predictions suggest that the single-scale model is likely to result in major errors of omission and commission across much of the study area. These differences in spatial prediction suggest that restricting the analysis to 1 spatial scale not only alters variable selection and model performance but also affects the ability of the model to predict marten occurrence.

Limitations of Our Analysis

We acknowledge that there are 2 important limitations of our analyses. First, optimizing the scale at which each habitat variable was calculated in the multiscale model may result in overfitting of the model. Although the multiscale model predicted habitat relations that were strongly supported by previous research, the generality of this model should be tested with novel data through cross-validation. In a similar multiscale analysis of marten habitat selection in Oregon and Washington, we found that scale-optimized models were highly predictive of marten occurrence based on k-fold cross-validation (A. Shirk, unpublished data). Second, stronger relations may occur at larger spatial scales than those included in our analyses. The multiscale model revealed that martens avoided areas with high road densities at the largest radius we evaluated (1080 m); however, had we included larger radii, we might have found a unimodal peak of support for a single optimum scale at which roads influence marten occurrence.

Implications for Managing American Marten Habitat

Decades of research on marten habitat relations, including this study, indicate that martens select habitat based on the cumulative influence of many factors acting across a hierarchy of spatial scales, from the landscape to microsites. We have demonstrated that constraining an analysis to a single spatial scale may result in false inferences, weaker models, and poorer spatial predictions of marten occurrence. Conducting multiple single-scale analyses of the same occurrence data (e.g., evaluating 3 separate models constructed at the patch, stand, and home range scales) does not eliminate these risks because such an approach does not produce a single model that simultaneously accounts for the potential influences of habitat variables occurring at multiple spatial scales.

For resource managers, the consequences of false inferences could be significant. For example, in this study, inferences from the single-scale model that marten habitat selection at the patch scale was negatively related to canopy cover and positively related to road density might be interpreted as providing empirical support for forest practices (i.e., intensive harvest and clear-cutting) and road development that have been shown in other studies to cause declines in marten populations. In addition to stronger inferences, the greater predictive power of the multiscale approach provides a more accurate depiction of the areas within a landscape that are likely to support martens. This improved spatial understanding of marten habitat can help guide conservation efforts by enabling managers to better identify habitat areas important for population persistence and for maintaining habitat connectivity.

Furthermore, the multiscale approach explicitly seeks to optimize the scale at which each habitat variable is related to occurrence, and in a way that does not constrain variables to 1 spatial scale, or that ignores the potential influence of habitat conditions occurring at other scales. This attention to spatial scale and flexibility in optimization provides resource managers with strong inferences regarding the scale at which martens perceive each habitat variable, enabling them to more appropriately match the scale of management activities to the scale at which the habitat variable influences marten occurrence. For example, in this study, the multiscale model suggests that, to improve habitat conditions for martens, managers should limit road density and forest fragmentation at the home-range scale while preserving patches of large western redcedars (found mainly in riparian habitats) at elevations centered around 1400 m.

Much of what we have learned about the hierarchical habitat relations of the American marten has been based on models that do not simultaneously account for the influence of habitat variables at all relevant scales. We suggest that future studies of marten habitat selection incorporate a multiscale approach to facilitate the identification of novel habitat associations, and produce better predictive models of marten occurrence that will help to elucidate our understanding of marten ecology and conservation.

Acknowledgments

We thank the editors and anonymous reviewers for their comments and help in preparing this manuscript. We also thank David Wallin, Western Washington University, and Jim Hayden, Idaho Department of Fish and Game, for their assistance. This work was supported by the USDA Forest Service Rocky Mountain Research Station and Pacific Northwest Research Station.

13

The Use of Radiotelemetry in Research on *Martes* Species

Techniques and Technologies

CRAIG M. THOMPSON, REBECCA A. GREEN, JOEL SAUDER, KATHRYN L. PURCELL, RICHARD A. SWEITZER, AND JON M. ARNEMO

ABSTRACT

Radiotelemetry was first used on a *Martes* species in 1972, when 5 American martens (*Martes americana*) captured incidentally during a snowshoe hare (*Lepus americanus*) research project in Minnesota were radio-collared. Since then, at least 128 research projects have used radiotelemetry to investigate various aspects of the ecology of *Martes* species worldwide. The most common application of radiotelemetry to research on *Martes* species has been a ground-based study of American marten home range or habitat use in North America using VHF collars. Other telemetry-based projects have included studies involving all but 1 *Martes* species that were conducted in 20 additional countries, used Global Positioning System (GPS) or Argos telemetry, and addressed a broad spectrum of research questions such as survival, density, diet, community interactions, and responses to disturbance. To better understand the application of radiotelemetry to research on *Martes* species, we summarize the use of ground, aerial, and satellite-based telemetry techniques and outline the strengths and limitations of each. We also review the use of alternative attachment techniques such as breakaway devices and intraperitoneal implant transmitters in research on *Martes* species, and provide recommendations for minimizing the risk of adverse effects while radio-tracking *Martes* species.

Introduction

Radiotelemetry was first used on a *Martes* species in 1972, when D. Mech and L. Rogers collared 5 American martens (*Martes americana*) that were captured incidentally during a snowshoe hare (*Lepus americanus*) research project in Minnesota in the United States (Mech and Rogers 1977; D. Mech,

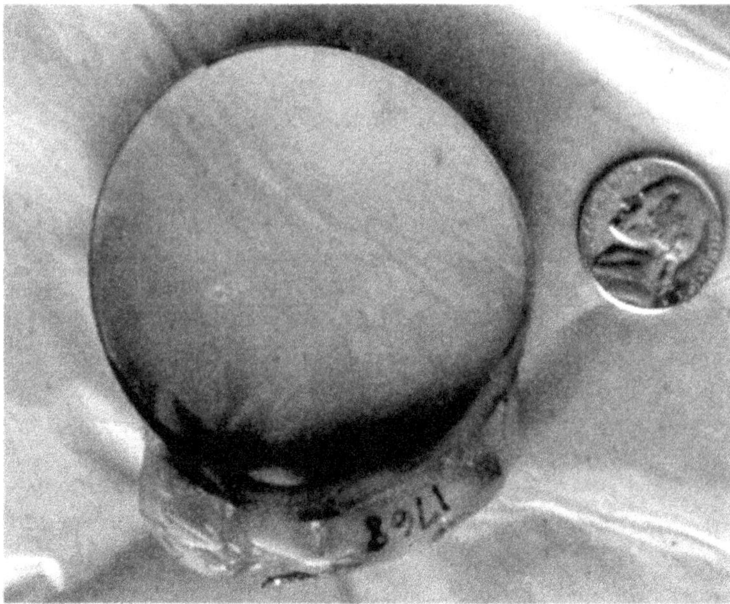

Figure 13.1. First-generation radio-collar used on American martens and fishers in 1972 (photo R. Powell).

U.S. Geological Survey, personal communication). Later that year, R. Powell began tracking 4 radio-collared fishers (*M. pennanti*) captured in the same area (R. Powell, North Carolina State University, personal communication). In 1973, researchers with the Ontario Ministry of Natural Resources in Canada began tracking 16 American martens in Algonquin Provincial Park to determine home range sizes and movement patterns (Taylor and Abrey 1982). The earliest collars, distributed by the AVM Instrument Co. (Champaign, Illinois, USA), consisted of a brass strip that doubled as the antenna and was either bolted or welded around the animal's neck (Figure 13.1). The transmitters, coated in dental acrylic, weighed 20–110 g, and had a lifespan of approximately 2 months and a range of 0.5–1.0 km (Buck 1982; D. Mech, personal communication; R. Powell, personal communication).

Following these early projects, the use of radiotelemetry to study American martens and fishers increased rapidly in the mid-1970s. In 1973, R. Powell began radio-tracking fishers in Michigan (USA) to investigate hunting patterns and ecological energetics (Powell 1977) and, in 1974, G. Kelly began following fishers in the White Mountains of New Hampshire (USA) to study movement patterns (Kelly 1977). In 1975, T. Campbell radio-tracked American martens in the Rocky Mountains of Wyoming (USA) to evaluate the impacts of timber harvesting (Campbell 1979), and M. Davis used telemetry to monitor the movements of American martens translocated into Wisconsin

(USA) (Davis 1983). From 1977 through 1979, additional telemetry studies were initiated on American martens in Maine and California in the United States, and the Yukon in Canada (Steventon 1979; Spencer et al. 1983; Archibald and Jessup 1984) and on fishers in California (Buck et al. 1979; Mullis 1985).

In Europe, radiotelemetry was first used on stone martens (*M. foina*) in 1980, when M. Herrmann began tracking 14 animals in Germany to study home range use and habitat preferences (Herrmann 1994), and on European pine martens (*M. martes*) in 1984, when P. Marchesi monitored the movements of 20 pine martens in Switzerland (Marchesi 1989). From 1986 to 1987, H. Kruger monitored the movements of both species near Gottingen, Germany, to determine the degree of ecological and spatial overlap (Kruger 1990). In Asia, M. Tatara began radio-tracking Japanese martens (*M. melampus*) in 1987 (Tatara 1994), and J. Ma and colleagues used telemetry in the 1990s to look at the seasonal activity patterns of sables (*M. zibellina*) in the Daxinganling Mountains of China (Ma et al. 1999).

To summarize the use of radiotelemetry in research on *Martes* species, we reviewed 178 published papers, book chapters, dissertations, theses, and unpublished reports, ultimately identifying 129 distinct research projects that used radiotelemetry to investigate various aspects of *Martes* ecology (Figure 13.2). We considered all work done in 1 study area over the same time span to be 1 research project, because the number of publications or reports varied greatly by project. Most studies focused on a single species, but 5 studies involved 2 species. We contacted *Martes* researchers worldwide and requested copies of unpublished reports, regional articles, and contact information for other relevant researchers. Where possible, we included preliminary results (e.g., posters, symposium abstracts) not published elsewhere. Regrettably, we undoubtedly missed some unpublished reports, theses and dissertations, or articles published in non-English journals that were not readily available. These 129 projects were conducted on 7 of the 8 recognized *Martes* species (no radiotelemetry research has been conducted on the Nilgiri marten [*M. gwatkinsii*]) and represented 22 countries (Figure 13.3). There was a clear bias toward North American species, with 58 research projects focused on American martens, 42 on fishers, 14 on European pine martens, 12 on stone martens, 4 on sables, 2 on Japanese martens, and only 1 on yellow-throated martens (*M. flavigula*). Although this bias is certainly inflated by our resources and search techniques, it was corroborated by numerous *Martes* researchers worldwide. Habitat use is the most widely reported ecological characteristic investigated (54% of studies), followed by home range (49%), movement patterns (21%), activity patterns (16%), and spatial organization (15%). Least reported are community interactions (2%), responses to disturbance (10%), density (10%), and survival (11%; Table 13.1). A list of the documents we reviewed is available from the authors on request.

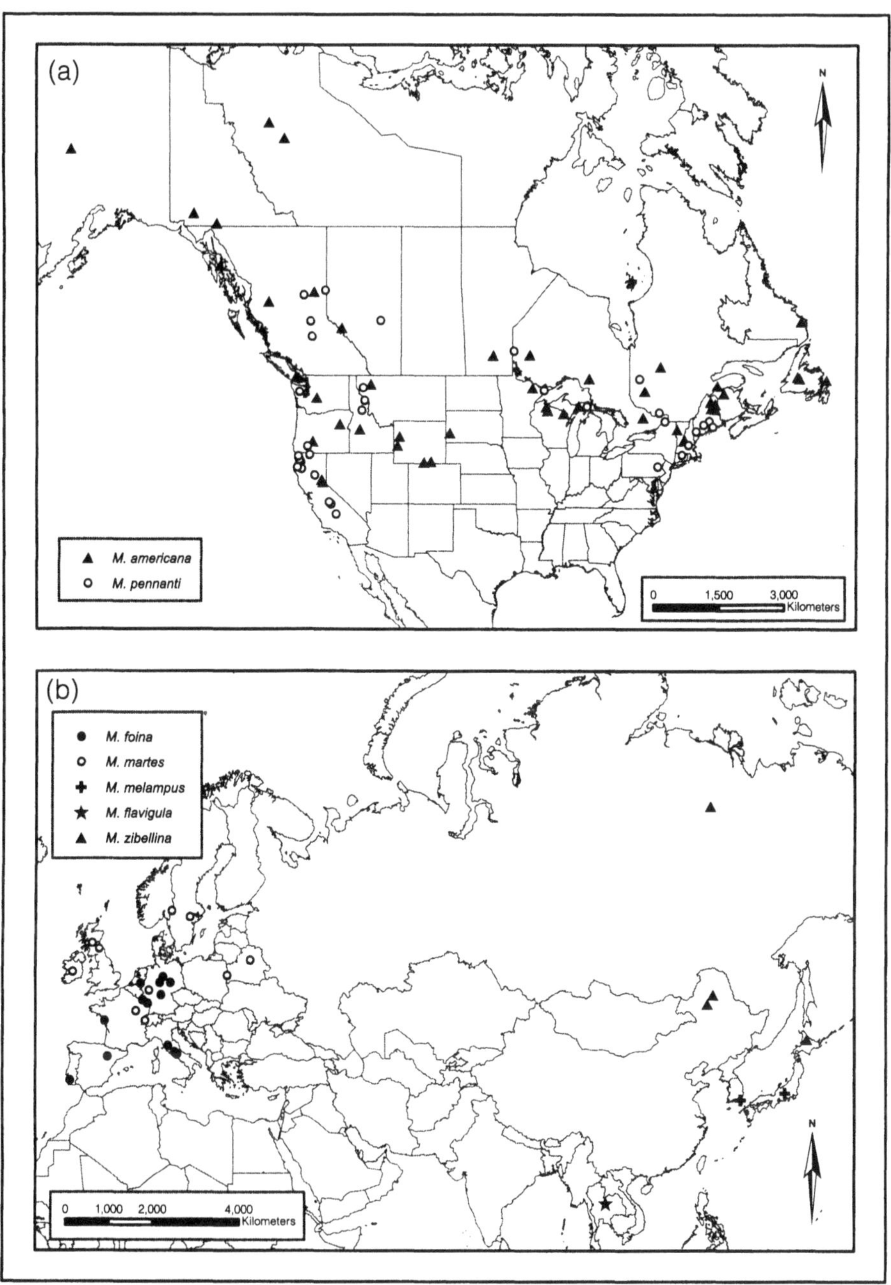

Figure 13.2. Distribution of telemetry-based *Martes* research projects in (a) North America and (b) Europe and Asia.

Figure 13.3. Radio-transmitter attachments on *Martes* species. *Top:* radio-collared yellow-throated marten (photo P. Grassman); *center left:* VHF collar on a Japanese marten (photo T. Nakamura); *center right:* ARGOS collar on a female fisher (photo J. Sauder); *lower left:* GPS collar on a male American marten (photo K. Moriarty); *lower right:* insertion of an implant transmitter into the body cavity of a fisher (photo R. Weir).

We collected methodological information from 119 research projects to evaluate data-acquisition rates, location accuracy, and efficiency. To estimate the relative cost per location for different telemetry techniques, we also interviewed many current *Martes* researchers. We summarized the results of these comparisons in Table 13.2.

Table 13.1. Summary of telemetry-based research on *Martes* species

Species	Total number of studies	Home range	Move-ment	Repro-duction	Density	Spatial organi-zation	Habitat use	Survival	Activity patterns	Diet	Distur-bance	Reintro-duction	Com-munity
M. americana	58	31	12	4	7	4	39	9	7	3	10	2	—
M. flavigula	1	1	1	—	—	—	—	—	1	—	—	—	1
M. foina	12	10	1	2	—	5	5	—	5	3	1	1	2
M. gwatkinsii	0	—	—	—	—	—	—	—	—	—	—	—	—
M. martes	15	8	2	1	1	7	7	—	2	3	1	—	1
M. melampus	2	2	—	—	—	1	2	—	—	1	—	—	—
M. pennanti	42	16	11	9	4	5	25	5	5	4	2	4	—
M. zibellina	4	1	1	1	1	—	2	1	1	1	—	—	—

Note: Spatial organization refers primarily to inter- or intraspecific territoriality

Table 13.2. Comparison of telemetry techniques used to study *Martes* species

Parameter	Ground	Aerial	Ground and aerial	Argos	GPS
Number of studies	78	14	23	1	3
Mean number of study animals	18 ± 19 (*n* = 67)	39 ± 52 (*n* = 7)[a]	24 ± 15 (*n* = 15)	~20	ongoing
Mean number of locations	1990 ± 2897 (*n* = 14)	1961 ± 2707 (*n* = 3)	447 ± 236 (*n* = 3)	~3000	ongoing
Efficiency (locations per animal)	132 ± 185[b] (*n* = 15)	25 ± 8[a] (*n* = 3)	31 ± 13 (*n* = 3)	~50	~123
Cost per location[c]	$8.26–$43.90[d]	$26.50	—[e]	$16.75	$21.06
Estimated average location error	102.6 m (*n* = 8)	212.4 m (*n* = 13)	—[e]	68% class 3 (<250 m)	12.4 m
Transmitter mass[f]	20–40 g	20–40 g	20–40 g	75–120 g	45–75 g
Estimated battery life[f]	2–3 yr	2–3 yr	2–3 yr	1–2 yr	3–5 months

Note: Mean values are followed by standard deviations when appropriate

[a] Three large research efforts using aerial telemetry are currently ongoing in Washington, California, and New York (USA); summary statistics from these studies are not yet available

[b] One research project was excluded from this summary because of the large number of locations collected on a small number of animals: Posillico et al. (1995) collected >7500 locations on 3 stone martens in rural Italy; inclusion of this study results in an average of 291 ± 641 locations per study animal for ground-based telemetry projects

[c] Cost per location was roughly estimated by several researchers based on equipment, contractor services, and labor costs as well as sample size

[d] Cost per ground telemetry location varies widely depending on terrain and access

[e] Data on the distribution of aerial vs. ground locations were not presented; values will reflect some combination of ground and aerial cost and accuracy

[f] In all cases, mass and life span depend on battery size and reflect programmable factors such as pulse rate, range, fix interval, etc.; values reflect those typically selected by current *Martes* researchers

Primary Telemetry Techniques

Ground-based VHF Telemetry

Background

Of the 129 discrete research projects we reviewed, 86 used ground-based telemetry as the primary tracking technique. These 86 studies included 7 *Martes* species: American marten (38), fisher (28), European pine marten (10), stone marten (7), sable (4), Japanese marten (2), and yellow-throated marten (1); 4 studies included both American marten and fisher (Raine 1981;

Gilbert et al. 1997; Zielinski et al. 1997a; Belant 2007). Seventeen of these projects were supplemented by aerial telemetry.

A wide variety of questions have been addressed via radio-tracking using ground-based techniques, but the majority (81%) examined habitat use. Many of these habitat-oriented studies compared animal locations or home ranges with random locations, but others specifically addressed resting (40% of studies examining habitat), denning (18%), and traveling habitat (6%). Traveling habitat was generally addressed using snow tracking of radio-collared individuals. Resting-site studies examined the effects of season, sex, weather, and site availability on resting-site selection. Denning-site studies have examined the timing and duration of denning, as well as the characteristics of cavities and surrounding habitat conditions.

More than half the studies (58%) estimated home-range size and examined variation in home ranges due to sex, season, reproductive status, and habitat. Studies addressing questions related to the spatial organization of local populations were also common (27%), as well as studies that estimated density (19%). Other research objectives that have been addressed with ground-based telemetry include rates of movements or dispersals (30%) and the timing of seasonal or daily activities (20%). Thirteen percent of studies examined some aspect of the effects of disturbance, including logging, road construction, fire, sensitivity to fragmentation, and trapping. Studies done in areas with fur-trapping assessed harvest rates, responses to trapping pressure, and the sustainability of harvest practices. Seven studies evaluated reintroductions, addressing questions related to optimal release dates, release methods, post-release movements, and the fates of translocated animals.

Tracking Techniques

Three primary techniques were used to locate radio-marked animals: omnidirectional antennas, directional antennas, and homing. Omnidirectional antennas mounted on vehicles were often used to obtain approximate locations before obtaining more-precise locations with directional antennas.

Most studies used hand-held directional H, or 2-, 3-, or 4-element hand-held Yagi antennas. Locations were triangulated from points on the ground, usually taking advantage of road systems, and considering topography, accessibility, and known home ranges of animals. Most studies that used triangulation techniques specified the following: (1) a minimum number of bearings (2–7), (2) a maximum time period (usually 10–30 min) to help ensure the animal had not moved far during the interval, and (3) a minimum angle (generally 30°) between azimuths; these guidelines help to reduce errors associated with animals moving too far, while maximizing the accuracy of locations. To further increase accuracy, some studies attempted to be within a given distance of the animal (≤50 m to ≤1 km, depending on the study objectives). Directional antennas can be mounted on vehicles (Powell 1977; Gilbert et al.

1997; Pereboom et al. 2008) or used by observers on snowmobiles, skis, snowshoes, mountain bikes, or foot (Hauptman 1979; Simon 1980). Directional antennas can also be mounted on towers to increase detection range. Powell (1977) used directional antennas on two 14-m towers permanently mounted in the study area, and Taylor and Abrey (1982) used 3 permanent towers situated at high points approximately 1.5 km apart. Each of the latter towers was fitted with an omnidirectional antenna, an 8-element Yagi antenna, and a communications antenna to maintain continuous contact among observers. Observers at all 3 towers located animals simultaneously, removing any error associated with the focal animal's movement.

Typically, researchers used homing techniques when animals were inactive to locate resting and denning sites and to locate animals with mortality signals. Variations in signal strength and pulse rate are often used to determine activity, though activity sensors have recently been incorporated into some collars. Hand-held directional telemetry antennas and receivers were used to follow the radio signal to the point of origin, following the increasing strength of the signal. When close to an inactive animal, the external antenna of the receiver was often disconnected to assess proximity, based on whether a signal could be detected with only the internal antenna of the receiver (the "box signal"). When sufficiently close, observers generally reduced the gain and used the external antenna to identify the exact location. In a few cases, automated telemetry receivers were used near resting and denning sites to monitor attendance (Buskirk et al. 1989; Ruggiero et al. 1998), or near noninvasive monitoring stations (e.g., track plates, hair snares) to evaluate research effectiveness (Ivan 2000).

The interval between locations of individual animals has varied greatly among studies, depending on their objectives. Intervals between successive locations ranged from 3 min for studies examining activity cycles and movement patterns, to >1 week in others. A few studies focused on movement patterns and space use plotted sequential locations using a grid system (Taylor and Abrey 1982; Balharry 1993; Herrmann 1994). The size of the grid was determined on the basis of the accuracy of telemetry-based locations (Balharry 1993). Most studies focused on diurnal locations, but species that are strictly nocturnal (stone marten) or primarily nocturnal (European pine marten) were located at night unless daytime resting locations were of interest (Schröpfer et al. 1997; Herr 2008; Pereboom et al. 2008). A few studies noted the proportion of locations taken at night, and several studies obtained locations over time periods as long as 24–48 h to monitor activity (Fuller and Harrison 2005; Santos and Santos-Reis 2010; E. Manzo, Ethoikos, Convento dell' Osservanza, unpublished data).

Accuracy

The accuracy of locations acquired via ground-based triangulation is highly variable and depends on the terrain, observer experience, animal movement, proximity to the animal, number of bearings taken, and the angle of intersection between subsequent bearings (White and Garrott 1990). Acceptable error depends on the objectives of the study and the precision of locations needed to address those objectives. Homing to locate resting and denning sites generally results in accurate locations because errors are limited to Global Positioning System (GPS) or mapping errors. Excluding studies where homing was used to locate structures and those that relied solely on visual sightings, approximately 45% of the studies reviewed described methods for assessing accuracy or estimates of accuracy. However, given the use of rugged terrain and cavities by *Martes* species, as well as their capacity for rapid movements, some consideration of triangulation accuracy and data screening is necessary, depending on project objectives.

Location error reported in 12 studies of *Martes* species ranged from ≤5 m in a study of Japanese martens (Tatara 1994) to 848 m for fishers (Tully 2006). Mean bearing error varied from 2° for fishers (Johnson 1984) to 12.6° for American martens (Dumyahn et al. 2007). Roy (1991) reported a 95% error arc of 16° for his study of fishers in Montana (USA). Few studies reported the mean size of error polygons. Self and Kerns (2001) calculated error polygons of approximately 2.5 ha, but they also noted that actual locations of their collared fishers were often outside the polygon. Payer and Harrison (2003) estimated a mean error-polygon size of 3.7 ha. More often, authors discarded observations when the associated error polygon exceeded some predetermined size. Weir and Corbould (2008) assigned precision classifications based on the size of the error polygon and then used various subsets of the data, depending on the precision needed for analyses at various spatial scales. Poole et al. (2004) calculated a mean error ellipse of 9.8 ha and used the 90th percentile ellipse (75 ha) as the upper limit for locations used in data analyses. Koen (2005) calculated a 95% confidence ellipse of 99 ha. Dumyahn et al. (2007) used only locations with an error ellipse ≤20 ha. Caryl (2008) calculated both error ellipses and mean location error, then used the larger estimate.

Koen (2005) provided an extensive evaluation and review of telemetry accuracy and triangulation error for fishers. Location and angular errors were estimated using a combination of blind tests and known positions of dead fishers. The author evaluated the effect of animal movements on location precision by having volunteers carry a transmitter while walking along forest trails in occupied fisher habitat at an average speed of 0.6 m/sec. Average angle error was 6.4 ± 14.5°, and average location error was 268.6 ± 270.9 m, with an average distance of 1014 m between the observer and transmitter. Error did not vary significantly based on the statistical approach used (geometric mean vs. maximum likelihood, $P = 0.068$), but was significantly correlated with the distance between observer and transmitter ($P < 0.001$). Location error was 49% greater for mobile compared with stationary locations. See Koen (2005) for a detailed explanation of analytical techniques, discussion of error, and guidance on screening bearings.

Finally, the recent development of digitally encoded VHF signals allows an observer to monitor multiple animals on a single frequency. A unique digital "tag" is incorporated into the VHF signal, and signals are then sorted by a digital receiver. This capability greatly increases the efficiency of scanning receivers because only a single frequency is monitored, reducing observer fatigue and eliminating the risk that a brief appearance by a study animal will be missed by either a fast-moving (e.g., aerial) scanner or an automated monitoring system (Lotek Wireless, http://www.lotek.com/radio.htm).

Aerial VHF Telemetry

Background

The unique ability of observers in aircrafts to rapidly search and locate radio-collared animals over large and inaccessible areas, while allowing for nearly line-of-sight reception between transmitter and receiver, makes aerial radiotelemetry an attractive research technique (Gilmer et al. 1981). It is particularly appropriate for studying *Martes* species that occur in remote areas with limited access (Weir and Corbould 2008). Wildlife biologists have used aircraft for monitoring American martens and fishers in North America since the early to mid-1970s (Mech 1974). Kelly (1977) used a fixed-wing airplane to monitor the movements of fishers in the White Mountain National Forest of New Hampshire and Maine, and Mech (1974) reported detection distances from fixed-wing airplanes of 3.2 and 8 km for martens and fishers, respectively. In comparison, current projects report detection distances of 11–15 km in the direction of flight using a forward-mounted 3-element Yagi antenna, and 8–11 km using side-mounted directional H-antennas (R. Sweitzer and R. Barrett, University of California at Berkeley, unpublished data). Helicopters have been used infrequently for monitoring American martens (Latour et al. 1994; Potvin et al. 2000; Hearn 2007), and 1 project used an ultralight aircraft (M. Cheveau, Université du Québec en Abitibi-Témiscamingue, unpublished data). Notably, we were unable to locate any studies of *Martes* species in Europe or Asia that reported using aerial telemetry.

The most common use of aerial telemetry for studies of *Martes* species has been for assessing movements and home range use. Other common uses include assessment of habitat use/selection (Potvin et al. 2000) and survival (Koen et al. 2007). Typically, aerial telemetry has been used in association with ground-based telemetry, most often as a backup method for finding animals that were difficult to locate from the ground. Because of the high cost of flight time, few projects were able to conduct aerial surveys more than once per week. Of the 37 studies we identified that used aerial telemetry to track study animals, 14 used it as the primary monitoring technique. The remaining 23 studies used aerial monitoring to supplement or inform ground-based telemetry efforts.

Tracking Techniques

The most common method for tracking animals from fixed-wing airplanes involves the use of H-style antennas mounted on both the right and left wing struts, linked to a switchbox and receiver. Once a signal has been detected, the biologist begins alternating which antenna is active/audible with the switchbox and works with the pilot to orient the flight path toward and then over the animal; the pilot then circles back, and the biologist marks the estimated location using a handheld or aviation GPS receiver based on peak signal strength (Gilmer et al. 1981; Seddon and Maloney 2004). Although helicopters are much more expensive to operate than fixed-wing airplanes, they have 2 distinct advantages: the ability to fly at slow speeds and to hover. Gilmer et al. (1981) recommend searching for radio-collared animals from helicopters outfitted with a forward-pointing antenna at an elevation of about 200 m, then obtaining signal directionality on approach by occasionally yawing left and right; however, excessive noise and windwash disturbance is a concern with low-level helicopter (<100 m) approaches to radio-collared animals. In addition, although the accuracy of positions may be improved by helicopter-based radiotelemetry, the relatively short cruising radius will limit the number of animals that can be tracked and positioned, especially in remote areas (Gilmer et al. 1981).

Accuracy

A large number of factors influence location error when tracking radio-collared animals from aircraft. In general, location error increases with flight speed, elevation above ground level (AGL), and signal reflection in rugged topography, and decreases as the experience levels of the pilot and biologist increase. Problems associated with signal reflection in steep topography or around large expanses of exposed rock are a common issue for both ground and aerial telemetry. One advantage of locating animals from an airplane in those situations is that the area can be thoroughly examined until a direct line-of-sight signal is acquired (Gantz et al. 2006).

We identified 34 different studies that used some type of fixed-wing aircraft to study American martens or fishers in North America. Fourteen of the studies provided estimates of location error. Two of these studies simply estimated location error at <10 m (Arthur et al. 1989) and <500 m (Slough 1989), without providing details on how the errors were estimated. Two studies reported location error in terms of error ellipses: 2.7 ha (Fuller and Harrison 2005) and 3.5 ha (Gosse et al. 2005). Eleven studies reported error in terms of the distance between the estimated and actual location of test transmitters, which averaged 207 ± 188 m (range: 30–729 m).

Although helicopters have been used extensively for tracking other wildlife species, we found only 3 studies that used helicopters for tracking radio-collared *Martes* species. Latour et al. (1994) used a Bell 206B helicopter to monitor the movements and home ranges of American martens in the Mackenzie Valley of the Northwest Territories, Canada. They reported an average location error of 242 ± 100 m. Potvin (1998) used both a Bell 206B helicopter and a Cessna 185 fixed-wing airplane to track American martens in western Quebec, Canada, and reported average errors of 9 m for the helicopter and 223 m for the airplane. Finally, Hearn (2007) used both a Bell 206B and an Aerospatial A-star helicopter to track American martens in Newfoundland, Canada. Although he did not report an estimate of location error, he did report hovering <10 m above the canopy and visually observing collared martens (Hearn 2007).

Although it may be possible to improve the accuracy of some locations by operating aircraft closer to the ground, as a general rule, it should not be necessary to fly a fixed-wing aircraft below 150 m AGL. Indeed, most federal and state agencies and the U.S. Federal Aviation Administration (FAA) require that aircraft not operate below 150 m, except in special circumstances (Gilmer et al. 1981). Safety must remain the highest priority when using helicopters or light aircraft to track radio-collared wildlife; there is no situation when improving the accuracy of positions for radio-collared animals is more important than the safety of the pilot and biologist. In our experience, the optimal elevation AGL for locating radio-collared fishers is between 150 and 230 m; lower-level flying significantly detracts from the concentration needed to interpret variation in signal strength, and it is more difficult to pinpoint locations when flying much above 450 m.

Efficiency

Seven studies provided sufficient information to evaluate the efficiency of aerial telemetry. Studies that relied exclusively on aerial telemetry averaged 1961 total locations on 39 study animals, or 50.3 locations per animal ($n = 4$). Projects that used aerial telemetry to supplement ground efforts averaged 447 total locations on 24 animals, or 18.6 locations per study animal ($n = 3$).

Current and Future Applications

The primary benefit of aerial telemetry is the ability to consistently locate a large number of study animals and to monitor long-distance movements (e.g., dispersals, male breeding activities) in inaccessible terrain. In North America, *Martes* species exist at relatively low densities in rugged terrain, use large areas, and are capable of making rapid long-distance movements—all characteristics that make ground-based telemetry difficult. If flights are frequent, detecting and locating mortalities quickly can help establish the cause of death. The primary weaknesses of aerial telemetry are reduced location accuracy, the loss of additional insights and data gleaned by tracking animals on the ground, and the high cost of flight time. Koen (2005) compared the accuracy of aerial and ground-based telemetry and found that aerial locations were 57% less accurate, making them largely unsuitable for investigating habitat use or selection at fine spatial scales. Aerial telemetry also precludes the identification of specific habitat elements, such as resting or denning sites. For this reason, most studies combine aerial and ground-based monitoring. Although aerial telemetry facilitates the monitoring of larger numbers of study animals, this often translates into fewer locations collected per study animal as a result of reduced tracking efforts (e.g., weekly flights vs. daily ground-tracking).

A relatively recent innovation in wildlife radiotelemetry is the use of ultralight aircraft. The ultralight classification includes motorized aircraft weighing <115 kg with a power-off stall speed of 25 knots (Quigley and Crawshaw 1989). Advantages of using ultralight aircraft compared with larger fixed-wing airplanes or helicopters include lower cost and extended flight time (better fuel economy), flexibility of use (e.g., transport into remote areas), and relatively slow flight speeds, which can translate into more precise locations. Disadvantages of ultralight aircraft include relatively high sensitivity to weather conditions (especially wind) and the potential for reduced safety compared with fixed-wing airplanes (Quigley and Crawshaw 1989).

For locating radio-collared jaguars (*Panthera onca*) from an ultralight, Quigley and Crawshaw (1989) used 2 H-antennas mounted approximately 2 m from each wing tip and angled downward at 45°. M. Cheveau (unpublished data) used an ultralight aircraft equipped with skis to track radio-collared martens in a remote area of northwestern Quebec. Cheveau mounted one 4-element Yagi antenna pointing forward on the front of an ultralight aircraft, with a second antenna oriented sideways on the left side of the ultralight. Using this arrangement, she reported locating study animals daily with an average location error of 33 m (median = 12 m) based on blind tests.

GPS Telemetry

Background

In 1988, the Ontario Ministry of Natural Resources unveiled an ambitious program to double the number of moose (*Alces alces*) in the province over the next 20 years through a combination of timber management and habitat conservation (Ontario Ministry of Natural Resources 1988). As part of this program, they reviewed all available and potential technology for wildlife tracking and settled on the (then) new NAVSTAR GPS satellite constellation as the most efficient and cost-effective way to meet their monitoring needs (Rodgers et al. 1996). In 1992, Lotek Engineering (Newmarket, Ontario, Canada) was contracted to design and develop a GPS-based wildlife telemetry system (Rodgers 2001), and by 1994, the first 1.8-kg GPS collar was available commercially (Rodgers et al. 1996). Since then, steady advancements in GPS telemetry technology such as the miniaturization of

GPS chips, the addition of remote-download and solar-recharging capabilities, and improvements in antenna design have resulted in collars as small as 5–7 g.

Currently, the most appropriate GPS units for tracking *Martes* species range from 40 to 75 g in weight, which includes the battery, epoxy, and collar material. This means that current GPS transmitters weigh approximately 1–8% of a study animal's body weight, depending on the species and sex of the animal. Solar-powered units, while significantly smaller, are generally considered inappropriate for research on *Martes* species because of their preference for areas of dense canopy, tendency to rest in tree cavities or other protected structures, and nocturnal or crepuscular behavior.

The first deployment of a GPS collar on a *Martes* species occurred in February 2009, when both the USDA Forest Service, Pacific Southwest Research Station and the University of California at Berkeley, Sierra Nevada Adaptive Management Project (SNAMP) began field-testing units manufactured by Telemetry Solutions (Concord, California, USA) on fishers in the southern Sierra Nevada of California. To date, 45 miniature GPS collars have been deployed on fishers by these 2 research organizations. One additional GPS collar was deployed on an adult male American marten in 2010 on the Sagehen Experimental Forest in the central Sierra Nevada (K. Moriarty, Oregon State University, personal communication).

Tracking Techniques

Most miniaturized GPS collars contain "store-on-board" transmitters, meaning that data are stored in the unit until it can be retrieved, after the animal either sheds the collar or is recaptured. A VHF beacon operating on a separate power supply allows the collar to be found if it is shed by the animal. Because electronic drop-off mechanisms add 20–40 g of weight, most researchers have used homemade inserts in the collars to facilitate eventual drop-off (see below). Store-on-board units are a risky proposition for most *Martes* researchers, because the collar must be recovered to retrieve the data. Should the collar be dropped in a tree cavity, underground burrow, or roof space where the VHF beacon is suppressed, or the collar is not retrieved prior to failure of the VHF signal, all data are lost. For this reason, researchers typically add the remote-download feature to their collars. This feature operates from the GPS battery and allows a researcher to wirelessly connect with the collar to download the data or reprogram the collar when within 200–400 m of the animal; for example, data can be downloaded from the base of a rest or den tree when it is occupied by a study animal. If the animal is using an area that is difficult to access, tends to flee at the researcher's approach, or is part of a larger project with multiple animals, stored data can be downloaded from an aircraft properly outfitted with a UHF antenna by either circling or hovering at 150–230 m AGL (R. Sweitzer, personal communication).

Accuracy

Accuracy and efficiency are significant concerns when applying GPS technology to tracking *Martes* species. These species typically live in structurally complex habitats where GPS accuracy and fix rates may be compromised. The 2 primary research projects using miniaturized GPS collars on *Martes* species have evaluated their accuracy in 2 ways. The USDA Forest Service, Pacific Southwest Research Station's Kings River Fisher Project (KRFP), in the Sierra Nevada of California, tested collar accuracy by placing a GPS collar on a scat-detector dog during surveys. Scat-detector dogs are trained to locate the scats of a target species (see Long and MacKay, this volume), and over the course of a month, the dog and a handler could survey the majority of the study area. To test the accuracy of the GPS collar while moving through brush and varying forest conditions, researchers compared locations collected by the collar with a tracklog generated by a standard Bluetooth GPS unit carried by the handler. The average distance between the GPS collar and the handler was 12.4 ± 18.6 m (n = 165) (C. Thompson and K. Purcell, unpublished data); 90% of all errors were <28 m.

Researchers with the University of California at Berkeley SNAMP project placed collars in multiple fixed locations and allowed them to collect locations for up to 24 h (R. Sweitzer and R. Barrett, unpublished data). They compared recorded locations with those taken from handheld GPS units, categorizing errors by canopy cover, slope, and aspect. Average error varied from 12 to 42 m, with no obvious relation to cover or topography.

Efficiency

To date, 46 miniaturized GPS collars have been deployed on 13 fishers in the SNAMP project, 8 fishers in the KRFP project, 24 fishers in a USDA Forest Service, Pacific Southwest Research Station fuel-reduction project, and 1 American marten in the Oregon State University, Sagehen Creek study. Because collars are user-programmed, fix rates are highly variable and data presented here should be taken as suggestive of potential collar performance. Based on multiple remote-download attempts for all collars, the average performance of 75-g collars was 152 points collected over an average of 58 days. The average performance of 50-g collars was only 16 points collected over 27 days. The average fix success for both collar types was 28%. Two 75-g collars were successfully removed from recaptured adult male fishers after the GPS batteries were drained, allowing for an accurate assessment of actual lifespan and productivity. One unit lasted 77 days, attempting 711 fixes during that time and successfully acquiring 182 locations (25.6% success rate). The second lasted 85 days, attempting 766 fixes and successfully acquiring 281 locations (36.7% success rate; C. Thompson and K. Purcell, unpublished data). A 45-g collar placed on a male American marten collected 179 fixes at

5-min intervals over 2 days, for a success rate of 39% (K. Moriarty, personal communication). Comparatively, manufacturer specifications indicate that collars collecting 1 location per day should last 331 and 114 days for the full-AA (75 g) and half-AA (45 g) configurations, respectively.

Current and Future Applications

GPS-tracking technology has now advanced to the point where it is appropriate for the largest *Martes* species and individuals, namely fishers and males of other species >2 kg in weight; however, GPS collars are still too heavy and bulky for use on smaller animals. Their use can greatly expand research capabilities through increased accuracy, potentially greater numbers of locations in remote terrain, greater spatial coverage in rugged or roadless areas, collection of locations during times when aerial or ground telemetry may not be feasible (e.g., at night, during severe weather), collection of sequential locations indicating routes of travel, and elimination of the need for an on-the-ground researcher who could potentially disrupt the behavior of study animals. Nonetheless, use of GPS technology has many limitations; for example, data on the cause of mortality and other real-time information are unavailable without additional effort, because the VHF frequency is not monitored consistently. Another concern of particular relevance to *Martes* researchers is the possibility of habitat-related biases resulting from unequal satellite reception within a study area. *Martes* species typically move rapidly through complex landscapes, and a careful assessment of potential biases is necessary. Finally, GPS collars currently have much shorter lifespans than VHF collars, typically on the order of 2–3 months vs. 2–3 years.

Significant advances in GPS technology will undoubtedly be made in the near future that provide greater reliability and longevity. GPS chips will continue to decrease in size, though this will likely mean more storage and functionality, rather than lighter collars, because chips are already an insignificant component of transmitter weights. Lighter collars are most likely to result from improvements in battery technology, which is currently the limiting factor for collar weight, lifespan, and fix attempts. Another area where further advancement is possible is antenna technology; a $4 GPS chip can achieve greater accuracy than a $19,000 GPS receiver, if the antenna used is of significantly higher quality (van Diggelen 2010).

One active focus area for current research and development is the transmission of data back to the user. Currently, collars suitable for use on *Martes* species must either be retrieved or their data downloaded remotely from within 400 m. With the explosion of cellular technology worldwide, new and innovative options are emerging. Currently, features are available to remotely download data from larger GPS collars whenever an animal comes within range of a cellular tower using GSM (Global System for Mobile communications) technology, and it is only a matter of time before this technology is incorporated into smaller collars. Although this may not be useful in North America, where most *Martes* populations inhabit remote terrain with limited cellular phone coverage, it could vastly improve tracking capabilities in Europe or Asia, where *Martes* species inhabit rural and urban areas, and cellular coverage is widespread. Similar proposals have been made to use either unmanned aerial vehicles (Schulte and Fielitz 2001) or other collared animals (Markham and Wilkinson 2008) as download relay stations. At least 1 manufacturer is currently working on a remote-download station that, when placed in conjunction with a bait station, will download data from any collared animal that comes within 100 m (Q. Kermeen, Telemetry Solutions, personal communication).

The most likely area of rapid technological advancement will be through GPS+, a reference to supplementation of the GPS satellite constellation with additional satellite or terrestrial networks. In 1993, Russia announced the functionality of its version of the GPS satellite constellation, known as GLONASS, and China, Japan, and the European Union are developing similar networks (Compass, QZSS, and Galileo systems, respectively). The creation of products capable of communicating with multiple satellite systems will greatly increase coverage and reduce the time-to-fix, resulting in more successful locations and overall power conservation. Similarly, technology is available to supplement GPS location data with data from terrestrial Wi-Fi or GSM networks. These networks provide supplemental location information and increase the speed of communications between the satellite and collar, resulting in more successful locations and overall power conservation. Although this technology may be slow to work its way into wildlife applications, its prevalence in navigation and cellular-phone technology ensures its development and eventual availability.

Argos Telemetry

Background

Originally conceived as a collaborative project between the United States and France to improve meteorologic and oceanic information (CLS America 2011), the Argos Satellite System has expanded markedly since its inception >30 years ago. Currently, movements of >300 species are monitored by >6500 Argos Platform Terminal Transmitters (Argos PTTs; D. Stakem, CLS America, personal communication). Although many biologists are familiar with the mechanics of GPS telemetry, Argos satellite telemetry works quite differently. Unlike GPS telemetry units that receive signals from orbiting satellites and internally calculate a location, Argos units broadcast a signal into space that is picked up by polar-orbiting satellites. Signals from the PTT received by the satellite are relayed to a ground processing station; if ≥4 signals are received by a satellite during a pass, a location and an estimate of that location's accuracy are calculated using the Doppler-shift principle. For more-detailed descriptions of the Argos system, the principles behind the calculation of locations, and current data options, see Fancy et al. (1988) and CLS America (2011).

Much of the initial research into Argos telemetry concluded that the moderate location accuracies of this system limited its suitability to studies of species that moved long distances (Harris et al. 1990; Keating et al. 1991; Britten et al. 1999). Accordingly, much of the wildlife research that has used Argos telemetry has focused on large-scale movements, particularly of birds (e.g., Higuchi et al. 1998; Soutullo et al. 2006), marine mammals (e.g., Baumgartner and Mate 2005; Blix and Nordoy 2007) and large terrestrial mammals (e.g., Walton et al. 2001; Mauritzen et al. 2002; Ito et al. 2006). Over the past 10–15 years, however, the evolution of transmitter design, satellite configuration, and data-processing techniques have significantly improved Argos satellite telemetry performance (J. Sauder, unpublished data). Recent research suggests that Argos telemetry can be suitable for smaller-scale analyses of space use and habitat selection by species that do not move large distances, if appropriate techniques and units of analysis are used (Moser and Garton 2007).

Tracking Techniques

One of the unique characteristics of Argos telemetry is that data collection is entirely automated and coordinated through CLS America. Signals received by the satellites are relayed to a ground-based processing station, where a location and accuracy assessment is calculated. These data are made available to the researcher in multiple formats. Data are accessible in near real-time on the Argos Users website (http://www.argos-system.org), where tools for mapping, downloading, and working with the data can be found. Alternatively, data can be delivered daily or at other intervals to the owner via e-mail. This facilitates data collection in traditionally difficult research situations (e.g., at night, in wilderness areas, during inclement weather), though at significantly increased cost. The initial cost of Argos PTTs can exceed $2500, and additional data-delivery fees are charged by CLS America, based on the number and timing of signals received by the satellites.

The weight and size of Argos PTTs designed for deployment on *Martes* species are mainly a function of battery and collar configurations. Collars built for fishers have generally had about 15 mm-wide nylon-belting collar material and were powered by 1, 2, or 3 AA batteries, resulting in collar weights of approximately 75, 97, and 120 g, respectively (K. Lay, Sirtrack Ltd., personal communication). Lifespan is a function of 4 parameters: repetition rate (how often the transmitter broadcasts a signal), duty cycle (hours per day the transmitter is functioning), output wattage (strength of transmission), and battery size. For a 3 AA-battery collar with a repetition rate of 45 seconds, duty cycle of 12 h "on" in each 72-h period, and output wattage of 0.5 watts, the manufacturer's projected lifespan is approximately 400 days. For identically configured transmitters with 2 and 1 AA batteries, the projected life span is approximately 265 and 135 days, respectively. In practice, however, effective collar life may not meet the manufacturer's estimate. Low temperatures (<10 °C) are known to diminish the lifespan of Argos collars. Also, at the end of a transmitter's life, performance often becomes more variable, because the unit's voltage may hover around the minimum required to emit a signal.

Ground- or aircraft-based tracking of Argos collars is possible but difficult, requiring specialized equipment. Argos transmitters broadcast their signals at approximately 401.65 MHz, a frequency that most VHF receivers do not pick up. Police scanners have been used to locate PTTs (Bates et al. 2003), but the slow repetition rate of signals makes homing-in on the signal difficult. Recently, specialized receivers with highly directional antennas have been designed to detect PTTs. Some of these Argos PTT receivers have an internal meter that indicates signal strength audibly. Using such a receiver makes recovering stationary collars fairly simple, but locating a collar still on a study animal is difficult because the 45 sec between signals is long enough for the animal to leave the area.

Accuracy

When 4 or more signals from an Argos PTT are detected by a satellite during 1 pass, a location and accuracy assessment are calculated. The accuracy assessment assigns the location to 1 of 4 "location classes" (LC) ranging from 0 (least accurate) to 3 (most accurate). The Argos Users Manual states that 68% of locations assigned to LC 3 have an error radius of ≤250 m, increasing to ≤500 m for LC 2, ≤1500 m for LC 1, and ≥1500 m for LC 0 (CLS America 2011). Independent testing of these assessments has shown them to be fairly accurate (J. Sauder, unpublished data). If fewer than 4 signals are received by the satellite during a pass, a location may be calculated with auxiliary data processing; these are assigned to location classes A, B, or Z, but no accuracy assessment is provided by Argos for these locations. Such data points are of variable accuracy and probably of little value to *Martes* researchers, with the possible exception of using them to document long-distance dispersals. In addition to the location class, CLS America recently began providing an error ellipse, an angle of orientation, and a metric called geometric dilution of precision (conceptually similar to positional dilution of precision for GPS telemetry). This additional information on location error may improve the accuracy of error evaluations for analyses of Argos telemetry data. The 95% error ellipses for LC 3, 2, and 1 locations from collars deployed on fishers were 6.0, 23.1, and 186.1 ha, respectively (J. Sauder, unpublished data).

ferrets (*Mustela nigripes*; Biggins et al. 2006); in both projects, the researchers concluded that animals easily removed the harness unless it was attached so tightly that it caused severe neck irritation. One researcher experimented with gluing transmitters between the shoulder blades of 2 stone martens, but the transmitters were shed within 24 h (Herr 2008).

Risks to Study Animals from Telemetry-based Research

Before telemetry-based projects are implemented, researchers should weigh the potential risks associated with capture, handling, and radio-tagging of animals against the benefits of the data to be collected and the status of the population of interest (Kenward 2001; Casper 2009). The attachment of radio transmitters to any wildlife species can potentially lead to injury or mortality during the trapping and handling process or after release (Casper 2009). However, appropriate training in animal handling, a thorough investigation of previous related studies, and some forethought about how transmitters might interact with the morphology and behavior of individual species can, in most cases, minimize adverse effects on study animals (Millspaugh and Marzluff 2001). Limitations on collar weight as a percentage of body weight (3–5%) are commonly adhered to for most mammals (Kenward 2001; Gannon et al. 2007), but the potential effects of the shape and flexibility (especially ability to stretch) of radio-collars relative to the focal species' body shape and habits has received relatively little attention for medium-sized mammals. Because of their long and slender bodies, affinity for tight spaces (e.g., tree cavities, subnivean passages, rock crevices, roof spaces), and tendency to move through vegetation that could snag collars, *Martes* species are likely to be more susceptible than others to collar-related injuries. In this section, we discuss the potential negative effects of transmitters on *Martes* and related species, options to minimize injuries, and results of alternative transmitter attachments to date.

There are 3 primary ways that radio-collars can negatively influence animal safety and research objectives: (1) collars can get caught on external objects, (2) collar fit may change over time, and (3) collar attachment can alter behavior and result in biased data. Although options exist for dealing with these concerns, each involves a trade-off of some kind; thus, the most appropriate choices for each project will vary depending on the species of interest, research goals, study area, time frame, and budget. When initially planning a telemetry-based study on *Martes* species, researchers should consider these 3 points, determine whether concerns about transmitter attachment differ among sex or age groups, talk to other biologists about their experiences, and then decide which techniques are most appropriate for their situation. Finally, if injuries or mortalities occur in association with transmitter attachments, researchers should take the time to review the details of the incident, attempt to improve their methodology, and share the outcome with the larger scientific community to help prevent similar problems for other researchers.

The first concern with attaching radio-collars to *Martes* species is that they can potentially get caught on objects (e.g., sticks, splinters, wire fencing) or wedged in confined spaces (e.g., rock crevices, tree cavities) in natural or human-altered environments. These situations could result in injury, asphyxiation, or vulnerability to predation for individual animals. One American marten in Oregon died after getting its collar caught on a splinter inside a hollow tree (Bull and Heater 2001). Evidence at the scene of 2 separate fisher mortalities in Canada suggested that individuals died as a result of collars getting hung up in branches associated with debris piles (R. Weir, Artemis Wildlife Consulting, personal communication).

A second potential problem with radio-collars is that they are not generally constructed to adjust to changes in neck size associated with age, weight gain or loss, or season. As a result, collars may cause discomfort, lesions, injuries to skin or muscle, and even death if the neck becomes constricted or infected. For most (if not all) *Martes* species, subadult males and juveniles of both sexes present the greatest challenges for balancing the trade-offs between fitting a collar loosely enough to allow for growth, but secure enough that it will not be easily shed or snagged. Neck injuries associated with tight-fitting collars have been documented for fishers on several projects (Weir 1999; Mazzoni 2002; C. Thompson and K. Purcell, unpublished data; M. Higley, Hoopa Tribal Forestry, personal communication). In the reported instances, collars of recaptured individuals were found embedded in the neck (Figure 13.4). The animals were treated and released without collars, and at least 3 had recovered fully by the time they were recaptured. To reduce the risk of neck abrasion, one researcher substituted flexible nylon webbing for the typically stiff collar material but found that the webbing became stiff and abrasive with the accumulation of oil and dirt, contributing to the death of 4 fishers (Heinemeyer 1993). Similar neck injuries associated with radio-collar attachment have been recorded for other carnivores, including the American mink (Zschille et al. 2008) and black bear (*Ursus americanus*; Koehler et al. 2001). Because recapture and the removal of radio-collars from all study animals on a project is often not feasible, the documentation of injuries associated with long-term collar wear is undoubtedly underrepresented.

Finally, transmitter attachments can present problems for study animals and researchers if the device alters behavior, limits habitat-related choices, or increases the risk of predation. For example, a radio-collar with a bulky shape may limit the size of cavity openings that females can use during the denning season, which is not the case with implanted transmitters (R. Weir, unpublished data). Often, these potential adverse effects are difficult to assess

Figure 13.4. Ingrown collar removed from a juvenile male fisher. Animal was treated, released, and later recaptured showing no signs of the injury (photo J. Garner).

or quantify for species that are infrequently seen in the field; however, consideration of the species' ecology and morphology can help inform the selection of collar shape, size, material, and color. For example, species that use circular cavities may be least impeded in their movements by small cylindrical canisters, and dark-colored collars are likely the best choice for animals with dark pelage. Two primary options exist for minimizing problems associated with attaching radio-collars to *Martes* species: collar alterations and intraperitoneal implants.

Collar Alterations—Break-aways and Spacers

In this chapter, we refer to a *break-away* as a weak point on a radio-collar that is designed to break under pressure or deteriorate over an extended period of time, or both. A *spacer* represents an expandable portion of the radio-collar that provides some allowance for growth, weight gain, or increase in muscle mass around the neck. These options are not mutually exclusive and can be combined in a single collar. These types of alterations have been most commonly designed for use with large mammals, such as white-tailed deer (*Odocoileus virginianus*; Diefenbach et al. 2003), elk (*Cervus canadensis*; Smith et al. 1998), and black bears (Koehler et al. 2001), but the same general ideas are appropriate for use with *Martes* species. To our knowledge, Paragi (1990) was the first to insert a small piece of leather into a collar to ensure that it would eventually drop off. Weir (1999) used cotton firehose spacers in fisher radio-collars with mixed results; the cotton material on 2 collars rotted away after only 4 months, but remained intact on another collar that had become embedded in the tissue of a male fisher's neck. On the Hoopa Reservation in northern California, a piece of soft leather was added to some collars to provide a degree of stretch and a long-term breaking point (S. Matthews, Wildlife Conservation Society, unpublished data).

Two ongoing research projects in California's southern Sierra Nevada have incorporated the objectives of break-aways and spacers into their VHF and GPS radio-collar designs for use on fishers (C. Thompson and K. Purcell, unpublished data; R. Sweitzer and R. Barrett, personal communication). After testing several designs, the Kings River Fisher Project researchers settled

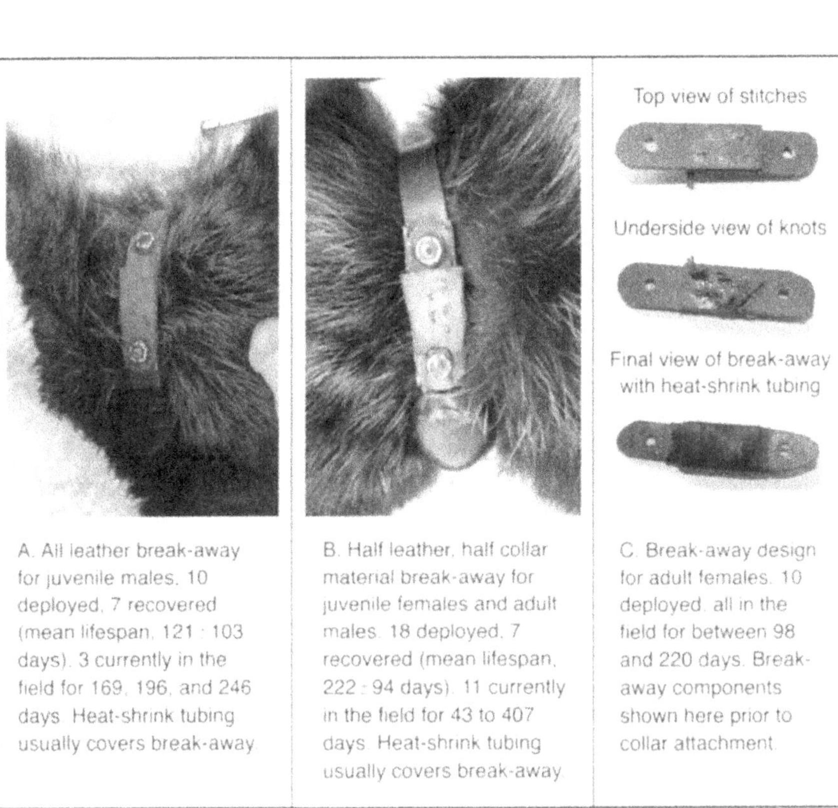

Top view of stitches
Underside view of knots
Final view of break-away with heat-shrink tubing
A. All leather break-away for juvenile males. 10 deployed, 7 recovered (mean lifespan, 121 ± 103 days). 3 currently in the field for 169, 196, and 246 days. Heat-shrink tubing usually covers break-away.
B. Half leather, half collar material break-away for juvenile females and adult males. 18 deployed, 7 recovered (mean lifespan, 222 ± 94 days). 11 currently in the field for 43 to 407 days. Heat-shrink tubing usually covers break-away.
C. Break-away design for adult females. 10 deployed, all in the field for between 98 and 220 days. Break-away components shown here prior to collar attachment.

Figure 13.5. Examples of break-away devices being used on the U.S. Forest Service's Kings River Fisher Project (Thompson et al. 2010). Break-aways guarantee that the animal will be able to shed the collar eventually, if it is not recaptured, and provide a weak point for breaking loose if the collar gets entangled.

on 1 pattern and 3 combinations of materials to accommodate animals in different sex and age groups (Figure 13.5; R. Green, unpublished data). The basic design involves 2 short pieces of material stitched together by hand. The stitches are then coated with nail polish and covered with heat-shrink tubing. The break-aways are made in advance, bolted to VHF collars during handling, and can be replaced easily if an animal is recaptured. Materials used in making this break-away include soft leather, neoprene-impregnated polyester straps (used in radio-collars), cotton thread, cotton-polyester thread, nail polish, and heat-shrink tubing. The combination of materials used depends on the likelihood of the animal's neck increasing in size. Two pieces of soft leather allow for the greatest amount of stretch (best for juvenile males; Figure 13.5A), 1 piece of polyester strap combined with 1 piece of soft leather will permit some stretching (suitable for juvenile females, subadult males, and adult males; Figure 13.5B), and 2 pieces of polyester strap are unlikely to stretch (appropriate for adult females; Figure 13.5C). Although these break-away/spacer designs may not prevent all possible injuries associated with collars in *Martes* species, they are inexpensive and easy to make, can be adapted to different collar designs and materials, provide some room for growth, and decompose over time. Furthermore, although it is rarely possible to document the success of a break-away insert, there have been at least 2 instances where collars were recovered from sites where entanglement was possible (e.g., exposed root mass, narrow rock crevice) and the break-away was found split open, as anticipated.

Intraperitoneal Implant Transmitters

History

Implanted transmitters (implants) were first used on *Martes* species in the early 1980s, when Skirnisson and Feddersen (1984) implanted transmitters in 9 stone martens in Germany. The authors compared the function of the intraperitoneal implants with 5 radio-collars, and found that the implants worked without malfunction, whereas the collars all failed prior to the end of the manufacturer's predicted lifespan. They also reported that 1 implanted animal died after the sutures tore loose, but 1 implanted female successfully gave birth to and raised 3 kits while carrying the implant.

The next reported use of implants with *Martes* involved American martens reintroduced into the Yukon between 1984 and 1987 by the Yukon Department of Renewable Resources. Thirty-nine martens were monitored via radiotelemetry, 12 with collars and 27 with intraperitoneal implants. During the initial 30-day monitoring period, 17 (63%) of the implanted animals died from a variety of implant-related causes, whereas all collared animals survived (Slough 1989). During approximately the same period, Buskirk et al.

(1989) implanted 16 g temperature-sensitive transmitters in American martens in Wyoming to evaluate winter rest-site requirements, and reported no ill effects. In 1989 and 1990, 37 fishers were translocated into the Cabinet Mountains of Montana: 26 were radio-collared and 11 had intraperitoneal implants (Roy 1991; Heinemeyer 1993). Four of the collared animals, fitted with collars made of lightweight webbing (described above), developed severe neck lesions and died. Another 2 animals showed hair loss and irritation when the collars were removed, but no adverse effects of the implants were reported.

To our knowledge, implants were not used again on *Martes* species until 1998, when researchers in British Columbia (Canada) began implanting fishers (Weir and Corbould 2006) as a result of concerns about the safety and reliability of radio-collars. Since 1998, researchers in British Columbia have implanted 32 fishers on 41 occasions with no negative consequences (R. Weir, personal communication). Recently, implanted transmitters have been used in fisher reintroduction efforts in Washington (J. Lewis, personal communication) and California (A. Facka, personal communication).

Function

Early generations of implant transmitters suffered from reduced signal strength. As stated by Biggins et al. (2006: 182), "Relatively poor reception range is a well-known attribute of implantable transmitters, in part because of the compromises necessary with transmitter antennas, which can translate into reduced precision and accuracy of data." Green et al. (1985) compared the performance of implants and collars on coyotes (*Canis latrans*) and kit foxes (*Vulpes macrotis*) and found that collars had 5–10 times greater range and a 30% longer lifespan. Koehler et al. (2001) reported that implanted black bears were more difficult to locate than collared bears because of reduced signal strength, and recommended that implants be used on animals with smaller or more dependable home ranges. These problems appear to have been resolved in current generations of implantable transmitters, and researchers currently using implants in fishers report no difference in signal strength or range (R. Weir and J. Lewis, personal communications). Most researchers also report comparable lifespans between implants and collars (J. Copeland, USDA Forest Service (retired), personal communication; R. Weir personal communication; J. Persson, Swedish University of Agricultural Sciences, personal communication), though problems with malfunctioning mortality switches in implants have led to premature transmitter failures in some cases (J. Lewis, personal communication). Additional research on brown bears (*Ursus arctos*) in Sweden has suggested that the failure of mortality switches in implant transmitters may be associated with the accumulation of moisture inside the transmitters, and subsequent corrosion

(J. Arnemo, Norwegian School of Veterinary Science, unpublished data; see below).

Potential Risks

Transmitters can be implanted either subcutaneously or into the peritoneal cavity. Many studies have found that intraperitoneal implants are preferable for mammals, because subcutaneous implants often cause abscesses, and the rate of transmitter loss or damage is high (Agren et al. 2000; Echols et al. 2004; Biggins et al. 2006).

The greatest risk associated with implanted transmitters involves the surgical procedures and the 24–48 h recovery time while the incision begins to heal. Implant surgeries often take place in uncontrolled environments where the risk of infection is high, and procuring an experienced veterinarian willing to work under field conditions can be challenging (J. Copeland, personal communication). Eagle et al. (1984) implanted 32 American minks with intraperitoneal transmitters, and attributed 3 mortalities within 2 months to infection and inexperienced personnel. Copeland (1996) reported a single implanted wolverine (*Gulo gulo*) dying after chewing out the sutures. Likewise, Zschille et al. (2008) implanted 14 American minks and reported 1 male dying after chewing out the sutures; subsequently they recommended that implanted animals be held for 72 h after surgery to ensure a full recovery.

Once an animal has recovered from surgery and the incision has healed, the risk of mortality from the implant appears to be low. We found no reports of *Martes* species dying from implant-related causes, once the sutures had healed. Van Vuren (1989) performed 300 implant surgeries on 183 yellow-bellied marmots (*Marmota flaviventris*) and reported only 2 mortalities in which the animals expired in burrows and the cause of death was unknown. Guynn et al. (1987) reported that 1 of 18 implanted American beavers (*Castor canadensis*) died after the intraperitoneal implant lodged in the intestinal tract, and Herbst (1991) reported that 1 of 18 implanted nine-banded armadillos (*Dasypus novemcinctus*) died from a similar blockage. Recently, extensive use of intraperitoneal transmitters on sea otters (*Enhydra lutris*) and Vancouver Island marmots (*Marmota vancouverensis*) has resulted in no consistent complications (Ralls et al. 2006; M. McAdie, British Columbia Ministry of Environment, personal communication).

Evidence accumulated to date suggests that intraperitoneal transmitters have no effect on reproduction. Although this is nearly impossible to prove because of their low and variable reproductive rates, numerous accounts have been published of implanted females of various *Martes* species giving birth and successfully raising kits. Weir and Corbould (2008) found similar reproductive rates between implanted and collared female fishers; 5 of 8 implanted and 7 of 10 collared females successfully reproduced. More extensive evaluations of the impacts of implant transmitters on reproduction have been done on other species, with most reporting no noticeable effect. Herbst (1991) found that female nine-banded armadillos carrying intraperitoneal implants reproduced successfully, and Van Vuren (1989) reported that both growth rates and reproductive rates were similar between implanted and non-implanted yellow-bellied marmots. Reid et al. (1986) evaluated the ability of river otters (*Lontra canadensis*) to reproduce while carrying implanted transmitters and found that all stages of the reproductive cycle were successful. The authors' only caution was that surgery on adult females during pregnancy or lactation should be avoided, because many drugs used for anesthesia cross the placental barrier and may induce anesthetic effects or affect development in fetuses. This concern is complicated by delayed implantation, although the use of anesthesia while collaring female *Martes* study animals is not known to impact the development of offspring in the subsequent year.

Long-Term Implications

Very few data are available on the long-term implications of implanted transmitters for study animals. Implants are coated in inert materials and hermetically sealed, so it is generally assumed that they pose no long-term health risks for study animals, beyond the limited risks described above. However, because of the difficulties associated with locating and retrapping study animals after a project is completed, very few researchers have removed intraperitoneal transmitters, inspected them for wear or damage, or assessed long-term side effects on study animals that carry them throughout their lifespans.

To our knowledge only 1 research project has attempted to evaluate these long-term implications. Between 2004 and 2010, the Scandinavian Brown Bear Research Project (http://www.bearproject.info/) surgically removed 64 intraperitoneal implant transmitters (Telonics IMP/400/L) carried by brown bears for 3–9 years (Arnemo et al. 2007; J. Arnemo, unpublished data). In 30 cases (47%), the transmitter was encapsulated by connective tissue in the omentum, and histopathology showed a typical reaction to a foreign body. A detailed inspection was conducted on 49 of the implants. Thirteen of the transmitters (27%) showed visible damage, such as discoloration, wear, or melting of the exterior wax coating. One implant (2%) had cracks in the Plexiglas cylinder holding the transmitter components, and the end cap on the Plexiglas cylinder was loose or open in 6 implants (13%). Humidity levels inside the supposedly waterproof implants ranged from 5 to 20%, and condensed moisture was visible inside the cylinder in all cases. In 30 of the implants (63%), moisture had resulted in battery corrosion and 3 implants (6%) had subsequently short-circuited. Of greatest concern was the fact that 2 of the 3 short-circuits resulted in battery corrosion and breaking of the end cap of the Plexiglas cylinder. In one case, a 5-year-old implant suffered internal corrosion, short-circuit, and breakage of the end cap, which caused significant leakage from the battery into the abdominal cavity. The implant was encapsulated by a thick wall of connective tissue and removed. Upon release, the female bear survived and had 3 cubs the following year before being legally shot by a hunter. In 2010, 10 years after implantation, another adult female brown bear was found dead. Necropsy showed that the batteries had short-circuited, the end cap of the Plexiglas cylinder had broken off, and a metal wire (the antenna) had perforated the stomach. The cause of death was peritonitis with subsequent sepsis (J. Arnemo, unpublished data).

These researchers concluded that the implant transmitters were not waterproof, were not physiologically inert, and had a high rate of technical failure attributed to the gradual accumulation of moisture inside the transmitter (Arnemo et al. 2007). Although they acknowledged that implant transmitters were a viable method for minimizing the adverse effects of radio-collars on study animals, they recommended that all implants be surgically removed during the operational life of the batteries. These concerns have yet to be addressed directly in relatively short-lived animals, such as *Martes* species, although transmitters replaced after 3 years in Olympic marmots (*Marmota olympus*) in Washington and British Columbia, and 2 years in wolverines in Sweden did not show similar levels of deterioration (M. McAdie and J. Persson, personal communication).

Conclusions

Researchers tend to have strong opinions about the viability of implanted transmitters as a research technique, based on their personal experiences. It is undeniable, however, that radio-collars add some unknown degree of risk to the safety of study animals and may influence their behavior in unknown ways. Available evidence indicates that implanted transmitters provide a viable alternative to radio-collars and, in some situations, may pose less overall risk to the study animal, if they are used responsibly. For example, in translocation efforts where animals are already being held and can be monitored after surgery, implanted transmitters may be preferable to collars. Additional guidance on implant procedures is available from studies on other species. After extensive use of implanted transmitters on Vancouver Island marmots, researchers at the British Columbia Ministry of Environment recommended that personnel implanting transmitters should (1) use liquid disinfectants instead of gas sterilization, which can damage the wax coating and result in adhesions, (2) flush transmitters well with sterile saline solution after removing them from the disinfectant, (3) avoid temperature extremes, which can damage the wax and lead to adhesions, (4) visually inspect all wax-coated transmitter packages for any surface defects or cracks prior to implantation, and (5) handle transmitters gently with sterile gloved hands to prevent contaminating the transmitter—the use of forceps or other instruments creates small inconsistencies in the wax coating, which can lead to focal adhesions (M. McAdie, personal communication).

More information is needed on the long-term risks and potential deterioration of implanted transmitters within the body cavity. In the interim, we recommend that at the end of each study, researchers should recapture study animals and remove implanted transmitters, despite the additional cost, time, and effort this requires.

Summary

Radiotelemetry has provided essential information on the ecology of *Martes* species, and has been used extensively to study many aspects of American marten and fisher ecology. Questions addressed include demographic rates, habitat use, responses to disturbance, diet, activity patterns, movements, and spatial organization. Despite the significant contributions of individual projects (e.g., Brainerd et al. 1995; Genovesi and Boitani 1995, 1997a,b; Zalewski 1997a,b, 2000; Herr 2008), little telemetry-based data on either the European pine marten or the stone marten are available. Much of the work that has been done on these species remains unpublished, although some information is available in non-English articles and reports (e.g., Skirnisson 1986; Clevenger 1993a; Seiler et al. 1994; Simon and Lang 2007). Recent and ongoing research will help fill this void, once it becomes widely available (e.g., E. Manzo, Ethoikos, Convento dell'Osservanza, Siena, Italy; M. Mergey, Centre de Recherche et de Formation en Eco-Ethologie, Boult-aux-Bois, France; and D. O'Mahoney, Ecological Management Group, Belfast, Ireland, unpublished data); however, many questions that are best addressed using radiotelemetry, such as density or causes of mortality, remain unanswered. Telemetry-based research on the yellow-throated marten, Japanese marten, and sable remains sparse, with only 1, 2, and 4 projects conducted, respectively; none have been conducted on the Nilgiri marten.

We have seen remarkable advances in technology since telemetry was first used on *Martes* species. As technological advances such as the miniaturization of ARGOS and GPS collars open exciting new opportunities for data collection and mesocarnivore research, it is important to clearly identify research objectives and carefully select the most appropriate tools for the questions being addressed. Although the number of accurate locations generated by a properly functioning miniature GPS collar is impressive when compared with the effort required to obtain a similar sample using ground-based telemetry, this new technology and its relatively high failure rates make it a risky choice for short-term or underfunded projects. Similarly,

ARGOS transmitters are currently appropriate only for the largest *Martes* species, the fisher.

The use, efficiency, cost, and accuracy of different monitoring techniques and technologies are summarized in Table 13.2. Ground-based telemetry, the most common technique used in research on *Martes* species, typically provides >100 locations per study animal during the course of a typical research project, but the overall cost will vary considerably, depending on terrain. In comparison, projects that used aerial telemetry monitored more animals, but collected fewer locations per animal, as a result of the reduced frequency of monitoring (e.g., weekly flights vs. daily ground-tracking). Satellite-based monitoring techniques have not been widely used, but they appear to provide higher data-collection rates and potentially reduced costs; however, there are potential trade-offs in reliability, weight, longevity, and the accuracy of locations obtained.

Selecting the appropriate technology and monitoring technique requires careful consideration of research objectives and available resources, as well as the morphology, behavior, and ecology of the focal species. When radiotelemetry is the most appropriate choice and animals must be handled, researchers should maximize available opportunities by collecting data on age, sex, reproduction, physiological state, and other covariates that may influence resulting data. Furthermore, if break-away collars are not used, researchers should consider recapturing animals to remove collars or implants while the transmitters are still functioning, even though it can be challenging, time-consuming, and expensive.

Despite its limitations and associated concerns about the invasive nature of radiotelemetry research, during the past 4 decades we have seen remarkable growth in the application of this technology to ecological research questions. Handmade beacons of the past have developed into miniature processors capable of collecting physiological, meteorological, or proximity data, and the research questions that can be addressed have evolved accordingly. The next few decades promise to be an exciting period for mesocarnivore researchers, as technological advances result in the development of ever smaller and lighter transmitter packages.

Acknowledgments

We are grateful to the numerous researchers worldwide who provided copies of manuscripts, presentations, reports, proposals, and many other valuable sources of data. In particular, Jan Herr, Scott Brainerd, Kazuki Miyoshi, Andrzej Zalewski, and Sandro Lovari not only provided their own data but also assisted in locating other researchers and unpublished reports. Roger Powell and L. David Mech provided a historical perspective. Jeff Copeland, Jens

Persson, Malcolm McAdie, and Rich Weir all provided insights into the use of intraperitoneal transmitters. Jeff Lewis and Jeff Larkin agreed to quantify their program cost and efficiency during the creation of Table 13.2. We are grateful to Martin Raphael and 2 anonymous reviewers for editorial comments, and to the editors of this volume for the opportunity to share what we have learned.

14

Noninvasive Methods for Surveying Martens, Sables, and Fishers

ROBERT A. LONG AND PAULA MACKAY

ABSTRACT

The study of secretive, low-density, and wide-ranging carnivores is inherently challenging. Although track and scat surveys have long been used by carnivore researchers, these methods have been limited in their ability to provide data for estimating population parameters. Over the last decade, however, advances in both remote-camera technology and DNA laboratory techniques have opened the door for researchers to collect valuable information about occupancy, population size, relatedness, genetic structuring, hybridization, recolonization, and behavior with increased resolution and efficiency. Such advances have, in turn, provided the impetus to further develop methods for collecting DNA and photographic data and for analyzing these data. We present an overview and review of the methods currently available for the noninvasive study and monitoring of *Martes* species.

Introduction

Nearly 2 decades have passed since Raphael (1994) published a review of available methods for monitoring American martens (*Martes americana*) and fishers (*M. pennanti*). In this relatively short time period, the number and sophistication of mechanical, molecular, and statistical tools for studying *Martes* species and other carnivores have expanded greatly. Noninvasive survey methods have become particularly important in recent years (Piggott and Taylor 2003; Waits and Paetkau 2005; Long et al. 2008). At the time of Raphael's (1994) review, film-based remote cameras suitable for field research were just coming into use, and the analysis of DNA extracted from feces (scat) or hair was not yet on the horizon. Zielinski and Kucera's (1995) survey manual for forest carnivores placed noninvasive survey methods into the hands of carnivore biologists and introduced the first field-ready remote-camera systems (Kucera et al. 1995), but still preceded hair collection as a pivotal research technique. This chapter is designed to provide *Martes* researchers and managers with an update of the noninvasive methods currently available for surveying martens, sables (*M. zibellina*), and fishers. Although not an exhaustive literature review, it introduces readers to the most prevalent techniques being used worldwide and presents the information necessary to evaluate and choose among them.

Taxonomic and Geographic Breadth of Our Review

To the extent that information was available, our review includes all *Martes* species. To provide a complete review of relevant noninvasive survey methods, we also reference other species groups, especially other carnivores. Most noninvasive surveys for *Martes* species have focused on American martens, fishers, and European pine martens (*M. martes*). Although many of the methods applicable to these species are potentially useful for other *Martes* species, we caution that small differences in species-specific physical characteristics (e.g., coarseness of hair) or behavior (e.g., propensity to enter cavities of certain sizes and shapes) can have major effects on the success of a survey.

What Are Noninvasive Methods?

The scientific literature presents divergent views of what constitutes a noninvasive survey method (Garshelis 2006; MacKay et al. 2008b; Pauli et al. 2009b), and even whether any method can be considered truly noninvasive (Garshelis 2006). Here, we define *noninvasive survey methods* as those that do not require target animals to be observed directly, handled by the surveyor (MacKay et al. 2008b), or restrained or detained when the surveyor is absent. Although noninvasive methods may involve devices that pluck hairs and follicles from the skin, they do not require penetration of the skin or the removal of tissues beyond hair follicles (e.g., as with biopsy darting, Pauli et al. 2009b). Nonetheless, noninvasive methods impact study animals to varying degrees, and camera flashes, scent lures, baits, and the presence of researchers or scat-detection dogs can potentially contribute to changes in the behavior of the species under study (MacKay et al. 2008b). In most cases, classifying a method as noninvasive is self-evident, and wildlife researchers now commonly use the term to refer to both the remote collection of DNA samples (e.g., scats, hair) from free-ranging animals (e.g., Morin and Woodruff 1992), and non-DNA based wildlife surveys using other methods discussed in this chapter (e.g., remote cameras, Moruzzi et al. 2003; snow tracking, Jiang et al. 1998).

Benefits of Noninvasive Research Methods

Modern noninvasive survey methods provide many benefits to wildlife researchers. Because these methods do not require study animals to be captured or observed directly, they minimize risks of injury and disturbance and help reduce sampling bias. Station-based survey devices can be deployed over vast areas with minimal upkeep and long intervals between revisits, often resulting in relatively low costs per data point. Noninvasive methods can also often be used by technicians or volunteers without the rigorous training and certification required to handle wildlife. Consequently, with a given amount of funding, noninvasive surveys can provide better information about population size, distribution, and occurrence from a larger area and from more animals than surveys using more traditional methods (e.g., telemetry, live-trapping).

Martens, sables, and fishers often occur at low densities and are elusive and wide-ranging, making them difficult to study with observational or capture-based methods. Most *Martes* species are curious, patrol and scent-mark their territories, and travel along traditional routes, which are traits that lend themselves well to noninvasive survey methods and the use of noninvasive devices that can be deployed for extended periods, or tracking methods that require only wildlife sign to be successful.

Telemetry and other capture-based methods continue to offer advantages for meeting certain objectives (Mech and Barber 2002; C.M. Thompson et al., this volume), such as documenting long-distance animal movements (e.g., Inman et al. 2004), determining the causes of mortality, and assessing habitat use (Garshelis 2006; but see section on "Track and Scat Surveys" below). However, technological advances have narrowed the gap between the types of objectives that can be met with capture-based methods and those achievable with noninvasive techniques. MacKay et al. (2008b: 3) effectively summarized the choices faced by researchers: "The bottom line is that researchers conducting any wildlife survey must weigh the tradeoffs associated with methods that will allow them to achieve their goals. For some surveys, the target species and primary objectives will lend themselves well to one or more affordable and effective noninvasive methods. We presume that such methods will be the obvious choice when this is the case. In other situations, the required data may only be obtainable via more traditional capture- or observation-based methods."

Classes of Noninvasive Survey Methods

In this chapter, we review 5 broad classes of noninvasive methods designed to collect survey data from *Martes* species: tracking- and scat-based methods, track stations, remote cameras, hair collection, and scat-detection dogs. To some extent, each of these methods is capable of fulfilling a complete range of potential survey objectives, from simple detection-nondetection assessments to the identification of individuals and estimates of abundance and density, but certain methods lend themselves better to some objectives than others. For example, individuals of most *Martes* species have few obvious markings (e.g., stripes) and are therefore difficult to distinguish morphologically. If the unique identification of individuals is required (as with most mark-recapture estimators), then remote cameras may be of limited value (but see Rosellini et al. 2008). On the other hand, if only detection-nondetection data are necessary for successful monitoring, remote cameras should be sufficient in many cases.

New Technologies

The contribution of novel technologies to contemporary noninvasive survey and monitoring techniques cannot be overstated. For example, a radical shift from film to digital photography has opened many new avenues for the use of remote cameras in research. The digital format permits thousands of photos to be stored on a single media card, resulting in significant cost savings because cameras can be deployed for weeks at a time between revisits and there are no film-developing costs. Further, the increasingly reliable nature of modern sensor/camera units, coupled with the minimal lag time between the sensing of an individual and photo capture, permits the detection of rapidly moving individuals.

Advances in the field of molecular ecology—most notably in the ability to effectively extract and amplify DNA from hairs and scats (e.g., Taberlet et al. 1996, 1999)—have dramatically increased the number and types of questions that noninvasive field studies can address (Waits and Paetkau 2005; Schwartz and Monfort 2008; Schwartz et al., this volume). Indeed, scat- and hair-collection methods that, 20 years ago, could largely generate only detection-nondetection data, often of poor quality, are now able to yield full multi-locus genotypes for capture-recapture analyses and population-genetic applications (Broquet et al. 2006b; Beja-Pereira et al. 2009). Such methods are also increasingly used to verify anecdotal occurrence and sighting data for carnivores (Aubry and Jagger 2006; McKelvey et al. 2008). The ability to extract valuable information from DNA at relatively low cost has, in turn, encouraged new innovations in hair- and scat-collection techniques and applications. For example, hair samples collected along snow trails can provide important information that tracks themselves cannot (McKelvey et al. 2006), and scats can now be positively identified to the species and, sometimes, to the individual from which they came. Species-specific hair-snagging devices are becoming more common (DePue and Ben-David 2007; Kendall and McKelvey 2008), and the use of scat-detection dogs has increased rapidly in recent years; specially trained dogs are now capable of finding larger quantities of scats with higher success and less bias than can human researchers.

Finally, computing advances have enabled the development of quantitative and statistical approaches that permit researchers to address complex and previously challenging questions about habitat use and associations (e.g., Buskirk and Millspaugh 2006; Baldwin and Bender 2008), occurrence (e.g., MacKenzie et al. 2002; Nichols et al. 2008; Slauson et al., this volume), abundance (e.g., Petit and Valiere 2006; Royle et al. 2009; Gardner et al. 2010), behavior (e.g., Grassman et al. 2006a), and conservation (e.g., Zielinski et al. 2006a; Carroll et al., this volume). Royle et al. (2008) present a comprehensive introduction to new methods of inference for noninvasive survey data.

Method Descriptions

In this section, we discuss each method in detail, refer readers to appropriate publications, and highlight those methods commonly used to survey for *Martes* species.

Track and Scat Surveys

Exploring the behavior and confirming the presence of animals by observing animal tracks and sign (e.g., scats, dens) date back to the first human hunters (Liebenberg 1990), and the relatively young field of wildlife management has a rich history of obtaining biological insights from natural sign. Tracking, in particular, has provided valuable knowledge about where animals travel, how they use habitat, and their relative abundance, whereas scat collection and analysis has often been used to study diet and species occurrence.

Tracking

Tracking in its most basic form is simply finding, following, and documenting animal tracks in earth, sand, mud, or snow. Expert trackers can infer a great deal of useful information from a trail of tracks, including species identity, the speed at which the animal was moving, the animal's behavior and interactions with conspecifics and prey, and occasionally, the animal's sex and individual identity (Rezendes 1992; Zalewski 1999). Tracking was once among only a handful of methods available for collecting information about animal movements (Heinemeyer et al. 2008), and it has long played a role in the study of *Martes* species (e.g., Thompson 1949; de Vos 1951c; Douglas and Strickland 1987).

Although more recently developed noninvasive methods, such as the deployment of remote cameras and hair-collection devices, have largely replaced traditional tracking techniques for documenting occurrence, snow tracking is still highly effective and often used. In fact, snow tracking is often the method of choice for detecting *Martes* species (Halfpenny et al. 1995; Proulx and O'Doherty 2006), given their large home ranges, the forested landscapes they inhabit, and their year-round activity patterns. In addition to its utility for confirming presence or permitting the inference of absence, snow tracking remains the only noninvasive technique that can be used to investigate micro-habitat use effectively (e.g., Andruskiw et al. 2008). Further, when used in an appropriate sampling design (Thompson et al. 1989; Helle et al. 1996; Helle and Nikula 1996), snow tracking can also be an efficient method for monitoring population trends (Hellstedt et al. 2006) of wide-ranging mustelids, and for estimating the species richness of carnivore communities (Pellikka et al. 2005). Proulx and O'Doherty (2006) identify a number of benefits of snow tracking compared with other monitoring or survey methods, including the lack of dependence on scent lures or baits, the precise location and travel information provided by snow tracks, and the opportunity to infer behavior from animal trails.

Snow tracking has been used to address a wide variety of objectives for *Martes* species, including confirming presence and inferring absence (Moruzzi et al. 2003); assessing abundance (de Vos 1951c; Zalewski et al. 1995); evaluating habitat selection (Jiang et al. 1998; Proulx 2005b; McKague 2007); investigating behaviors, such as locomotion, foraging (Goszczyński et al. 2007), and hunting success (Andruskiw et al. 2008); and assessing the effects of forestry practices (Buskirk et al. 1996b) and human disturbance (Robitaille and Aubry 2000) on their populations. Snow-tracking surveys repeated over many years as part of a long-term monitoring program can also yield useful information about trends in species abundance (Long and Zielinski 2008; Dhuey 2009; Woodford 2009b). Finally, when combined with scat collection, snow tracking has been used to investigate competition between sexes and the diet of European pine martens (Zalewski 2007).

With the advent of advanced molecular methods, snow tracking—coupled with hair (McKelvey et al. 2006) or scat collection (McKelvey et al. 2006; Goszczyński et al. 2007)—can now provide researchers with accurate species identification and information about individuals, even when trails are relatively old and track details are lacking. This is an important development, because misidentification of species from snow tracks is one of the primary drawbacks to the use of snow tracking to study carnivore populations. Modern genetic methods allow the quantification of misidentification rates and enable surveyors to determine whether trails in close proximity were from the same individual. Despite the effectiveness of pairing snow tracking with DNA collection, the availability of appropriate snow conditions is still a limiting factor for the successful application of this method.

Scat Surveys

Scat surveys for *Martes* have long been used to assess distribution, index abundance, and examine food use and diet. Although some diet studies have relied on scats collected at the capture or rest sites of radio-collared animals (e.g., Zielinski et al. 1999), noninvasive scat surveyors typically search for scats along trails and roads, natural landscape features (e.g., streams, ridgelines), or snow tracks. Despite the common use of scat collection and analysis historically, this method has been criticized in light of recent genetic studies that revealed high species-misidentification rates based on scat morphology (e.g., Davison et al. 2002). Birks et al. (2004) present a strong argument against using scat surveys as they are commonly practiced for assessing the distribution and population status of European pine martens. These authors identify 4 major problems with scat surveys: (1) they are often biased because they rely on trail- or latrine-based survey designs; (2) scat identification without DNA confirmation can be inaccurate; (3) assessments of presence are unreliable for low-density populations; and (4) the relation of scat density to abundance is unknown or inconsistent. Additionally, new calls for obtaining verifiable occurrence data for carnivores (Aubry and Jagger 2006; McKelvey et al. 2008) require more than morphologically based scat identification.

Modern molecular techniques have made it feasible and cost-effective to genetically confirm the identification of scats to the species level (e.g., Foran et al. 1997a; Domingo-Roura 2002; Dalén et al. 2004), rendering morphologically based species identification from scats largely obsolete. Species confirmation can be especially critical for surveys of *Martes* species because confusion is common among scats of sympatric *Martes* species (e.g., European pine martens and stone martens [*M. foina*]), or of *Martes* and other mesocarnivores (e.g., Kurose et al. 2005).

Notably, a number of recent scat-based studies and monitoring efforts have used new DNA markers and approaches (Francesca et al. 2004; Livia et al. 2006; Pilot et al. 2007) to explore differences between sables and Japanese martens (*M. melampus*; Sugimoto et al. 2009) and also between pine and stone martens—including differences in diet and trophic patterns (Posluszny et al. 2007), foraging and locomotion (Goszczyński et al. 2007), and distribution (Rosellini et al. 2008). Further, scat surveys combined with genetic testing have been used to assess the distribution (O'Mahony et al. 2006; Balestrieri et al. 2008) and metapopulation dynamics (Jansman and Broekhuizen 2000) of pine martens. And relatively recent advances in endocrinology now make it possible to measure various hormones in scats (e.g., Monfort et al. 1993), enabling the evaluation of health, stress levels, and reproductive status in free-ranging wildlife (Schwartz and Monfort 2008). Such research approaches, combined with effective field methods (e.g., see section on "Scat-Detection Dogs" below), are likely to play increasingly important roles in noninvasive wildlife research.

Many studies of *Martes* species have conducted scat surveys without the benefit of DNA confirmation; recent examples include Gosse and Hearn (2005), Zalewski (2007), and Skalski and Wierzbowska (2008). Although this approach may be valid in certain circumstances (e.g., if there are no species with morphologically similar scats in the study area), researchers should be careful not to overestimate their ability to accurately differentiate among sympatric species (Davison et al. 2002; Birks et al. 2004; Heinemeyer et al. 2008). Interestingly, no recent studies of American martens or fishers have been based on scat surveys by human searchers, perhaps because other methods (e.g., remote cameras, hair snares) can now provide more information with less ambiguity.

Experimental approaches have attempted to identify optimal methods for scat preservation in the field and the effects of various factors on DNA extraction and genotyping (e.g., Murphy et al. 2000; Piggott 2004; Rutledge et al. 2009). Despite such efforts, there is no consensus about the best method for preserving scat, and recommended approaches vary by species, regional attributes (e.g., wet climates vs. dry climates), and objectives (e.g., species identification vs. individual identification). We therefore recommend that researchers discuss scat-preservation options with the genetics lab used for DNA analyses.

Track Stations

Track stations include both track plots and track plates (Ray and Zielinski 2008). Track plots are prepared areas of soil or other media designed to record the tracks of target species; they can be baited to attract target animals to the site or deployed along active trails. Baited track plots are often referred to as *scent stations* and have been used primarily for canids and felids (Wood 1959; Roughton and Sweeny 1982; Zielinski 1995). Because these methods are most appropriate for medium and large species (Conover and Linder 2009), and are rarely or never used for surveying *Martes* species, we will not discuss them further.

In contrast, track plates have been used extensively for studies of American martens and fishers since the 1980s (e.g., Zielinski 1995; Zielinski and Stauffer 1996; Ray and Zielinski 2008). Track plates—metal plates coated with fine soot or chalk to retain a negative record of tracks—were first deployed by Barrett (1983) to detect American martens. His original design was subsequently improved by Fowler and Golightly (1994) to include a sticky, track-receptive surface that would provide a permanent record of a visiting animal's sooty print. Track plates are typically deployed within an enclosure (e.g., cubby box) and can record very high-resolution, permanent track records. Enclosed track plates deployed as part of a comprehensive survey protocol (Zielinski 1995) have been the primary method used to detect and monitor American martens and fishers in the western United States (e.g., Mowat et al. 2000; Zielinski et al. 2005; Slauson et al. 2007), but have been used less often in other regions or for other *Martes* species (Barea-Azcón et al. 2006).

Data from track-plate surveys have been used extensively for habitat-related studies of American martens and fishers (e.g., Carroll et al. 1999; Zielinski et al. 2006a; Kirk and Zielinski 2009). This method has also been the source of much statistical exploration and modeling related to survey effort, power, and survey design (Zielinski and Stauffer 1996; Hamm et al. 2003; Smith et al. 2007). Track-plate impressions are typically highly effective for differentiating between visiting species—even those that are closely related (e.g., Zielinski and Truex 1995; Zielinski and Schlexer 2009)—and can potentially be used to identify individuals (e.g., Herzog et al. 2007). The narrow set of conditions required for the latter application, however, have precluded any large-scale study from using track plates for this purpose.

The first track-plate enclosures were constructed of plywood (Zielinski 1995), which made them heavy, difficult to set up in the field, and prone to absorbing water. Modern enclosures are typically composed of single sheets of corrugated polypropylene—a light, flexible, and durable material that can be folded for transport in the field. Whereas soot has typically been the tracking medium of choice in North America (Zielinski 1995), researchers monitoring stoats (*Mustela erminea*) in New Zealand have successfully used ink-soaked cards in tracking tunnels (King and Edgar 1977; Choquenot et al. 2001). Chalk has also been used effectively (Belant 2003b; S. Reed, Colorado State University, personal communication) but may be compromised in some circumstances by excessive moisture (Zielinski 1995). More recently, Hooper and Rea (2009) experimented successfully with a foam tracking medium.

Advances in remote cameras and hair-collection techniques may soon render track plates obsolete for many field applications. Track plates require that stations be revisited often (typically at 2-day intervals) to avoid over-tracking of the sticky contact paper when multiple animals visit the station. Cameras and hair snares, on the other hand, are often deployed for 2 weeks between checks and therefore demand far less personnel time for site revisits. Cameras and hair-collection devices are also easier to deploy than track plates since they do not require the application of a tracking medium, and are generally more resistant to moisture problems. Remote cameras and hair-collection devices, which generally yield more information than track plates (e.g., high-resolution species or individual identifications), are now commonly used alone or in conjunction with track plates. Nonetheless, track plates may still be the best option for some monitoring and research efforts for *Martes* species. These include efforts for which sufficient field personnel are readily available, access to deployment locations is fairly easy, and the costs associated with DNA analysis or remote cameras are prohibitive. Track plates can also be very effective when multiple species are targeted (Manley et al. 2006), because they enable the detection of a broad array of species, especially those too small or fine-haired to be detected by remote cameras or hair-collection devices, respectively. Ray and Zielinski (2008) provide a more complete overview of track-station methods for carnivores, including detailed discussions of design and deployment issues.

Remote Cameras and Video

Field-ready remote cameras capable of reliably collecting data from free-ranging wildlife have been part of the noninvasive toolbox since the mid-1990s (Jones and Raphael 1993; Kucera et al. 1995). However, the shift from film to digital technology, combined with other advances in the design of remote cameras over the past 5–7 years, has revolutionized the use of this method for carnivore research (Kays and Slauson 2008). One advance of particular importance is the long time frame over which cameras can now function without revisits by investigators. With a fresh set of batteries, today's camera/sensor units are capable of capturing many thousands of images before the memory card requires changing. For most applications, this number of photos translates into many days or even weeks of monitoring.

Another critical technological advance is the reduction in the wake-up time required for a digital camera to capture an image after the sensor detects an animal. Early digital cameras required wake-up times of 2–5 s, during which animals could easily move out of the frame before being photographed. This was especially problematic for trail-based or other studies that did not use baits or scent lures to detain animals within the camera frame. Most high-quality remote cameras now feature wake-up times of less than a second and can capture even rapidly moving species, such as birds and bats.

Newer cameras typically possess a number of additional features that further benefit wildlife monitoring and research. These include infrared illumination that reduces the need for intrusive flashes; "stealth" models have illuminators that operate beyond the range of light visible to many mammals, including humans. Such options can be useful in locations where there is a risk of detection and theft of cameras by people. Other characteristics of modern remote cameras include near-silent or completely silent operation, no moving parts, compact and weatherproof cases, and lightweight materials.

Together, this suite of technological improvements has transformed the field of noninvasive carnivore research and is especially relevant to surveys for *Martes* species. Specifically, detection rates have increased as a result of large photo capacity and short wake-up times, and costs associated with maintaining a large number of cameras in remote locations have decreased because cameras remain functional for long periods without the need for replacement of batteries or storage media. Although unbaited trail sets are less common for surveys of *Martes* species than for some species, such as tropical felids (but see Grassman et al. 2006a and Chen et al. 2009), short wake-up times are nevertheless important in situations where target species are particularly shy or unlikely to loiter at a baited site.

High detection rates and the ability to deploy cameras over large areas enables not only surveys designed to detect rare species with high confidence, as when assessing reintroductions (Moruzzi et al. 2003), but also the formal evaluation of occupancy (MacKenzie et al. 2002, 2006; Slauson et al., this volume) and the indexing of abundance (Rovero and Marshall 2009). In the foreseeable future, it may also be possible to estimate abundance in some *Martes* species via detection data from camera stations. Researchers studying uniquely marked individuals of other carnivore species have successfully used sight-resight methods to estimate abundance (McCarthy et al. 2008; Sarmento et al. 2009). Despite the fact that most *Martes* species possess few obvious markings to permit individual identification, the "bibs" of pine and stone martens have been used to differentiate species and individuals (Rosellini et al. 2008). Creative camera deployment designed to photograph these bibs—as with the gular patch in wolverines (*Gulo gulo*; Magoun et al. 2011a,b)—may one day enable sight-resight estimates of abundance. In addition, new statistical approaches (e.g., Rowcliffe et al. 2008) may ultimately allow the estimation of abundance without the requirement of unique markings for some *Martes* species. Finally, researchers have successfully combined live-trapping and marking with noninvasive remote-camera resightings as a method for estimating fisher abundance (Fuller et al. 2001; Jordan 2007).

The design of most remote digital cameras permits these units to record many frames with a single trigger of the sensor, resulting in a series of images that provide a pseudo-video effect when viewed in quick succession. Pseudo-videos allow researchers to examine certain aspects of animal behavior that film cameras are rarely able to capture, such as interactions with detection devices or between multiple individuals in the camera frame (R. Long, unpublished data; Figure 14.1). Indeed, "den cams"—both pseudo-video and true-video based—have been used to successfully monitor aspects of the daily activities of *Martes* species, including denning (Lewis and Happe 2008; R. Sweitzer, University of California at Berkeley, personal communication; R. Weir, Artemis Wildlife Consultants, personal communication). Pseudo-videos can also facilitate more accurate identification of species and individuals from photos because multiple frames of each animal detected are available for review. Given these benefits, we suggest that researchers choose remote-camera systems that capture multiple frames per second and ≥10 frames per trigger of the sensor.

Remotely triggered video cameras have also been used to monitor and study mustelids (Stevens and Serfass 2005), including *Martes* species (Aubry et al. 1997; J. Gilbert, Great Lakes Indian Fish and Wildlife Commission, unpublished data). Because high-definition equipment records fine detail, it can provide behavioral insights that are generally not possible with pseudo-video-equipped cameras. Nonetheless, because many, if not most, research and monitoring objectives can be addressed with less complex and more af-

Figure 14.1. An American marten interacts with a tree-mounted hair-collection device (photo R. Long).

fordable digital still cameras, true-video applications may become less attractive for wildlife applications in the future. Alternatively, many "still" remote cameras now permit video clips to be recorded, making it possible to record true videos with simpler, still cameras.

As with the film-based remote cameras that preceded them, today's remote digital cameras use either an active or a passive sensor system; researchers must evaluate which type is more useful for meeting their particular survey or monitoring goals. Active systems require an animal to break an infrared beam projected between 2 sensor components, resulting in a very narrow sensor region. Passive systems detect the movement of an object that differs in temperature from its background with a single, cone-shaped sensor region.

This type of sensor is generally more flexible in its use and, with proper deployment, passive systems can monitor large and loosely defined target areas. Although it is now possible to purchase digital remote cameras at very low cost (e.g., $150–200), many inexpensive camera units are poorly designed for rigorous scientific research and often experience mechanical or electrical problems within the first season of use (R. Long, unpublished data). Indeed, our experience has been that only a few high-quality camera brands perform sufficiently well for use in professional wildlife research, and that attempting to minimize project costs by purchasing less expensive, lower quality camera units may have substantial negative effects on data quality (e.g., resulting from slow wake-up times, excessive failure rates, slow frame-rates). Further, given equipment-replacement costs, the loss of use during repairs, and the trade-offs of manipulating and interpreting poor data, inexpensive camera units may end up costing a project more in the long run. An extensive summary of modern remote-camera applications can be found in Kays and Slauson (2008), and others have explored logistical, design, and statistical aspects of modern camera-based surveys (Kelly 2008; Morrison et al. 2011; Rowcliffe et al. 2011).

We expect that remote cameras will be an important part of future wildlife research and monitoring efforts. Cameras will no doubt continue to increase in resolution, ruggedness, weather resistance, and image capacity, while decreasing in size, weight, and cost. Camera-related features and functions currently being tested and refined, and likely to become commonplace in the near future, include the remote downloading of images from both still and video equipment (MacNulty et al. 2008), wireless sensors (currently available from some companies), and the application of camera networks to address spatio-temporal aspects of mammal activity (Kays et al. 2009).

Hair Collection

Prior to a decade ago, there was little scientific reason to systematically collect hair from free-ranging carnivores. Although keys were available to help with species identification (e.g., Moore et al. 1974), the lab methods required were laborious and often resulted in high-confidence identification only to the genus or family level. Raphael (1994) summarized a number of hair-snaring efforts for *Martes* species (e.g., Nelson 1979; Barrett 1983; Jones et al. 1991) and highlighted the benefits of hair snares when compared with other detection methods. He also noted, however, that most efforts had low detection rates, and that many hairs are difficult to identify morphologically, especially if they are not dorsal guard hairs (Raphael 1994). The huge leaps in molecular ecology that occurred not long after 1994 (Goossens et al. 1998; Schwartz and Monfort 2008) now make it possible to extract useful DNA from hairs, and also to identify the species and individual from which the sample came. These dramatic advances have made hair collection a fundamental component of many field studies and monitoring efforts of *Martes* species.

DNA extracted from hair has become increasingly valuable for detecting genetic structuring (Jansman and Broekhuizen 2000), estimating abundance (Mowat and Paetkau 2002; Gardner et al. 2010), and detecting rare species (McDaniel et al. 2000). In response, researchers have created novel methods for collecting hair from carnivores, including devices designed specifically or in part for martens and fishers (e.g., Foran et al. 1997b; Belant 2003a; Zielinski et al. 2006b). Further, powerful molecular markers have been developed that enable researchers to easily and affordably determine species, sex, and individual identification for *Martes* species from relatively small hair samples (e.g., Riddle et al. 2003; Lynch and Brown 2006; Jordan et al. 2007). Such advances have been particularly helpful in regions where sympatric *Martes* species are difficult to distinguish based on the morphology of hairs or scats (Livia et al. 2006; Ruiz-González et al. 2008).

The growing interest in hair collection has encouraged the concurrent development of advanced statistical approaches, such as abundance estimators designed specifically for noninvasively collected data (e.g., Schwartz et al. 1998; Royle et al. 2008; Gardner et al. 2010). Not surprisingly, hair-snaring efforts are now becoming commonplace for studies of *Martes* species that involve assessing occurrence and distribution (Tóth 2008; R. Truex and W. Zielinski, U.S. Forest Service, personal communications; J. Gosse, Terra Nova National Park, personal communication), detecting dispersal (Jordan 2007), indexing abundance (Lynch et al. 2006), estimating abundance and censusing populations (Mowat and Paetkau 2002; Williams et al. 2009; Mullins et al. 2010), and conducting landscape-genetic analyses (Wasserman et al. 2010; Schwartz et al., this volume).

Hair-collection devices tend to be designed specifically for a species of interest, because they generally snag hair by compelling an animal to squeeze by or rub against a hair-snaring object (e.g., sticky tape, glue, barbed wire, gun brush; Kendall and McKelvey 2008). Consequently, the device must be large enough that the species can enter the device, but small enough to force contact with the hair snare. One early design used sticky tape for snagging hair (Foran et al. 1997b). Although similar designs are still in use (e.g., Williams et al. 2009), tape or glue is difficult to remove from hair samples and can lose effectiveness when wet (Zielinski et al. 2006b; Kendall and McKelvey 2008). Other designs use barbed wire strung across the opening of a standard track plate (Zielinski et al. 2006b; Jordan 2007) or use spring mechanisms (Lynch et al. 2006) to collect hair samples. Gun-cleaning brushes are the most recent and easy-to-use mechanism for snagging hairs from martens and fishers (Kendall and McKelvey 2008; R. Long, unpublished data), and may be the most effective. These were first deployed at both ends of a triangular cubby constructed from corrugated polypropylene (M. Schwartz, U.S. Forest

Figure 14.2. (a) A tree-mounted hair snare designed for American martens and fishers, with (b) the view from below showing the location of chicken legs nailed under the snowshield; and (c) a close-up of a gun-cleaning brush with marten hairs attached. Note different methods for attaching brushes, including (b) electrical lugs with tap screws and (c) threaded T-nut attachments, which are simpler to install and help facilitate brush insertion and removal (photos R. Long).

Service, personal communication), with a piece of chicken fastened in the center for bait.

Recent designs for a tree-mounted mesocarnivore hair snare (K. Slauson, U.S. Forest Service, personal communication, as adapted from a design by P. Figura and L. Knox, California Department of Fish and Game), include a snowshield and collar—both embedded with gun brushes—fastened to the bole of a tree (Figure 14.2). The benefits of this design include the following: (1) hair samples on the brushes are largely protected from DNA-degrading precipitation and UV radiation; (2) shields can be folded and easily transported; (3) brushes can be removed and replaced with little effort; and (4) bait is located such that it does not contaminate the device, which is important when removing devices from the field.

A final consideration when deploying hair snares with scent lure or bait is the potential for attracting multiple individuals to the device. In such cases, gun brushes or other mechanisms may collect hair samples from a number of individuals or species. Although mixed samples can still enable the identification of species, DNA from these samples will typically be useless for individual identification (Roon et al. 2005). Some hair-collection devices have therefore been designed to allow only a single individual to leave a hair sample (e.g., Bremner-Harrison et al. 2006; DePue and Ben-David 2007; Pauli et al. 2008).

As with scat collection, those wishing to extract DNA from hairs should take care to preserve and store samples carefully. Hair samples are typically easier to preserve and store than scats, and suggestions are available from a number of sources (e.g., Roon et al. 2003; Schwartz and Monfort 2008). Generally, once hair samples are dry, they should be stored at a stable temperature and in a dry location in paper envelopes. Plastic bags or containers should not be used unless a desiccant is included, because plastic can trap moisture, causing mold to grow on the sample and leading to degradation of DNA. Plastic vials containing desiccant can be useful in the field, because brushes, barbs, or other hair-snagging devices can be deposited directly into the vial. Because cleaning such samples prior to analysis can be laborious, however, some molecular labs discourage direct contact between desiccants and samples. Researchers should discuss strategies for preserving and storing hair with the genetics lab that will be processing their samples.

Scat-Detection Dogs

As with the detection of drugs, explosives, avalanche victims, agricultural contraband, and other odors of human interest, domestic dogs (*Canis familiaris*) are being used increasingly in wildlife field research (e.g., Wasser et al. 2004; Smith et al. 2005; Long et al. 2007b; Figure 14.3). Advances in genetic and endocrine techniques during the 1990s now enable researchers to extract a wealth of biological information from scats (Putnam 1984; Kohn and Wayne 1997; Schwartz and Monfort 2008), which can be used to investigate distribution and movements, abundance, population density, habitat use, home range size, physiological condition, and other ecological measures of interest (see MacKay et al. 2008a). Given the outstanding olfactory abilities possessed by most dogs, wildlife biologists have realized that many of the criticisms directed at scat surveys conducted by people (e.g., Birks et al. 2004) can be overcome with canine searchers.

During the last decade, detection dogs have been used to systematically search for scats from several dozen carnivore species in the ursid, canid, felid,

Figure 14.3. A scat-detection dog awaits her reward for detecting a marten scat (foreground) in the Cascade Range in Washington (photo P. MacKay).

and mustelid families (MacKay et al. 2008a). In addition, scat-detection dogs have been used for fine-scale discrimination in a number of applications. For example, dogs have been trained to distinguish the scats of closely related species (Hurt et al. 2000; Smith et al. 2003; Harrison 2006) and to identify individual animals from their scats (Kerley and Salkina 2007; Wasser et al. 2009).

Scat-detection dogs can offer several advantages over other noninvasive survey methods. First, dogs can survey for multiple species over extensive areas (e.g., Wasser et al. 2004; Beckmann 2006; Long et al. 2007b), and are proficient at recovering small and cryptic scats (e.g., Smith et al. 2001; Long et al. 2007b). These attributes are potentially valuable to the study of *Martes* species, given their wide-ranging, elusive nature and the small size of their scats. In addition, scat-detection dogs have demonstrated greater success at locating scats from some carnivores than researchers relying on visual detection alone (Smith et al. 2001; Long et al. 2007b), with presumably less sampling bias. Furthermore, scat-detection dogs may provide less spatially biased data than detection devices that require attractants to draw target animals to a site (see section on "Attracting *Martes* Species to Survey Stations" below), and can potentially confirm occupancy more quickly and accurately than other survey methods (Long et al. 2007a). Finally, the charismatic nature of this method should not be underestimated; dogs have broad public appeal and are compelling ambassadors for carnivore research and conservation.

The use of scat-detection dogs has its limitations, however. As with any scat-based method, the persistence of scats in the environment is key to detection, and scat decomposition or consumption by wildlife is a potential source of sampling bias (Sanchez et al. 2004; Livingston et al. 2005). Conversely, the potential long-term persistence of scats can violate assumptions of closure for some abundance estimators unless the age of scats can be controlled for (e.g., by attempting to clear all target scats prior to conducting the survey). Further, although many studies have shown scat-detection dogs to be highly efficient at locating target scats (e.g., Smith et al. 2003; Wasser et al. 2004; Long et al. 2007b), accuracy rates can vary between dogs, revisits, and sites. Thus, DNA confirmation of species identification is generally recommended. Laboratory expenses may add significantly to the cost of scat-dog surveys, and despite advances in genetic techniques, scats can still be an unreliable source of DNA for individual identification or even species confirmation. Finally, detection rates can vary between dog/handler teams (Smith et al. 2003; Wasser et al. 2004; Long et al. 2007b), which underscores the importance of proper training and a study design that allows detectability to be both estimated and corrected for. MacKay et al. (2008a) provide a comprehensive overview of the application of scat-detection dogs to carnivore surveys.

Early surveys conducted with scat-detection dogs targeted ursids (Wasser et al. 2004; Long et al. 2007b) and kit foxes (*Vulpes macrotis mutica*; Smith et al. 2003, 2005). Although the application of this method to research on *Martes* species has been less prevalent (but see Long et al. 2007b, 2011), several recent scat-detection dog surveys in the western United States were focused on martens or fishers. Dogs are currently being used in 2 American marten studies in California; one focused on assessing summer habitat use and the other investigating population genetics (K. Slauson, unpublished data). The Kings River fisher project in California is using both scat-detection dogs and telemetry techniques to investigate habitat use, population density, stress hormone levels, matrilineage, and prey analysis (C. Thompson, U.S. Forest Service, unpublished data). In addition, researchers in Montana have been using scat-detection dogs to collect DNA for a landscape-genetic analysis for fishers (M. Schwartz, personal communication). Results from these studies will help evaluate the effectiveness of scat-detection dogs for locating the scats of *Martes* species in densely forested landscapes, including areas where multiple *Martes* species occur.

Combining Multiple Methods

Using multiple methods can increase the likelihood of meeting project objectives (e.g., by increasing detectability), expand the number and types of objectives that can be addressed (Campbell et al. 2008), and increase the number of species for which the survey is effective (O'Connell et al. 2006;

urban Krakow, Poland, found chicken eggs to be the most effective bait, with other types of baits (including various meats, and eggs from other species) being less effective (I. Wierzbowska, Jagiellonian University, personal communication). Most surveys for American martens and fishers in North America, especially those using track-plate or hair-collection methods, have used 1–2 pieces of chicken (typically drumsticks) in conjunction with skunk-scented lure.

Important Considerations for the Use of Attractants

Ideally, attractants are strong enough to draw the target species, but not so strong that they invalidate the assumptions required for achieving the survey objective. For example, researchers wanting to detect a rare species may not be concerned if an individual is drawn from a substantial distance, but those engaged in studies of habitat selection may require that animals be attracted from only nearby locations, if at all. The attractive quality of a bait or scent lure should be long lasting, thus requiring fewer revisits if the survey design and analysis methods permit longer periods between revisits. If periods between rebaiting are extensive, a scent lure combined with bait may provide longer term attraction.

Deciding whether to use reward or non-reward bait will depend on the target species, survey objectives, and ethical or safety considerations for feeding wildlife. For example, most recent surveys of ursids (e.g., Proctor et al. 2005; Cushman et al. 2006; Kendall et al. 2008) have used non-reward liquid scent lures, presumably to minimize the chance of food habituation. As mentioned above, many surveys for American martens and fishers have used standardized reward baits along with a commercial scent lure (e.g., Caven's Gusto, Minnesota Trapline Products, Pennock, MN). This combination results in sufficient attraction while minimizing visits by multiple individuals because the bait is removed during the first visit. Reward baits, however, are also available to nontarget species, which can reduce the effectiveness of surveys. For example, an American marten study in California experienced a reduction in detectability in areas where black bears (*Ursus americanus*) were common because bears quickly removed the chicken bait (Slauson et al., this volume). In cases where survey stations are likely to be visited by nontarget species, or when using bait with remote digital cameras capable of capturing many thousands of photos, it may be advantageous either to place bait in a wildlife-proof container that allows odor to escape, or to suspend it out of reach of wildlife and make it a non-reward bait (Schlexer 2008).

Other bait-related considerations are primarily logistical, such as bait size. One important advantage of chicken legs as bait is that they can be easily transported in the field; larger baits, such as deer parts or beaver carcasses, are difficult to carry long distances from either roads or snowmobile trails.

This becomes especially relevant when a research team must deploy multiple survey stations located far from roads or trails.

A final consideration for choosing baits and lures is the season during which the survey will be conducted. Meat baits can rot very quickly during summer months, resulting in reduced attraction or even aversion. Similarly, liquid scent lures can desiccate and lose odor in hot, dry environments. Such compromised baits and lures may still be effective, but researchers should plan for more frequent revisits to rebait the site. Where bears are present, it may be advantageous to conduct surveys for *Martes* species during winter to reduce the probability of bear-inflicted bait loss or equipment damage. Indeed, winter surveys may alleviate or reduce many of the problems associated with the use of baits, including visits by nontarget species, bait decomposition, and human intrusion. Because baits and scent lures used in below-freezing temperatures may have reduced olfactory attractiveness, it is advisable under such conditions to use antifreeze agents in liquid scent lures (Schlexer 2008).

It has long been noted that carnivores are more likely to approach baits and scent lures at traps and survey sites during winter, the (untested) assumption being that prey availability at this time is more limited and less diverse (Carman 1975). Indeed, lower detection rates during summer than in winter have been observed for American martens and fishers throughout their ranges when baits and scent lures are used to attract animals to noninvasive devices (J. Gilbert, personal communication; P. Jensen, New York Department of Fish and Wildlife, personal communication; R. Long, unpublished data; W. Zielinski and K. Slauson, personal communication). Other hypotheses for this pattern include the following: (1) densities of *Martes* species are lower or home ranges are smaller in summer than in winter; (2) during summer, *Martes* species move away from areas that experience high winter activity; or (3) survey devices requiring bait are less effective in the summer than in winter as a result of more rapid bait deterioration or removal by nontarget species (e.g., bears).

Only 1 project that we know of has attempted to test seasonality hypotheses pertaining to noninvasive *Martes* surveys (Slauson et al., this volume). Initial results from this study in California suggested that, in areas where black bears were common, removal of chicken bait by bears decreased the probability of marten detections at remote cameras in summer and fall (W. Zielinski, personal communication). This effect could not explain all the observed variability between results for winter vs. summer surveys, however, and that project is now exploring other hypotheses. Notably, in a limited number of surveys recently conducted in Washington, detection rates of American martens were lower during summer than in winter, despite low rates of bait removal by bears (R. Long, unpublished data). If detection rates are dramatically different among seasons, investigators should consider analyzing data and making inferences based on separate seasonal datasets.

When attractants are used, it is possible that a visit to a survey station by a given individual can affect whether another individual of the same species (or a different species) visits at a later time. For example, in a live-capture study in Wisconsin, American martens were never captured at locations where fishers were previously captured, if the trap and site were not thoroughly cleaned (J. Gilbert, personal communication). In the same study, female martens would not enter traps that had previously captured male martens. Although such situations may be difficult to address, researchers should be aware of the issues involved so they can be explored later or taken into consideration during analysis.

Final Thoughts and Future Directions

Advances in molecular laboratory methods, digital-camera technologies, and statistical analyses have paved the way for modern noninvasive survey methods. The application of these survey methods to the study of *Martes* species has enabled researchers to ask questions that were impossible to address in the past (e.g., from which population did a given individual originate?), and the framing of such questions has, in turn, driven the development of new and better field methods. At the same time, tried-and-true methods like tracking and scat collection continue to contribute to research on *Martes* species, and in some cases, these methods have become more valuable when combined with modern molecular and statistical techniques. Technological and methodological advances will no doubt continue to move the field forward. Faster and higher resolution remote cameras, remote image downloading and viewing, increasingly sensitive and affordable genetic tools, and more creative uses of hair-collection devices and scat-detection dogs are already being applied in many field situations. In the future, we also expect that on-site DNA confirmation of species (and perhaps individual) identity and the aging of animals via the analysis of hair samples may be possible, which would have profound effects on the way surveys are designed and conducted. Finally, advances in computing capability and new statistical approaches will continue to improve inferences from limited and imperfect data.

Acknowledgments

We thank the organizing committee of the 5th International *Martes* Symposium for their dedication to the production of this important resource, the *Martes* research community for their generous contribution of information, the valuable input of reviewers, and finally, the coeditors and contributors of the book *Noninvasive Survey Methods for Carnivores* for their guidance.

15

Occupancy Estimation and Modeling in *Martes* Research and Monitoring

KEITH M. SLAUSON, JAMES A. BALDWIN, AND WILLIAM J. ZIELINSKI

ABSTRACT

Occupancy estimation and modeling was developed to account for variation in detectabilities in noninvasive survey data. This type of modeling involves the estimation of 2 parameters, *P* and ψ, from the sequences of detections and nondetections obtained at each survey unit (site). *P* is the probability of detecting a species when it is present at the site, and is a function of the survey protocol used (e.g., number of stations, survey duration), and ψ is the probability that ≥1 individual of the species occupies the site, which is related to covariates at the sites surveyed (e.g., habitat structure, community composition). In this chapter, we present details from 3 case studies involving fishers (*Martes pennanti*) and American martens (*M. americana*) to illustrate key issues in occupancy estimation and modeling for *Martes* research, management, and monitoring. For each case study, we used data from standardized survey protocols for fishers and martens to estimate detection probabilities and investigate factors causing detection heterogeneity. We found 5 significant sources of detection heterogeneity: (1) detection dependencies within a survey unit, (2) variation in detection probabilities during a survey, (3) variation among seasons, (4) variation between sexes, and (5) variation due to the presence or absence of bait. Incorporating covariates to account for detection heterogeneity significantly increased the accuracy of parameter estimates and the appropriateness of the fitted models. Probability of detection varied by survey-occasion, and increased over the survey duration. Fishers had a significantly lower probability of detection during the summer (July–September) compared with all other seasons. The same was true for American martens, but not at all locations. Custom models created to estimate *P* always outperformed standard models available in software packages. Overall, noninvasive surveys are effective tools for *Martes* research and monitoring; however, observed results were typically biased low as a result of the combined effects of protocols with detection probabilities <1, and the bias and additional variability that detection heterogeneity can have on parameter estimates based on these types of data. Proper application of occupancy estimation can remedy or at least reduce the effects of these problems, leading to improved understanding of survey results and inferences derived from them.

Introduction

Survey methods that provide detection/nondetection data have been used in *Martes* research for decades. Snow-tracking was one of the first survey methods used for *Martes* species, whereby the habitat characteristics of sites containing American marten (*Martes americana*) or European pine marten (*M. martes*) snow tracks were compared with those without snow tracks (e.g., de Vos 1951b; Pulliainen 1981a; Raine 1983). Devices capable of remotely detecting the presence of *Martes* species are being used increasingly in *Martes* research, monitoring, and management, including track plates (Ray and Zielinski 2008), camera traps (Kays and Slauson 2008), and hair snares (Kendall and McKelvey 2008). The newest addition to this suite of techniques is the use of scat-detection dogs to locate the fecal droppings (scats) of target species, which can then be identified to species, sex, and, in some cases, individual with genetic analyses (MacKay et al. 2008a). Collectively, these noninvasive survey methods offer a diverse array of options for addressing research and management questions for *Martes* species (Long and MacKay, this volume). Because all detection devices suffer from the problem of imperfect detection, occupancy estimation and modeling enable researchers to account for variation in detectability when interpreting survey results. In this chapter, we focus on the use of occupancy estimation and modeling in *Martes* research and monitoring, and present 3 case studies to illustrate the use of occupancy analysis and the interpretation of resulting data. We briefly review relevant conceptual and statistical-design issues for occupancy analysis, but refer readers to Long and Zielinski (2008) and Royle et al. (2008) for more in-depth treatments of these important topics.

Modern Occupancy Estimation and Modeling

The first step in designing any ecological investigation or monitoring program is the formulation of clear objectives (Nichols and Williams 2006). In most cases, 1 or more population parameters (e.g., occurrence, abundance) are compared over time or space (Pollock et al. 2002). Proper estimation of selected parameters and inferences about their variation over time or space requires careful consideration of spatial variation in animal abundance and detectability (Williams et al. 2002). In most cases, investigators cannot survey the entire area of interest and must select a sample area to survey. The sample area should be selected with an unbiased approach that permits valid inferences to the larger, unsurveyed area (MacKenzie et al. 2006). A key consideration is that no survey method detects target species with 100% certainty. Occupancy estimation addresses this problem explicitly by using probabilities of detection to account for unexplained variation in survey results (MacKenzie et al. 2006).

Why Estimate Occupancy?

Occupancy estimation and modeling involves collecting binary occupancy data (detected, not detected) at a collection of survey units (sites) to estimate occurrence-based population parameters, such as site occupancy, proportion of area occupied, or a species' distribution (Royle et al. 2008). Any parameter that can be measured in a binary (0/1) or small number of states (e.g., male/female/both/unknown) can be used for occupancy analysis. Importantly, spatial and temporal covariates can be included in the modeling framework to investigate their effects on occupancy. Furthermore, research or monitoring objectives can include parameters that are impractical to measure (e.g., changes in the density of breeding adult females over large areas). In such cases, occupancy status can provide useful surrogates for the population parameters of interest.

When deciding whether to conduct noninvasive surveys, careful consideration should be given to whether binary occupancy data will be useful. In their simplest use, noninvasive surveys provide data on locations where a species does or does not occur. The strongest inference from these data is to contrast survey units that are occupied with those that are not occupied, and identify key ecological differences between them. Surveys that are repeated among seasons or years can distinguish locations that are occupied seasonally or occasionally from those that are never or always occupied, and provide opportunities to explore environmental features associated with each state.

Despite the benefits, there is an inherent limitation in the information content of binary data from occupancy surveys. For example, in a single-season survey, all occupied locations are considered to be of equal value to the target species. This can be misleading when there is spatial variability in the importance of site covariates. For example, sites occupied by nonbreeding juveniles have different effects on estimates of population parameters than sites occupied by breeding females. Furthermore, sites that are occupied consistently should be interpreted differently from sites where occupancy varies from year to year (MacKenzie et al. 2003). And, if detections are primarily males, but this is unknown, the occupancy results, as well as habitat models built from these results, will be biased to the extent that occupancy by males differs from that by females.

Another important consideration when evaluating noninvasive surveys is the relation between occupancy data and true population parameters. Intrasexual territoriality is common in *Martes* species, whereby larger male home ranges usually overlap >1 female home range. Home-range size is also likely to vary with habitat quality (Thompson 1994). Thus, if occupancy is used as a surrogate for abundance, it may not accurately reflect fluctuations in true population size. Furthermore, if habitat quality and population size change over time, occupancy rates can remain relatively stable because only 1 individual needs to be detected to indicate that the site is occupied. Changes in home-range size will also interact with occupancy rates, such that the same number of individuals can potentially yield very different occupancy estimates if their home ranges vary in size over time. Despite their potential severity, these pitfalls can be minimized by using the appropriate sampling design (MacKenzie and Nichols 2004; Long and Zielinski 2008). If an appropriate survey design and protocol are used, and assumptions about the relation between occupancy and true population parameters are valid, analytical techniques are available for estimating relative or absolute abundance (Royle and Nichols 2003; MacKenzie and Nichols 2004; Kerry et al. 2005). Furthermore, partitioning detections by sex or individual via DNA analyses provides opportunities to evaluate the validity of assumptions, and can significantly improve population inferences from occupancy analysis (Smith et al. 2007; see Case Study 3).

One of the primary benefits of noninvasive surveys is their low cost per unit area (e.g., Long et al. 2007a), resulting in large datasets and opportunities for broad geographic inferences (e.g., Hines et al. 2010). These attributes provide substantial advantages for noninvasive survey approaches compared with more traditional approaches, such as capture-mark-recapture and radiotelemetry.

Modern Occupancy Analysis: An Overview

If noninvasive surveys have been selected to meet research or monitoring objectives, the next step is to address the key problem with using survey data—failure to detect the target species when it is present. A nondetection can represent either a true absence (i.e., the species does not occur in the site surveyed), or a false or sampling absence (i.e., the species occurs in the site surveyed, but was not detected). If not corrected, the resulting dataset will contain a combination of both, which will confound resulting interpretations.

Occupancy analysis addresses this problem by estimating 2 parameters, P, the probability of detecting a species in a survey unit when it is present, and ψ, the probability that a species occurs at a site, where probability refers to the chance that a randomly selected site from a collection of sites will be occupied. Both parameters are estimated from a detection history; that is, the sequences of 1s and 0s that indicate detection or nondetection, respectively, at each site during a survey. P is influenced strongly by characteristics of the survey protocol, such as the number of stations and survey duration, whereas ψ is influenced strongly by covariates of the site (e.g., habitat structure, presence of other carnivores). Thus, P is generally considered a nuisance parameter that must be estimated properly to derive an accurate estimation of occupancy (ψ), which is the parameter of interest. Nonetheless, P plays an important role in the design of future studies, where it can be used to allocate effort efficiently and obtain estimates of ψ at desired levels of precision.

The relation between P and the survey protocol depends on the allocation of survey effort. For example, suppose that a survey is conducted for fishers using a survey unit containing 6 track-plate stations placed in a 6.4 km^2 area, and the stations are visited by a technician every 2 days for 16 consecutive days, resulting in a total of 8 visits. Each visit (v) represents a replicate sampling occasion. Thus, the probability of detecting a fisher using this survey protocol, is $P = 1 - (1 - p)^v$, where p is the per survey-occasion probability of detecting a fisher when it is present, Survey-occasions are independent, and the probability of detection is the same at each survey-occasion. Clearly, P increases as both p and v increase, allowing for the evaluation of alternative survey protocols. Probability p is influenced by the number of stations within a survey unit, their spatial distribution, the number of days elapsed per survey-occasion, and the presence or absence of bait.

The probability that a species occurs at a site (ψ) can be used to investigate habitat covariates that influence occupancy. It can also be estimated repeatedly to monitor changes in occupancy over time. The probability of occupancy is the product of ψ and P for sites where the species is not detected, but in cases where $P < 1$, the occupancy status of sites where an animal is not detected is confounded by imperfect detectability. Thus, the proportion of sites where the target species has been detected is considered the naïve estimate of occupancy, which will be negatively biased (Bailey et al. 2004) when P is not known or is estimated poorly.

When a survey unit comprises multiple stations, they are not independent; consequently, detection status is assigned for the collection of stations, not for each station individually. Thus, a survey unit surveyed on $v = 1, 2, \ldots,$ V occasions yields the detection history for survey unit i: $y_i = (y_{i1}, y_{i2}, \ldots, y_{iV})$ where $y_{iv} = 1$ if a detection was made on survey-occasion v and $y_{iv} = 0$ otherwise. To estimate p and ψ, the site-specific detection histories can be summarized by

$$y_{i.} = \sum_{v=1}^{V} y_{iv}$$

where the probability distributions of the total number of detections, y_i, can be specified as a function of the parameters p and ψ (Royle et al. 2008). Then, estimates of p and ψ can be obtained by maximizing the likelihood function (MacKenzie et al. 2006).

Sample size factors into occupancy analysis in 2 ways: the number of observations per survey unit (hereafter, survey-occasions) used to estimate P with a desired precision, and the number of survey units needed to estimate ψ with a desired precision. This presents a trade-off in the application of survey effort—is it better to survey fewer sites more often, or to survey more sites less often? The decision on how to allocate survey effort should be guided by the objectives of the survey and with data from pilot studies or previous surveys (Bailey et al. 2007). In general, when the objective is to confirm presence or absence at a particular location, the number of survey-occasions should be large enough that P approaches 1. For surveys designed to estimate distribution or to monitor changes in the proportion of sites occupied over time, once P becomes large enough, investigators can usually reduce the number of repeat survey-occasions at each site and increase the number of sites surveyed. In that case, the number of survey-occasions required will depend on the desired level of precision for estimates of ψ (MacKenzie and Royle 2005). Furthermore, more survey-occasions will typically be required when employing methods that use baits or scent lures to attract the target species to detection devices than when the survey employs search methods (e.g., scat-detector dogs, snow-track surveys). This results from the inherently lower probability of attracting an individual to a few fixed locations and detecting it there, compared with searching broadly to find evidence that the species occurs in the survey area.

Assumptions of Occupancy Analysis

Occupancy analysis involves 4 key assumptions (MacKenzie et al. 2006): (1) the occupancy status at each site is constant during the survey (i.e., sites are "closed" to changes in occupancy); (2) the probability of occupancy is either constant among sites, or differences are modeled using covariates; (3) the probability of detection is constant among both sites and surveys, or is a function of site- and/or survey-specific covariates (i.e., there is no unmodeled heterogeneity in detection probabilities); and (4) the detection of species and the detection histories at each site are independent. If these assumptions are not met, estimators may be biased and inferences about factors that influence either occupancy or detection may be incorrect. We consider each assumption below as it relates to the ecology of *Martes* species.

Assumption of Closure

The typical structure and spacing of *Martes* populations is such that individuals usually maintain the same annual home range over time and exclude same-sex conspecifics while tolerating those of the opposite sex (Powell

1994b). This pattern suggests that the assumption of closure is probably met in most cases; however, there are circumstances where violations may occur, such as during the breeding or dispersal season, or when seasonal changes in habitat use occur. To minimize these potential violations, surveys can be designed to avoid potentially confounding periods of the year. Furthermore, for multiyear comparisons, sampling should occur at the same time each year so that seasonal effects do not confound estimates of among-year changes in occupancy.

Assumption of Constant Probability of Occupancy

The impact of unmodeled variation on the probability of occupancy (hereafter, occupancy heterogeneity) is relatively unknown compared with other model assumptions. In cases where 2 groups of sites exist with different occupancy probabilities, but detection probabilities among sites in each group are equal, estimates of occupancy will reflect the average level of occupancy between the 2 groups, but reported variances will be too large (MacKenzie et al. 2006). Investigators can predict how the probability of occupancy may differ as a result of particular site characteristics, and explicitly test for occupancy heterogeneity by including appropriate site covariates (Royle and Nichols 2003).

Assumption of Constant Probability of Detection

The probability of detection has not been constant for most datasets on martens and fishers that we have analyzed (see Case Studies 1–3), requiring modeling of heterogeneity in the probability of detection (p). The probability of detection has varied significantly by season (see Case Study 1), elevation (K. Slauson, unpublished data), sex (K. Slauson, unpublished data), time periods during a survey (e.g., seasons; see Case Studies 1–3), and abundance (Smith et al. 2007). If left unmodeled, heterogeneities in detection probabilities will often result in underestimates of true occupancy rates. Detection probabilities may also vary because of season, environment (e.g., temperature, precipitation), and social behavior. Bias due to detection heterogeneity may be further exaggerated when sample sizes for either survey units or survey-occasions are small. During both the design and analysis phases, anticipating detection heterogeneity and minimizing its effects are essential for generating reliable occupancy models (MacKenzie et al. 2006).

Assumption of Spatial Independence of Detection and Detection Histories

Lack of independence among survey-occasions can arise when an individual of the target species can be detected at >1 survey unit. When this occurs, the precision of occupancy estimates is usually overstated. As a general rule, if the spacing between survey units is less than the diameter of a typical home range for a male (the sex with the larger home range), this assumption is likely to be invalid. In these instances, the *effective sample size* (the number of independent sites or detection histories) is actually smaller than the number of sites surveyed, and the estimated standard errors obtained from the model are too small (MacKenzie and Bailey 2004; MacKenzie et al. 2006). Although this problem should be addressed during the design phase, standard errors can be adjusted after the fact with an estimated variance-inflation factor (MacKenzie et al. 2006).

Lack of independence in detection histories (i.e., when a detection increases or decreases the probability of a subsequent detection at that site) will also create bias in parameter estimates. This is more likely to be a problem for methods that use bait or lure, because they can increase the probability of subsequent detections. This issue also has the potential to significantly influence parameter estimates, because subsequent detections will most likely be of the same individual. Case Study 1 provides an example of evaluating a dataset for temporal dependency in detections and accounting for its effects.

Design Considerations

Designing effective occupancy surveys is a complex topic that deserves considerably more attention than is possible here. We refer those who are interested in more-detailed considerations of this topic to Bailey et al. (2007), Long and Zielinski (2008), and Long and MacKay (this volume).

Survey Objective

Surveys are typically conducted (1) to provide spatially explicit information about the occurrence of a species in a particular location (i.e., "single-location" surveys), (2) to estimate the proportion of survey units occupied by a species, or (3) to create maps that display the relative probability of occurrence. Single-location surveys are often used to determine whether a species is present in an area where a management activity is proposed. For objective 1, investigators should use a survey design and protocol that provides a very high probability of detection ($P \approx 1$). For objective 2, the goal is to accurately estimate the proportion of survey units that are occupied. This can generally be accomplished with less effort per survey unit and a lower P, because effort can be optimized between increasing the number of sites surveyed, which will increase the precision of occupancy estimates, and including sufficient survey-occasions to maintain an acceptable level of variance in P (MacKenzie et al. 2003). For most objectives, a simple estimate of the proportion of survey units occupied will not be adequate, and further information about the characteristics of sites that are likely to be occupied will be of interest. The probability of site occupancy can be estimated using site covariates occurring in or around survey units (e.g., forest structural attributes), and, if there is good fit to the data, site covariates can be used to determine the probability of occupancy in each survey unit in the study area.

Survey Design

Survey design should be guided by the survey objectives and include the following design elements: (1) size of the survey area (the spatial scope of inference); (2) size of the survey unit (the unit of inference); and (3) the timing of surveys (the temporal scope of inference). Design element 1 also includes the geographic location of the survey, and how the survey area will be sampled (e.g., systematic vs. stratified sampling) to best meet survey objectives. Spacing of survey units should be sufficient to provide independence between adjacent survey units. Design element 2 typically involves decisions on both the size of the survey unit and how many sampling stations to include in each survey unit. If an objective is to detect both male and female fishers, the size of the survey unit should be between the average sizes of male and female home ranges and include enough survey stations to ensure that >1 station will occur in each home range. Design element 3 involves consideration of how the timing of surveys will determine which portion of the population is being surveyed effectively and may affect the detectability of the target species (see Case Study 1). For example, fall surveys are likely to detect higher proportions of dispersing young than of other segments of the population, whereas summer surveys are more likely to detect lower proportions of breeding females, because of their concentrated use of areas around dens within their home ranges.

Survey Unit

The survey unit is the smallest spatial resolution, or grain, at which occupancy data are collected and varies with detection method. Baited survey methods involve the use of fixed points or stations in the survey unit where detection devices are deployed. Unbaited survey methods typically involve the use of fixed-distance transects within the survey unit (Helle et al. 1996; MacKay et al. 2008a). In either case, results for all stations or survey distances in each survey unit are aggregated by survey-occasion to provide a "1" (≥1 detection) or a "0" (no detections) in the detection history. Detection data for each survey unit can also include additional information about the quantity of detections (e.g., number of scats or photos obtained), the sexes detected, or other characteristics of interest (see Case Study 3).

The number of stations (for baited methods) or the length of transects (for unbaited methods) will depend on the target level of P. In general, P will increase as the number of stations (or transect length) or the duration of the survey increases. The effects of increasing the number of stations or the length of transects on P can be evaluated via simulation, if previous survey data are available. The number of stations can also influence the robustness of the survey unit to inoperabilities, such as when bait is removed by a nontarget species or when batteries fail in a remote camera system. Simulations on the effect of station inoperabilities on detection probability for fishers indicated that 6-station versus 2-station survey units were more robust to inoperability effects, whereby a 20% decrease in survey duration can result in a >0.15 decrease in P for a 2-station survey unit, and a <0.05 decrease in P for a 6-station survey unit (K. Slauson, unpublished data).

Survey Protocol

The survey protocol includes (1) survey duration (the total amount of time or distance that surveys will be conducted in each survey unit); (2) number of survey-occasions (the number of times each site is visited to replenish bait and lure, maintain the device, and collect evidence of detection, or to conduct snow-tracking or scat-detection surveys); (3) visit interval (the length of time between visits); and (4) rules for locating sampling devices or efforts, if station or transect locations are not fixed. To be clear, we use the term *visit* to refer to a visit either by technicians to detection devices or by surveyors (e.g., snow trackers or scat-detector dog teams) to survey units. For baited detection devices, such as track-plate boxes and remote cameras, a greater number of visits increases the number of survey-occasions, usually resulting in higher values for P (see Case Studies 1–3). For unbaited survey methods, such as snow-tracking or scat-detector dogs, 1 or several survey-occasions may be sufficient to attain high values for P as a result of the high value of p for each survey-occasion (Long et al. 2007a; Heinemeyer et al. 2008). Care should be taken when determining the survey-occasion interval, because it can create problems of temporal dependency (see section on Dependency within Detection Histories below).

Detection Histories and Developing Candidate Models

Once a survey has been completed, the first task is to prepare the detection data for analysis. A detection history is constructed for each survey unit; thus, if the survey unit has >1 station, results from all stations in the survey unit are aggregated. For each survey-occasion at that survey unit, a "1" means that ≥1 station(s) detected the target species, a "0" means it was not detected at any station during that survey-occasion, and another symbol indicating missing data (which can vary by software program) means that station inoperabilities or difficulties in checking a unit arose as a result of weather, personnel, or other issues. Careful construction of detection histories is important to avoid introducing bias in occasion- and day-specific estimates of p.

Another issue when baits are used is whether a survey-day or survey-occasion should be considered valid when the bait is missing. If bait is missing, it may alter detection probabilities and create additional detection heterogeneity (see Case Study 2). For devices other than cameras, it will not always be possible to know whether the bait was present between checks; however, the presence or absence of bait should be included as a covariate whenever possible.

After a detection history has been created for each survey unit, the next step is to consider sources of detection heterogeneity. Recall that P is a nuisance parameter with respect to estimating ψ, and that occupancy estimation assumes no unmodeled heterogeneity in detection probabilities. Thus, at this point in the process, the investigator should thoroughly explore and account for potential detection heterogeneity in the dataset. More complex models can typically be used to estimate P than ψ, because the sample size for modeling p is equal to the number of sites surveyed multiplied by the number of survey-occasions or survey-days (for remote cameras). A good first step in evaluating detection heterogeneity is to plot detection histories by the proportion of detections occurring on each sequential camera-day or survey-occasion (see Figure 15.1). This step will reveal important patterns in the detection histories, such as whether p is constant (Figure 15.1a) or increases during the survey (Figure 15.1b), or whether groups of sites have similar values for p (Figures 15.1c,d).

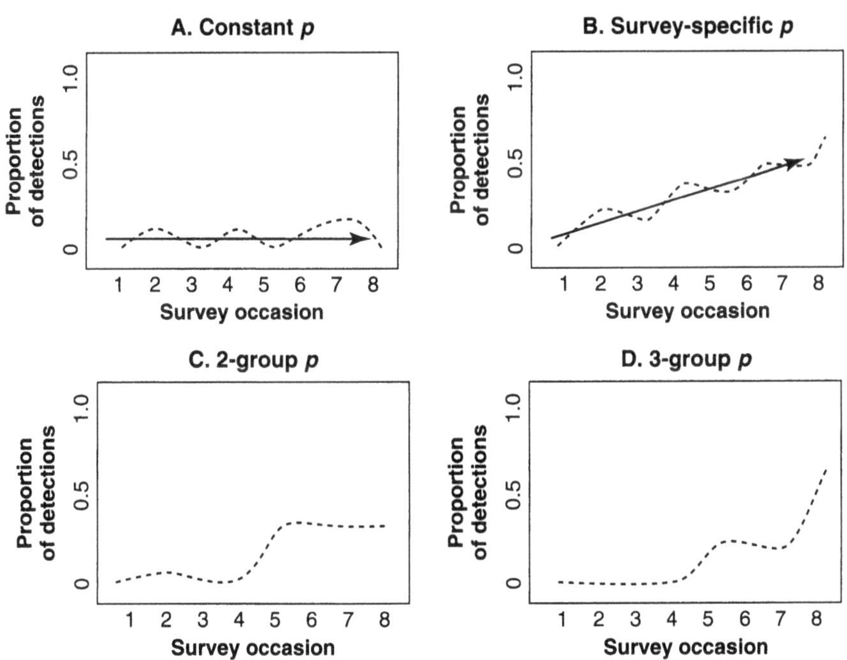

Figure 15.1. Conceptual examples of variation in p during surveys for *Martes* species.

Once alternative functions for p have been identified, they can be paired with additional covariates to account for heterogeneity. Additional sources of detection heterogeneity include, but are not limited to, temporal variation during the overall survey period (e.g., by month or season), changing weather patterns, elevational variation, and differences among detection devices (when >1 type is used). Using scat-detector dogs creates other potential sources of detection heterogeneity, including the dog, the handler, characteristics of scats, wind speed, wind direction, and average temperature. Snow-track surveys can have similar sources of heterogeneity, including intraobserver variation in track detection and identification, and time since the last snowfall.

In most cases, detection histories will not be uniform at each site, nor in the distribution of detections during the survey. In these cases, overdispersion is a concern if patterns in detection histories are not random with respect to a binomial distribution. Instead, 1 or more types of detection histories (i.e., "groups") may be more common than others; e.g., histories that include many detections (e.g., 01111010) or few detections (e.g., 00001000) may be more common than expected (Royle and Nichols 2003). Whether this pattern has a biological basis or is simply a statistical artifact, it is important to account for this type of heterogeneity. Grouping variables can be used to estimate detection probabilities for ≥2 distinct groups of survey sites, resulting in a more precise estimate of ψ (Royle and Nichols 2003). Grouping variables can be created to model a specific source of heterogeneity (e.g., survey month; see Case Study 1) or to account for unknown sources of overdispersion (e.g., 2-group and 3-group variables in Program PRESENCE).

Dependency within Detection Histories

Dependency refers to changes in detection probabilities resulting from detection status during a previous survey-occasion; i.e., a departure from statistical independence. Researchers should anticipate and test for dependency in detection histories, particularly when an attractant is used. Baits reward species detected and will likely increase the probability of detection on subsequent survey-occasions (positive dependency; see Case Study 2); Case Study 1 provides an example of testing and accounting for dependency. Most remote cameras include time and date stamps, such that sequential detections can be separated in time; however, the investigator must decide how much time is needed to ensure that detections are independent. Despite a lack of quantitative support, many studies have considered each 24-h period to be independent; although this assumption may be valid for unbaited camera stations, it may not be when bait is used. Furthermore, if this interval is not defined by temporal behavior patterns of the study species, additional dependency can be introduced. For example, for a largely nocturnal species, the 24-h camera-day should begin and end at noon, not midnight. Careful examination of temporal patterns in the dataset will help investigators select a survey period that minimizes dependency issues. Dependency issues can also arise when unbaited survey methods are used; for example, Royle et al. (2008) argued that repeat surveys using scat-detector dogs may suffer from reduced detection probabilities (negative dependency) if scats are removed on initial surveys.

Types of Occupancy Models and Their Inferences

Several classes of models exist for estimating occupancy and occupancy dynamics, each with its own considerations for design, analysis, and inference.

Single-Season Model

The most basic occupancy model is the single-season model, where detection/nondetection data from a single survey (usually with multiple survey-occasions) are used to estimate occupancy in the area surveyed (see Case Studies 1–2). Site covariates (e.g., seral stage, canopy closure) can be evaluated for their influence on the probability of occupancy, and their effects are interpreted similarly to those for covariates in logistic regression (Talancy 2005).

Multiseason Model

The multiseason model allows for the exploration of temporal occupancy dynamics when surveys are repeated ≥2 times (MacKenzie et al. 2003). We distinguish multiple survey periods, which occur among seasons or years, from multiple survey-occasions, which occur within a single survey. Two additional parameters, the probability of extinction (ε) and the probability of colonization (γ), can be included in multiseason models to account for changes in occupancy status between time periods of interest (1 to 0 and 0 to 1, respectively). Probabilities for no change in occupancy status (i.e., 0 to 0 and 1 to 1) are calculated as $1 - \gamma$ and $1 - \varepsilon$, respectively. Site covariates can be used to test hypotheses about factors that may influence extinction and colonization probabilities.

Scale is also important when applying multiseason models to detect changes in the population using occupancy data. To increase the confidence that changes in occupancy reflect a true population fluctuation, the spacing between survey units needs to be large enough, such that each individual is likely to be detected in only 1 survey unit. Furthermore, surveys designed to detect population fluctuations should have repeat surveys during the seasons of interest (see Case Study 3) to minimize the influences of changes in home-range sizes or movement patterns between seasons.

Multistate Models

The multistate model allows for including information other than occupancy status. For example, if hair snares were used and species and sex could be identified from DNA, the occupancy state could be: (1) male, (2) female, (3) both male and female, or (4) no target species detected. Site covariates and temporal dynamics can then be evaluated relative to each occupancy state (Nichols et al. 2007; see Case Study 3). The multistate model can be used as either a single- or multiseason model (MacKenzie et al. 2009); however, when used in a multiseason model, the number of parameters quickly increases. For a single-season model, the number of parameters is twice the number of possible states; for a 2-season model, the number of parameters is 6 times the number of states.

Multispecies Models

An important new advance in occupancy modeling is the inclusion of more than 1 species (MacKenzie et al. 2004; Bailey et al. 2009), enabling the investigator to examine interspecific effects on occupancy rates and detection probabilities. To evaluate whether species interactions produce static or dynamic occupancy outcomes, multispecies models can also be single- or multiseason (MacKenzie et al. 2006). Carnivore interactions and dynamics have been extremely difficult to study. We expect that because *Martes* populations typically occur in carnivore assemblages that include both larger- and smaller-bodied species (e.g., Campbell 2004), interspecific interactions will affect occupancy dynamics. Like multistate models, multispecies models require relatively large sample sizes, because of the number of parameters being estimated.

Multiscale and Multiple-Detection-Method Models

Occupancy models have recently been extended to incorporate the inclusion of multiple detection methods (Nichols et al. 2008). These models incorporate the simultaneous use of all data obtained from multiple methods to make inferences about detection probabilities for each method used. In addition, this modeling approach allows for the estimation of occupancy at 2 spatial scales: a larger scale corresponding to occupancy of the survey unit, and the smaller scale of the station or transect (Nichols et al. 2008).

Spatial and Temporal-Dependency Models

For surveys with spatial or temporal dependency, detection probabilities will be overestimated, and estimates of occupancy will be negatively biased (Hines et al. 2010; see Case Study 1). For example, spatial dependency is likely to occur if subunits of a transect are treated as independent survey-occasions. Temporal dependency is likely to occur when an animal is not equally detectable during the survey, or when an animal becomes "trap-happy" or "trap-shy." If either type of dependency is suspected, these models should be included in the candidate set to explicitly test for such dependencies and, if found, to adjust for negative bias in occupancy estimates.

Software for Occupancy Estimation and Modeling

Program PRESENCE (http://www.mbr-pwrc.usgs.gov/software/presence.html) was developed specifically for occupancy analysis (MacKenzie and Hines 2006) and Program MARK also supports occupancy analysis (http://warnercnr.colostate.edu/~gwhite/mark/mark.htm). Both PRESENCE and MARK support the main model types (e.g., single- and multiseason, multistate, multispecies) and include easy-to-use interfaces, with the ability to include both standard models for p and ψ, as well as the ability to create custom models and include covariates. However, PRESENCE has recently been updated with new model types (multimethod, multiscale, trap-response) not yet available in MARK. Bailey et al. (2007) developed the program GENPRES to enable comparisons of design trade-offs in occupancy analyses (http://www.mbr-pwrc.usgs.gov/software/bin/genpres.zip). Other statistical software packages that have been used for custom programming in occupancy analyses include SAS (SAS 2008; http://www.sas.com), R (R Development Core Team 2009; http://www.r-project.org), and WinBUGS (Lunn et al. 2000; http://www.mrc-bsu.cam.ac.uk/bugs/winbugs/contents.shtml). These 3 programs require custom programming for all occupancy estimation and modeling.

Case Studies

We present 3 case studies that provide examples of applying occupancy estimation and modeling to a variety of research and management questions for *Martes* species.

Case Study 1: Single-Season Model and Detection Heterogeneity

Source

Zielinski et al. (2005); K. Slauson, unpublished data.

Study Area

>30,000 km^2 in the historical range of the fisher in California (USA; Figure 15.2).

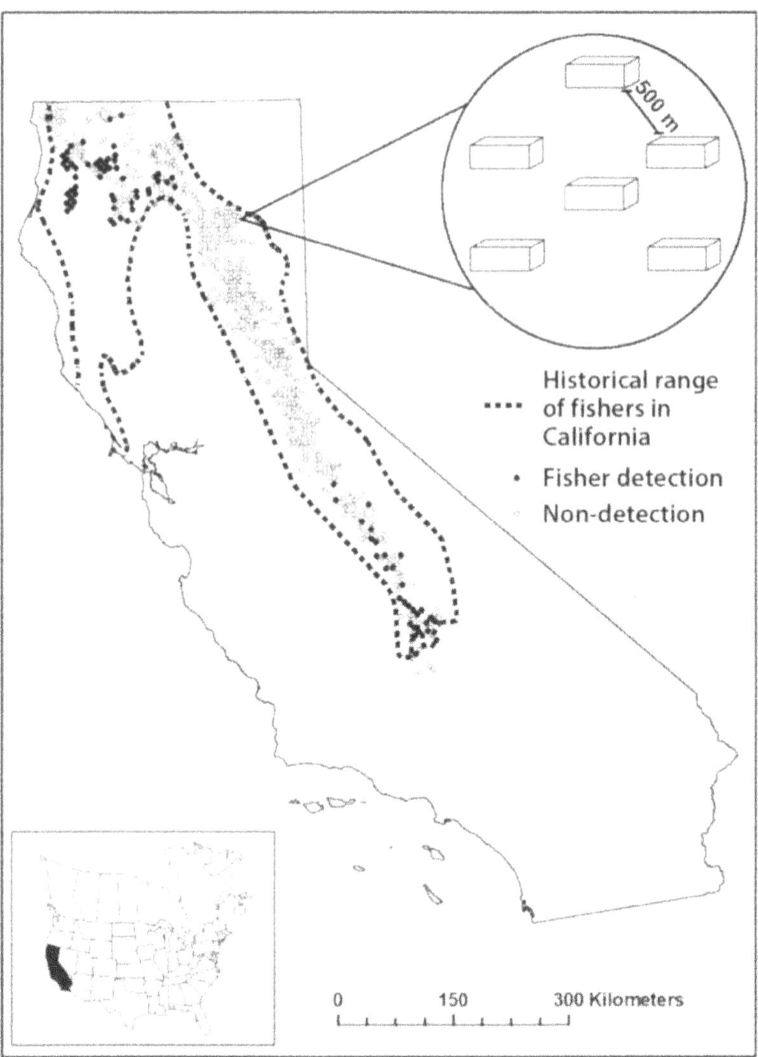

Figure 15.2. Case Study 1: results of systematic surveys for fishers throughout their historical range in California from 1996 to 2006 using track-plate boxes.

Survey Objectives

To estimate the contemporary distribution of fishers within their historical range in California and estimate detection probabilities for fishers using standard survey protocols.

Survey Design

Survey units were distributed at ~10 km intervals throughout the historical range of the fisher in California (Figure 15.2). The ~10-km spacing was chosen to provide statistical independence between adjacent survey units.

Survey Unit

Each survey unit comprised 6 track-plate stations in a pentagonal array with 1 station in the center and 5 around the perimeter, with 500–600-m spacing between adjacent stations (Figure 15.2).

Survey Protocol

Each survey unit was run for 16 consecutive days and checked by field technicians every other day for a total of 8 survey-occasions. Each station was baited with a single drumstick-sized piece of chicken, and a commercial trapping lure (Gusto, Minnesota Trapline Products, Pennock, Minnesota, USA) was applied to a nearby tree when each station was established and reapplied after 8 days if a marten or fisher had not yet been detected. During each visit by technicians, bait was replaced and tracks removed.

Survey Results

A total of 530 survey units were surveyed, primarily from June to October 1996–2006. Fisher detections were obtained at 84 survey units (Figure 15.2), providing a naive occupancy estimate of 0.16.

Modeling P and ψ

A plot of the proportion of fisher detections by survey-occasion suggested that p was best modeled either by using separate detection probabilities for each survey-occasion or by fitting a trend in detection probabilities across the occasions. Eight potential sources of detection heterogeneity were evaluated using univariate tests, including survey month, northern versus southern fisher population, distance to the coast as a resource-productivity gradient, elevation, whether the survey unit was on the periphery or core of the species range in that area, habitat suitability, detection dependency, and station inoperability. Only survey month and detection dependency contributed significantly to estimates of P. Survey month and detection dependency in combination with several models for how p increased during the survey (survey-specific p) were used to build 8 custom models. In addition to the 8 custom models, 3 "standard" models were included in program PRESENCE for a total of 11 candidate models. Fisher detection dependency was modeled using a first-order Markov-chain persistence factor (Barton et al. 1962) that accounted for the dependence in fisher detections between subsequent survey-occasions.

Results and Conclusions

- The top model (lowest ΔAIC_c value) contained the 3 variables with the highest importance weights (a measure of the relative contribution of a variable to all models in the model set): Markov-chain persistence factor, survey-specific p, and 2 seasons. This model had a very high

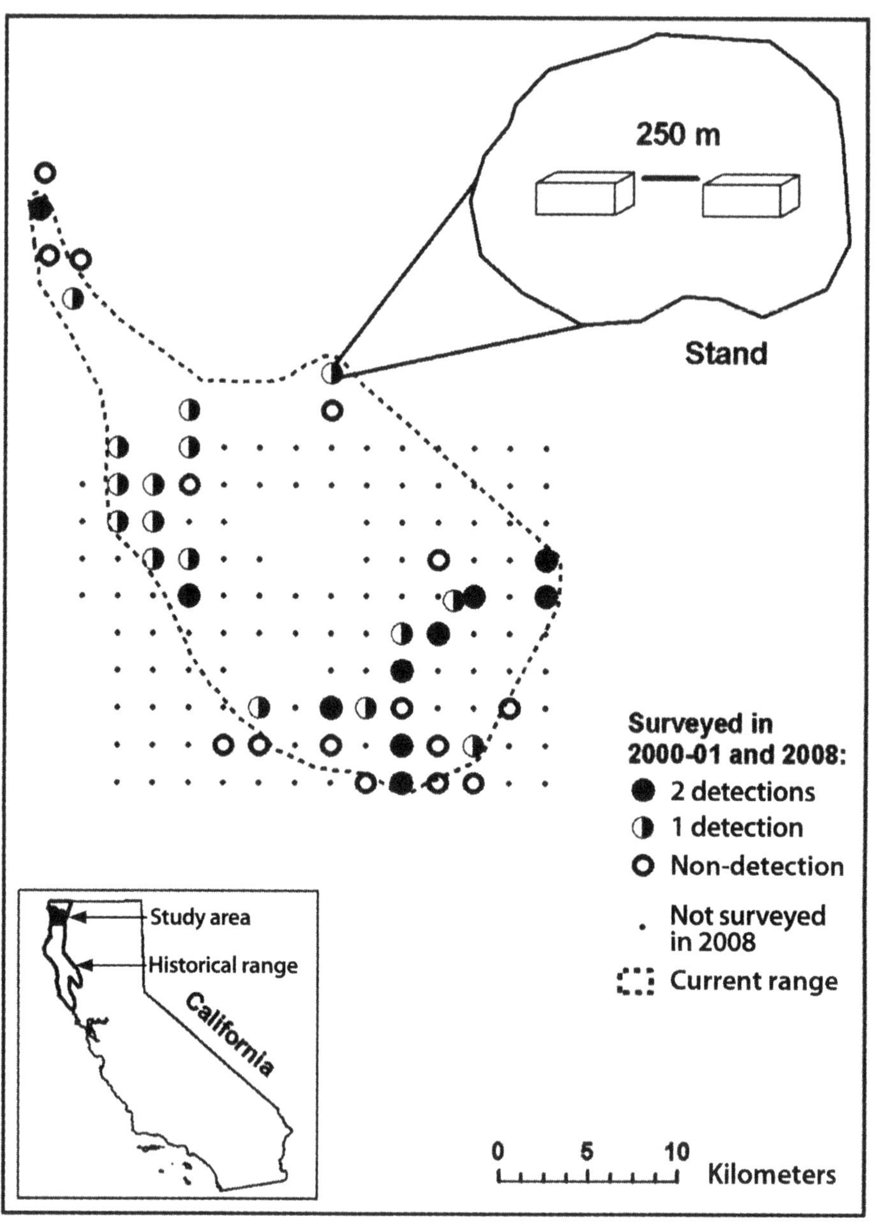

Figure 15.4. Case Study 3: results of repeated surveys for Humboldt martens in a portion of their historical range in California in 2000/2001 and 2008 using track-plate boxes.

Survey Objectives

Determine whether changes in distribution occurred from 2001 to 2008 for the only known population of the Humboldt marten in California.

Survey Design

During 2000 and 2001, the general vicinity of the study area was surveyed to document distribution and habitat use of the Humboldt marten (Slauson et al. 2007). The original survey included 164 survey units, 28 of which had marten detections; however, most survey units were not located in areas that contained the key habitat characteristics associated with occupancy (e.g., large patches of late-successional forest or serpentine habitat with dense, spatially extensive shrub cover). A prospective power analysis was conducted with the original survey data to determine the number of survey units that would need to be resurveyed in 2008 to detect a >30% decline in occupancy with 80% power. This analysis determined that 7 additional surveys units were needed for a total of 35 (all of which were part of the original survey).

Survey Unit

Each survey unit consisted of 2 track-plate stations, 1 located on the original 2-km systematic-survey grid point, and the second 250 m away, but in the same forest stand as the grid point (Figure 15.4).

Survey Protocol

Each station was run for 16 consecutive days and checked every 2 days to replace bait and remove tracks, resulting in 8 survey-occasions. A commercial lure (Gusto) was used and reapplied after 8 days if a marten or fisher had not yet been detected.

Survey Results

In 2008, only 30 of 35 survey units were surveyed because of an active wildfire, reducing the number of directly comparable survey units to 30. In 2000/2001, marten detections were observed at 23 of the 30 units (naïve occupancy = 0.77) compared with 14 of 30 (naïve occupancy = 0.47) in 2008 (Figure 15.4). Based on sex identification from tracks (Slauson et al. 2008), male-only marten detections were observed at 14 and 7, females only at 11 and 3, and both sexes at 3 and 4 survey units in 2000/2001 and 2008, respectively.

Modeling P and ψ

A 2-step process was used to develop candidate models. First, a set of candidate models including 4 functions for p (occasion-specific, occasion-group, week 1 vs. week 2, constant p) and 1 detection covariate (year) were evaluated to identify the best model for estimating P. The second step was to use the best model(s) in an information-theoretic framework (Burnham and Anderson 2002) to develop a set of competing models containing habitat covariates to distinguish between sites with stable vs. unstable occupancy. Eight landscape-configuration covariates were used to develop 17 a priori competing models to identify factors that influence the probability that a survey unit becomes unoccupied (i.e., the extinction probability (ε). Multiseason occupancy modeling was used in Program MARK.

Multistate Modeling

The dataset was further explored by conducting a multistate analysis in which a detection could have 1 of 4 true states: (1) nonoccupancy, (2) male-only occupancy, (3) female-only occupancy, or (4) male and female occupancy. Limitations in sample sizes permitted only single-season modeling. Several candidate models were developed and evaluated for each "season" of data (2000/2001 and 2008) to explain variation in sex-specific detection probabilities using the same methods described in the previous section. Custom modeling for the multistate modeling was conducted in Program SAS, but Program PRESENCE was used for standard occupancy calculations.

Results and Conclusions

- Six models were highly competitive (<2 ΔAIC_c units) and together had 72% of the Akaike weight. Each of these models included the occasion-group variable to model *p*, which fit the data better than the 4 other functions used to model *p*.
- Because there was no clear top model, model averaging was used to generate estimates for each parameter in the models. Detection probability (P) was 0.92 (SE = 0.02) in 2000/2001 and 0.95 (SE = 0.01) in 2008, and did not differ significantly between time periods ($t = -1.31$, df = 12, $P = 0.207$).
- The occupancy estimate ($\hat{\psi}$) in 2000/2001 ($\hat{\psi} = 0.79$, SE = 0.09) was significantly higher than in 2008 ($\hat{\psi} = 0.46$, SE = 0.10; McNemar's Chi-square Test: $\chi^2 = 45.3$, df = 31, $P = 0.046$). The change between 2000/2001 and 2008 at the 30 survey units marks a significant decline in occupancy for the survey units, equaling a $\hat{\psi} = 0.58$ (SE = 0.13) or a 42% decline in occupancy during the 7-year period. It follows, therefore, that the estimated probability of extinction ($\hat{\psi} = 0.49$, SE = 0.12) was higher than the probability of colonization ($\hat{\psi} = 0.29$, SE = 0.23).
- Multistate model results revealed that female-only occupancy from 2000/2001 to 2008 ($\hat{\psi}_{f,01} = 0.36$, SE = 0.09 and $\hat{\psi}_{f,08} = 0.06$, SE = 0.06, respectively) declined more severely than male-only occupancy ($\hat{\psi}_{m,01} = 0.36$, SE = 0.09; $\hat{\psi}_{m,08} = 0.26$, SE = 0.06); sites occupied by

both sexes remained relatively constant ($\hat{\psi}_{m/f,01} = 0.15$, SE = 0.07; $\hat{\psi}_{m/f,08} = 0.16$, SE = 0.08). This suggests that the major factor in the decline from 2000/2001 to 2008 was a decrease in the number of females.

Summary

Occupancy estimation and modeling provides a valuable tool for improving the use of noninvasive surveys for research and monitoring of *Martes* species, because survey results may not reflect occupancy accurately. The case studies provided examples of the ways in which results of single-season surveys can vary during the survey season, and the frequent occurrence of detection heterogeneities in *Martes* survey data. To minimize the effects of detection heterogeneities on parameter estimates, potential sources should be anticipated and addressed in the survey design and analysis phases. Key sources of detection heterogeneity that should be considered carefully by *Martes* biologists include: detection dependence, bait status, time periods within the survey season, sex-specific detection probabilities, and the functional form of p in each dataset. For situations where both baited and unbaited methods are used, the potential effect of bait or lure on behavioral responses to the survey devices should be considered carefully. There are undoubtedly additional sources of detection heterogeneity not listed here that investigators should attempt to identify and account for. Furthermore, for small datasets, it may not be possible to model all potential sources of detection heterogeneity; in these cases, it is important to base survey designs on previous survey efforts with known sources of heterogeneity so that their effects can be mitigated.

Noninvasive surveys are attractive options for addressing research and monitoring objectives because of their generally low costs and the large spatial scale at which they can be deployed (see Long and MacKay, this volume); however, they must be designed and implemented carefully. Use of survey data to address research or monitoring objectives involves making assumptions about the relation between occupancy status and the presence of various segments of the population (e.g., individuals, sexes, age classes) at each survey unit. If the true population parameter and the observed occupancy status are not closely related, resulting inferences and management recommendations will likely be incorrect. We therefore recommend that a secondary method (e.g., sex or individual identification) be used at a subset of sites whenever possible to validate implicit assumptions about the relation between observed occupancy status and the population parameters of interest.

Occupancy estimation and modeling has the potential to expand our understanding of *Martes* ecology in at least 2 specific areas. Multiseason and multistate frameworks allow for the investigation of long-term temporal patterns of occupancy dynamics and provide opportunities to test hypotheses about population dynamics, stability, and change. In addition, multispecies occupancy models provide a template for evaluating how *Martes* species interact with other members of the carnivore community. Because of their relatively low cost, noninvasive surveys can often be conducted at larger temporal and spatial scales than other research methods. Thus, occupancy estimation and modeling provides very practical and informative tools for exploring new frontiers in *Martes* research.

Acknowledgments

We thank the *Martes* Working Group, and specifically the organizing committee of the 5th International *Martes* Symposium for the opportunity to share this information. We thank Tom Kirk and Jan Werren for GIS support.

SECTION 5

Conservation of *Martes* Populations

16

Martens and Fishers in a Changing Climate

JOSHUA J. LAWLER, HUGH D. SAFFORD, AND EVAN H. GIRVETZ

ABSTRACT

Average global temperatures are projected to rise between 1.1 and 6.4 °C by the end of the century. Coupled with changes in precipitation and increasing atmospheric carbon dioxide concentrations, these increases in temperature will alter species distributions and phenologies, with cascading effects on ecological communities and ecosystem functions. In this chapter, we investigate how projected changes in climate are likely to affect 4 *Martes* species: American marten (*M. americana*), fisher (*M. pennanti*), stone marten (*M. foina*), and European pine marten (*M. martes*). We review both recent trends in climate and projected future climate changes throughout the geographic ranges of these 4 species and project shifts in their distributions under multiple climate-change scenarios. To provide insights at a finer spatial scale, we describe analyses that have explored the potential effects of climate-driven changes in fire regimes and fisher habitat in the southern Sierra Nevada Mountains in California. We projected that both North American species will experience northward range shifts during the coming century. Projected climate-driven shifts in the potential ranges of the 2 European species were more complex and more variable. The finer-scale analyses for California fisher populations revealed potential changes in forest composition from conifers to mixed hardwoods in the north, and to mixed-hardwood forests and grasslands and shrublands in the south. Increases in fire frequency and intensity were also projected for the southern Sierra Nevada population. Finally, more-detailed simulations resulted in projected losses of large conifer and hardwood trees as fire severity increased. For fishers in the southern Sierra Nevada, we recommend protecting old-forest habitats through targeted forest-fuel treatment, and applying more liberal fire-suppression policies to naturally ignited fires during moderate weather conditions. Overall, our results suggest that martens and fishers will be highly sensitive to climate change and, as with many species, will likely experience the largest climate impacts at the southernmost latitudes and lowest elevations within their ranges. Furthermore, several marten populations currently considered at risk of extirpation will likely experience future climates outside the range of current climatic conditions.

Introduction

During the last 100 years, average annual global temperatures have risen 0.7 °C (IPCC 2007b), but temperatures have been increasing more rapidly in the recent past. For example, during the last 50 years, average global temperatures have risen twice as fast as over the previous 50 years. This trend in warming is projected to continue into the future and will likely be accompanied by changes in precipitation patterns. Global average surface temperatures are projected to rise between 1.1 and 6.4 °C by 2100 (IPCC 2007b). The largest increases are projected for high-northern latitudes, where average annual temperatures may increase by more than 7.5 °C. There is far less agreement among projections of future precipitation patterns. In the winter months, climate models generally agree that there will be an increase in precipitation in the mid- to high-northern latitudes. In the summer months, most land masses are projected to experience less precipitation; however, there is less confidence in these projections than for winter projections. Among summer projections, there is more confidence in drying trends projected for Europe, the Mediterranean region, southern Africa, and the northwestern contiguous United States (IPCC 2007b).

These changes have had, and are projected to continue to have, profound effects on many other physical processes. For example, recent increases in the size and frequency of fires in the western United States have been linked to changing moisture regimes (Westerling et al. 2006). Portions of that area are projected to experience a 2- to 4-fold increase in the frequency of large fires during the coming century (McKenzie et al. 2004). Changes in temperature and precipitation have already had impacts on hydrology (IPCC 2007a). Future changes will likely have even more profound effects on snowpack, timing and frequency of stream flows, and water availability for many plant and animal species.

Recent trends in climate have also been linked to numerous ecological changes (Parmesan 2006). The best documented of these changes are shifts in species distributions and changes in phenologies (Parmesan and Yohe 2003; Root et al. 2003). Many species have shifted their ranges in accordance with changing temperatures, often resulting in poleward or upward shifts in species distributions (e.g., Moritz et al. 2008). Similarly, changes in the timing of important annual life-history events have occurred that generally correspond with recent trends in climate. Flowering in plants, migratory patterns and egg laying in birds, and mating in amphibians have all begun to occur earlier in the season for some species (Parmesan 2006).

In this chapter, we provide an overview of some of the potential impacts of future climate change on the American marten (*Martes americana)*, fisher (*M. pennanti*), stone marten (*M. foina*), and European pine marten (*M. martes*). We chose these 4 species because of the availability of detailed data on their current and historical distributions; because of limited data, it was not possible to model potential climate-change impacts on all 8 *Martes* species. We begin by summarizing recent historical and projected future changes in climate across the ranges of these 4 species. We then use bioclimatic models to project broad-scale shifts in their potential geographic distributions. We conclude by using a set of finer-scale forest-growth and fire models to examine potential changes in habitat for the fisher in California.

Recent Changes in Climate

We assessed trends in mean annual temperatures and total annual precipitation from 1950 to 2002 across the current ranges of these 4 species using the Climatic Research Unit (CRU) TS (time series) 2.1 monthly climate dataset (Mitchell and Jones 2005). This dataset has a 0.5-degree spatial resolution (grid cells approximately 50 km per side, depending on latitude). We calculated linear trends in both temperature and precipitation with the Climate Wizard climate-analysis tool using a restricted maximum-likelihood linear-regression analysis method (Girvetz et al. 2009).

American Marten and Fisher

For the American marten and fisher, the largest changes in temperature have occurred in Alaska (marten) and northern Canada (both species). In parts of these regions, mean annual temperatures have increased at a rate of 0.4 °C per decade. Changes have been less dramatic in the eastern and southernmost portions of the species' ranges, where temperatures have generally increased by less than 0.2 °C per decade. California, northeastern Quebec, and Newfoundland showed no significant changes in temperature over this 52-year period.

Patterns of recent trends in total annual precipitation within the ranges of the 2 North American species are far less consistent than those of temperature trends. For the American marten, the largest significant increases in precipitation were in the northeastern and north-central parts of its range. Only a few areas in the range (e.g., eastern Alaska and south of James Bay at the southern tip of Hudson Bay) showed significant decreases in precipitation. Decreasing trends in precipitation in other parts of its range were not significant. For the fisher, precipitation has decreased in the extreme northwestern part of its range, and in portions of the north-central part of the range.

Stone Marten and European Pine Marten

From 1950 to 2002, temperatures also increased over most of the ranges of stone and pine martens. The strongest trends were in southern France and Spain for the stone marten and in northern Spain, France, Estonia, and Latvia for the pine marten, where, in some places, temperatures increased at rates exceeding 0.4 °C per decade. Conversely, there were very few places within the ranges of either species that experienced significant trends in precipitation. Both species experienced increases in precipitation in central France and decreases in Slovenia. The pine marten experienced increases in precipitation in Scotland, northern Scandinavia, Estonia, Latvia, Lithuania, and southern Sweden.

Projected Changes in Climate

We assessed projected changes in mean annual temperature and total annual precipitation for all 4 species at the end of the 21st century. We calculated differences between climate averaged over a recent 30-year period (1970–1999) and projected climate averaged over a future 30-year period (2070–2099) using the Climate Wizard climate-analysis tool (Girvetz et al. 2009). Projected future temperature and precipitation values were downscaled from 16 general circulation models (GCMs) run for a mid- to high-CO_2 emissions scenario (SRES A2) (Nakicenovic et al. 2000). This scenario provided the closest approximation of the current emission trajectories for the 3 scenarios for which all 10 GCMs were run. We chose to explore the range of projections from multiple GCMs and a single scenario because there is more variation in the output of the GCMs than from using the 3 scenarios. We calculated median changes in temperature and precipitation for all 16 models within the ranges of each of the 4 *Martes* species.

American Marten and Fisher

We projected that both the American marten and the fisher will experience warming throughout their ranges by the end of the 21st century (Figures 16.1a,c). We projected that average annual temperatures will increase by more than 5 °C in far northern and western Alaska and in the area directly south of Hudson Bay. The ensemble of 16 projected future climates showed the least change for the 2 *Martes* species in coastal portions of British Columbia and California, where we projected that temperatures will increase by less than 3 °C by the end of the century. It is important to note, however, that these temperatures were the medians of the 16 projections; thus, some of the models project larger, and some smaller, increases in temperature. No models projected decreases in temperatures throughout the species' ranges.

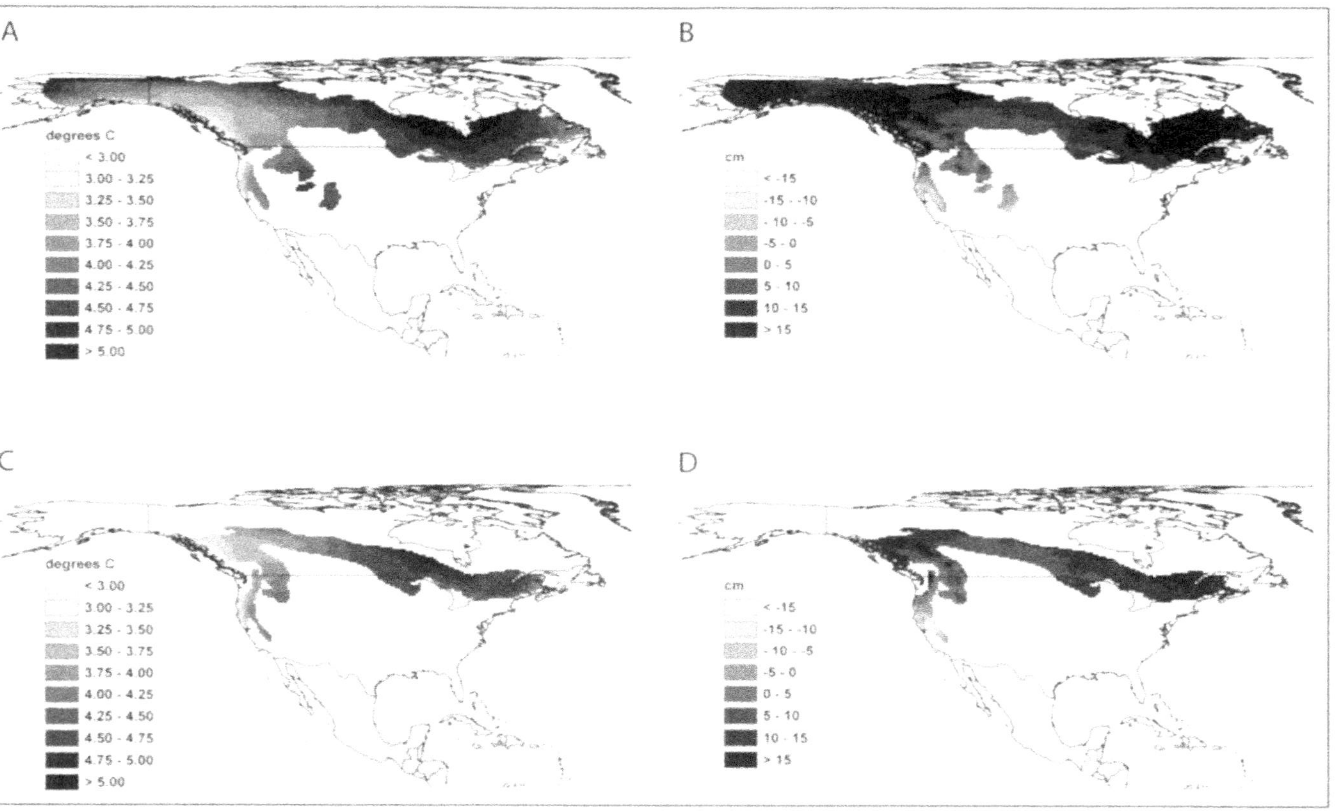

Figure 16.1. Projected changes in temperature (A and C) and precipitation (B and D) for the American marten (A and B) and the fisher (C and D) in North America. Changes are the differences between climate averaged over a recent 30-year period (1970–1999) and projected climate averaged over a future 30-year period (2070–2099) and are the median of projected climates from 16 different general circulation models (GCMs). Geographic ranges for both species are based on digital range maps in Patterson et al. (2003).

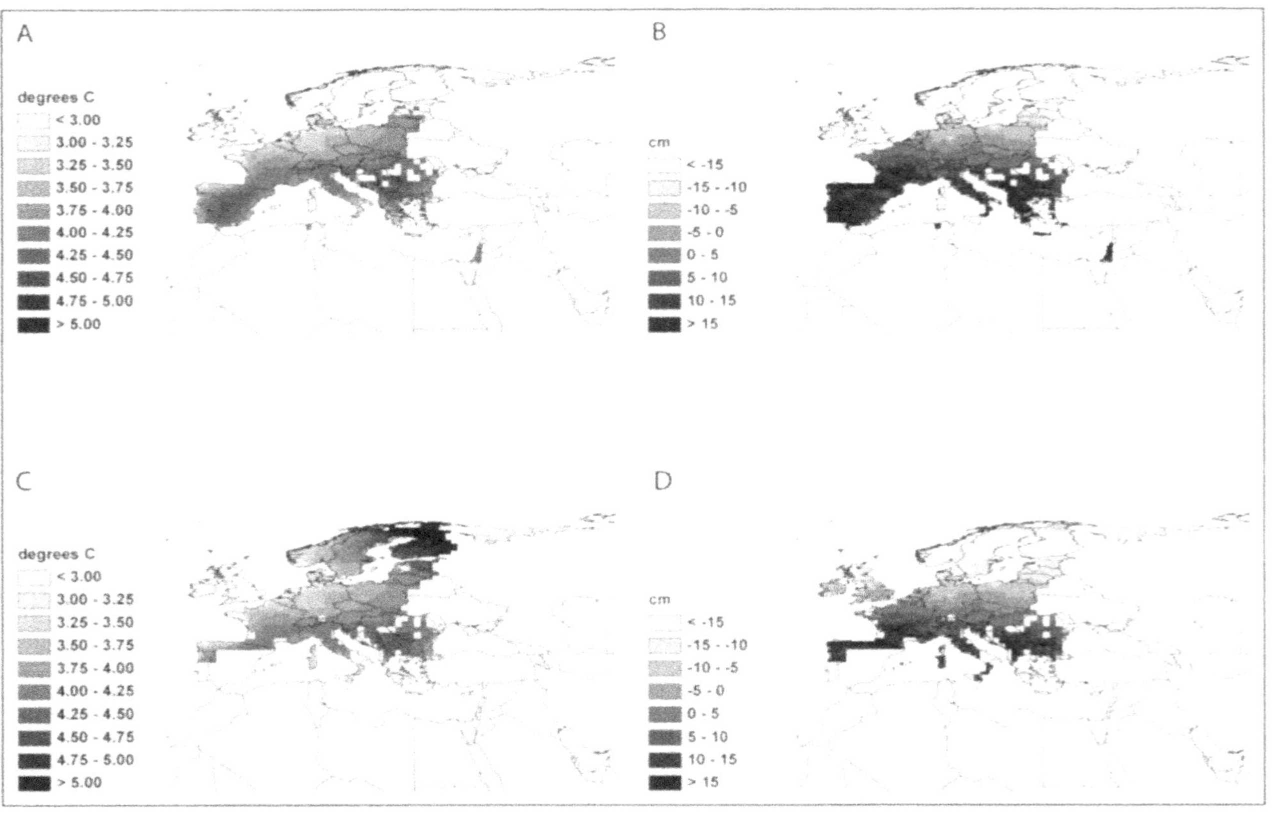

Figure 16.2. Projected changes in temperature (A and C) and precipitation (B and D) for the stone marten (A and B) and the European pine marten (C and D) in Europe. Changes are the differences between climate averaged over a recent 30-year period (1970–1999) and projected climate averaged over a future 30-year period (2070–2099) and are the median of projected climates from 16 different general circulation models (GCMs).

We projected that total annual precipitation will increase throughout most of the ranges of the 2 North American species (Figures 16.1b,d). Projected increases were largest in the far northwest and northeast, where total annual precipitation is projected to increase by >20 cm. In the southern portions of the species' ranges, we projected that total annual precipitation will decrease by 5–20 cm.

Stone Marten and European Pine Marten

For both European species, we projected that temperatures will increase throughout their ranges (Figures 16.2a,c). For the stone marten, the greatest increases in temperature were in central Spain, where we projected that average annual temperatures will increase by 4–5 °C. For the pine marten, we projected that the greatest increases in temperature will be in the far northeastern part of its range in Finland and northern Sweden. We projected that the smallest changes in temperature for both species will occur in western France, the United Kingdom, and Ireland.

In general, we projected that total annual precipitation will increase in far northern Europe and decrease in central and southern Europe (Figures 16.2b,c); thus, we projected a decrease in precipitation across much of the stone marten's range, and for the pine marten, we projected a decrease across the southern half of its range, and an increase across the northern half. We projected that the stone marten will experience the largest decreases (>20 cm) in total annual precipitation across the western Balkan and Iberian Peninsulas and the largest increases (10–15 cm) at the northern extent of its range. We projected the pine marten to experience increases in precipitation of >15 cm in most of Scandinavia and Scotland.

Projected Range Shifts

We projected climate-driven changes in the ranges of each of the 4 species using bioclimatic models (Pearson and Dawson 2004). Bioclimatic models, often referred to as climate-envelope models, are based on correlative relationships between the current distribution of a species and current climate. The models are applied to projected future climate data to estimate the future potential range of the species. These correlative models have their shortcomings; they do not account for behavioral or evolutionary changes, they do not directly model interspecific interactions, they cannot adequately account for future non-analogue climatic conditions, and they assume that the range of the species being modeled is at equilibrium (i.e., not expanding or contracting). Despite these shortcomings, these models provide a useful first approximation of potential changes in habitat conditions under various climate-change scenarios.

We obtained data for building the models from several sources. We derived distribution data for the 2 North American species from digital range maps of their breeding ranges (Patterson et al. 2003). We mapped these data on a 50 × 50-km cell grid, whereby we considered each species to be "present" or "absent" in each grid cell based on whether that cell fell within or outside the mapped range of the species. Because the map of the fisher's current range that we used to build our model did not include areas where fishers had been extirpated, but have been reintroduced relatively recently, we built a second model for the fisher based on a digitized map of its presumed historical range (Powell 1982). If differences between the fisher's historical and current ranges are due to extirpation (either directly through trapping or indirectly through habitat loss), building a bioclimatic model based on the current distribution would result in an overestimate of the potential impacts of climate change on the fisher's range. However, differences between current and historical ranges could have resulted from recent climate changes (see Krohn, this volume); consequently, building a model based on historical range and current climate may underestimate the potential impacts of climate change. Accordingly, we explored potential climate-driven range shifts for the fisher by using both current and historical ranges to build our models. We obtained distribution data for the 2 European species from the Global Biodiversity Information Facility data portal (http://data.gbif.org), gridded to 50 × 50-km cells. These occurrence data included both museum specimens and observational data; there were >1000 occurrences for each species.

We built models for the North American species based on recent climate data from the 30-min CRU CL (average climatology) 1.0 (New et al. 1999) and 10-min CRU CL 2.0 datasets (New et al. 2002). We downscaled these data to our 50-km grid and averaged them for a 30-year period from 1961 to 1990. This downscaling involved a topographically informed spatial interpolation. Thus, the downscaled layer takes into account the effects of elevation and aspect on temperature and precipitation patterns. To build the bioclimatic models, we used a set of 37 bioclimatic variables (e.g., evapotranspiration, total annual snowfall) derived from the temperature, precipitation, and sunshine data in the CRU datasets. See Lawler et al. (2009) for complete descriptions of downscaling methods and bioclimatic variables. We derived recent climate data for the European species for the same time period (1960–1990) from the same dataset used to summarize historical changes in climate for the European species (see previous section on Recent Changes in Climate). We based models for the European species on 34 variables, including monthly, seasonal, and annual mean temperatures, and total precipitation values. Although we could have used the same global dataset and variables for both North American and European *Martes* species, we wanted to take advantage of the more-detailed climate dataset available for North America for modeling the American marten and fisher.

We used a 50-km grid to balance the relatively coarse nature of the species distribution data and the need to capture relatively fine-scale climate variation (e.g., the effects of topography). Although the climate datasets do capture the effects of large mountain ranges, they do not capture finer-scale topographic relief. Thus, projections based on the bioclimatic models do not address the relatively fine-scale effects of slope, aspect, and elevation, nor do they capture finer-scale effects of climate on potential habitat.

We built bioclimatic models for all species using the approach described in Lawler et al. (2009). We built all models using random-forest predictors (Breiman 2001; Cutler et al. 2007). Random forests are a machine-learning, model-averaging approach based (in our study) on classification trees. Random-forest models have been shown to perform better than many other approaches for modeling species distributions (Lawler et al. 2006; Prasad et al. 2006; Cutler et al. 2007). We built the models using 80% of the data, reserving the remaining 20% for a semi-independent model validation.

For the North American species, we used projected future climate data from 10 different GCMs run for a mid-high (SRES A2) greenhouse-gas emissions scenario. We obtained future climate projections from the World Climate Research Programme's Coupled Model Intercomparison Project phase 3 multi-model archive (http://www-pcmdi.llnl.gov/ipcc/about_ipcc.php). We down-scaled projected future temperature anomalies to the North American 50-km grid by spatially interpolating the values for the coarser GCM grid. We added these anomalies to the current climate variables and then recalculated the 37 bioclimatic variables. We then calculated future climate data as a 30-year average for the period 2071–2099. For the European species, we used projected future climates from 16 different GCMs run for the mid-high (SRES A2) greenhouse-gas emissions scenario. We downscaled anomalies from these projections to the same European grid, and added them to the current climate data to generate the 34 variables for the period 2071–2099.

American Marten and Fisher

Models for both species accurately predicted their current distributions. The model for the American marten correctly predicted 96% of the test data that were within the range, and 98% of the test data that were outside the range. Models built with current and historical ranges for the fisher predicted 91 and 92%, respectively, of the test data within the range correctly, and both models correctly predicted 99% of the test data outside the range. Models for both species projected potential climate-driven range expansions to the north and contractions across much of the southern extent of the ranges (Figure 16.3). We projected that both species will lose most of their climatically suitable range in the contiguous United States by the end of this century. For the American marten, projected contractions were greatest in the central western

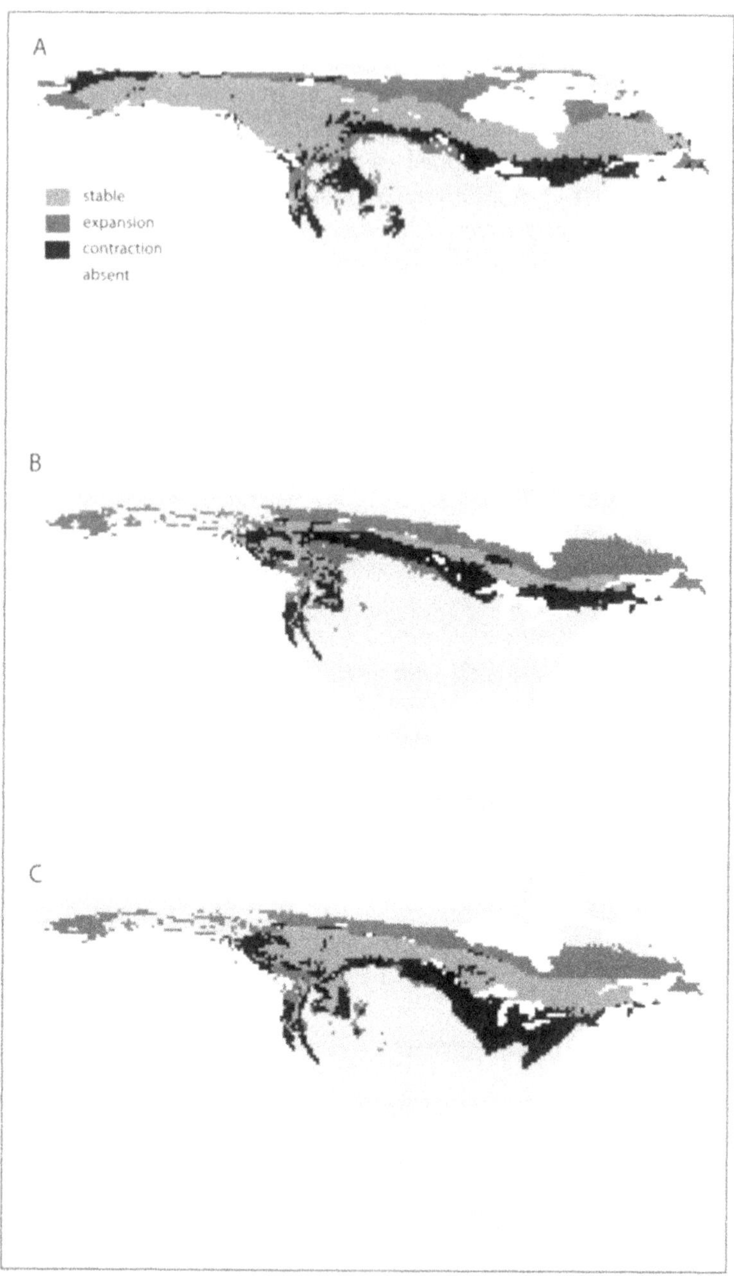

Figure 16.3. Projected climate-induced range shifts in North America for (A) the American marten and the fisher based on (B) current and (C) historical ranges. Areas of potential range expansion and contraction (the darker shades of gray) represent areas where at least 1/10 projected future climates resulted in a range expansion or contraction. Projections are for an averaged period from 2071 to 2099. Geographic ranges for both species are based on digital range maps in Patterson et al. (2003).

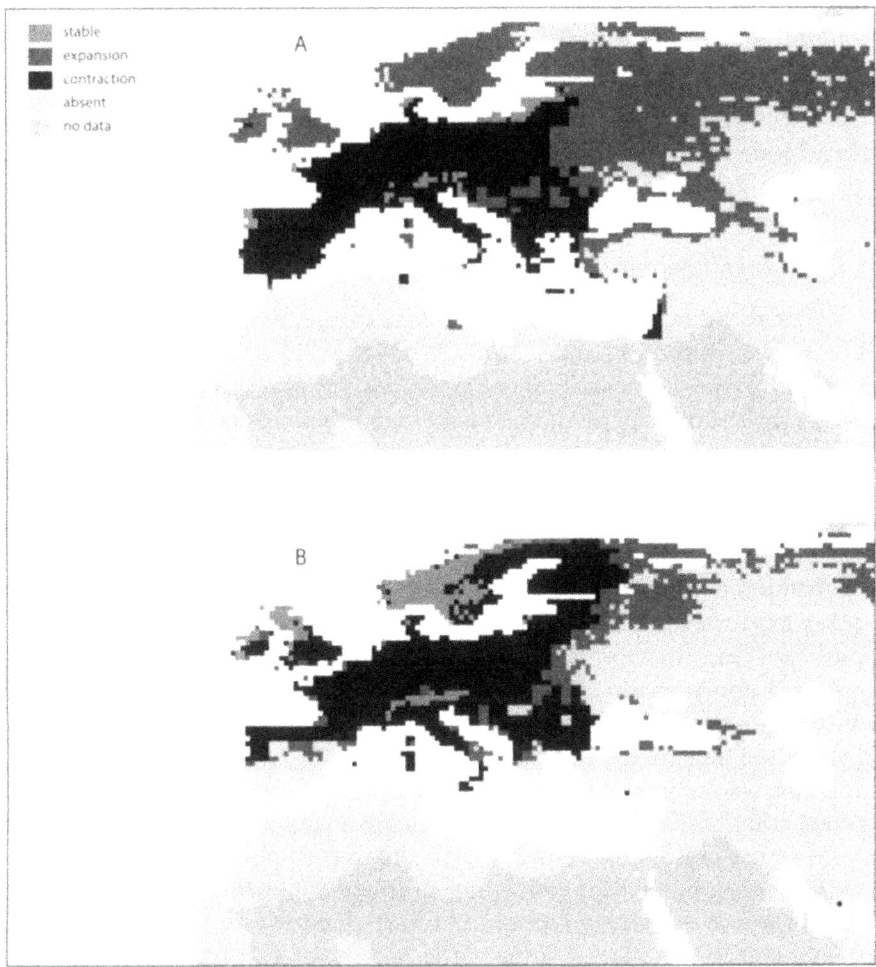

Figure 16.4. Projected climate-induced range shifts in Europe for (A) the stone marten and (B) the European pine marten. Areas of potential range expansion and contraction (the darker shades of gray) represent areas where at least 1/16 projected future climates resulted in a range expansion or contraction. Projections are for an averaged period from 2071 to 2099.

United States and around the Great Lakes. The largest area of range expansion for the fisher was in the north-central and northeastern parts of its range. For the American marten, Hudson Bay prevented an expansion in the central part of its range, but we projected large expansions to the east and west of the Bay. Models built using both current and historical ranges for the fisher produced similar projected range shifts (Figures 16.3b,c).

Stone Marten and European Pine Marten

Models for both stone and pine martens correctly predicted 97% of presences in the test data, and 97 and 98% of the absences, respectively. Both models projected potential range contractions across much of Europe (Figure 16.4). For the stone marten, there were very few places where all 16 projected future climates resulted in a projected stable range (Figure 16.4a); however, there was only moderate agreement among model projections for the contractions projected in most of its range. The areas with the most model agreement for range contractions were in southern Europe, particularly the southern half of Spain. Similarly, although the models projected a large area of potential range expansion for the stone marten (Figure 16.4a), there was relatively little model agreement across most of this zone of potential expansion. The areas with the greatest model agreement for expansions were in southern United Kingdom, Ireland, and Scandinavia.

For the pine marten, a slightly larger but still relatively small portion of the range was projected to be stable under all 16 projected future climates (Figure 16.4b). There was relatively high model agreement for range contractions across much of the southwestern and eastern portions of the range and low model agreement for contractions throughout most of the species' range in continental Europe. Projections of future climates resulted in relatively few areas of potential range expansion, and very few areas with high model agreement for such expansions.

Potential Climate Impacts on Fisher Habitat in California

The southernmost fisher populations in North America are in the Klamath Mountains and Sierra Nevada of California (Figure 16.5). These 2 populations are believed by many to represent disjunct remnants of a formerly contiguous range in California (e.g., Wisely et al. 2004). Fisher conservation is a major management concern in California, especially in the Sierra Nevada, where lawsuits challenging proposed actions by federal land management agencies in both occupied and unoccupied fisher habitat are common.

In California, fisher home ranges are characterized by forest structural characteristics strongly skewed toward mid- to late-seral stands with high canopy cover and large cavity-forming trees that are required for resting and

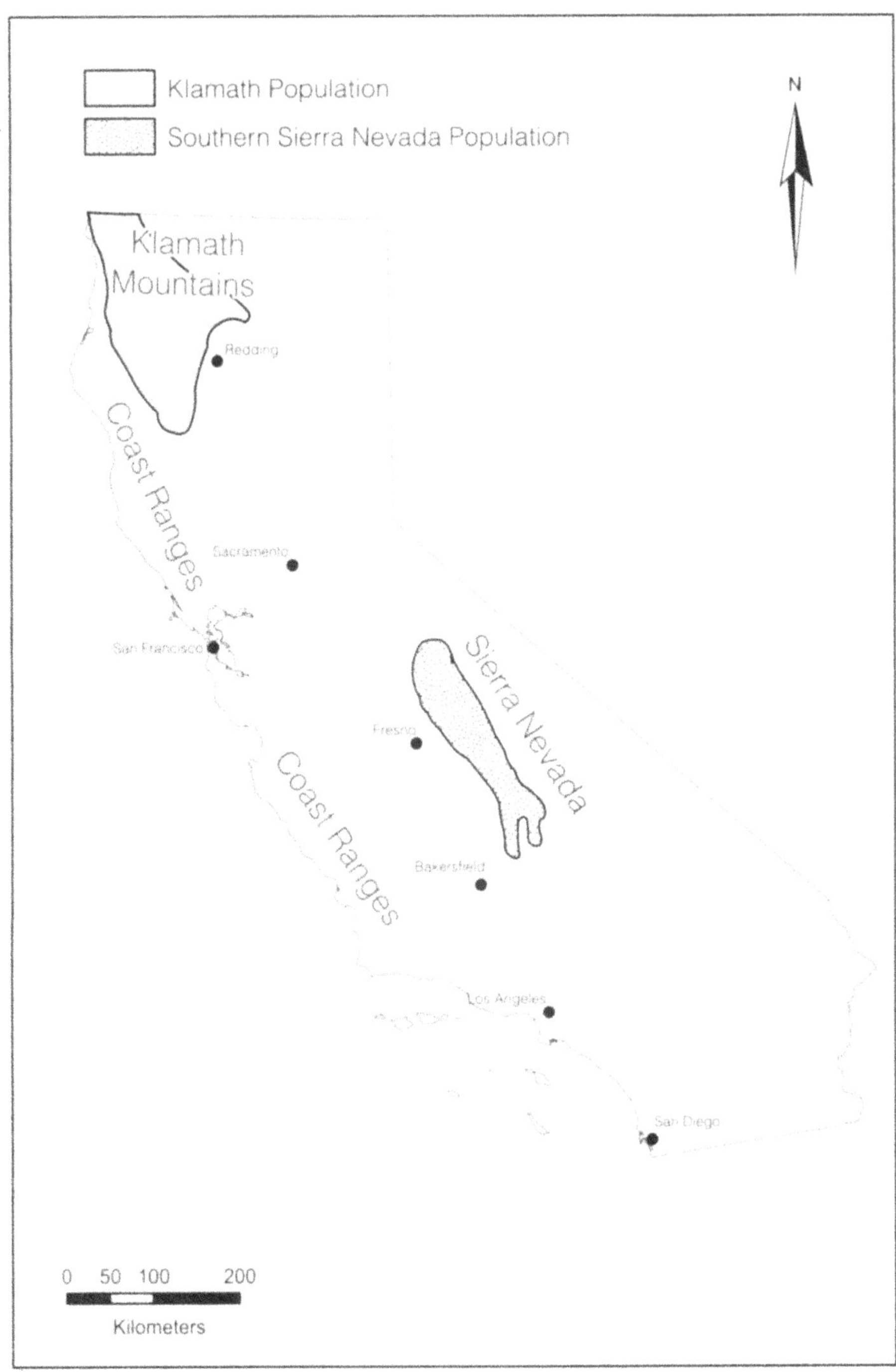

Figure 16.5. Locations of the 2 fisher population areas in California.

denning (Zielinski et al. 2004b; Yaeger 2005). Oaks, primarily black oak (*Quercus kelloggii*), are a key component of California fisher habitat, particularly in the Sierra Nevada (Zielinski et al. 2004a; Davis et al. 2007). In California, forest-structural conditions associated with fisher occurrence are correlated with areas of complex topography, steep (often north-facing) slopes, and proximity to water, especially in the Sierra Nevada, where precipitation is lower and summertime drought more pronounced than in the Klamath Mountains (Zielinski et al. 2004b; Carroll 2005a).

Potential climate-change impacts on the fisher and its primary habitat in California have yet to be evaluated quantitatively. Climate change is likely to affect fisher habitat in California by altering species and structural composition directly through changes in climatic conditions and competitive relationships, and indirectly by altering fire and other disturbance regimes. Climate-driven changes in tree-species composition may either decrease or increase the amount of habitat available to fishers for foraging, denning, or resting. Climate-driven changes in fire regimes are likely to drive much of the short-term response in vegetation floristics and structure (Flannigan et al. 2000; Dale et al. 2001). If longer or more severe fire seasons result from climate change, the probability of losing local populations of species associated with late-seral habitat conditions will increase (McKenzie et al. 2004).

The Landscape Scale: GCM-based Projections of Future Fire Regimes and Vegetation Types for Fishers in California

In this section, we combine previously published GCM-based model projections of fire regimes and vegetation response in California (Lenihan et al. 2003a,b, 2008) with stand-scale fire and forest-growth modeling outputs to explore potential future climate-driven changes to fisher habitat in California. Lenihan et al. (2008) projected potential future changes in fire regimes and vegetation across California using outputs from 2 different GCMs. The GFDL CM2.1 model (medium climate sensitivity of about 3 °C for a doubling of CO_2 above pre-industrial levels) and the NCAR PCM model (lower sensitivity of about 1.8 °C for CO_2 doubling) were both run from 1895 to 1995 using observed trends in greenhouse gases, and into the future using the SRES A2 and B1 emission scenarios (IPCC 2007b). Climate outputs from these models were input into the dynamic global vegetation model MC1, which models changes in vegetation and fire and other ecosystem properties (Lenihan et al. 2003a,b, 2008).

We overlaid outer range boundaries for the 2 California fisher populations on maps of projected vegetation-type distributions in Lenihan et al. (2008) for the period 2071–2100, summed the number of pixels of each vegetation type occurring within the current population areas, and compared the outcomes with the current (modeled) condition. We also overlaid the fisher range boundaries on maps of projected changes in fire frequency and intensity presented in Lenihan et al. (2003a). The 2003 effort was similar to the 2008 effort but used different GCMs (Hadley CM2 and an earlier version of NCAR PCM) and did not incorporate the IPCC (Intergovernmental Panel on Climate Change) emissions scenarios (Lenihan et al. 2003a,b).

In is important to note that the Lenihan et al. (2003a,b, 2008) MC1-based modeling efforts were conducted on a 10 × 10 km spatial grid. This is a high resolution for this type of model, but fisher habitat quality depends primarily on vegetation and landscape features occurring at finer spatial scales than 100 km^2. Thus, the MC1-based results we summarize below represent broad, landscape-scale patterns of hypothesized change whose influence on fisher habitat will be filtered by variability in topography, vegetation, and other factors occurring outside the resolution of the MC1 model. In addition, the fire module in MC1 was calibrated by Lenihan et al. (2003a,b, 2008) to restrict the occurrence of simulated fires to extreme events. They explain that "Large and severe fires account for a very large fraction of the annual area burned . . . and these events are also . . . least constrained by heterogeneities in topography and fuels that are poorly represented by relatively coarse-scaled modeling grids" (Lenihan et al. 2003a: 1669). Although fires in the California fisher population areas (especially in the southern Sierra Nevada) may have been dominated by smaller fires of low- and mixed-severity effects historically (and thus poorly represented by the MC1 fire module), fire-exclusion policies result in the immediate suppression of almost 99% of ignitions on national forest lands in California. Increasing forest fuels and warming climates have led to a situation where uncontrolled fires in the mixed-conifer zone are becoming larger and burning at increasingly severe levels (Miller et al. 2009). We conclude that the MC1 fire module is probably a reasonable representation of future fire activity, at least at the broad scale of application.

Lenihan et al. (2008: S226) projected increases in the annual area burned by wildfire throughout most of California by the end of the 21st century. In the Sierra Nevada fisher population area, annual area burned by wildfire was projected to increase on about 10–15% of the landscape and remain stable on 85–90% of the landscape, depending on the GCM-emissions scenario. Lenihan et al. (2008) projected much greater changes in burned area in the Klamath fisher population area, with increases occurring on 20–27% of the landscape, but decreases occurring on 2–5% of the landscape. In both population areas, the GFDL-A2 scenario (much warmer and drier than today) produced the greatest increases in burned area.

Lenihan et al. (2003a,b) included maps of projected changes in mean fire frequency and intensity under the Hadley CM2 and NCAR PCM GCM scenarios (with no specific emissions assumptions). The Hadley CM2 model projected much higher future precipitation in California, now considered unlikely by most climate experts (Cayan et al. 2009). Like most of the current GCMs, the NCAR PCM model used by Lenihan et al. (2003a,b) projected gradual decreases in annual precipitation during the next century, but their temperature projections are among the more optimistic scenarios, with only about 2.4 °C warming. We report the NCAR PCM-based outputs here. For the Klamath Mountains fisher population area, Lenihan et al. (2003a,b) projected more frequent fires in about 50% of the landscape, with most of those areas experiencing decreases of 10–50% in average fire-return intervals (FRIs) (i.e., 10–50% more fires). About 45% of the landscape was projected to have fire frequencies similar to those of today, and 5%, to have decreased fire frequencies. In contrast, in the Sierra Nevada population area, they projected more frequent fires in about 30% of the landscape, with fire frequencies remaining similar to current values on most of the remaining landscape. In both population areas, most areas with greatly increased fire frequencies are associated with areas where conifer forest transitions to hardwood-dominated mixed forest and woodland. Projected transitions from woody vegetation to grassland also drive (and are driven by) higher fire frequencies in the Sierra Nevada (Lenihan et al. 2008).

Lenihan et al. (2003a,b) projected that average fire intensities will remain unchanged or decrease in 90% of the Klamath Mountains fisher population area, primarily because of widespread transitions from conifer to hardwood-dominated forests and woodlands. In contrast, in the southern Sierra Nevada, they predicted increases in fire intensity in about 35% of the fisher population area and decreases on 15% of the landscape; fire intensities were projected to remain similar to current values on about 50% of the landscape. Lenihan et al.'s (2003a,b) projected changes in fire intensity are due primarily to changes in fuels growth resulting from climatic factors, changes in fire frequencies, and vegetation-type shifts to more or less flammable fuels. For example, transitions from woody vegetation to grassland are accompanied by increased fire frequencies but decreased fire intensities, and a similar pattern occurs where conifer forest transitions to hardwood-dominated vegetation. Most locations projected to remain in the same vegetation type will experience only moderate or no changes in fire regime.

Interactions between climate and fire generate significant changes in projected vegetation cover in both California fisher population areas (Lenihan et al. 2008; Figure 16.6). In the Klamath Mountains, all the GCM-emissions scenarios projected losses of >50% in conifer-forest cover by 2100, with increases of about 100% in mixed (hardwood-dominated) forests, and 100–400% in mixed woodlands. Shrubland and grassland cover were projected to change only minimally. In the Sierra Nevada, the different GCMs projected 25–75% losses in conifer forest (Figure 16.6). The response of mixed forest depends strongly on the GCM, but mixed woodland was projected to increase in area by at least 40%; overall, the area of hardwood-dominated veg-

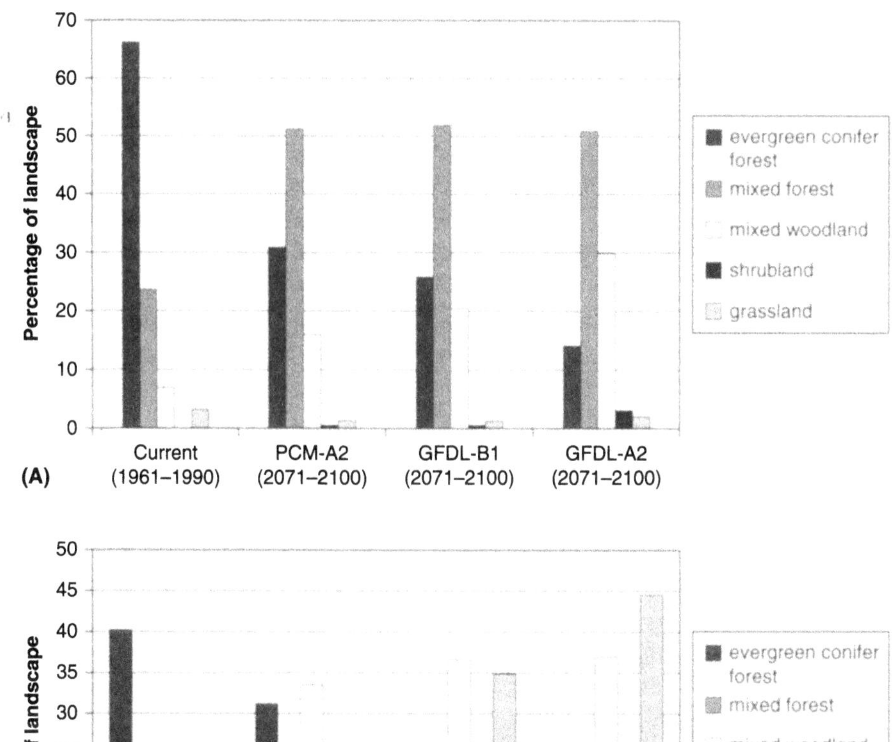

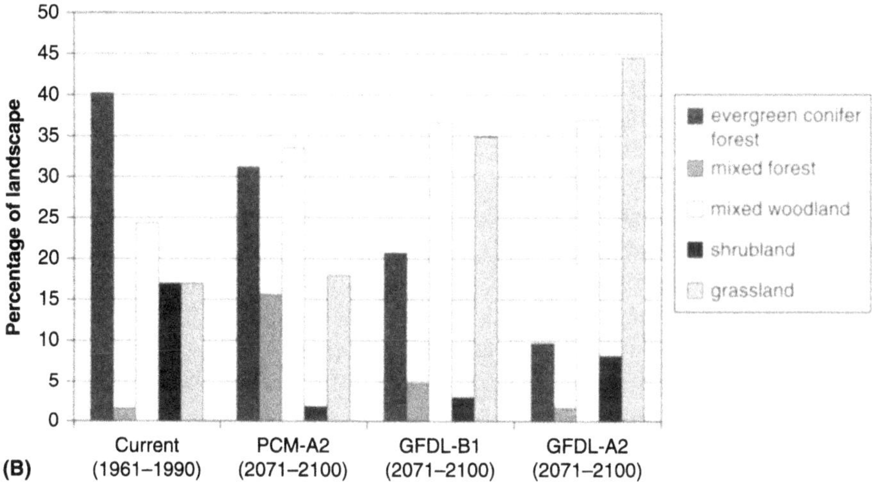

Figure 16.6. Current (mean during 1961–1990) and predicted future (mean during 2071–2100) projected vegetation types for (A) Klamath Mountains and (B) southern Sierra Nevada fisher populations under 3 future climate-emissions scenarios. Data represent the percent of landscape within each vegetation type. Evergreen conifer forest includes montane and subalpine forests; mixed types are dominated by hardwood species. The PCM-A2 scenario is moderately warmer than today but with similar precipitation; GFDL-B1 is moderately warmer than today but with less precipitation; GFDL-A2 is much warmer than today and much drier. See Lenihan et al. (2008) for more information on these climate-emissions scenarios. Data from Lenihan et al. (2008).

etation was projected to increase by 45–75%. Grassland increases substantially, especially in the drier GFDL scenarios (Figure 16.6). In the most extreme (GFDL-A2) scenario for the Sierra Nevada population area, grassland and shrubland cover >50% of the landscape by 2100, and conifer-dominated forest <10%. Note that, except for the last scenario described (GFDL-A2 for the southern Sierra Nevada), all the simulated vegetation outcomes for both fisher population areas predicted that the overall extent of forested landscape will not change appreciably during the 21st century. The biggest projected change is a transition to hardwood-dominated forest types from conifer-dominated types (Figure 16.6). In areas where mixed woodland is predicted to rise substantially in importance, we can expect a decrease in canopy cover within these forested stands.

The Stand Scale: Simulations of Fire Effects on Forest Structure for Fishers in California

To explore the potential impacts of projected changes in fire regimes at the landscape scale on fisher habitat at the scale of the forest stand, we simulated the effects of fire regimes at different frequencies and severities on forest structure in the southern Sierra Nevada population area. We used the Fire and Fuels Extension to the Forest Vegetation Simulator (FVS-FFE) (Reinhardt and Crookston 2003) to model fire effects on the structure of late-seral conifer forest in the southern Sierra Nevada. To parameterize the model, we used Forest Inventory and Analysis (FIA) plots sampled on the Sierra and Sequoia national forests. We used the fisher resting-habitat-quality index developed by Zielinski et al. (2006c) to rank FIA plots on both national forests, then chose the 40 plots with the best fisher habitat. Most plots were in the mixed-conifer type, and many had a significant component of oak and other hardwoods.

We used FVS simulations to grow trees in each plot for 90 years, beginning in 2010 and ending in 2100, and ran fire simulations for 3 different fire-weather scenarios between June 15 and September 15 (high = 97th percentile weather; moderate = 80th percentile weather; low = 10th percentile weather [essentially corresponding to a prescribed-fire scenario]) at FRIs of 10, 20, 30, 50, and 100 years. We assessed variables at 10 decadal time steps. We derived fire-weather scenarios from the Shaver Lake "RAWS" climate station (elevation 1710 m), which is in the middle of the elevational and latitudinal range of the fisher in the Sierra Nevada, using the fire-weather ranking system in the Fire Family Plus program (Main et al. 1990). As a control, we also simulated the same stands without fire. We tracked the following variables, which are correlated with occupied fisher habitat in California and elsewhere (Powell 1993; Zielinski et al. 2004a,b, 2006c): tree canopy cover; number of large trees (dbh ≥ 76 cm) per acre, number of hardwood trees per acre, number of large snags (dbh ≥ 76 cm) per acre, and number of down logs ≥30 cm diameter

Table 16.1. Mean values for forest-stand variables, averaged across the 10 decadal time steps of the simulation

	Fire scenarios						
Variable	Control	20yr_mod	20yr_low	50yr_high	50yr_mod	100yr_high	100yr_mod
Large trees/ha	33.0c	11.5a	26.3bc	8.7a	20.7b	12.5a	25.2b
Canopy cover (%)	85.9b	69.7a	78.1ab	92.6bc	77.6ab	96.8c	88.4ab
Large snags/ha	8.6a	16.2c	12.2abc	15.6bc	13.7abc	13.0abc	10.5ab
Coarse woody debris (large down logs/ha)	64.0	63.6	64.5	79.4	75.3	85.6	70.8

Notes: Each column represents a different fire scenario, including return interval and level of severity. Values with different letters are significantly different among fire scenarios at $P \leq 0.05$ (1-way ANOVA, Tukey HSD).

per acre. We report results from the following 7 fire-regime scenarios: control (no fire); 20-year moderate and low severities; 50-year high and moderate severities; and 100-year high and moderate severities. We report these scenarios for the following reasons: (1) because moister, mixed-conifer stands in the pre-European era in the Sierra Nevada are thought to have burned at mean FRIs of 10–20 years, primarily at low and moderate severities (Sugihara et al. 2006; Stephens et al. 2007); (2) because of historical fire-suppression policies, most areas in the Sierra Nevada have not burned in the last century, hence the 100-year FRI; and (3) the 50-year FRI represents a 50% increase in fire, which is the upper bound of likely increases in fire frequency for most places in the California fisher population areas projected to experience more fire by the end of the century (Lenihan et al. 2003a,b, 2008).

We began the FVS-FFE simulations with a mean of 26 large trees per hectare, but different fire-regime scenarios produced highly variable responses by the end of the century (Figure 16.7a). At the end of the 90-year period, the mean number of large trees per hectare was highest in the control plots, followed by the 100-year moderate-severity scenario and the 20-year low-severity scenario. Taking the average number of large trees in the simulated plots over the entire time period (i.e., the mean of the 10 decadal time steps), the same 3 scenarios supported the most large trees, whereas the 50-year high, 100-year high, and 20-year moderate scenarios supported the fewest large trees per hectare (Table 16.1). For all hardwoods ≥6 cm dbh, final (i.e., in 2100) and overall mean hardwood tree densities were not significantly different among scenarios, but the mean density of large hardwoods (≥20 cm dbh) was about twice as high in the control and 20-year low scenarios as in any of the other scenarios ($F = 13.37$, $P < 0.0001$). Tree canopy cover from all size classes (including seedlings and saplings) was highest in the 100-year and 50-year high-severity scenarios, both at the end of the scenario and as an average of the entire simulation period. The lowest canopy cover was in the two 20-year fire scenarios (Table 16.1).

By the end of the 21st century, the number of large snags per hectare was highest in the two 20-year scenarios and in the 50-year moderate-severity scenario. The lowest number of large snags was in the 100-year high-severity scenario (Figure 16.7b). When averaged across the simulation period, the highest large-snag densities were in the 20-year moderate-severity simulation, although they were statistically equivalent to the 20-year low, both the 50-year, and the 100-year high scenarios (Table 16.1). The lowest large-snag density occurred in the control plots. When averaged across the simulation period, coarse woody debris was not statistically different between fire regimes (Table 16.1), but at the end of the simulation, the highest number of down logs was in the 50-year moderate-severity scenario, and the lowest number was in the 20-year moderate-severity scenario.

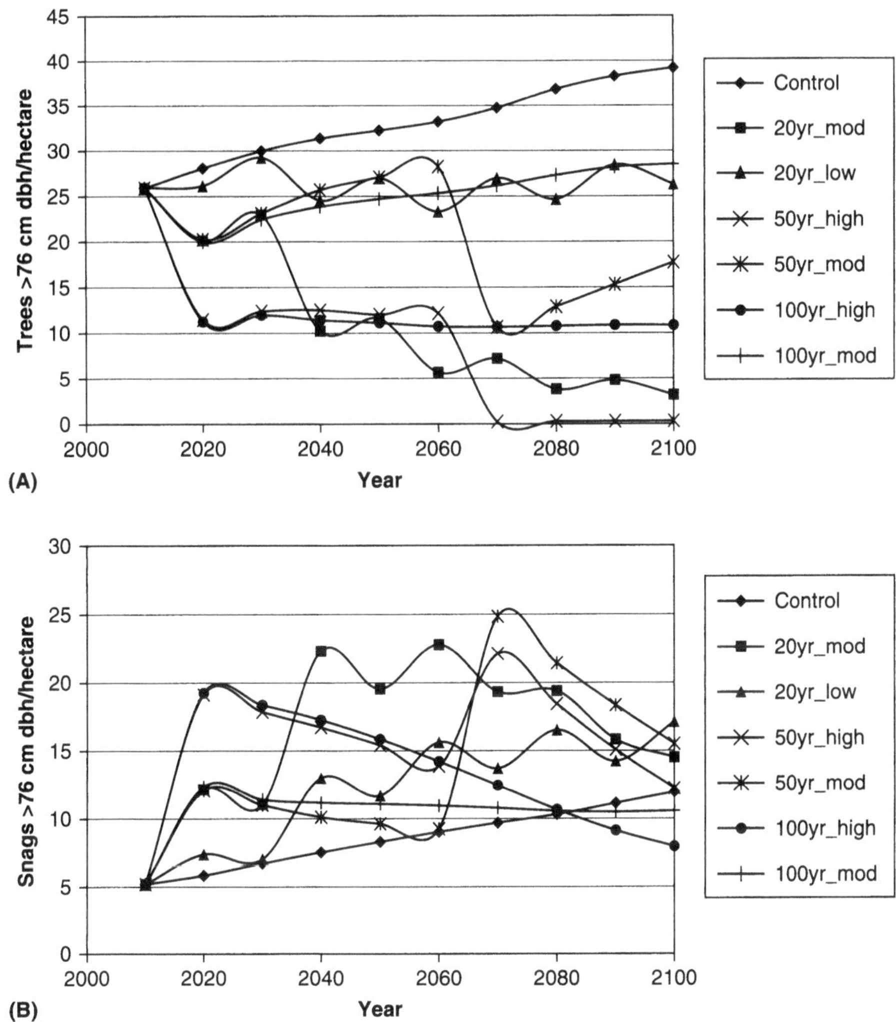

Figure 16.7. Outputs from FVS-FFE fire and forest-growth simulations for 2 key fisher habitat variables in the Sierra Nevada of California: (A) number of large trees (≥76 cm dbh) per hectare, and (B) number of large snags per hectare. See text for details.

We also compared forest structure in the low fire-severity scenarios versus the high-severity scenarios for each FRI class (Figure 16.8). The densities of large trees and hardwood trees decreased by 50–75% as fire severity increased, but the density of large snags increased as fire severity increased. For the 10- and 20-year scenarios, canopy cover and coarse woody debris both decreased as fire severity increased, but both variables increased with severity for the 50- and 100-year scenarios, although canopy cover increases were due

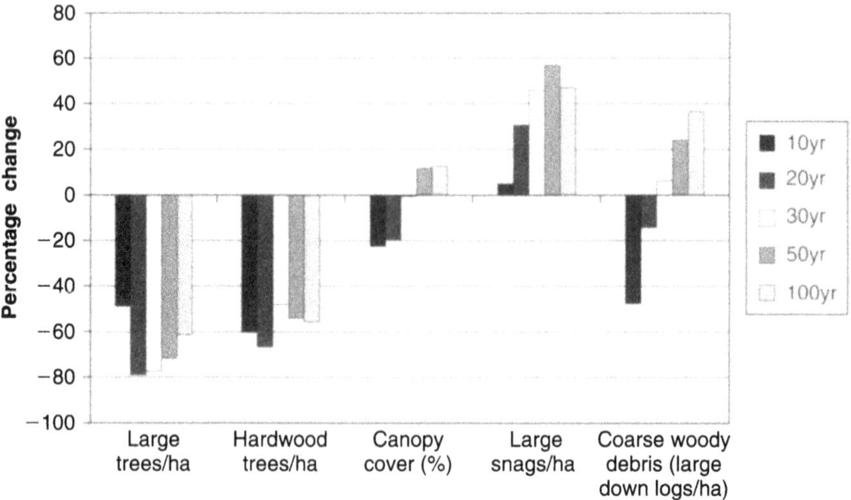

Figure 16.8. The effects of changing fire severity from low to high on key fisher habitat variables in the Sierra Nevada of California, for 5 fire-frequency classes, measured as percent change. Data from FVS-FFE fire and forest-growth simulations. See text for details.

entirely to small- and medium-sized trees; large trees were almost entirely absent from the high-severity stands. Because 50- and 100-year FRIs rarely if ever burn at low fire-weather conditions, we also compared moderate- and high-severity scenarios for both FRIs. Patterns were similar to those in Figure 16.8, except for coarse woody debris, which reversed direction and decreased by 16.6% from the moderate- to high-severity scenarios.

Management Implications for Fishers in California

The Lenihan et al. (2003a,b, 2008) models suggested that both fire frequency and annual burned area are likely to increase in substantial portions of both fisher population areas during the next century. This generally corroborates findings from other modeling efforts (Miller and Urban 1999; Flannigan et al. 2000; Hayhoe et al. 2004) and extends increasing trends in fire frequency and size for both the Klamath Mountains and Sierra Nevada (Miller et al. 2009; Miller et al. 2012). If these trends continue, fisher habitat will be affected. Our stand-scale modeling effort sought to predict what may happen to key fisher habitat attributes under different future fire scenarios. These results showed that fires occurring only rarely (every 50–100 years) under extreme conditions (i.e., a continuation of current and projected trends) generate high densities of snags and coarse woody debris, but substantially reduce the number of large conifer and hardwood trees, and maintain canopy-cover values only through the recruitment of dense stands of smaller trees (Figures 16.7–16.8; Table 16.1). In general, we found that shorter FRIs under low and moderate fire-weather conditions were best able to balance the resilience of forests with the creation of legacy forest structures such as snags and down logs.

There is substantial controversy surrounding current and future management of fisher habitat in California, especially in the critically small Sierra Nevada population area (Spencer et al. 2008). There is general agreement that fire should be reintroduced as a key ecosystem process in the Sierra Nevada, but disagreement as to whether pre-fire fuels treatment is required to protect old-forest habitats. We agree with Spencer et al. (2008) that current and predicted future trends in fire activity warrant fuel-reduction treatments in high fire-risk areas to reduce the probability of losing fisher habitat to fire. At the same time, we also believe that a comprehensive review of fire-management policies should be undertaken throughout the Sierra Nevada. The strong increase in fire size and severity seen in Sierra Nevada mixed-conifer forests during the last quarter century is tied to increasing forest fuels and warming climates (Miller et al. 2009), but it is also linked to current fire-management practices, which extinguish nearly all fires burning in moderate weather conditions, except those on national park lands. These are precisely the fires that are most likely to both preserve and create habitat for old-forest associates like the fisher. More than 30 years of let-burn policies in Yosemite and Sequoia-Kings Canyon national parks have returned many areas in these parks to nearly pre-suppression era conditions (Collins and Stephens 2007). Even moderate relaxation of fire-suppression policies and practices outside the parks (e.g., to allow for some naturally ignited fires under moderate weather conditions, especially after forest fuels were reduced in high-risk areas) would be a significant step toward sustaining naturally functioning forests and the wildlife they support.

Conclusions

Our bioclimatic models indicate that climate change will likely result in relatively large changes in the distribution of both North American and European *Martes* species. In North America, multiple climate-change scenarios consistently projected northward range shifts for both the American marten and fisher. In Europe, greater variability in climate projections resulted in much less agreement about the nature of range shifts, but suggested the potential for large range contractions for both species, and relatively small range expansions for the pine marten.

These projected changes have potential implications for the management of several populations in each of the 4 species. There are 7 states and provinces in which American martens are protected: California, Nevada,

Newfoundland, New Hampshire, Nova Scotia, South Dakota and Wisconsin (Proulx et al. 2004: 52–54). Populations in South Dakota, Wisconsin, New Hampshire, and Nova Scotia are all likely to experience shifts in climate that will make those areas less suitable for martens as a result of increased temperatures, loss of snowpack, or both. Although the 2 most endangered populations (the subspecies *M. a. atrata* in Newfoundland, and *M. a. humboldtensis* in north-coastal California) are not projected to experience a complete loss of suitable climates, they will likely experience warmer temperatures. Warmer temperatures and decreases in snowpack may have detrimental effects on the Newfoundland subspecies; however, because the Humboldt marten does not occur in areas where snowpacks accumulate, it may experience fewer negative effects of climate change. Projected potential range contractions in the southeastern portion of the fisher's range in California raise the question of how much effort should be put into reintroductions in that region or into maintaining reintroduced populations in West Virginia, Pennsylvania, or New York, should they begin to decline. We are not suggesting that additional reintroductions are unwarranted; however, range-shift projections can be useful when deciding among areas for reintroductions or between putting effort into maintaining a population in an area projected to remain climatically suitable vs. reintroducing fishers into an area projected to become climatically unsuitable.

Given that models for the 2 European martens did not include distribution data for the eastern portions of the ranges, we are more cautious about the management implications of projected range shifts. Nonetheless, these models can be used to anticipate potential future population declines, particularly in populations already at risk. For example, protected populations of the stone marten in Albania, Bulgaria, Hungary, Portugal, and Spain are likely to experience increased risks of extirpation due to increasing temperatures.

One of the largest shortcomings of bioclimatic models such as those reported here is their correlative nature. These models do not address many of the mechanisms that affect species distributions directly. Mechanistic models applied within the fisher's current range in California, however, provide some support for range contractions projected by the bioclimatic models. The increase in mixed-evergreen forest projected for both the Klamath and southern Sierra Nevada fisher population areas suggests that floristic conditions for fisher survival may be enhanced by climate warming, as long as annual precipitation remains near or above current levels, and as long as direct temperature (or other) effects on metabolic rates for the fisher and its prey do not result in significant range shifts upward in elevation. The drier climate-change scenario resulted in the conversion of large expanses of forested area to grasslands and shrublands. Furthermore, although our simulations indicate that projected increases in fire frequency, area, and intensity will likely lead to increases in large snags and down logs, such increased fire activity will likely also lead to decreases in the density of large trees and the canopy cover they provide. Given that overstory cover and the density of large trees are among the factors most closely correlated with fisher occurrence in California (Powell 1993; Zielinski et al. 2004a,b), the net result of these simulated changes in forest structure may ultimately be a decrease in fisher habitat.

Other studies have come to similar conclusions about the potential effects of future climate change on the distribution and habitat of at least 2 of the 4 species we discuss here. Based on historical trends, Krohn (this volume), concluded that increased warming will likely lead to further contractions at the southern edge and expansions at the northern edge of the ranges of both the American marten and the fisher in the eastern United States. Additionally, he concluded that the fisher will likely continue to be distributed further south than the marten, a pattern we also found in our model projections. Carroll (2007) projected a 40% decline in American marten populations in the far eastern portion of their range as a result of reduced snowpack resulting from projected changes in climate. Burns et al. (2003a) used regression models of vegetation associations of animal species and climate-change effects on vegetation in a subset of U.S. national parks to predict mammalian species turnover by the end of the 21st century. They found that the fisher was one of the most climate-sensitive carnivores and predicted its loss from both national parks in California where it occurs today. Miller et al. (2009) found that trends in both fire severity and size since the early 1980s in the Sierra Nevada were strongly positive, particularly in those forest types that support old-forest associated species like the fisher. Miller and Urban (1999) conducted a long-term (500–800 years) simulation for the Sierra Nevada and found that fires were projected to be more frequent under future climate, and that forests at lower elevations would be replaced by grasslands and more-open woodlands. Flannigan et al. (2000) projected that seasonal fire-severity ratings would increase by 10–30% by the middle of the 21st century in most of the areas occupied by fishers. These results are similar to the climate-, fire- and vegetation-change projections reported by Lenihan et al. (2003a,b, 2008).

Although our models capture some of the potential effects of climate change on American martens and fishers, there are several other ways in which climate change will likely influence these species. For example, temperature extremes (which were not captured in any of our modeling efforts) may be more important than mean values in driving species distributions. Hayhoe et al. (2004) found that the probability of extreme heat events in California will undergo a marked increase as temperatures rise. They estimated that there will be 22 more days with maximum temperatures >32 °C by the year 2100. Although extreme high temperatures may have direct effects on fishers, they may also indirectly affect fishers through their effects on prey species. Likewise, changes in fire frequency and forest structure will also influence prey availability.

Climate change is also likely to result in a number of other complex and largely unpredictable effects on *Martes* species. For example, the complex relationships between rising temperatures and drought, water stress, insect and disease occurrence, and fire appear to be already driving rapid ecosystem response to climate change in western forests (Dale et al. 2001; Breshears et al. 2005). Such feedback effects among multiple disturbances are thought to represent causal factors for recent large-scale geographic shifts in the distributions of ponderosa (*Pinus ponderosa*) and piñon (*P. edulis*) pines in the southwestern United States (Allen and Breshears 1998). In many cases, insect- or disease-caused mortality among conifers may benefit fishers by creating resting, denning, and foraging habitat. Large inputs of standing and down dead wood, such as occurs during pine beetle (*Dendroctonus* spp.) outbreaks, however, may also increase the probability of stand-replacing fires and the resulting loss of critical habitat structures (Shaffer and Laudenslayer 2006; U.S. Fish and Wildlife Service 2006).

There are, of course, several other factors that neither our broad- nor finer-scale bioclimatic models take into account. For example, human land use, habitat alteration, and both predator-prey and competitive relationships (among other things) will likely interact with climate change in complex ways. An example of an unforeseen outcome is the recent discovery that pine martens are rapidly expanding into agricultural areas in southern Europe and displacing stone martens, which are competitively subordinate and usually occupy lower-quality habitat (e.g., urban and agricultural areas) in areas where the 2 species are sympatric (Balestrieri et al. 2010). Another example of a more nuanced effect of climate change is the potential for reductions in snowpack to alter competitive relations between martens and fishers. Snow has the potential to mitigate competitive interactions between the species, because martens have a higher foot-loading than fishers (increasing their energetic efficiency) and, unlike fishers, are capable of hunting subniveally (Krohn et al. 1995, 1997, 2004).

Not surprisingly, our results suggest that populations of *Martes* occurring at more southerly latitudes and at lower elevations will be most affected as temperatures continue to rise As our finer-scale modeling for the fisher revealed, future climate-change impacts are likely to be more complex. As discussed above, both the broad- and finer-scale models presented here can be used to guide management actions at multiple spatial scales. Nonetheless, all models are imperfect representations of reality. Given that future trends can never be projected perfectly, we recommend that those charged with managing *Martes* populations and their habitats implement a variety of management practices that can be monitored and assessed in an experimental fashion, and apply lessons learned in an adaptive framework. Climate change will undoubtedly challenge our ability to manage and sustain populations of many species. It is our hope that analyses such as those presented here, coupled with adaptive-management approaches and flexible planning frameworks, will help managers and planners meet that challenge.

Acknowledgments

We thank Denis White, Peter Kareiva, Ron Neilson, Bill Zielinski, Cheryl Carrothers, Diane MacFarlane, Dave Schmidt, and Andy Taylor for useful discussions and their contributions to the various modeling efforts. We also thank Douglas Kelt and an anonymous reviewer for helpful comments on the manuscript. Portions of this paper were originally written for the interagency Southern Sierra Nevada Conservation Assessment for the California Fisher. We acknowledge the Program for Climate Model Diagnosis and Intercomparison and the WCRP's Working Group on Coupled Modelling for their roles in making available the WCRP CMIP3 multi-model dataset. Support of this dataset is provided by the U.S. Department of Energy.

17

Conservation Genetics of the Genus *Martes*

Assessing Within-Species Movements, Units to Conserve, and Connectivity across Ecological and Evolutionary Time

MICHAEL K. SCHWARTZ, ARITZ RUIZ-GONZÁLEZ, RYUICHI MASUDA, AND CINO PERTOLDI

ABSTRACT

Understanding the physical and temporal factors that structure *Martes* populations is essential to the conservation and management of the 8 recognized *Martes* species. Recently, advances in 3 distinct subdisciplines in molecular ecology have provided insights into historical and contemporary environmental factors that have created population substructure and influenced movement patterns of several *Martes* species. Intraspecific phylogenetics has allowed us to understand the role of large-scale historical events, such as the last glacial maxima and their associated refugia, in the ecology of at least 5 *Martes* species in North America, Europe, and Asia (*M. americana*, the American marten; *M. martes*, the European pine marten; *M. melampus*, the Japanese marten; *M. pennanti*, the fisher; *M. zibellina*, the sable). In addition, population genetics has examined how *Martes* populations are connected within species across space and, in some cases, how this level of connectivity has changed over recent time. These studies have been conducted on *M. americana*, *M. martes*, and *M. pennanti*. More recently, several landscape genetic analyses, including graph-theoretic and least-cost-path approaches, have been used to evaluate the correlation between landscape features and genetic relatedness among individuals across a landscape. These new approaches are showing promising results for understanding the ways in which multiple habitat features at multiple scales promote or reduce connectivity. Different forms of this landscape-genetics approach have been applied to *M. americana*, *M. martes*, and *M. pennanti* in portions of their ranges. In this chapter, we review the intraspecific phylogenetic, population genetic, and landscape genetic studies conducted on *Martes* populations; discuss commonalities found among species; and identify knowledge gaps for understanding movements and substructuring in the genus *Martes*.

Introduction

Understanding the biotic and abiotic forces that influence the movements of animals has been a central focus of wildlife management for nearly a century. This topic has come to the forefront of wildlife biology in recent years, as the perils of habitat fragmentation and climate change are becoming clearer and more pronounced. Habitat areas that were once extensive and connected are now becoming small, degraded, or completely isolated. In fact, we can consider these habitat changes on a gradient: from those that completely eliminate the movement potentials of animals, to those that marginally limit the probability of a successful dispersal of an individual, to those that have no effect. For example, urbanization of a once-forested area may act as a complete barrier to movements and isolate populations, whereas various forest-thinning treatments may remove the cover necessary for *Martes* species to disperse, thus exposing them to predation risks that they did not historically face in unmanaged forests.

Some habitat changes are tied to natural cycles occurring on a temporal scale of centuries or millennia, while others are functions of short-term natural or human-induced landscape changes. Understanding whether the contemporary distributions and substructure patterns of animal populations result from long-term influences, such as glaciations, or have been caused by more recent landscape uses is critical for managing and conserving wildlife. Only by disentangling historical from contemporary factors can we determine whether barriers to movements are part of the natural history of the species or are caused by recent human activities. From these understandings, we can also determine whether management actions, such as corridor protection or habitat improvement, will increase animal movement and gene flow within a species' range.

In this chapter, we first provide a basic primer of the principles, methods, and tools of molecular ecology. Then, we discuss recent findings on movements and substructure for each species based on intraspecific phylogenetic studies, which have been conducted on only 5 of the 8 currently recognized *Martes* species: *M. americana*, the American marten; *M. martes*, the European pine marten; *M. melampus*, the Japanese marten; *M. pennanti*, the fisher; and *M. zibellina*, the sable. In general, such information is obtained by extracting information from mitochondrial DNA sequences, which mutate at a much slower rate than many of the nuclear DNA regions commonly used (e.g., microsatellites). Thus, changes in the sequence occur at a slower rate and differences among sequences reflect more-ancestral splits (Avise et al. 1987).

Table 17.1. List of intraspecific phylogenetic, population genetic, and landscape genetic studies conducted on *Martes* species

Species	Intraspecific phylogenetic	Population genetic	Landscape genetic
M. americana	Carr and Hicks 1997; Stone and Cook 2002; Stone et al. 2002; Dawson 2008; Slauson et al. 2009	McGowan et al. 1999; Kyle et al. 2000; Kyle and Strobeck 2003; Small et al. 2003; Broquet et al. 2006a,b; Swanson et al. 2006; Swanson and Kyle 2007; Williams and Scribner 2007	Broquet et al. 2006b; Wasserman et al. 2010
M. flavigula			
M. foina			
M. gwatkinsii			
M. martes	Davison et al. 2001; Pertoldi et al. 2008b; Ruiz-González 2011	Kyle et al. 2003; Mergey 2007; Pertoldi et al. 2008a; Ruiz-González 2011	Mergey 2007; Ruiz-González 2011
M. melampus	Hosoda et al. 1999; Kurose et al. 1999; Murakami et al. 2004; Sato et al. 2009b; Inoue et al. 2010		
M. pennanti	Williams et al. 2000; Drew et al. 2003; Vinkey et al. 2006; Schwartz 2007; Knaus et al. 2011	Williams et al. 2000; Kyle et al. 2001; Wisely et al. 2004; Carr et al. 2007a,b; Hapeman et al. 2011	Carr et al. 2007b; Garroway et al. 2008
M. zibellina	Hosoda et al. 1999; Kurose et al. 1999; Murakami et al. 2004; Inoue et al. 2010; Malyarchuk et al. 2010; Rozhnov et al. 2010		

Notes: To date, there have been no genetic studies conducted on *M. flavigula*, *M. foina*, or *M. gwatkinsi*, and no detailed population or landscape genetic studies on *M. melampus* or *M. zibellina*. The table is shaded to indicate areas where there has been much research (white), some research conducted in limited geographic areas (light gray), and no research conducted (dark gray). Many of these studies are reviewed in the text.

Intraspecific phylogenetic information is used to compare among populations or different parts of the geographic range of a single species, often to make phylogeographic inferences regarding the importance of geographic features for structuring populations in evolutionary time scales. This is in contrast to *interspecific* phylogenetic information, which evaluates the relations among species. We do not include interspecific phylogenetic information that has helped reveal the evolutionary history of the genus *Martes*, as this subject is well covered by Koepfli et al. (2008) and Hughes (this volume). Next, under each species subheading, we follow our intraspecific synthesis by reviewing published population and landscape-genetic studies for the 3 species that have been studied in this regard: *M. americana*, *M. martes*, and *M. pennanti*. To date, no genetic studies have been conducted on *M. flavigula*, the yellow-throated marten; *M. foina*, the stone marten; or *M. gwatkinsii*, the Nilgiri marten (Table 17.1). Population genetic methods use highly variable nuclear DNA markers for evaluating contemporary patterns of gene flow and genetic variability. An extension of the population genetics analysis is the landscape genetics approach (Manel et al. 2003), which combines landscape ecology methods and population genetics data to examine, at a finer resolution, how specific landscape features structure extant populations. Finally, we conclude the chapter with a synthesis of the lessons molecular data has taught us about the genus *Martes*.

The Role of Conservation Genetics

Genetic Diversity and Inbreeding

During the last 2 decades, the role of genetics in conservation biology, and in ecology in general, has been steadily growing (for reviews, see Frankham 1995, 2005; Allendorf and Luikart 2007; Pertoldi et al. 2007; Ouborg et al. 2010a), partly because changes in genetic diversity can help quantify the status of endangerment of a given population, species, or group of species. The assessment and monitoring of genetic diversity in endangered animals is now pervasive (Schwartz et al. 2007), because powerful DNA-analysis methods have become increasingly available to infer the causes of the spatio-temporal dynamics of populations, and to estimate genetic diversity within populations and the organization of genetic diversity among populations (Allendorf and Luikart 2007).

Genetic diversity (also called genetic variability) can be quantified in several ways; we will focus on genetic diversity quantified by molecular genetic methods. These are expressed as the proportion of polymorphic loci (or proportion of polymorphic sequence sites), the proportion of heterozygous loci, and the number and frequency of alleles at these loci. An implicit assumption often made as a first principle of conservation genetics is that of the causal relation between genetic variability and both the short- and long-term persistence of a population or species (Ouborg et al. 2010a,b). One such measure of genetic variability, called expected heterozygosity (H_E), can provide an indication of the immediate evolutionary potential of the population, although this measure has no deterministic relation to its future value.

Given an initial pool of unrelated founder genes, the potential change and loss of genetic diversity can be assessed by the increase in relatedness. The initial genetic variability of a population is reduced as relatedness among individuals increases. The decline is proportional to the reduction in heterozygosity or increase of average inbreeding coefficient *(F)* of the parents, caused by inbreeding and genetic drift (change in gene frequencies in a population between generations due to random sampling). The degree of inbreeding within an isolated population is quantified by F_{IS} (the proportion of the variance in the subpopulation [S] contained in an individual [I]), which varies from –1 to 1, with 0 representing a non-inbred population with random mating (Holsinger and Weir 2009). F_{IS} is greater than 0 when observed heterozygosity (H_O, the level of heterozygosity measured in a sample) is smaller than the level of heterozygosity that would be observed with the population in genetic equilibrium. This occurs when populations are inbred. Inbreeding depression, the reduction in a fitness trait due to inbreeding, is one of the principal concerns of conservation geneticists.

On the contrary, increased genetic divergence between the parents may be an advantage to the offspring by increasing their heterozygosity, which could cause a *heterosis* or *hybrid-vigor* effect, but only until the genetic distance between the parents reaches a limit. Beyond this limit, the divergence and differences in co-adaptation between the parents may reduce fitness in the offspring because of outbreeding depression (Tallmon et al. 2004). Therefore, genetic divergence between the parents of an individual lies on a continuum, with varying fitness consequences for the offspring.

All metrics described above are estimated using genetic markers (also called loci) that are unlikely to be under selection pressure (neutral markers). Another form of genetic variation is the quantitative variation that results from natural selection, which can produce, for example, a continuous distribution of a phenotypic trait (e.g., size, weight). Quantitative genetics is correlated to the population's adaptive potential, or its capacity to adapt via changes in allele frequencies in the gene pool. The correlation between neutral and quantitative variation is weak and, therefore, divergence among populations at neutral loci is potentially uninformative (Lynch 1996). Given that we are unaware of any quantitative genetic studies on wild *Martes* populations, this topic will not be covered extensively in this review, although we expect that new molecular-genetic approaches will make the study of genes under selection an important pursuit in the near future (Ellegren 2008; Ouborg et al. 2010a,b).

Connectivity and Genetic Diversity

Generally, small fragmented populations are genetically depauperate. This loss of genetic variability has 2 potential consequences: (1) low genetic variability can reduce their adaptive potential to changing environmental conditions, including disturbed habitats, and (2) small, fragmented, and isolated populations can suffer from inbreeding depression due to increasing relatedness among individuals. One of the most common practical conservation strategies to offset concerns about inbreeding depression is to increase the level of connectivity and, thus, gene flow among populations; however, high levels of gene flow can also reduce or impede the capacity for adaptation to a stressor (Lenormand 2002; Tallmon et al. 2004; Postma and van Noordwijk 2005). The actual degree of adaptation is a dynamic interaction between the selective pressures acting on the population and gene flow.

Gene flow can be estimated using methods based on genetic differentiation among populations, quantified with the use of neutral (non-adaptive) regions of the genome. One common metric is F_{ST}, which is a measure of genetic divergence among subpopulations (S) compared to the total population (T) (Allendorf and Luikart 2007; Holsinger and Weir 2009). F_{ST} is a higher hierarchical level than F_{IS}, mentioned earlier, which compares the individual to the subpopulation. F_{ST} ranges from 0 (no genetic differentiation among populations) to 1 (complete genetic differentiation or no gene flow). Alternatively, there are suites of metrics that assign genotypes to populations based on the frequency of alleles in each population (Manel et al. 2005). These metrics have been used in several *Martes* species to quantify movement rates and to identify immigrants (e.g., Kyle et al. 2001; Carr et al. 2007b).

Population genetic data have been used to delineate substructure, identify isolated populations, and define units of conservation. By definition, these data rely on group or population statistics. This poses a challenge when species appear to be distributed continuously across a landscape, and groups are not readily apparent. Although some elements of population genetics, such as measures of between-population genetic distance, are inherently spatial, they do not specifically take the landscape into account. The field of landscape genetics is an extension of population genetics that uses either individual or population genetic data, explicit spatial information, and associated covariates (e.g., elevation, forest type, distance to roads) to identify environmental variables that influence the species' movement patterns.

Landscape genetic approaches are relatively new, but since the term was coined in 2003 (Manel et al. 2003), >500 papers have been published that reference or use these methods. The most common landscape genetics approach is to compare ecological distances among individuals or populations to a matrix of genetic distances (or the inverse, genetic relatedness). These ecological distances are measured along streams, and through forest cover, riparian zones, nonhuman habitations, savanna, steppe, or any other environmental variable deemed important to the organism's life history, survival, and ability to disperse. This approach becomes more complicated when the landscape is a mosaic of habitat patches, and there is not a continuous path within the ecological covariate of choice, forcing populations or individuals to move through nonoptimal habitats to interact. Here the standard landscape genetics approach has been to impose cost values on habitats of different quality and type, and conduct least-cost-path modeling to derive a matrix of least-cost paths among individuals (or populations). Given that a specific cost per habitat type is rarely known, multiple models with different cost penalties are often created. These multiple models are then evaluated by comparing the many matrices of least-cost paths to the matrices of genetic distances (using Mantel tests as described above). In more complex models, these resistance values can be an aggregation of costs imposed by multiple variables or can be evaluated using multiple matrix regression models, where each covariate's influence on genetic relatedness can be evaluated (Balkenhol et al. 2009).

In addition to least-cost-path modeling, there have been several graph-theoretic approaches developed for landscape genetic analyses. These approaches allow identification and prioritization of important locations and populations for maintaining connectivity. The most widely used graph-theoretic approach is one based in electrical circuit theory and incorporated into the program CIRCUITSCAPE (McRae and Beier 2007; McRae and Shah 2009; Schwartz et al. 2009). This model simultaneously considers all possible paths connecting individuals or populations based on resistance distances. This approach is similar to least-cost-path modeling, but it can provide different results because it simultaneously evaluates contributions for multiple dispersal pathways, which can identify areas where connectivity is most tenuous (i.e., "pinch-points"; McRae and Shah 2009).

Genetic Monitoring

Recently, many research efforts have used molecular markers to monitor wild populations of fish, wildlife, and plants (Boulanger et al. 2004; Schwartz et al. 2007; Grivert et al. 2008; Jacob et al. 2010; Palstra and Ruzzante 2010). These methods use diagnostic molecular markers either to monitor changes in estimated parameters such as abundance, using traditional wildlife biology tools, or to monitor changes in population genetic metrics (Schwartz et al. 2007; McComb et al. 2010). These approaches may be particularly useful for the study of *Martes* species, given the animals' secretive nature. Deciding on the best strategy for monitoring will depend largely on the number of markers available for the species of interest and the power associated with various metrics. To detect trends in population numbers, Tallmon et al. (2010) have shown that monitoring population genetic metrics may be as powerful as, or potentially more powerful than, monitoring changes in abundance.

One common population genetic metric to monitor is effective population size, or N_E, formally defined as the size of an ideal population with the same rate of change of allele frequencies or heterozygosity as the observed population. N_E has been considered the most important and critical surrogate parameter to describe the status of small populations. In populations with an N_E smaller than a few hundred individuals, natural selection is not very effective and is easily overpowered by genetic drift; hence, a small N_E reduces the population's potential to adapt to environmental changes. Populations with large N_E have the potential to react to the selective pressures generated by environmental changes, if their genetic variability is high enough and if the speed of the environmental changes is not too high (i.e., if the rate of adaptive evolution at least matches the rate of environmental change; Allendorf and Luikart 2007). Therefore, the N_E of a population can predict its capacity to survive in a changing environment more reliably than the population size; furthermore, N_E determines the speed at which genetic variability is lost (Schwartz et al. 1999; Luikart et al. 2010; Hare et al. 2011).

To predict the long-term persistence of animal populations, accurate estimates of population size are also necessary. Thus, abundance has been a commonly monitored metric. Census methods based on direct counts can be inaccurate if individuals are difficult to detect. New molecular techniques for the analysis of noninvasive genetic samples (feces or hairs) typed for diagnostic genetic markers allow counts of individuals in a population by determining the number of unique genotypes in the population (Luikart et al. 2010; Marucco et al. 2011). This possibility has created a relatively new discipline, noninvasive genetics, which is a set of field, laboratory, and analytical techniques that enable researchers to study the biology of natural populations without observing or capturing individuals (Long and MacKay, this volume).

Phylogenetic Inference within Species

One important aspect of conservation genetics is the use of molecular genetic data to understand the ways in which geographic features influence long-term connectivity among groups of individuals or clades (where a *clade* is an organism and its descendants). This is the field of phylogeography (Avise 2004). By understanding phylogeographic relations, it is now possible to investigate geographic variation using different molecular markers and to deduce the phylogenetic relations of populations within *Martes* species (e.g., Davison et al. 2001; Stone et al. 2002; Pertoldi et al. 2008a,b; Ruiz-González 2011). Intraspecific phylogenetic approaches also allow us to identify the genetic legacy of species translocations (e.g., Drew et al. 2003; Vinkey et al.

2006), elucidate taxonomic uncertainties or the validity of subspecies (e.g., Drew et al. 2003; Knaus et al. 2011), delineate conservation units (e.g., Stone et al. 2002; Sato et al. 2009b; Slauson et al. 2009), reconstruct postglacial colonization histories (e.g., Davison et al. 2001; Ruiz-González 2011), understand range expansions and contractions, and even explore temporal changes in genetic variation through the use of historical DNA (e.g., Pertoldi et al. 2001; Schwartz 2007; Pertoldi et al. 2008a).

Molecular Markers

Many ecological questions can be answered with molecular genetic data; however, no single molecular tool (i.e., molecular marker) is best for all questions. The choice of the molecular marker depends largely on the question being addressed, available laboratory facilities, and prior research conducted with that category of marker on the target species. For example, most studies of intraspecific phylogenetics are based on variation in mitochondrial DNA (mtDNA) because of its properties, which include maternal transmission, extensive intraspecific variation, general lack of interspecific variation, and absence of genetic recombination. Thus, intraspecific studies conducted to date on *Martes* species have largely involved analyses of mtDNA sequences (Table 17.2). Because mtDNA has a relatively fast rate of nucleotide divergence, it is well suited to examining events that occurred during the last few million years.

Two mitochondrial molecular markers have been used mainly on intraspecific phylogenetic studies of *Martes* species: the control region or displacement (D-) loop and cytochrome *b* (cyt *b*). Historically, the sequence length examined for both of these markers has been short (300–500 bp) due to logistical constraints, although this is rapidly changing (Morin et al. 2010). A few studies have used both cyt *b* and the control region to obtain greater resolution, whereas others have complemented analysis of a single region with data on restriction fragment length polymorphisms (RFLPs) or with additional mtDNA markers, such as the NADH dehydrogenase subunit 2 gene, with several transfer RNA (tRNA) markers, or with internal spacer regions of the nuclear ribosomal DNA (rDNA). Several earlier studies on the intraspecific phylogenetics of *Martes* species have not detected clear structuring, possibly because they used a small fragment of mtDNA that did not provide enough resolution to identify intraspecific patterns.

Each DNA sequence has its own genealogy, and these genealogies may evolve at different rates. Furthermore, various methods of analysis probe different aspects of molecular and spatial histories. Consequently, to reconstruct a species' phylogeographic history, one would ideally use a range of sequences (e.g., nuclear, cytoplasmic, sex-linked, autosomal, conserved, neutral, high and low mutation-rate DNA fragments) and apply a suite of pertinent

Table 17.2. Information on the molecular marker and length of DNA sequences examined for understanding intraspecific phylogenetic relationships in *Martes* species

Species	Intraspecific phylogenetic data	Molecular marker (number of basepairs)	References
M. americana	Yes	Cyt *b* (441 bp)	Carr and Hicks 1997
		Cyt *b* (1140 bp); ald C (241 bp)	Stone and Cook 2002
		Cyt *b* (441 bp) and complete cyt *b* (1140 bp) in combination with RFLPs	Stone et al. 2002
		1428 bp cyt *b* (1140 bp); tRNA-Pro (25 bp); D-loop (263 bp)	Slauson et al. 2009
		mtDNA; 14 microsatellites	Dawson 2008
M. flavigula	No	—	—
M. foina	No	—	—
M. gwatkinsii	No	—	—
M. martes	Yes	D-loop (321 bp)	Davison et al. 2001
		D-loop (350 bp)	Pertoldi et al. 2008b
		Cyt *b*; tRNA-Thr; tRNA-Pro; complete D-loop; rDNA 12S (1608 bp)	Ruiz-González 2011
M. melampus	Yes	Cyt *b* (402 bp); RFLPs of rDNA	Hosoda et al. 1999
		Cyt *b* (1140 bp)	Kurose et al. 1999
		Cyt *b*; tRNA-Thr; tRNA-Pro; D-loop (521–524 bp)	Murakami et al. 2004
		Cyt *b*; D-loop; NADH-2 (2814 bp)	Sato et al. 2009b
		D-loop (535–537 bp)	Inoue et al. 2010
M. pennanti	Yes	Isoenzymes	Williams et al. 2000
		D-loop (301 bp)	Drew et al. 2003
		D-loop (301 bp); cyt *b* (428 bp)	Vinkey et al. 2006; Schwartz 2007
		Entire mitochondrial genome	Knaus et al. 2011
M. zibellina	Yes, parts of range	Cyt *b* (402 bp); RFLPs of rDNA	Hosoda et al. 1999
		Cyt *b* (1140bp)	Kurose et al. 1999
		Cyt *b*; tRNA-Thr; tRNA-Pro; D-loop (521–524 bp)	Murakami et al. 2004
		Cyt *b* (702 bp)	Malyarchuk et al. 2010
		D-loop (535–537 bp)	Inoue et al. 2010
		D-loop (495 bp)	Rozhnov et al. 2010

analytical approaches (Sato et al. 2009b; Balloux 2010). New techniques are becoming available to sequence entire genomes (Morin et al. 2010), promising a multitude of new data and greater insights into the intraspecific relations of *Martes* species (Knaus et al. 2011).

Microsatellites, which are regions of the genome with high mutation rates, have been widely used in conservation genetics investigations, and many software programs are available for analysis of these data (Excoffier and Heckel 2006). Microsatellites, considered neutral genetic markers, have been useful for estimating N_E, abundance, gene flow, hybridization, and genetic diversity in natural populations. They are highly variable in nearly all vertebrates, which enables the identification of individuals in a population. They also have many advantages compared with other genetic markers. First, microsatellite loci are codominant, which means alleles from both the chromosome pairs in diploid organisms (e.g., mammals) are detected. Next, microsatellites are believed to be largely selectively neutral, thus conforming to many theoretical population genetic models. Third, the ability to inexpensively develop microsatellites for a particular species or adapt those already developed for related taxa has made microsatellites popular for many investigations of wildlife genetics (Schwartz and Monfort 2008). In fact, there have been nearly 20 studies of *Martes* species that used microsatellites to infer gene flow or estimate abundance and effective population size (e.g., Kyle et al. 2003; Wisely et al. 2004; Hapeman et al. 2011).

Recently, a surge of new molecular genetic tools will enable researchers to examine molecular markers under selection. One of the most promising is the use of single nucleotide polymorphisms (SNPs). SNPs represent the most widespread source of sequence variation within genomes (Brumfield et al. 2003; Chen and Sullivan 2003; Ellegren 2008; Wang et al. 2009) and have the potential to significantly expand our ability to survey both neutral variation and genes under selection in natural populations, because SNPs have the very advantageous characteristic of being detectable throughout the genome. The use of SNPs involves amplification of very short fragments of DNA, which makes them particularly attractive for noninvasive genetic monitoring projects, where DNA is often degraded (Seddon et al. 2005). Furthermore, SNP genotypes based on single nucleotide changes are universally comparable and do not require standardization across laboratories, which can be a problem when comparing other genetic data (e.g., microsatellites) produced by different laboratories (Vignal et al. 2002). The enhanced opportunities for collaboration among laboratories located in different countries or continents will aid our ability to understand the population dynamics of *Martes* species and the influence of those dynamics on genetic variation. The main obstacle remaining for the use of SNPs is the difficulty of identifying them in nonmodel organisms (Smith et al. 2004; Ryynänen et al. 2007), but that situation is changing rapidly.

The Conservation Genetics of *Martes* Species

In this section, we synthesize the results of intraspecific phylogenetic, population genetic, and landscape genetic studies on each *Martes* species. We review species from west to east, starting with *M. americana* with its western extent in Alaska (USA), and moving to *M. melampus* in Japan.

Martes americana—American Marten

The intraspecific phylogeny of *M. americana* has been debated for several decades, with clear doubts about subspecific and specific status. As many as 14 subspecies of *M. americana* have been described (*M. a. abieticola*, *M. a. abietinoides*, *M. a. actuosa*, *M. a. americana*, *M. a. atrata*, *M. a. brumalis*, *M. a. caurina*, *M. a. humboldtensis*, *M. a. kenaiensis*, *M. a. nesophila*, *M. a. origenes*, *M. a. sierra*, *M. a. vancouverensis*, and *M. a. vulpina*; reviewed in Dawson and Cook, this volume). Traditionally, these have been placed into 2 morphologically distinct groups, *americana* and *caurina* (Merriam 1890; Clark et al. 1987). Although several studies (e.g., Merriam 1890; Anderson 1970; Hall 1981; Clark et al. 1987; Carr and Hicks 1997; Stone and Cook 2002; Stone et al. 2002; Small et al. 2003; Dawson 2008) have corroborated the separation of *M. americana* into these 2 groups, the level of distinctiveness between them has been vigorously debated. Before 1953, these 2 groups of martens were recognized as distinct species: *M. americana* and *M. caurina*, the Pacific marten (Merriam 1890). Yet, on the basis of intergradation in British Columbia (Canada) and Montana (USA), Wright (1953) proposed they be considered a single species. Since then, they have been synonymized under *M. americana*, which is reflected in currently accepted taxonomy (e.g., Wilson and Reeder 2005); however, the presence of 2 distinct clades has continued to be acknowledged by many researchers. Preliminary molecular data corroborated the distinction of *caurina* and *americana* as 2 monophyletic mitochondrial clades (Carr and Hicks 1997; Stone et al. 2002), but subsequent studies gave only subspecific status to the *caurina* and *americana* clades (Stone and Cook 2002; Stone et al. 2002).

These 2 lineages are largely allopatric. The *americana* clade is widespread from interior Alaska south to Montana and eastward to Newfoundland (Canada) and New England (i.e., northwestern, north-central, and northeastern North America) with little or no geographical structure present among populations. In contrast, the *caurina* clade occurs in western North America, extending from Admiralty Island in southeastern Alaska south through 2 large peninsular extensions, the first in the Cascade and Coast ranges to California (USA), and the other through the Rocky Mountains to Colorado (USA; Wright 1953; Hall 1981; Carr and Hicks 1997; Stone et al. 2002). Within *caurina*, there is strong structuring throughout its distribution, with several haplotypes confined to single populations (Stone et al. 2002). These 2 lineages appear to have diverged as a result of isolation in distinct southern glacial refugia; one in eastern and the other in western North America (Stone et al. 2002). The authors hypothesized that the individuals belonging to the *caurina* clade represent an early Holocene colonization northward along the west coast as coastal ice receded at the end of the last glaciation, whereas *americana* populations represent a later colonization from continental source populations that expanded through river corridors and traversed the coastal mountains. Interbreeding between these 2 lineages at contact zones has been shown (Stone and Cook 2002; Stone et al. 2002).

Microsatellite studies corroborated the patterns of population structure derived from sequences of the mitochondrial cyt *b* gene fragment (Carr and Hicks 1997; Stone et al. 2002), the nuclear aldolase C gene fragment (Stone and Cook 2002), and earlier morphological comparisons (Merriam 1890; Anderson 1970). Consistent with the occurrence of 2 distinct species of martens in North America, Small et al. (2003) showed that northern populations of *M. caurina* have greater genetic differences among populations and lower within-population genetic diversity than northern populations of *M. americana*, likely caused by the longer periods of isolation in coastal forests that were fragmented during the early Holocene period. The lack of differences among *M. americana* populations has been attributed to either continued gene flow or a more recent expansion throughout the range (Small et al. 2003). These results are consistent with a previous study that used randomly amplified polymorphic DNA markers (RAPD) to examine substructure among martens in Canada from southern British Columbia (*caurina*), Northwest Territories (*americana*), Labrador (*americana*), and the island of Newfoundland (*americana*; McGowan et al. 1999). They found that genetic distances were small among *M. americana* populations throughout Canada, yet large among all comparisons between *americana* and *caurina*, supporting the phylogenetic findings of Carr and Hicks (1997). This was again confirmed with larger sample sizes and microsatellite DNA by Kyle et al. (2000), who showed little substructure among populations of *M. americana* from the Yukon to the Northwest Territories in Canada.

Most recently Dawson (2008) and Dawson and Cook (this volume) reviewed previous molecular studies and developed a more-detailed view of genetic differentiation throughout the range of North American martens. These authors concluded that there are 2 distinct clades in North American martens that are consistent with species-level differences, and that *M. americana* and *M. caurina* are valid species that parallel their original taxonomic descriptions.

The subspecific status of several marten populations in the *caurina* clade from Oregon (USA) and California was investigated by Slauson et al. (2009). These authors evaluated the subspecific identity of a rediscovered population of martens in northern California that was within the historical range of a subspecies presumed to be extinct—*M. a. humboldtensis*, the Humboldt marten. They compared the mtDNA (1428 bp) sequence diversity of contemporary specimens within the presumed historical range of *M. a. humboldtensis*, including samples from neighboring populations of *M. a. caurina* and *M. a. sierrae*, and a historical museum specimen of *M. a. humboldtensis*. The museum specimen shared 1 haplotype with martens from both the rediscovered population in California and from coastal Oregon. This result suggests that the rediscovered population descends from a relictual population that previously existed in coastal California, Oregon, or both. They also concluded that the subspecific boundary between *M. a. humboldtensis* and *M. a. caurina* is questionable, because the historical haplotype from *M. a. humboldtensis* was shared with contemporary populations in both coastal Oregon and coastal California; consequently, extant marten populations in these regions should be managed collectively. One additional finding from Slauson et al. (2009) was that *M. a. sierrae* differed substantially from both *M. a. humboldtensis* and *M. a. caurina*, suggesting that marten populations were not a single large, genetically homogeneous population throughout the Pacific states (Washington, Oregon, and California) historically, and that this divergence may have occurred in separate glacial refugia.

Kyle and Strobeck (2003) studied the effects of habitat configuration on genetic variability and differentiation among *M. americana* populations in Canada. In agreement with previous studies, they observed little genetic structure in the northern regions, where habitat is homogeneous and few barriers to dispersal are thought to exist, compared with the more-fragmented southern region. Contrary to their expectations, no strong breaks in gene flow were found between any of the 35 sampled regions, regardless of the degree of fragmentation, with the exception of the insular Newfoundland population (*M. a. atrata*). This lack of genetic structure suggests a very large N_E for these populations and that, at larger spatial scales, dispersal by *M. americana* is not limited by landscape features, as was believed previously.

Although marten populations on the mainland appear to have large N_E values, there has been much concern about the N_E and genetic variation of *M. americana* on large islands. The abundance of *M. a. atrata* decreased from an estimated 630–875 animals in 1986 to only 300 animals in 1995 (Snyder 1986; Forsey et al. 1995). Such a drastic and rapid population decline raised concerns that inbreeding could affect the average fitness of this population (i.e., lead to inbreeding depression; Forsey et al. 1995). Population genetics studies on Newfoundland are consistent with this bottleneck, because *M. a. atrata* had the lowest mean number of alleles per locus and expected heterozygosity among the 25 populations sampled throughout Canada (Kyle et al. 2003). Whether this is a natural property of insularity (similar to *M. martes* in Ireland and Scotland; Kyle et al. 2003) or the result of recent declines in abundance or habitat fragmentation is unclear. It will be important to con-

Figure 17.1. Geographic range of *Martes pennanti*, the fisher, adapted from Knaus et al. (2011) and Reid and Helgen (2008). The area in light gray is the presumed current range; labels indicate locations discussed in the text.

translocations suggested historical gene flow among populations in British Columbia, Washington, Oregon, and California (Drew et al. 2003). The authors concluded that anthropogenic impacts have greatly reduced and isolated extant populations in Oregon and California and, consequently, that British Columbia would be the most appropriate source population for future translocations to recover *M. pennanti* in Washington and some localities in Oregon and California. This conclusion was confirmed by Warheit (2004) using the same molecular markers, as reported by Lewis and Hayes (2004).

Recently, more-detailed studies of *M. pennanti* have been conducted in the U.S. Rocky Mountains. It was assumed that *M. pennanti* occurring in the U.S. Rocky Mountains in the late 20th century were all descended from reintroduced stocks. However, Vinkey et al. (2006) reported that mtDNA (428 bp of cytochrome *b* and 301 bp of D-loop) haplotypes found only in *M. pennanti* from west-central Montana were likely derived from a relict population that persisted despite intensive fur harvesting in the early 20th century. Using the same molecular markers as Vinkey et al. (2006), Schwartz (2007) compared *M. pennanti* in west-central Montana with samples from north-central Idaho and found no differences. One museum specimen, collected in north-central Idaho in 1896, before any known translocation, had the same haplotype as the "native Montana haplotype" discovered by Vinkey et al. (2006). Thus, *M. pennanti* in north-central Idaho and west-central Montana are the only confirmed native populations in the U.S. Rocky Mountains, although many of these individuals have likely interbred with translocated animals. *Martes pennanti* from Idaho and Montana are not all descendants of translocated individuals, but are also the descendants of those that survived early 20th-century trapping.

A recent study by Knaus et al. (2011) has re-examined some of the results reported by Drew et al. (2003) and Vinkey et al. (2006), using the complete mitochondrial genome (16,290 bp). The most striking result was that the full-genome analysis identified patterns that were obscured by using only the control region; for example, Drew et al. (2003) showed that both northern and southern California shared a common haplotype, suggesting gene flow, yet the full-genome analysis revealed that these geographic areas each had unique haplotypes, concordant with microsatellite data (Wisely et al. 2004) and consistent with long-term isolation. Furthermore, similar to findings about *M. martes* and *M. americana*, Knaus et al.'s (2011) work on *M. pennanti* has shown expansion from refugia following the last glacial maximum; they suggest that *M. pennanti* expanded from an Eastern refugium and radiated westward within the past 16,700 years (range: 9000–31,300 years ago). The full-genome analysis also confirmed the uniqueness of the native Montana haplotype (Knaus et al. 2011).

Kyle et al. (2001) used microsatellite DNA to show relatively high levels of genetic structuring ($F_{ST} = 0.14$; range: 0.028–0.261) of *M. pennanti* in Canada compared with that of *M. americana* ($F_{ST} = 0.020$). Despite this high level of substructure, the populations maintained high genetic variability ($H_E = 62\%$). The greater amount of genetic structure in *M. pennanti* could be a reflection of philopatry and the large demographic changes that affected many populations after European settlement. Wisely et al. (2004) and Aubry et al. (2004) also found high levels of genetic substructure among *M. pennanti* populations (F_{ST} varied from 0.11 to 0.60) from southern British Columbia to the southern Sierra Nevada in California. Unlike Kyle et al. (2001), however, Wisely et al. (2004) found much lower values of H_E for *M. pennanti* in the fragmented populations along the Pacific coast (H_E range: 0.16–0.42) associated with relatively high estimates of F_{IS}, suggesting inbreeding. This pattern of reduced heterozygosity follows a north-south gradient, with *M. pennanti* populations in the southern part of the Sierra Nevada having the lowest levels of genetic variation in western North America (Wisely et al. 2004), a pattern that Wisely et al. (2004) suggest is due to the peninsular shape of the distribution in the Pacific states.

Population substructure in *M. pennanti* has been observed in eastern North America in both nuclear (microsatellite) and mitochondrial DNA (288 bp section of D-loop). MtDNA indicated 4 refugial populations in the northeastern United States and Canada: the Adirondack Mountains in New York, White Mountains in New Hampshire, the Moosehead Plateau in Maine in the United States, and the Cumberland Plateau in New Brunswick, Canada (Hapeman 2006). Microsatellite analyses provided evidence of the same 3 distinct populations in the United States (New Brunswick was not analyzed), corresponding to the last known remnants of *M. pennanti* in the east by the end of the 1930s. There is strong evidence, however, of range expansions and subsequent contact among these 3 populations (with limited gene flow) in the narrow corridor between Lake George and Great Sacandaga Lake near the New York–Vermont border (Hapeman et al. 2011). In addition, in the United States there are remnant genetic signals of reintroductions of *M. pennanti* from Maine to Vermont, from the Adirondacks to the Catskill Mountains in New York, and from Vermont and New Hampshire to Connecticut (Hapeman et al. 2011).

Population genetic data have also been used to evaluate the success of *M. pennanti* reintroductions. Williams et al. (2000) showed that older reintroductions had significant allele-frequency differences from their source populations. Although some of these differences may be due to initial sampling error because reintroductions typically involve relatively few individuals, the fact that recent introductions of *M. pennanti* show no significant allele-frequency differences from source populations suggests that the differences are more likely due to genetic drift (Williams et al. 2000). Drift can occur rapidly in small populations, especially for species such as *M. pennanti* that exhibit a polygynous mating system (Allendorf and Luikart 2007).

Several new landscape genetics approaches have been used to evaluate a recolonizing population of *M. pennanti* in southern Ontario (Carr et al. 2007a,b; Garroway et al. 2008). Initial research tested the idea that Algonquin Provincial Park was the source population for this expanding population of *M. pennanti* by examining microsatellite profiles of 35 sites (groups of samples, or "populations") surrounding the park (Carr et al. 2007a). The authors found that these 35 sites could be clustered into 5 discrete genetic groups, suggesting multiple origins for *M. pennanti* in Ontario; thus, the origin of *M. pennanti* in Ontario was not Algonquin Park, as predicted initially, but rather remnant populations in Ontario and Quebec in Canada and New York in the United States. Carr et al. (2007a) also showed that these populations were rapidly homogenizing along their expansion fronts. Subsequent research used assignment tests to infer the proportion of immigrants into each of the 5 genetic clusters and relate the proportion of immigrants to habitat variables including snow depth, coniferous forest cover, deciduous forest cover, mixed-wood forest cover, and nonforest (Carr et al. 2007b). Carr et al. (2007b) showed a positive association between snow depth and the proportion of immigrants, and a negative association between the proportion of coniferous forest in the landscape and the proportion of immigrants. The best regression model included both snow depth and proportion of coniferous forest, suggesting that the most suitable landscapes for *M. pennanti* had low snowfall and large expanses of coniferous forest (Carr et al. 2007b).

Finally, this same dataset was used in a graph-theoretical framework to examine network structure for evaluating habitat quality, gene flow, and population substructure (Garroway et al. 2008). The graph-theoretical framework is a new approach for landscape genetics that can be used to evaluate complex systems of connectivity that lead to system-level properties not readily discerned by examining relations among populations. This analytical approach has been adopted in the fields of social-network analysis, neurobiology, and transportation efficiency-network analysis (Costa et al. 2007). Basically every complex network, in this case a network of connectivity among populations of *M. pennanti*, has very specific topological features that typify its connectedness and its responses to perturbations (Costa et al. 2007). Garroway et al. (2008) showed that the network for *M. pennanti* in Ontario displayed high levels of clustering, and short mean-path lengths connecting pairs of nodes (populations). Using the graph-theoretic approach also allowed the authors to explore the effect of removing populations (nodes) on system connectivity and resilience. Garroway et al.'s (2008) removal analysis suggested that trapper harvest (i.e., removal of nodes) is unlikely to affect genetic connectivity among *M. pennanti* populations, given current conditions. In addition, they demonstrated a negative association between measures of node connectivity and both the proportion of immigrants into a node and snow depth, confirming Carr et al.'s (2007b) previous results.

Martes martes—European Pine Marten

Martes martes is well distributed throughout Europe (Proulx et al. 2004; Figure 17.2). It is a habitat specialist confined to mature deciduous and coniferous forests (Domingo-Roura 2002; Ruiz-González 2011), has a limited dispersal ability compared with other mustelids (Kyle et al. 2000), and has a slow reproductive rate, potentially rendering it vulnerable to habitat changes (Bright 2000; Webster 2001). Traditionally, *M. martes* has been subdivided into at least 8 subspecies based on coat color and geographic range (*M. m. borealis, M. m. latinortum*, *M. m. lorenzi*, *M. m. martes*, *M. m. minoricensis*, *M. m. notialis*, *M. m. ruthena*, and *M. m. uralensis*; Amori et al. 1996; Mitchell-Jones et al. 1999), although support for recognizing all these subspecies may be limited.

The phylogeography of *M. martes* was investigated initially using a small fragment (321 bp) of the mtDNA control region (Davison et al. 2001). This study suggested that extant populations of *M. martes* in central and northern Europe are the result of colonizations from 1 or more glacial refugia and subsequent mixing. The fragment sizes of DNA used were too small though to identify the specific locations of the refugia, or the process of postglacial recolonization of central Europe. Moreover, the scarcity of samples from a suspected Mediterranean refugium (1 specimen from the Iberian Peninsula, 3 from Italy, and 2 from the northern Balkans) has left the recolonization hypothesis open. Interestingly, Davison et al. (2001) reported evidence of genetic introgression (i.e., widespread historical or contemporary hybridization) of *M. martes* with *M. zibellina,* the sable, in Fennoscandia, along with mtDNA and morphological evidence of introgression with *M. a. caurina* in England, which was later confirmed with microsatellite data (Kyle et al. 2003).

More recently, Ruiz-González (2011) investigated unresolved questions posed by Davison et al. (2001) and re-examined phylogeographic patterns of *M. martes* throughout its current range. With the advantage of newer technologies and methods, this study was more comprehensive in terms of the number of specimens included and the length of the mtDNA sequence examined (1600 bp). Sampling also covered a larger portion of the species range, including individuals from Scandinavia in the north, the Russian Federation in the east, and the Iberian Peninsula in the southwest. Ruiz-González (2011) revealed the presence of 69 haplotypes for *M. martes* (and 11 for *M. zibellina*), which are split into 2 major clades: the European-Mediterranean and Fennoscandian-Russian clades. The first clade, including all *M. martes* samples collected from throughout its current European range, is further subdivided into 2 subclades that connect haplotypes in central-northern Europe and the Mediterranean region. Surprisingly, haplotypes in the Mediterranean subclade apparently did not contribute to the postglacial recolonization of most of the Palearctic range of the species. It appears that central-northern

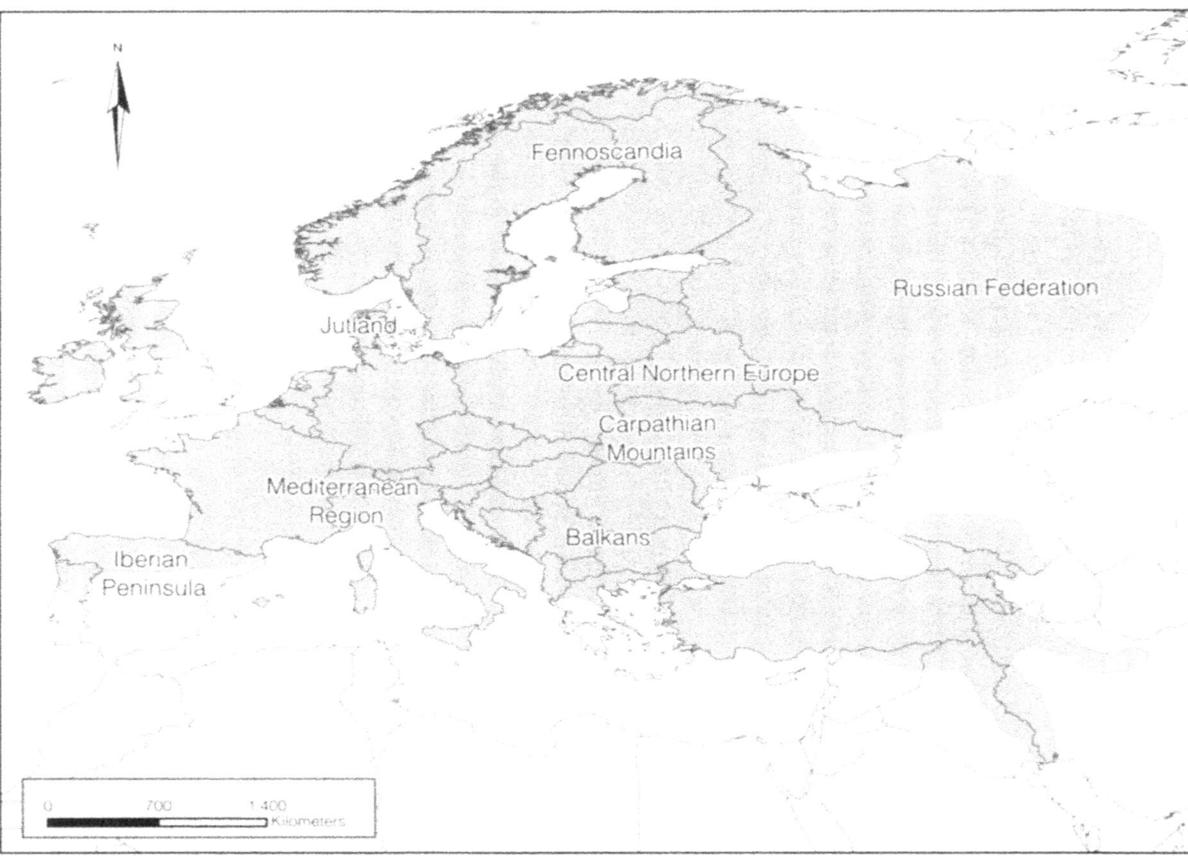

Figure 17.2. Geographic range of *Martes martes*, the European pine marten, adapted from Kranz et al. (2008). The area in light gray is the presumed current range; labels indicate locations discussed in the text.

Europe was recolonized by a population of *M. martes* that survived the last glaciation in an undetermined central-European refugium (possibly the Carpathian Mountains), as was suggested previously by paleontological data (Sommer and Benecke 2004). In addition to this complex recolonization of Europe, genetically differentiated populations of *M. martes* in Fennoscandia and Russia are introgressed with mtDNA of *M. zibellina*, highlighting the complex phylogeographic history of *M. martes* (Ruiz-González 2011).

In a more spatially restricted study, Pertoldi et al. (2008b) studied genetic differentiation of *M. martes* in 3 isolated geographic regions in northern Europe (Jutland and Sealand in Denmark, and southern Scania in southernmost Sweden) by sequencing the hypervariable region of the mtDNA D-loop (350 bp). Pertoldi et al. (2008b) found 8 haplotypes, with 2 shared by individuals from all 3 regions, yet with unique haplotypes found in all localities. This subdivision was likely due to the insular and peninsular nature of the northern European landscape. By comparing these data with previous haplotype analyses (Davison et al. 2001), Pertoldi et al. (2008b) confirmed the presence of 2 primary clades in central and northern Europe, with samples in southern Scania being well differentiated from those in central Sweden. Altogether, these studies point to 3 themes in the phylogeography of *M. martes*: (1) survival in multiple refugia during the last glacial period in the Mediterranean, central-northern European, and Fennoscandian-Russian regions; (2) postglacial recolonization of northern Europe by the central-northern European clade; and (3) recent genetic drift caused by isolating factors, such as major waterways and peninsulas.

Martes martes populations have been shown to have a higher level of genetic structure (with an overall F_{ST} value of 0.18, range: 0.016–0.330) and lower genetic variation (H_E range excluding the insular populations: 53.8–63.8%) than their North American sibling species, *M. americana*, the American marten, sampled throughout Canada (average H_E: 63.6% excluding the Newfoundland island population; Kyle and Strobeck 2003; Kyle et al. 2003). The level of genetic differentiation among *M. martes* populations is correlated with the geographic distance among populations ($r = 0.31$, $P = 0.11$; D_S: $r = 0.55$, $P = 0.007$, D_{LR}: $r = 0.91$, $P = 0.00006$; Kyle et al. 2003; Figure 17.3); thus, *M. martes* in Europe appears to have greater substructure per unit distance than *M. americana*. It is difficult to exclude more ancient processes (e.g., the influence of glaciations) as a cause of the differences observed, but it may be related to the greater level of persecution and habitat fragmentation experienced by *M. martes* (Kyle et al. 2003). At a more local level, *M. martes* in northern Spain are also highly substructured, with F_{ST} values of 0.057–0.172 in a 250-km^2 area (Ruiz-González 2011). Although the distribution of samples is nearly continuous, substructure in at least 1 of 3 genetically identified clusters corresponds to the presence of anthropogenic influences, such as reservoirs, high road densities, and urbanization (Ruiz-González 2011). Of

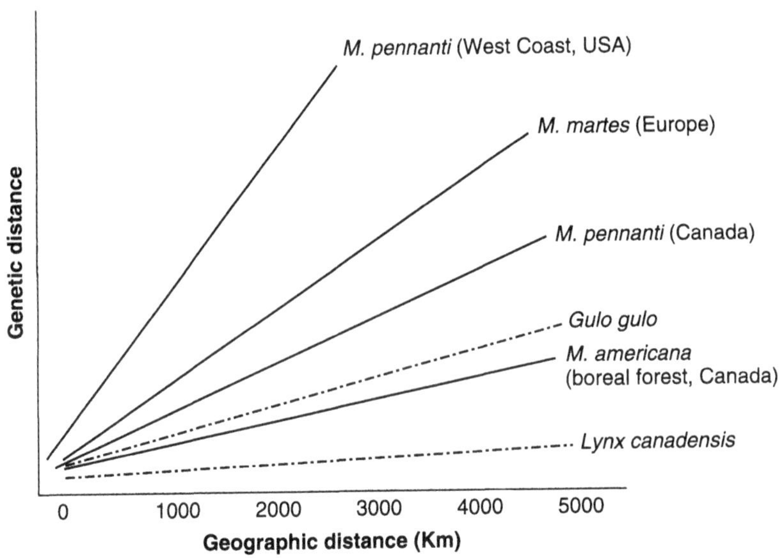

Figure 17.3. A schematic showing the approximate relation between genetic distance and Euclidean distance for several *Martes species*. Dotted lines show the approximate relation for other midsized carnivores as a reference. Adapted from Kyle and Strobeck (2003).

particular interest is the potential that 1 of the areas in northern Spain is structured by interspecific competition with *M. foina*, the stone marten (Ruiz-González 2011).

Ruiz-González (2011) also used landscape genetic approaches to evaluate how *M. martes* responds to various vegetation types and topographic and hydrologic features. He found that gene flow and connectivity were strongly reduced by croplands, wetlands, roads, urban areas, and reservoirs (Ruiz-González 2011). Intact forests and scrublands acted as corridors for connecting populations, whereas urban areas acted to reduce gene flow (Ruiz-González 2011). Similar work in Ardennes, La Bresse, and L'Isere in France has shown the importance of forest structure for dispersal of *M. martes* across large landscapes (Mergey 2007).

Martes zibellina—Sable

Martes zibellina exhibits substantial interpopulation variation in morphological characters and a multiplicity of local forms, complicating the study of intraspecific taxonomy (Monakhov 1976; Pavlinov and Rossolimo 1979). Based on phenotypic and geographic differences, as many as 16 subspecies have been designated for *M. zibellina* (*M. z. angarensis*, *M. z. arsenjevi*, *M. z. averini*, *M. z. brachyura*, *M. z. ilimpiensis*, *M. z. jakutensis*, *M. z. kamtschadalica*, *M. z. obscura*, *M. z. princeps*, *M. z. sahalinensis*, *M. z. sajanensis*, *M.*

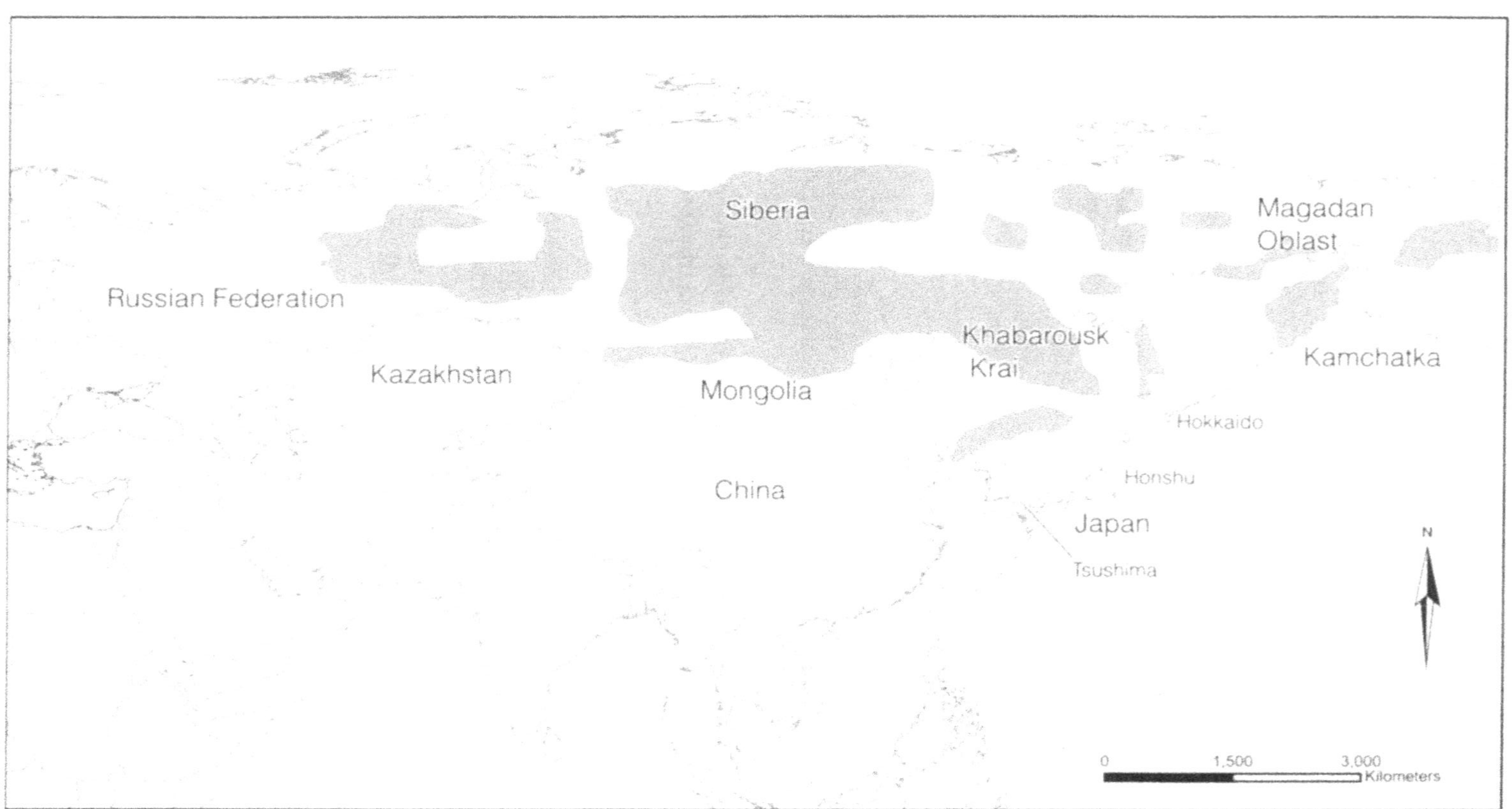

Figure 17.4. Geographic range of *Martes zibellina*, the sable, adapted from Abramov and Wozencraft (2008). The area in light gray is the presumed current range; labels indicate locations discussed in the text.

z. schantaricus, *M. z. tomensis*, *M. z. tungussensis*, *M. z. yeniseensis*, and *M. z. zibellina*; Ognev 1925; Kurose et al. 1999; Miyoshi and Higashi 2005; Wilson and Reeder 2005). Yet, untangling the intraspecific phylogenetic history of *M. zibellina* has been complicated by the massive number of human-mediated introductions and reintroductions throughout most of its range in Russia during the 20th century (Figure 17.4; Monakhov 2001; Powell et al., this volume).

Initial analyses of population-level genetic variation in the mtDNA cyt *b* gene fragment of *M. zibellina* from Russia pointed to the existence of high levels of intraspecific heterogeneity (Balmysheva and Solovenchuk 1999a,b; Petrovskaya 2007). Populations of *M. zibellina* from Siberia and the Far East showed the prevalence of 3 different haplotypes, probably representing 3 monophyletic clades (Balmysheva and Solovenchuk 1999a,b; Petrovskaya 2007). To confirm these findings, fine-scale analysis of mtDNA variation was recently conducted by Malyarchuk et al. (2010). This study focused on the analysis of phylogenetic relations using mtDNA cyt *b* sequences in 17 *M. zibellina* from Magadan Oblast, Kamchatka, and Khabarovsk Krai in Russia (Figure 17.4), and data published previously on *M. z. brachyura* from Hokkaido Island, Japan. Malyarchuk et al. (2010) identified 2 phylogenetic clades: the first was represented predominantly by *M. zibellina* from different regions of Northeast Asia, including Kamchatka, Khabarovsk Krai, and the Magadan Oblast, together with samples from Hokkaido, and the second consisted of haplotypes from Magadan Oblast and Khabarovsk Krai, without haplotypes represented in populations in Kamchatka or Hokkaido (Figure 17.4). Interestingly, a specimen of *M. martes* from Sweden clustered with this group, supporting evidence reported previously for historical introgression of *M. martes* with *M. zibellina* in Fennoscandia (Davison et al. 2001; Ruiz-González 2011), although contemporary hybridization has also been suggested (Rozhnov et al. 2010). In the Magadan Oblast, individual *M. zibellina* from both clades have been found, which has been explained by the introduction there of *M. zibellina* from Kamchatka and Khabarovsk, starting in the 1950s (Petrovskaya 2007; Malyarchuk et al. 2010).

Overall, there is strong evidence of at least 2 clades of *M. zibellina* in Russia, both with high sequence divergence that probably resulted from the impact of Pleistocene glaciations. During the Holocene, these 2 clades were reunited in a new contact zone (Malyarchuk et al. 2010). Interestingly, *M. zibellina* on the island of Hokkaido do not form a unique clade, which would be expected if Hokkaido served as a glacial refugium. Thus, it appears that divergence between the populations in Hokkaido and the Russian Far East has been a recent process, probably because Hokkaido was periodically connected to the mainland of Asia by a land bridge as recently as 10,000 years ago (Oshima 1990; Hosoda et al. 1999). Hokkaido may still have been a refugium for 1 of the *M. zibellina* clades during the last glaciation, with recent postglacial expansion into the mainland of Asia (Kurose et al. 1999). An examination of population genetic structure will be necessary to further understand the complex intraspecific phylogeography of *M. zibellina*, and how contemporary landscape features structure extant populations.

Martes melampus—Japanese Marten

Martes melampus is endemic to Japan, where it occurs on the main Japanese islands of Honshu, Shikoku, and Kyushu, but it is not endemic to Hokkaido Island (Masuda 2009), where *M. zibellina* occurs (Murakami et al. 2004; Murakami 2009). There are at least 3 recognized subspecies of *M. melampus*, based on differences in coat color: *M. m. coreensis*, *M. m. melampus*, and *M. m. tsuensis* (Anderson 1970). *Martes melampus* was introduced to Hokkaido from Honshu and is currently expanding its range in southern Hokkaido, whereas the native *M. zibellina* is distributed in central and eastern Hokkaido. The contact zone between the 2 species is in central Hokkaido (Murakami et al. 2004; Masuda 2009). *Martes melampus tsuensis* is restricted to Tsushima Island in the Korea Strait (Figure 17.4), and *M. m. corensis* is thought to occur on the Korean Peninsula, but its existence and identity are controversial (Hosoda et al. 1999).

Martes melampus has a complex taxonomic history; moreover, the presence of the closely related *M. zibellina* (Hosoda et al. 1997, 2000; Sato et al. 2003, 2009b; Koepfli et al. 2008) makes intraspecific assignment difficult. Several studies have focused on genetic relations within and between *M. melampus* and *M. z. brachyura* (Hosoda et al. 1999; Kurose et al. 1999). Hosoda et al. (1999) used the restriction fragment length polymorphism (RFLP) of rDNA spacer and the mitochondria1 cyt *b* (402 bp) gene-fragment sequences, and Kurose et al. (1999) sequenced the entire cyt *b* (1140 bp) gene to reveal the extent of intra- and interspecific variation in these 2 species. Both studies showed large genetic differences between the species, yet, in both studies, the clustering of haplotypes of *M. melampus* in phylogenetic trees did not correspond with expected geographic relations between populations on different Japanese islands. Only the Tsushima Island populations (*M. m. tsuensis*) showed geographically concordant genetic differentiation. These results suggest that mtDNA introgression between local populations of *M. melampus* may have resulted from incomplete geographic isolation on each island, limited interpretive power of the available sequence, or the fact that *M. melampus* may have recently expanded to the Japanese islands and the genetic signals have not yet had time to become geographically concordant.

One of the most comprehensive phylogenetic studies of *M. melampus* populations was recently published by Sato et al. (2009b). They conducted molecular phylogenetic analyses of 49 individuals sampled from throughout Japan, focusing on 3 mtDNA loci (cyt *b*, control region, and the NADH subunit 2 gene) and 1 nuclear gene (the growth-hormone receptor gene, including the polymorphic intron regions). Sato et al. (2009b) identified 9 intraspecific groups, not correlated with winter coat color, but consistent with the geography of the Japanese islands; in particular, they demonstrated the monophyly of *M. m. tsuensis*, the Tsushima marten, supporting the view that its genetic distinctiveness and uniformity resulted from a long history of isolation on small islands. This also confirmed earlier studies (and subspecies designations) proposing that the Tsushima Island population is an evolutionarily significant unit. Sato et al. (2009b) also provide support (although with a limited sample size) of the uniqueness of martens in the Iwate region of Honshu (Figure 17.4).

Overall, phylogenetic patterns of *M. melampus* are more complex and more difficult to resolve than for North American and European martens, which exhibit strong patterns of postglacial expansion. For *M. melampus*, the clearest signal is the isolation of island populations (e.g., Tsushima Island), probably since the Pleistocene. Other patterns are obscured by complex phylogeographic events, such as the dynamics of land bridges among islands (e.g., the Seto-Ohashi Bridge) and some recent translocations by the fur industry (Sato et al. 2009b). Additional research on the population genetic structure of *M. melampus* using variable nuclear markers will provide additional insights into these relations.

Synthesis

Each species in the genus *Martes* is a unique product of evolution with a distinct ecological niche, but we have revealed a few general patterns among these species. First, it is clear that the complex glacial histories of Europe, Asia, and North America created refugial populations that are only recently coming back into contact. In Europe, *M. martes* apparently persisted during the last ice age in well-established southern-Mediterranean refugia, and also in central-northern European and Fennoscandian refugium (Ruiz-González 2011). Genetic evidence suggests large-scale expansions across Europe from the central-northern refugium. *Martes americana* was restricted to multiple refugia, as well, with 2 distinct clades (*americana* and *caurina*) persisting in eastern and western forest refugia in the southern parts of North America during past glacial advances. These deep phylogeographic splits produced 2 clades with very distinct evolutionary histories (Stone et al. 2002), consistent with species-level differences. The Eastern clade, much like the central-northern clade of *M. martes*, expanded into a larger area after the last glacial period. The complex recolonization of ice-free forests in North America also produced separate clades below the species level in the *caurina* clade (Slauson et al. 2009). *Martes pennanti* also persisted in a North American refugium, which was likely in the midwestern or eastern United States (Knaus et al.

2011). Colonization of U.S. Rocky Mountain and California forests appears to be from a single source area, although more information is needed to confirm this hypothesis. More recent biogeographic processes have subsequently shaped this species, producing well-supported clades in the U.S. Rocky Mountains and Sierra Nevada.

Less is known about the intraspecific phylogenetic relations of *M. zibellina*, although there is support for at least 2 clades: one largely in the Russian Far East and on Hokkaido Island in Japan, and another farther west. In the areas between these 2 regions, haplotypes from both clades occur, suggesting recent contact. Of particular interest is that the island of Hokkaido was apparently not an isolated refugium, because *M. zibellina* on this island cluster well with samples from the Russian Far East (Malyarchuk et al. 2010). It is possible that Hokkaido Island may have been the refugium from which the Russian Far East samples originated. *Martes melampus* also has a complex evolutionary history, with support for glacial processes contributing to at least 2 clades: one on Tsushima Island in the Korean Strait (Sato et al. 2009b), and another on the Japanese islands (except Hokkaido). Overall, one of the challenges faced by molecular ecologists is that of untangling these deep phylogeographic signals that were created by complex postglacial histories and influenced by contemporary landscape changes.

Although intraspecific phylogenetic information is essential for providing a context for managers to understand the long-term dynamics of individual species, the field of population genetics has been instrumental in examining the relations among populations. Kyle and Strobeck (2003) plotted a linear regression of D_S (an index of genetic distance) and geographic distances for mainland populations of *M. americana*, *M. martes*, and *M. pennanti* (Figure 17.3). The highest levels of genetic structure per unit distance were found for *M. martes* and *M. pennanti,* followed by *M. americana* (Figure 17.3). This suggests that, outside the Canadian boreal forests, managers of both North American and European *Martes* species should not expect rapid recolonization of areas where populations are currently absent, probably because of life-history or dispersal constraints imposed by the more-fragmented southern habitats.

The field of landscape genetics, an extension of population genetics that examines the finer-scale movements of *Martes* species, is currently being used to provide more-detailed examinations of contemporary landscape features to understand the structuring of populations. This finer-scale genetic information will be most useful to managers of these species. One of the most important landscape-genetic findings to date is that habitat selection for daily requirements (e.g., shelter, food, mating opportunities) is different from habitat selection for dispersal (Wasserman et al. 2010). This provides a good example of the way information from molecular ecology studies can augment traditional field biology studies that use other tools, such as radio telemetry and habitat suitability analyses, to understand habitat use. These traditional studies excel at providing detailed habitat information, but because of the sometimes rare nature of dispersal, they have difficulty documenting the habitats used for long-distance dispersal (e.g., Aubry et al. 2004).

In a third application of landscape genetics, Carr et al. (2007b) showed that immigration of *M. pennanti* in Ontario was correlated with low snow accumulation and more coniferous forests, landscape elements that now appear key for maintaining connectivity among populations of this species. Similarly, Broquet et al. (2006a,b) showed that logging significantly changed the landscape for *M. americana* and reduced gene flow compared with unlogged landscapes. In Europe, Ruiz-González (2011) found analogous results using a landscape genetic approach to show that gene flow and connectivity for *M. martes* were restricted by croplands, wetlands, roads, urban areas, and reservoirs, and most facilitated by intact forests and scrublands.

In another type of landscape-genetics application, Garroway et al. (2008) examined the impact of removing a population on existing connectivity, and found that fur trapping is unlikely to affect genetic connectivity under current conditions. This type of approach can be readily used by managers to assess the relative impact of trapping regulations in one area compared with another for any *Martes* species.

Overall, we still have an incomplete taxonomic and evolutionary framework for a significant portion of the *Martes* species complex. Unfortunately, research on intraspecific phylogenetics of *Martes* species is limited and biased toward some species (see Tables 17.1 and 17.2). Most research has been conducted on *M. americana*, *M. martes*, *M. melampus*, and *M. pennanti*, whereas other species, including *M. zibellina* (Hosoda et al. 1999; Kurose et al. 1999; Murakami et al. 2004; Inoue et al. 2010; Malyarchuk et al. 2010), have been only partially or insufficiently studied (Tables 17.1 and 17.2). For *M. flavigula*, the yellow-throated marten; *M. foina*, the stone marten; and *M. gwatkinsii*, the Nilgiri marten, intraspecific phylogenetic studies are completely absent from the literature. Similar trends hold for population genetic and landscape genetic information.

With the recent explosion of genomic approaches, we expect significant amounts of molecular genetic data to be generated that will facilitate exploring the intra- and interpopulation dynamics of *Martes* species (Beja-Pereira et al. 2009). Along with these technological advances, concomitant developments are occurring in analytical approaches for inferring demographic histories and evolutionary relationships from molecular genetic data, and testing their statistical significance (Pertoldi and Topping 2004; Bouchy et al. 2005; Nomura 2005; Bach et al. 2006). Overall, we encourage the use of new genomic methods and associated analytical tools to further resolve intraspecific relations among clades within each species. Specifically, we anticipate the use of new approaches that examine genes under environmental selection. When we understand how these genes are structured across the landscape, we can better delineate appropriate units for conservation (e.g., species, subspecies, management units), better inform reintroduction efforts, and better predict how *Martes* populations will adapt as the climate changes. We also see the field of genomics providing unprecedented power to conduct novel and important population and landscape genetic studies.

Although these newer approaches will refine our understandings of the well-studied *Martes* species, our most urgent recommendation is to initiate sample collection and all levels of genetic studies on *M. flavigula, M. foina*, and *M. gwatkinsii*. There is a complete lack of knowledge about their phylogenetic histories and the factors that influence their movements on the landscape. Without this information, we will have no basis for understanding how these species will fare in a rapidly changing world (Hoffmann and Sgrò 2011).

Acknowledgments

Aritz Ruiz-González holds a PhD fellowship awarded by the Department of Education Universities and Research of the Basque Government (Ref. BFI09.396) and is a member of the research group "Systematic, Biogeography and Population Dynamics" of the University of the Basque Country, supported by the Basque Government (Ref. IT317–10; GUIC10/76). Michael Schwartz was partially funded to conduct this research by a Presidential Early Career Award for Science and Engineers and by the Rocky Mountain Research Station's Wildlife and Terrestrial Ecosystems Program. We thank Bill Zielinski, Keith Aubry, and Chris Kyle for comments on an earlier version of this manuscript.

18

Use of Habitat and Viability Models in *Martes* Conservation and Restoration

CARLOS CARROLL, WAYNE D. SPENCER, AND JEFFREY C. LEWIS

ABSTRACT

Conservation and management of *Martes* populations are increasingly informed by quantitative models that predict habitat suitability and population viability. Recent modeling efforts to support fisher (*Martes pennanti*) reintroduction planning in the state of Washington (USA) and conservation of an isolated fisher population in the southern Sierra Nevada (California, USA) have integrated results from empirical static habitat models, such as resource-selection functions, with those from dynamic population-viability and vegetation models. Additional methods have been developed to identify habitat linkages with potential importance for maintaining interpopulation dispersal. While such modeling frameworks can be useful in integrating data on species distribution, demography, and vegetation response to disturbance, the associated increased data requirements may also increase uncertainty regarding model projections to different places or times. The costs associated with reintroductions generally justify the use of such models to inform the planning process before substantial resources are committed. Given the challenges posed by increasing human demands on forest ecosystems, well-constructed quantitative models can be key tools for enhancing the success of wildlife conservation efforts, as long as model uncertainty is considered explicitly, and model results are used for informing decisions rather than predicting outcomes.

Introduction

Conservation and management of *Martes* populations are increasingly informed by quantitative models that predict habitat suitability and population viability. Statistical models of habitat relations can help increase understanding of the factors limiting a species' distribution, facilitate protection and enhancement of habitat, predict distribution in unsurveyed areas, and evaluate suitability of currently unoccupied areas for reintroduction. More complex population-viability models that integrate data on habitat and demography are also increasingly employed to evaluate the effects of alternative management options on population persistence.

Like many carnivores, North American *Martes* species (American marten [*M. americana*], and fisher [*M. pennanti*]) have undergone range contractions since European settlement (Krohn, this volume). Although both species retain large populations over much of their historical ranges, several regional subpopulations are of conservation concern (Krohn, this volume). Whereas overexploitation was a primary factor contributing to range contractions historically, current threats stem primarily from widespread habitat alteration due to intensive forestry and other factors (I.D. Thompson et al., this volume). Thus, conservation and management of existing populations may involve evaluation of both local- and regional-scale habitat suitability (Shirk et al., this volume). Planning at broader spatial scales may involve development of quantitative habitat suitability models from geographically extensive data, such as satellite imagery (Carroll et al. 1999; Spencer et al. 2011; Marcot and Raphael, this volume).

To reverse range contractions stemming from historical overexploitation, American martens and fishers have been translocated to many areas (Drew et al. 2003; Powell et al., this volume). Although early reintroductions often were based on qualitative assessments of the availability of suitable habitat, more recent efforts have typically involved quantitative modeling of habitat suitability (Lewis and Hayes 2004; Callas and Figura 2008). Reintroductions, especially in eastern North America, have frequently been successful in establishing populations (Aubry and Lewis 2003; Powell et al., this volume). However, at least 2 aspects of the demography and habitat associations of North American *Martes* species may create challenges for both reintroduction and conservation of small populations (Powell et al., this volume). *Martes* species generally have relatively large territory sizes for their body mass, and dispersal ability for the majority of species may be limited by their association with forest habitats (Powell and Zielinski 1994).

In this chapter, we review the types of habitat and viability models that have been used to guide conservation efforts for North American *Martes* species (Table 18.1). The types of habitat and viability models reviewed here fall into 4 distinct categories based on their output predictions: distribution, dispersal, persistence, and optimal reserves (Table 18.1). Each type of model can be characterized by its data structure (i.e., types of input and output data), treatment of uncertainty, and context in the planning process. We present 2 case studies for the fisher in western North America that provide guidelines for choosing and applying appropriate modeling tools, and may be useful for

Table 18.1. Types of habitat and viability models discussed in this chapter

Model type	Acronym	Input data	Output predictions	Reference
Habitat suitability index	HSI	Environmental	Distribution	Thomasma et al. 1991
Generalized linear	GLM	Environmental	Distribution	Carroll et al. 1999
Generalized additive	GAM	Environmental	Distribution	Davis et al. 2007
Hierarchical Bayesian spatial	HB	Environmental	Distribution	Carroll et al. 2010
Centrality	—	Habitat	Dispersal	Garroway et al. 2008
Non-spatial viability	—	Habitat, demographic	Persistence	Lacy and Clark 1993
Spatially explicit population	SEPM	Habitat, demographic	Persistence	Lewis and Hayes 2004; Spencer et al. 2011
Reserve selection	—	Distribution	Optimal reserves	Zielinski et al. 2006a

Martes conservation efforts in other regions. Such planning tools are particularly relevant for the fisher, which in 2004 was classified as a candidate species for listing as threatened or endangered under the U.S. Endangered Species Act (U.S. Fish and Wildlife Service 2004). This finding has motivated land-management agencies to evaluate important habitat areas for the species in the Pacific coastal states and provinces (Lofroth et al. 2010).

Models of Distribution and Resource Selection

The first habitat models used to inform *Martes* reintroduction efforts were conceptual models such as the Habitat Suitability Index (HSI), that allowed expert opinion on the value of various habitat types to be expressed formally as a quantitative index (Allen 1983). The maximum value in Allen's (1983) fisher HSI model occurs when canopy closure is >75%, mean diameter at breast height (dbh) of overstory trees is >40 cm, 3 or more tree species are present in the canopy, and 10–50% of the overstory is deciduous. Thomasma et al. (1991) used Allen's (1983) model to inform a fisher reintroduction effort in the Upper Peninsula of Michigan (USA). A comparison of the HSI model predictions with snow-tracking data revealed that, although the composite HSI score was a significant predictor of fisher habitat use, only 2 of the 4 component variables (mean dbh of overstory trees and percentage of overstory deciduous trees) were significant in the multivariate model. In addition, the shape of the response function was not monotonically increasing, but multimodal (Thomasma et al. 1991). In the last 2 decades, habitat-suitability models derived from statistical analysis of distributional datasets have increasingly supplanted the use of conceptual models. When applied in the context of broad-scale planning, wildlife-distribution models typically require input of presence-absence data (derived from systematic regional surveys or compiled from nonsystematic surveys) measured at the patch or site scale, and environmental covariates mapped across the landscape using a geographic information system (GIS). Uncertainty of a single model is assessed using standard metrics (McCullagh and Nelder 1989), whereas uncertainty inherent in the model-selection process is addressed using model averaging (Burnham and Anderson 2002). The robustness of distribution models can be assessed by using cross-validation techniques (McCullagh and Nelder 1989) or by testing against new data (Carroll et al. 1999). Distribution models play a key role in the planning process both as a tool to identify priority habitats and as input to more complex models (Table 18.1).

Distribution models are commonly based on a form of generalized linear model (GLM) such as logistic regression, which models the relations between a binary response variable (e.g., presence-absence) and environmental covariates (McCullagh and Nelder 1989) (Table 18.1). Generalized additive models (GAM) resemble GLMs, but allow more flexibility in the form of the relations between the response variable and the covariates (Hastie and Tibshirani 1990). Logistic regression and other GLMs are often used to fit resource selection functions (RSFs), which are models that yield values proportional to the probability of use of a resource unit (Boyce et al. 2002). For example, RSFs have been used to model fisher habitat relations in California (USA; Carroll et al. 1999; Davis et al. 2007; Spencer et al. 2011), the Rocky Mountain region (USA and Canada; Carroll et al. 2001), and British Columbia (Canada; Weir and Harestad 2003) at a range of spatial scales. Models derived from radiotelemetry data were applied to a 1500 km^2 study area (Weir and Harestad 2003), whereas those derived from "found" data (compilations of nonsystematic surveys; Carroll et al. 2001) or regional presence/absence surveys (Davis et al. 2007) were applied at the scale of states or provinces.

A limitation of presence/absence data is that observed nondetections can be attributed either to the true absence of the species or to failure of the survey protocol to detect a species that is present (false or sampling absence). Recently developed methods for occupancy modeling, which allow adjustment for incomplete detection, can increase the generality of models built from presence/absence data (MacKenzie et al. 2006; Slauson et al., this volume). Additionally, absence from a site may be due to factors such as barriers to dispersal, or local extirpation caused by predation or disease. Alternative modeling methods, such as Maxent (Phillips and Dudik 2008), that compare sites with presence to all available sites may be used when absence data is unavailable (e.g., for observational data).

Another limitation of typical RSF modeling techniques is that species distribution data often exhibit spatial structure caused by unmeasured environmental factors and population processes (e.g., dispersal, territoriality) that may result in aggregated distributions of individuals (Beale et al. 2010). If ignored, such spatial structure may lead to models that include spurious covariates and have poor predictive power (Beale et al. 2010). Spatial structure may be incorporated into a distribution model by addition of either a spatially structured covariate or an error term (Beale et al. 2010). The GAM developed by Davis et al. (2007) to model fisher habitat suitability in California incorporated a simple autoregressive term that estimated spatial effects through an additional covariate on the basis of observed presence-absence values at neighboring sites within a spatial neighborhood.

Rather than adding spatial structure as a covariate, Carroll et al. (2010) used a hierarchical Bayesian model to incorporate a spatial random effect term (ρ [rho]), which is estimated by means of an intrinsic conditional autoregressive (ICAR) model in which each spatial effect has a Gaussian distribution centered on the mean of the neighboring values (Latimer et al. 2006). ICAR models simultaneously produce estimates of the effects of environmental variables and spatial random effects, allowing the data to determine the best placement of spatial effects (Carroll et al. 2010). This approach also provides better estimation where response data are missing than is possible with simple autoregressive methods. If the spatial random effect is due primarily to population processes, such as dispersal limitation, evaluating predicted distribution without the spatial term can help identify areas that still contain suitable habitat, but from which a species has been extirpated due to historical overexploitation. In Carroll et al.'s (2010) model, predictions that include the spatial effect term highlighted areas currently occupied by fishers in northwestern California and the southern Sierra Nevada (California, USA). Predictions evaluated without the spatial term identified an additional area in the northern Sierra Nevada (Figure 18.1a).

A fisher reintroduction project currently (2011) under way in the northern Sierra Nevada near the potential reintroduction area identified above offers an example of the ability of distribution models to inform restoration efforts (Callas and Figura 2008). The reintroduction feasibility analysis compared predictions from a conceptual model, a GLM-based RSF model, and the spatial GAM model of Davis et al. (2007) to determine which portions of the proposed reintroduction area contained habitat most likely to support a fisher population. All 3 models generally agreed about which sites within the 1500-km^2 analysis area had the highest quality habitat (Callas and Figura 2008); however, even the best habitat within the analysis area was predicted to be of lower suitability than currently occupied areas elsewhere in the region.

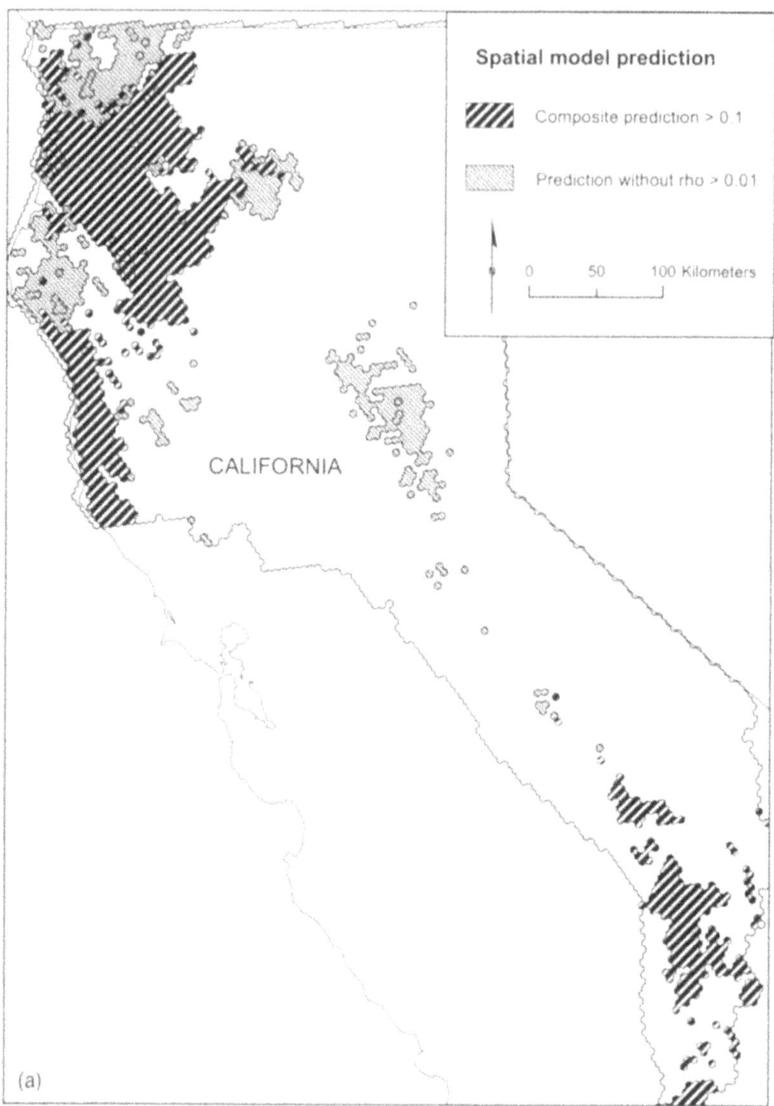

Figure 18.1. Use of a hierarchical spatial model and centrality analysis to evaluate fisher distribution and population connectivity in California. (a) Spatial model predictions of Carroll et al. (2010) identify currently occupied areas (composite prediction), as well as areas of potentially suitable habitat that may be unoccupied currently because of dispersal limitations (prediction without rho). (b) Shortest-path and circuit-flow betweenness centrality metrics were then used to predict habitat linkages that might provide dispersal routes between populations.

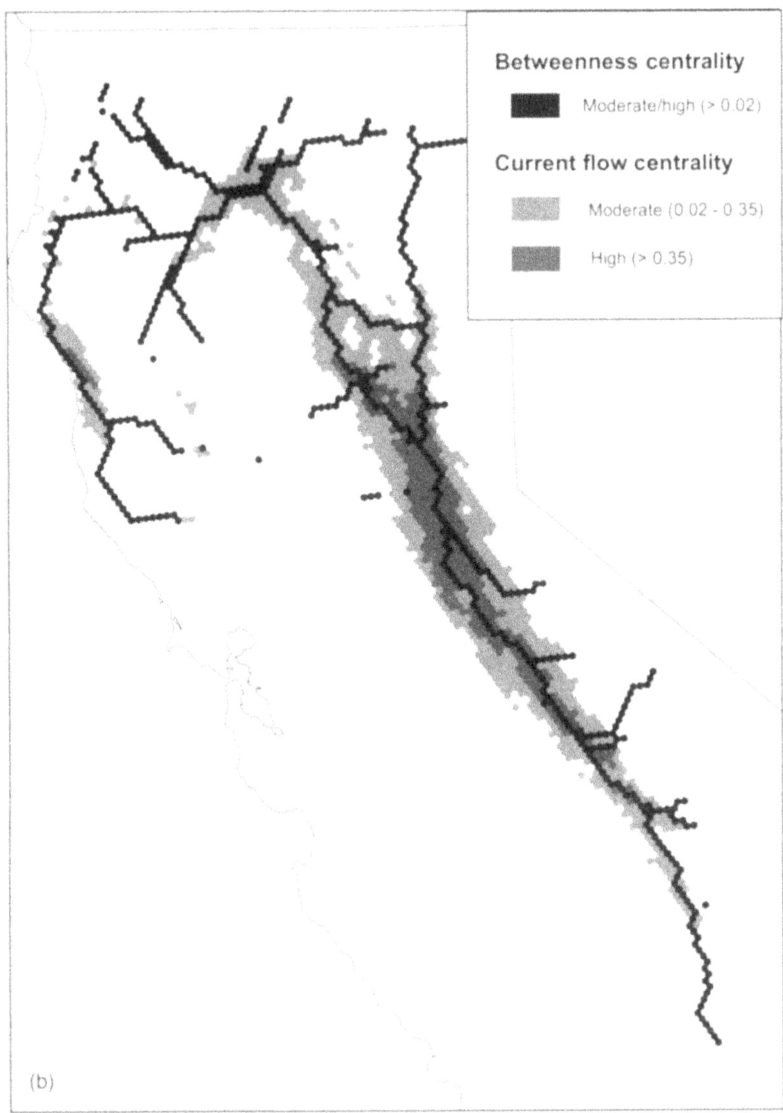

Although distribution models can provide useful guidance during the planning stages, reintroduction efforts are ultimately limited by practical constraints. For example, fishers were reintroduced to the northern Sierra Nevada onto industrial timberlands, where logistical and financial support was available (partly because of incentives for industrial timber companies to avoid listing of the species and attendant regulatory restrictions), rather than onto adjacent public forestlands with higher predicted suitability. The reintroduction provided an opportunity to test whether the contrast between modeled habitat suitability and current fisher distribution in the northern Sierra Nevada is due primarily to dispersal limitation or to the effects of unmeasured environmental variables such as forest type, prey abundance, or the presence and abundance of competitors and predators. Persistence of fishers in the northern Sierra Nevada, especially in areas predicted to provide high-quality habitat, would support hypotheses attributing their previous absence to historical extirpation. Conversely, if fishers do not persist or occupy atypical habitats, that would support hypotheses attributing their absence to lack of key habitat requirements that may be poorly represented in existing models.

Models of Habitat and Population Connectivity

Planners often wish to identify not only highly suitable habitat but also habitat linkages that may facilitate natural recolonization of suitable but unoccupied habitat without costly reintroductions. Models that predict dispersal or population connectivity often use distribution models as input (Table 18.1). This assumes that the habitat attributes that allow occupancy of an area will also facilitate dispersal. Connectivity analysis methods also make implicit assumptions about the movement process itself (McRae et al. 2008; Carroll et al. 2012). Although this model-based uncertainty is difficult to quantify, testing against genetic data can increase confidence in the predictions of habitat-based connectivity models (Schwartz et al., this volume).

Least-cost-path models are the most widely applied linkage-mapping method, in part because they are computationally simple and commonly available in GIS software (Singleton et al. 2002). The least-cost path can be modeled as a combination of the attraction to preferred habitats minus energetic and security costs (e.g., due to topography and exposure to roads, respectively) (Beier et al. 2008). Because habitat quality is expressed as the relative cost or distance to traverse a site, least-cost-path methods are also called shortest-path analyses. One limitation to early applications of the shortest-path method was that the single path identified might have the least total cost-distance but be unrealistic biologically (e.g., if segments of the path traversed developed areas). More recent applications of shortest-path analyses to coarse-scale planning problems have broadened the focus to identify a set of near-optimal paths rather than a single path (Theobald 2006; Beier et al. 2008).

Current-flow models, which are based on the design of electrical circuits, are similar in concept to shortest-path models but have a greater capacity to identify the contributions of multiple linkages (McRae et al. 2008; Newman 2010). These methods treat landscapes as conductive surfaces (i.e., networks of nodes connected by resistors) and represent individuals as random walkers that choose to move along edges at probabilities proportional to habitat-based edge weights. Because the importance of any one linkage in a network is affected by the availability of alternative pathways, a change in one portion of the landscape affects the inferred importance of all others.

Although both shortest-path and current-flow models are most commonly used to map linkages between a single pair of patches, centrality metrics can also be used to map linkages across the landscape. Centrality metrics consider the role of a node (i.e., a patch or site) in mediating flow between all other nodes in the landscape (i.e., as "gatekeepers" for functional connectivity) (Borgatti 2005). Because centrality metrics reflect potential linkages between all node pairs, they avoid the necessity for a priori identification of corridor endpoints. For example, "shortest-path betweenness-centrality" identifies the 1 or several shortest paths that connect each pair of nodes on a graph, and counts the number of such shortest paths in which a node participates (Newman 2010). The loss of a node lying on a large proportion of the shortest paths in the network would disproportionately lengthen distances or transit times between nodes. Like shortest-path corridor-mapping methods, shortest-path betweenness-centrality methods are based on the assumption that the dispersing individual has complete knowledge of the landscape and will choose optimal paths (Freeman 1977). Current-flow (or random-walk) betweenness-centrality assesses the centrality of a node by how often, summed over all node pairs, the node is traversed by a random walk between 2 other nodes (Newman 2010). Thus, this metric counts all paths between nodes, not just the shortest, and is analogous to the behavior of electrical flow in circuits (McRae et al. 2008).

To demonstrate the type of information provided by shortest-path and current-flow betweenness-centralities, we applied these metrics to analyze habitat connectivity for the fisher in California using the Connectivity Analysis Toolkit software (Carroll 2010; Carroll et al. 2012). The input habitat data were based on predicted fisher habitat suitability without the spatial-effect term from Carroll et al. (2010). Shortest-path betweenness-centrality identified a skeletal network of linkages connecting the range of the fisher in California, with a key linkage area connecting northwestern California to the northern Sierra Nevada (Figure 18.1b). Current-flow betweenness-centrality identified a "pinchpoint" of high centrality from the northern Sierra Nevada reintroduction area southward to currently occupied range in the southern Sierra Nevada (Figure 18.1b). This highlights the complementarity of shortest-path betweenness-centrality, which maps a complete network throughout both core and peripheral areas, and current-flow betweenness-centrality, which highlights only the areas where connectivity is important but challenging to maintain. The general outlines of the linkage are not surprising, given that fisher habitat in the Sierra Nevada lies primarily in a restricted band of mid-elevation forest along the western slope of the range (Spencer et al. 2011). The finer-scale pattern of current flow may be useful, however, for establishing conservation priorities within the mid-elevation zone.

When interpreting current-flow output, planners should note that areas with high current values represent pinchpoints of constrained connectivity (i.e., the best among limited options) and not necessarily highly suitable habitat. Such habitat-based linkage maps should be seen as hypotheses to be tested using empirical data on dispersal movements and genetic population structure (Schwartz et al., this volume). Alternatively, a graph network constructed using genetic distances can be tested against habitat data. For example, Garroway et al. (2008) constructed a graph based on genetic data from fishers in Ontario, Canada, and found that node-centrality was negatively related to both the proportion of immigrants in a node and snow depth, suggesting that central nodes were demographic sources located in areas with favorable snow conditions.

Spatially Implicit Dynamic Models

In the 1990s, the legal and policy debate over the northern spotted owl (*Strix occidentalis caurina*) focused attention on the effects of demographic and environmental stochasticity on long-term population viability (Lamberson et al. 1994). Static habitat-suitability models, such as those described above, may identify factors that affect viability deterministically (e.g., habitat loss), but typically do not provide information on viability thresholds below which a small population may go extinct due to stochastic factors, even when suitable habitat is available (Caughley 1994).

Nonspatial and spatially implicit population-viability models analyze such stochastic factors by combining demographic parameters with simplified representations of landscape pattern and structure (Lamberson et al. 1994). Although fishers and martens are relatively well-studied compared with most vertebrate species, demographic parameters in such models must often be estimated using data from other regions or with educated guesses. Nonetheless, nonspatial viability models assist planners by exploring the behavior of small populations and deriving conclusions that may be generalizable to other regions and contexts (Powell et al., this volume). For example, Lacy and Clark (1993) used VORTEX software to generate insights about viability thresholds for small marten populations. VORTEX can incorporate information on carrying capacity and patch isolation derived from the mapped distribution of available habitat, but lacks the topological information (e.g., patch shape) contained in spatially explicit population models (Lacy et al. 2005).

Schneider and Yodzis (1994) developed a pseudospatial model for the Newfoundland marten (*M. a. atrata*) in Canada that used the concept of "optimum territory size" to evaluate the potential influences of spatial dynamics (i.e., habitat quality and heterogeneity) on energy balance and reproductive output. As prey abundance, habitat area, or both decreased, martens in the model increased their territory sizes and associated energy costs, resulting in lower fecundity. Scenarios of population extinction resulted from both deterministic factors (e.g., negative growth rate, habitat loss) and stochastic risks to marginally viable populations. These results confirmed those from earlier nonspatial models (Lacy and Clark 1993), providing a means of linking changes in habitat area and pattern to demographic parameters. A subsequent spatially explicit model predicted marten distribution and viability under varying habitat scenarios (Schneider 1997).

Spatially Explicit Dynamic Population Models

Spatially explicit population models (SEPMs) are a class of simulation models that are both individual based and capable of retaining spatially explicit information on available habitat conditions (DeAngelis and Gross 1992). These models track the fates of many individuals through time as they move across a grid of cells, age, reproduce, and die. The behavior of large numbers of individuals collectively determines the aggregate characteristics that form the model output. SEPMs span a range of complexity, depending on the degree of biological realism and number of demographic parameters they incorporate. Model output may include the mean population size, mean time to extinction, or the percentage of suitable habitat occupied. Because these models can incorporate habitat-specific demographic parameters, the development of SEPMs has allowed data gathered from intensive demographic studies to be combined with GIS maps of landscape composition and pattern in dynamic models (Murphy and Noon 1992).

PATCH, the SEPM used in both case studies presented below (Lewis and Hayes 2004; Spencer et al. 2011), links the survival and fecundity of individuals or groups of animals to GIS data on mortality risk and habitat productivity at the scale of an individual or pack territory (Schumaker et al. 2004). Territories are allocated by intersecting the GIS data with an array of hexagonal cells. The different habitat types in the GIS maps are assigned weights based on the relative levels of fecundity and survival expected in those habitat types. Base survival and reproductive rates, derived from published field studies, are then supplied to the model as a population projection matrix. Data are rarely sufficient to build empirical models of the relations between habitat and demography; thus, these steps typically involve subjective judgments whose effect should be evaluated with sensitivity analyses. The model scales these base matrix values using the habitat weights within each hexagon, with lower means translating into lower survival rates or reproductive output. Each individual in the population is tracked through a yearly cycle of survival, fecundity, and dispersal events. Adult organisms are classified as either territorial or floaters. The movement of territorial individuals is governed by a parameter for site fidelity, but floaters must always search for available breeding sites. Movement decisions use a directed random walk that combines varying proportions of randomness, correlation, and attraction to higher quality habitat (Schumaker et al. 2004). HEXSIM, the successor to PATCH, provides additional analytical capabilities (www.hexsim.net; Heinrichs et al. 2010).

Parameterizing SEPMs requires the input of both predictions from a distribution model (habitat maps) and data on demographic rates (e.g., survival and fecundity in different habitat types). Although uncertainty in such models is difficult to quantify because of the large number of parameters involved, sensitivity analysis can be used to evaluate the robustness of model output to variation in input parameters (Spencer et al. 2011). If SEPM results are considered in the proper context, they have the potential to offer insights into both spatial and nonspatial factors that might influence the success of reintroduction and other conservation efforts. For example, SEPMs, like RSF models, may suggest which areas hold habitat that can support reintroduced populations and, like VORTEX, may suggest the minimum number of individuals to release to ensure a high probability of reintroduction success (Carroll et al. 2003b).

Case Study 1: Using a Dynamic Population Model to Evaluate the Feasibility of Reintroducing Fishers to Washington State

The feasibility assessment preceding the recent fisher reintroduction in Washington (USA) provides an example of the use of SEPM (Lewis and Hayes 2004). A conceptual habitat model was developed to map potentially suitable habitat throughout the Olympic Peninsula and Cascade Range in western Washington; both regions were within the historical range of the fisher and still supported extensive areas of late-successional conifer forests (Figure 18.2). Because the fisher was extirpated from Washington by the mid-1900s, little site-specific ecological or demographic information was available to quantitatively assess habitat availability for a reintroduced population. Consequently, the habitat model developed by Lewis and Hayes (2004) was based on fisher-habitat associations reported in the literature (e.g., Buskirk and Powell 1994; Powell and Zielinski 1994; Weir and Harestad 2003). The habitat suitability model included 4 variables (percent vegetation cover, percent conifer cover, quadratic mean diameter of overstory trees, and elevation) to identify relatively dense, late-successional forests at low and mid-elevations (Lewis and Hayes 2004). The study also identified suitable travel cover for fishers (low- and mid-elevation mid-seral forest) and mapped the distribution of suitable habitat and travel cover. The 3 largest blocks of interconnected

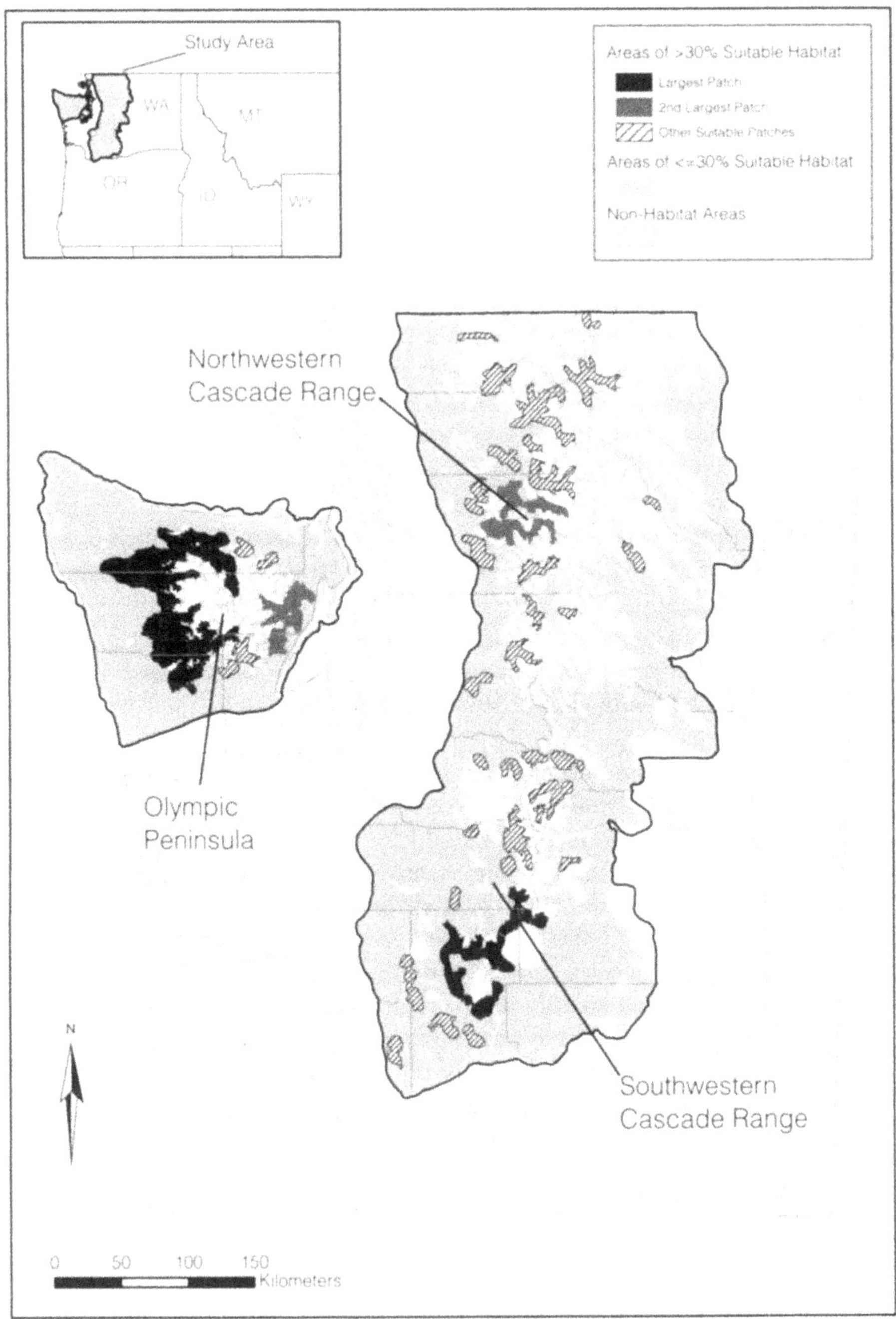

Figure 18.2. The Olympic Peninsula, northwestern Cascade Range, and southwestern Cascade Range were compared to evaluate their capability for supporting a reintroduced fisher population. Ninety fishers were released onto the Olympic Peninsula from 2008 to 2010.

suitable habitat and travel cover were identified as potential reintroduction areas: Olympic Peninsula, southwestern Cascade Range, and northwestern Cascade Range (Figure 18.2).

Lewis and Hayes (2004) used 6 criteria to rank the suitability of these 3 areas for fisher reintroduction: amount of suitable habitat, amount of suitable habitat on public land, amount of suitable habitat in protected areas (National Parks, National Monuments, and Wilderness Areas), amount of land with >50% suitable habitat, proximity to an existing fisher population, and estimated fisher carrying capacity. The authors used a GIS to calculate the first 5 criteria, and PATCH to estimate carrying capacity for each potential reintroduction area (Lewis and Hayes 2004). They used PATCH output to compare relative carrying-capacity estimates among areas, rather than to predict absolute carrying capacities. PATCH simulations incorporated a habitat map and suitability scores for 7 habitat types (Lewis and Hayes 2004). The authors based suitability scores on a scale of 0–10, with 0 being given to those cover types considered nonhabitat (e.g., water, urban) and 10 to cover types considered optimal habitat. Because fishers select late-seral forests at the stand scale and use structural elements typically found in late-seral stands for den and rest sites, the authors considered the late-seral forest cover type to be optimal habitat. The authors considered mid-seral forests to be intermediate in habitat suitability and early seral forests to have low suitability. Habitat scores also differed with elevation (above and below the Pacific silver fir [*Abies amabilis*] zone).

The authors assessed uncertainty in model outputs using sensitivity analyses performed with 3 estimates of dispersal distances and 3 Leslie matrices of survival and fecundity values (Lewis and Hayes 2004). They also used an alternative set of habitat scores in the sensitivity analyses that assigned lower values to suboptimal habitat types. Elasticity analyses indicated that results from all parameter sets (matrices with low, mean, and high values for survival and fecundity) were most sensitive to adult female survival, followed by juvenile survival, and then equally by subadult survival and adult female fecundity. PATCH simulation output was highly sensitive to assumptions concerning the contrast between optimal and suboptimal habitat types, but relatively insensitive to alternative assumptions about dispersal distances (Lewis and Hayes 2004).

The Olympic Peninsula was chosen as the best location for the first fisher reintroduction in Washington because it ranked first in 5 of 6 suitability criteria. Although the Olympic Peninsula ranked third in distance from an existing fisher population, the static habitat-suitability model suggested that the area supported the greatest amount, best configuration, and greatest protection of suitable habitat among the 3 potential reintroduction areas (Figure 18.2). The PATCH model simulations also indicated that the Olympic Peninsula supported a significantly greater carrying capacity than the 2 Cascade areas. Support from a key land-management agency (Olympic National Park) also motivated selection of this reintroduction area.

The 49 fishers released during late 2008 and early 2009 traveled extensively after being released and are currently distributed broadly across diverse landscapes, including lands previously classified as both suitable and unsuitable in the feasibility assessment (Lewis and Hayes 2004; Lewis et al. 2010). Altogether, 90 fishers were released from 2008 to 2010, but many translocated fishers have not yet established home ranges and formal analyses are not yet possible. Researchers are currently monitoring home-range establishment by translocated fishers on the Olympic Peninsula, but a comprehensive assessment of the accuracy of the suitability model will not be possible until longer-term occupancy of home ranges is assessed over the next decade.

Without region-specific data to evaluate the feasibility of a reintroduction, managers must rely on data from other areas. Additionally, like most geographically extensive models, the Washington habitat model did not include inputs for forest-floor structure (e.g., downed logs, slash and cull piles) or prey abundance and distribution, which are likely to be important predictors of fisher occupancy on the Olympic Peninsula. This points to the challenges of extrapolating habitat models developed from data on extant populations to new areas that may contain very different prey and vegetation communities. Although the assessment of suitable habitat for fishers was limited by an incomplete understanding of fisher habitat requirements in the coastal Pacific Northwest, the models provided useful tools to evaluate and draw informed conclusions about the feasibility of a successful reintroduction based on the best available information.

Case Study 2: Using Coupled Dynamic Habitat and Population Models to Evaluate Fisher Conservation Strategies in the Southern Sierra Nevada

An important limitation of most habitat models for *Martes* is that they predict suitability based on a static "snapshot" of current vegetation conditions. Consequently, some planners have sought to project how such vegetation attributes may change over time to better predict long-term habitat suitability. The accelerating pace of anthropogenic climate change (Lawler et al., this volume) is an additional motivation for analyzing habitat conditions in a dynamic model. To address these issues, Spencer et al. (2011) and Scheller et al. (2011) used coupled dynamic habitat and population models to project the potential effects of alternative management strategies (e.g., landscape-level fuels treatments) on the southern Sierra Nevada fisher population. This study was among the first to integrate many of the modeling approaches described above into a single decision-support process.

The analysis area supports the southernmost population of fishers, consisting of a few hundred individuals in the southern Sierra Nevada that are isolated from other populations by >400 km (Zielinski et al. 2005) (Figure 18.3). Several studies have projected future increases in wildfire intensity and extent in this region (Miller et al. 2009; Westerling et al. 2009), motivating proposals for vegetation management (e.g., forest thinning) to alter wildfire behavior (U.S. Department of Agriculture 2004). Such actions may also reduce fisher habitat quality and continuity, at least in the short term, because fishers select the densest forest stands as resting habitat (Zielinski et al. 2004a). To assess the status of the population and evaluate alternative vegetation-management approaches, Spencer et al. (2011) and Scheller et al. (2011) coupled a fisher SEPM with other spatially explicit models. The coupled models allowed the authors to test alternative hypotheses in a simulation environment, evaluate the relative risks of fuels treatments and wildfires to the population, and identify conservation actions that are likely to sustain and expand the population (Scheller et al. 2011).

Spencer et al. (2011) built a number of alternative GAM models using GIS data layers and fisher detection-nondetection data from a regional systematic monitoring program (Zielinski et al. 2006c). The best GAM model provided a strong fit to the fisher data using 3 predictors: latitude-adjusted elevation, annual precipitation, and total aboveground biomass of trees (kg/ha) averaged over 5 km^2 (see Spencer et al. 2011). The model suggested that fishers in the southern Sierra Nevada are associated with mid-elevation forests that experience relatively low annual precipitation and support abundant large trees. Forest biomass provided the best discrimination of fisher occupancy among the biotic variables tested, probably because it correlates strongly with many forest-structure variables known to predict fisher habitat selection at finer spatial scales (e.g., large trees, dense canopies, abundant dead-wood structures) (Zielinski et al. 2004a,b, 2006c).

Spencer et al. (2011) then parameterized the PATCH model using predicted probability of fisher occurrence from the GAM to estimate an equilibrium fisher population size (or carrying capacity) and to identify potential population source, sink, and expansion areas. Each territory was assigned a score from 0 to 10 in PATCH, based on mean occupancy scores within the territory. High-quality territories (territory score >7.5, reflecting an average within-territory occupancy >0.75) were assigned high fecundity and survivorship rates based on a 4-stage Leslie matrix and demographic rates from Lewis and Hayes (2004). Below the maximum score of 7.5, fecundity and survival rates were decremented linearly with territory score. The average resulting lambda (intrinsic rate of population growth) for territories scoring >7.5 was 1.19 (i.e., predicted source territories); lambda for territories scoring about 6.0–7.0 averaged about 1.0 (stable population); and lambda for territories scoring less than about 5.5 averaged <1.0 (predicted sink

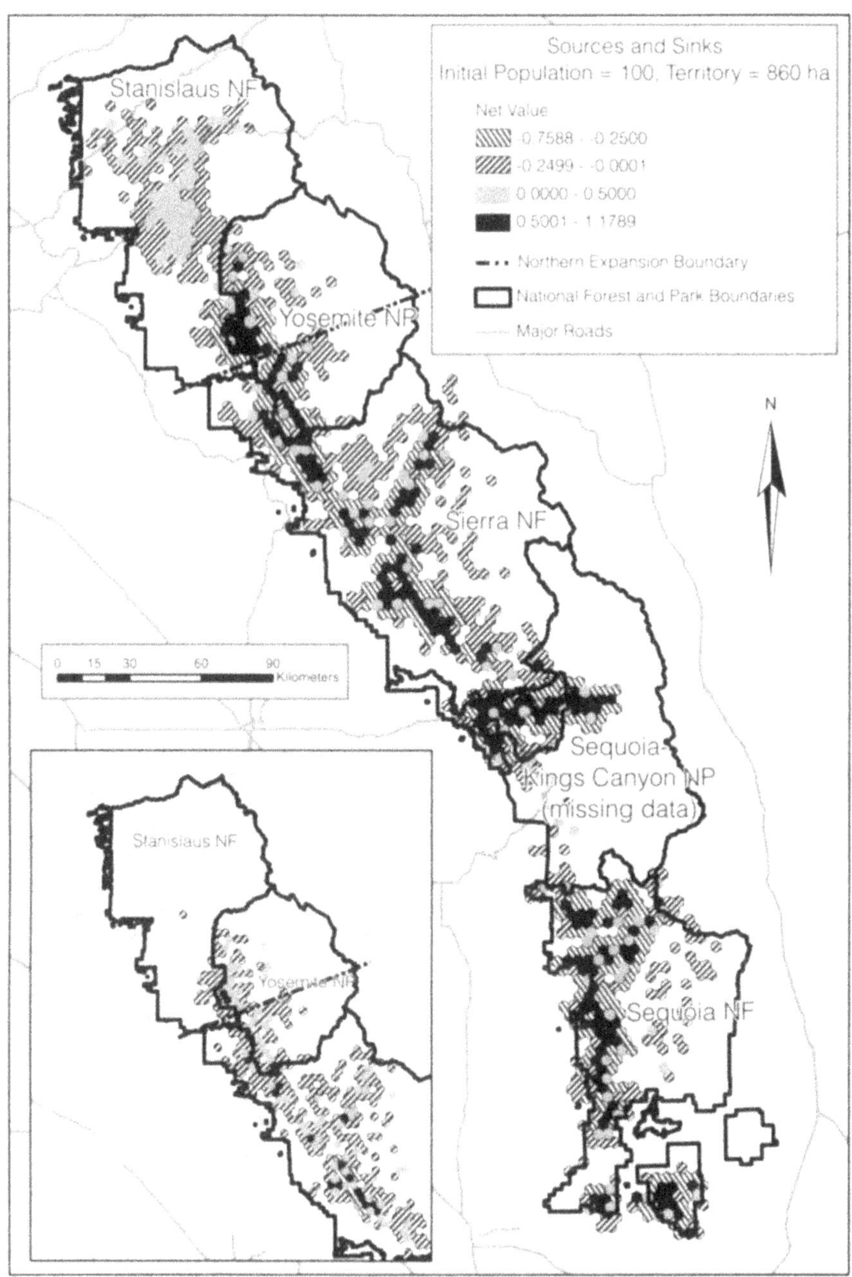

Figure 18.3. Influence of survival rate on modeled fisher source-sink dynamics and potential for population expansion north of the Merced River (dashed line) in California. Net value of territories (annualized births minus deaths) were averaged across 20 replicate simulations run for 200 years using intermediate parameter values. Source territories (births exceed deaths) are shaded, and sink territories (deaths exceed births) are hatched. The inset illustrates the reduced potential for northward expansion and establishment of new source territories with a 10% reduction in survivorship relative to baseline assumptions. Data gaps caused underrepresentation of territory value in Sequoia-Kings Canyon National Park. Adapted from Spencer et al. (2011).

territories) (Spencer et al. 2011). Despite assumptions about linear correlations between fisher vital rates and predicted occupancy (from the GAM), the SEPM parameters, when combined with appropriate sensitivity analyses, were judged useful for estimating potential equilibrium population sizes and for relative comparisons between alternative forest-management scenarios. Fishers were initially seeded into the highest-scoring territories and those where fishers were detected in the monitoring data. After the SEPM was run for 40 years to allow the model population to reach equilibrium on the landscape, the ending population size was recorded as a measure of carrying capacity for comparisons among different scenarios. The authors compared scenarios using equilibrium behavior (e.g., ending population size) rather than transient dynamics (e.g., temporal trends) because the latter metrics are typically more sensitive to uncertainty in model parameters (Grimm and Railsback 2005). Model uncertainty was assessed using a sensitivity analysis that varied vital rates, territory sizes, initial population sizes, maximum dispersal distances, and other model parameters. Population dynamics were highly sensitive to adult female survivorship and relatively insensitive to other age-specific vital rates. Territory size had a greater effect on estimated carrying capacity than did initial population size or dispersal distance (Spencer et al. 2011).

The southern Sierra Nevada fisher population is generally restricted to habitats south of the Merced River, despite some apparently suitable habitat to the north (Figure 18.3). Spencer et al. (2011) estimated carrying capacity in the area occupied currently by constraining the model fishers from establishing territories north of the river. To evaluate the potential for northward expansion under current habitat conditions, the authors then allowed fishers to establish territories north of the river. Because there is evidence of elevated human-influenced mortality rates in the northern portion of the occupied range (e.g., roadkill, diseases spread by cats and dogs), the analysis then decreased survivorship in 5% increments throughout the entire analysis area to observe effects on the modeled population expansion.

Using the most defensible range of input parameters, the SEPM results estimate the equilibrium population size at 73–147 adult females south of the Merced River (Figure 18.3). After weighing various uncertainties in these estimates and comparing them with several independent estimates (based on density extrapolations from fisher field studies and sampling theory applied to the regional fisher monitoring data), the authors estimated that the current population is most likely between 160 and 360 individuals, and includes only about 55–120 adult females.

When modeled fishers were allowed to cross the Merced River, the population expanded northward into mostly moderate-value habitat, establishing modest new population source areas north of the Tuolumne River. The expansion front stalled about 90 km north of the Merced River, where predicted probability of fisher occurrence generally dropped below 0.5. Predicted territory occupancy north of the Merced River dropped with decreasing survivorship rates, with no expansion north of the Tuolumne River when the survival rates were reduced by 15%, and no expansion north of the Merced River when survival rates were reduced by 20%.

To evaluate how fires, vegetation management, and other disturbances might affect the population, Scheller et al. (2011) coupled the spatially explicit habitat and population models with a stochastic, spatially explicit landscape-change model, LANDIS-II (Mladenoff 2004; Scheller et al. 2007). LANDIS-II simulates how succession, fires, forest management, and other processes may affect forest vegetation over time by tracking changes in forest biomass by tree species and age cohorts. Using a factorial experimental design with replication, the authors used LANDIS-II to simulate vegetation for 50 years in a variety of scenarios. These scenarios assumed 2 alternative future fire regimes (1 calibrated to match recent fire histories, and 1 that assumed more severe fire conditions in the future) and different types, rates, and locations of fuels treatments. For each decade and replicate, the forest-biomass outputs from LANDIS-II were used to update the habitat model on which PATCH runs. The model did not simulate direct mortality of fishers due, for example, to wildfire, but fishers could disperse to higher suitability areas in the event a fire or management action reduced the habitat value of their current territory. Model fishers unable to settle into unoccupied, high-quality territories following a disturbance event would experience relatively low survival and fecundity rates, as appropriate for the value of the particular territory where they settled. Scenarios were compared using total female population size at year 60. Structural-equation models were used to assess the net effects of fuels treatments on fishers, which can include direct negative effects (removal of forest biomass by logging) as well as indirect positive effects (by limiting fires that also remove biomass).

Model results suggested that, across a range of alternative fire regimes and vegetation treatments, the positive indirect effects of fuels treatments on fishers generally outweighed the direct negative effects. However, coarse-resolution landscape and fuels models, such as those used in this study, more accurately project changes in fire extent than in fire severity, yet the latter factor may have a greater impact on fisher habitat suitability. Spencer et al. (2011) and Scheller et al. (2011) did not use a formal population-viability analysis to estimate level of risk (e.g., probabilities of extirpation); however, the relatively small size of the southern Sierra Nevada fisher population, at <400 individuals, suggests that it may be at risk of extirpation due to stochastic threats. Northward expansion of the population appeared to have the greatest potential to increase the population's size and viability. Population expansion could be facilitated by (1) forest-management activities in the area north of the Merced River that increase fisher habitat quality and contiguity, and (2) adaptive management in the northern portion of the current range (in and immediately south of Yosemite National Park), to reduce fisher mortality sources and increase survivorship. The authors concluded that fuel treatments strategically located in and adjacent to areas with the greatest risk of severe crown fires, based on evaluation of human ignition sources and fuels characteristics, could also reduce risks to fisher habitat quality.

Multispecies Models

Although legal mandates often focus on conservation of individual species of concern, single-species conservation planning may often lead to less efficient solutions than an approach that simultaneously addresses the habitat needs of many species of concern (Marcot and Raphael, this volume). Several software packages (e.g., MARXAN [Ball et al. 2009], ZONATION [Moilanen et al. 2009]) are commonly used to analyze output from distribution models to find solutions (i.e., landscape designs) that optimally (e.g., in the minimum area) capture habitat needs for many species. Because most real-world optimization problems are too complex to be solved exactly, reserve-selection models identify both an approximation of the "best" solution and the degree of variation (or uncertainty) among alternative approximations (Ball et al. 2009). MARXAN has been used to assess the overlap and contrasts between the habitat requirements of *Martes* species and those of other taxa in the Rocky Mountain region and in California (Carroll et al. 2003a; Zielinski et al. 2006a).

Although these studies addressed spatial relations among habitat conditions for multiple species, it is more challenging to project how *Martes* populations will interact dynamically with populations of potential prey or predator species. Using the ZONATION software, Rayfield et al. (2009) developed a conservation plan for the American marten in an area of Quebec, Canada that considered the spatial interaction between the distribution of the marten and its 2 primary prey species. Areas were identified providing suitable habitat for all 3 species and of sufficient area and connectivity for marten population persistence. As yet, no studies have projected the indirect effects of climate change on *Martes* populations via changes in vegetation and disturbance patterns, although these have recently been completed for other taxa (Keith et al. 2008).

Use of Models within the Conservation Planning Process

The 4 types of models reviewed above (predicting distribution, dispersal, persistence, and optimal reserves; Table 18.1) have distinct roles within conservation-planning processes for *Martes*. Distribution models in some form are used to support almost all planning efforts attempting to integrate

Martes conservation into land-management practices (e.g., Spencer et al. 2011). The relevance of the remaining model types varies depending on available data and the goals of the planning process. Connectivity or dispersal is rarely addressed in planning for *Martes* or other wildlife populations because of limited availability of dispersal data and limited understanding of the influence of dispersal processes on population persistence. New techniques for analysis of genetic data may allow connectivity to be assessed more rigorously in the planning process (Schwartz et al., this volume).

Because the ultimate goal of *Martes* conservation is population persistence, one might anticipate broad use of viability models such as SEPM (Schumaker et al. 2004). The limited availability of demographic data, and the uncertainty associated with using demographic parameters from other populations, however, have limited the application of such complex models in planning processes. Recent examples (e.g., Spencer et al. 2011) where such models were developed, evaluated with detailed sensitivity analyses, and applied to management issues may encourage their broader use in planning. The key to appropriate use of such complex models is to assess what qualitative insights are robust to the many potential sources of uncertainty.

Conclusions

The relatively large territory sizes and habitat-limited dispersal ability of most *Martes* species suggest that their persistence may be influenced by large-scale population processes, as is true for other forest-associated "focal" species, such as the northern spotted owl (Thompson 1991; Lamberson et al. 1994). In contrast to findings about the spotted owl, several studies suggest that habitat requirements for the fisher in western North America are complex and poorly described by broad-scale GIS layers for such attributes as forest type and age class (Davis et al. 2007; Carroll et al. 2010). Fisher-habitat models with a relatively large proportion of unexplained variance, or that contain covariates that are surrogates for unmeasured factors, will have poor generality when extrapolated to other areas (Carroll et al. 1999). When this problem is combined with the challenge of representing large-scale population processes accurately in a simulation model, it is clear that the predictions of quantitative habitat and viability models will be highly uncertain. Additionally, the examples discussed in this review suggest that the link between the results of such models and management is often mediated by the many factors that constrain management decisions. Nevertheless, the costs associated with reintroductions and other broad-scale conservation efforts generally justify the use of such models to inform the planning process before substantial resources are committed. The complexity of conserving *Martes* populations, given increasing human demands on forest ecosystems, suggests that quantitative models will be increasingly relevant for enhancing the success of *Martes* conservation efforts, as long as model uncertainty is considered explicitly and model results are used to inform decisions rather than predict outcomes.

Acknowledgments

We thank the many biologists who assisted in collection of the fisher distribution data that supported the analyses described in this chapter. H. Rustigian-Romsos and John Jacobson helped prepare the figures. M. Raphael, K. Aubry, and 2 anonymous reviewers provided helpful comments.

19

Conservation of Martens, Sables, and Fishers in Multispecies Bioregional Assessments

BRUCE G. MARCOT AND MARTIN G. RAPHAEL

ABSTRACT

We review conservation strategies and guidelines for *Martes* species from several multispecies bioregional assessments throughout the world. We define a "multispecies bioregional assessment" as an evaluation of a species' status, habitat, ecology, and conservation needs at broad scales of geography and environmental conditions that integrates assessment and objectives for other species and ecosystem values. We review how conservation of *Martes* species is addressed in such assessments at broad landscape and regional scales, including descriptions of habitat conditions and patterns (e.g., forest structural and age classes, forest patch sizes, connectivity, and provision of key habitat elements) and integration with other regional-scale guidelines (including management for other species). Examples presented include bioregional assessments of sable (*M. zibellina*) in Far East Russia and northeastern China; Nilgiri marten (*M. gwatkinsii*) in south India; and Pacific marten (*M. caurina*), American marten (*M. americana*), and fisher (*M. pennanti*) in the Columbia River Basin of the western United States, Sierra Nevada of California (USA), U.S. Pacific Northwest, and southeast Alaska (USA). We summarize these examples and present steps that can provide a general framework for multispecies bioregional assessments, and identify *Martes* species and locations where such assessments are lacking and could be developed.

Introduction

The conservation of species deemed to be at risk or subject to loss from harvest often takes the form of single-species approaches to developing and implementing management guidelines. Nothing replaces local autecological research on species' life history, population status and dynamics, and responses to threats and stressors. For ensuring the long-term viability of at-risk species, however, a greater measure of success can be provided by embedding single-species assessments and guidelines into the broader context of multispecies bioregional assessments.

Martens, sables, and fishers (*Martes* spp.) occupy a diverse array of ecosystems throughout the world that vary greatly in their ecological communities, sympatric plant and animal species, and conservation challenges. Although species-specific conservation assessments and guidelines have been developed for a number of *Martes* species, few have been formally integrated into broader contexts of communities and ecosystems at the bioregional scale. The purpose of this chapter is (1) to review the state of such multispecies bioregional assessments and conservation guidelines pertaining to *Martes* species, (2) to explore the utility of such community-scale regional assessments for conservation of *Martes* species, and (3) to suggest new avenues for the conservation of *Martes* species. New avenues can include integrating the conservation of *Martes* species into general goals for biodiversity conservation, ecosystem management, sustainable natural-resource development, restoration of ecological communities, and other objectives.

What Is a Multispecies Bioregional Assessment?

We define a "multispecies bioregional assessment" as an evaluation of a species' status, habitat, ecology, and conservation needs at broad geographic and environmental scales in the context of management for other species. A multispecies bioregional assessment integrates assessment and management objectives for other species and ecosystem values (the "multispecies" approach), encompassing a significant portion of the species' range in the context of its broader ecological community (the "bioregion"). Other conservation objectives considered in a multispecies bioregional assessment can include evaluation and management of a broader suite of species or ecological communities; consideration of ecosystem-management goals, ecological disturbance regimes, and anthropogenic stressors; and improved international and other transboundary coordination. In this way, a multispecies bioregional assessment differs from, but can incorporate, the more traditional, single-species approach of assessing population viability, responses to harvest, and effects of threats and stressors. Thus, we use the term multispecies bioregional assessment to include not just an evaluation of the status, habitat, and ecology of a focal species in an ecosystem context, but also considerations for conservation or restoration guidelines in broader multispecies, community, and ecosystem contexts.

A bioregional scale, defined by broad geographic extent, can be of value for conservation assessments and management for several reasons. First, understanding how a given *Martes* species interacts with other organisms (e.g., as prey, predator, or competitor, as well as with vegetation structural elements and other substrates that provide key resting and denning sites) can provide information useful for devising effective conservation and restoration guidelines. Second, objectives and guidelines for conservation and resource use can often conflict. Considering up front how the species fits into a broader tapestry of administrative policies, legal mandates, and resource interests can help avoid unnecessary disputes and solve conservation problems among otherwise disparate interest groups and stakeholders. Third, bioregional assessments can provide the most meaningful context in which to consider overall population responses to conservation activities, including identifying needs, locations, and methods to provide core habitats and habitat linkages. Fourth, the multispecies management approach can improve the efficiency of conservation efforts by concomitantly addressing multiple at-risk species. Finally, the multispecies approach can provide greater efficiency in evaluating the potential responses of entire ecological communities and their component species to dynamic changes in their environment from anthropogenic stressors, major disturbance events, and climate change.

The use of bioregional approaches to conserve ecosystems and species is an increasingly common management strategy and one that has been implemented in a variety of circumstances (Johnson et al. 1999; Busch and Trexler 2003). For example, Hargiss et al. (2008) conducted an evaluation of seasonal wetlands in the Prairie Pothole Region in the northern Great Plains of the United States and Canada that resulted in the compilation of a comprehensive ecological dataset that could be used for mitigation, monitoring, inventory, and evaluation of ecological functions. Higgins et al. (2005) used a multispecies regional approach to integrate freshwater biodiversity into an evaluation of critical areas for conservation in the Columbia River Basin of the western United States and the Paraguay River in central South America. Large-mammal conservation, natural disturbances, and human influences were the centerpieces of a multispecies conservation assessment of the St. Elias region of Alaska (Danby and Slocombe 2005), and, in southern Appalachia (USA), Flebbe and Herrig (2000) conducted a bioregional evaluation of imperiled aquatic species. Mascarenhas et al. (2010) used a regional approach to identify a set of indicators of sustainable development in Portugal, concluding that such an approach provided a coherent assessment framework, prevented duplication of effort, and enabled the scaling-down of results to local areas. Collectively, these are advantages that could also pertain to incorporating *Martes* species in bioregional assessments.

In our review, we considered all 8 recognized species in the genus *Martes* (Table 19.1; Buskirk 1994; Proulx et al. 1997), and their global geographic distributions (Proulx et al. 2004). Based on recent evidence of species-level differences among American martens (*M. americana*), we also included the

Table 19.1. *Martes* species evaluated in this chapter, their geographic distributions, and whether they have been included in a bioregional assessment

Scientific name	Common name	Geographic distribution	Included in a bioregional assessment?
M. flavigula	Yellow-throated marten	Far East, southeast Asia	No
M. foina	Stone marten	Europe, Middle East, south Asia	No
M. martes	European pine marten	Eurasia	No
M. melampus	Japanese marten	Japan, Korean Peninsula	No
M. zibellina	Sable	Northern Asia, Far East	Yes
M. gwatkinsii	Nilgiri marten	South India (Western Ghats mountains)	Yes
M. americana	American marten	Northern North America	Yes
M. caurina	Pacific marten	Western U.S. and Canada	Yes
M. pennanti	Fisher	Northern North America	Yes

Pacific marten (*M. caurina*) of the western United States and Canada as a distinct species (Carr and Hicks 1997; Dawson and Cook, this volume).

Methods

We drew from a wide array of source materials for this review, particularly journal publications, published books, agency reports, Internet websites, and numerous personal communications (see Acknowledgments). Our search for source material focused on examples of multispecies and ecosystem-management assessments, conservation strategies, and resource-management plans in which *Martes* species had been addressed explicitly. We did not attempt to evaluate the many management plans for individual *Martes* species, such as local, state, provincial, or regional guidelines for trapping, hunting, or conservation, but we do cite a few examples of these. Rather, our review was focused on broader bioregional (multispecies and ecosystem-scale) assessments, as defined above. Although we intended our search for information to be comprehensive, we may have missed some pertinent assessments not highlighted in the source materials we obtained, especially in non-English sources.

Results

We report here on all instances where we discovered that a *Martes* species had been included in multispecies bioregional assessments and management guidelines (Tables 19.1, 19.2). These included: the sable (*M. zibellina*) in Far East Russia and northeastern China; Nilgiri marten (*M. gwatkinsii*) in south India; and American and Pacific martens and fisher (*M. pennanti*) in the Columbia River Basin of the western United States, Sierra Nevada of California (USA), U.S. Pacific Northwest, and southeast Alaska (USA).

Sable

Sables occur, and are harvested, in much of the Russian Commonwealth of Independent States, most heavily in the Far East. This species was included explicitly in at least 1 transboundary bioregional resource-planning effort in the 1990s that spanned the Russian Far East provinces of Primorski Krai and Khabarovsky Krai, and the northeastern China province of Heilongjiang in the Ussuri River watershed (ESD 1996; Marcot et al. 1997). This bioregion, characterized as the greater Ussuri River watershed, because that river forms a significant part of the Russian-Chinese border in the Far East, covered 262,000 km^2, an area larger than the United Kingdom. In that plan, the distribution of sable population centers (denoted by sable density) was mapped by Pacific Institute of Geography, Far East Academy of Sciences, Vladivostok, Russia, mostly throughout the Sikote-Alin Mountain Range. Potentially suitable habitat also was identified in the Wandashan and Laoling Mountains across the border in China by the Harbin Remote Sensing Centre.

Corresponding forest habitats were evaluated for the co-occurrence of other species of traditional and economic importance, including the brown bear (*Ursus arctos*), Asiatic black bear (*U. thibetanus*), Siberian tiger (*Panthera tigris altaica*), raccoon dog (*Nyctereutes procyonoides*), Amur wildcat (*Felis euptilura*), Siberian spruce grouse (*Dendragapus falcipennis*), and other rare and locally endemic wildlife species, as well as economically valuable plants such as ginseng (*Panax ginseng*), timber species such as Korean pine (*Pinus koraiensis*), and centers of endemic plant diversity. By overlaying all such distributions, including the sable's, the assessment team was able to delineate "hot spots" of biodiversity and culturally and economically important resources. The hot spots were delineated on maps as potential new protected areas, including several possible international peace parks based, in part, on the International Union for Conservation of Nature's (IUCN) protected area-management categories (IUCN 1994). The team provided a sustainable development plan with management guidelines to the Russian and Chinese governments that would help restore or conserve such forest habitats and their biota.

Table 19.2. Characteristics of multispecies bioregional assessments that include *Martes* species

Species	Location and size of assessment area	Salient features of the assessment pertaining to *Martes*	Source
M. zibellina	Greater Ussuri River watershed, Far East Russia and northeast China; 262,000 km^2	Included with a set of other regionally rare or endemic wildlife species, economically valuable plants, and centers of endemic plant diversity; mapped as hot spots of biodiversity and culturally and economically important resources spanning the border of Russia and China	ESD 1996; Marcot et al. 1997
M. gwatkinsii	Anaimalai Mountains, south India; 2338 km^2	Included as 1 of 7 regionally endemic wildlife species for conservation and restoration of native tropical forests and mitigation of adverse human activities, including antipoaching measures and delineation of habitat corridors among protected areas	Sajeev et al. 2002
M. caurina, M. pennanti	Interior Columbia River Basin, USA; 580,000 km^2	Included in wildlife-habitat databases; analyzed using Bayesian network models to determine degree of habitat reduction from historical (or to future) conditions; guidelines specified for old-forest source habitat including snags and large trees to also benefit a suite of associated wildlife species	Quigley et al. 1996; Marcot 1997; Wisdom et al. 2000; Raphael et al. 2001
M. caurina, M. pennanti	Sierra Nevada, USA; 69,560 km2	Identified for high-priority adaptive-management studies, along with 4 other focal wildlife species associated with old-forest ecosystems of conservation interest; guidelines for habitat conservation, den site protection, and habitat fragmentation included in multispecies planning approach	U.S. Department of Agriculture 2010
M. caurina, M. pennanti	Pacific Northwest, USA; 250,000 km2	Included in databases and habitat assessments, and for guidelines on late-successional forest reserves, riparian reserves, and habitat connectivity among reserves; viability status determined by expert panel	Forest Ecosystem Management Assessment Team 1993; U.S. Departments of Agriculture and Interior 1994
M. americana, M. caurina	Southeast Alaska, USA; 68,000 km^2	Part of an interagency, multispecies viability assessment, with guidelines for a network of old-forest reserves and specific protection measures for den and resting sites and beach fringe forest habitat	Suring et al. 1993; U.S. Department of Agriculture 2008

The sustainable-development plan included the delineation of key cross-boundary transportation routes for economic interchange and development between Russia and China, as well as consideration for border protection with a no-development buffer zone along the international boundary. It also presented general guidelines for potential management and restoration activities to provide for habitat conservation and human use of natural resources, including fishing, mineral mining, and logging of selected mixed conifer-hardwood forests of the region. Guidelines for natural-resource use took into account the need to conserve or restore hot spots of at-risk and economically important species, including the sable. Guidelines provided for protection of native woodland and forest environments, within hot-spot designations, from excessive amounts of timber harvesting, road clearing, and human settlement, but also allowed for limited resource consumption (e.g., pine nut harvesting by local indigenous peoples). Such guidelines were designed to provide for controlled resource use while maintaining forest-habitat conditions for suites of at-risk species, including the sable.

More specifically, the plan suggested designation of a formal Protected Territory of Traditional Use (based on IUCN Protected Area Category V; IUCN 1994) that would conserve mature, native conifer and hardwood forests in the heart of the Sikhote-Alin Mountains of Primorski Krai, Russia, along the middle and upper portions of the mostly undisturbed Bikin River watershed. In part, such protection would give the local Udege people access to productive and sustainable resources, including trapping of sables, fishing and hunting of game, and harvest of Korean pine nuts and other nontimber forest products, including edible and medicinal plants. Subsequently, the Primorski Krai government designated the area as a Territory of Traditional Natural Resource Use, which met the guidelines of the plan. In this way, both the protection and the trapping of sables were components of this broader resource and land allocation.

Since the plan was published, a Sino-Russian International Ussuri Commission was created to help resolve conflicts over transborder resource use to help sustain the regional economy, and to provide the basis for future joint establishment and management of shared parks and protected areas. Although the plan was at least partially adopted, the current status and trend of sables in the region are poorly known.

Nilgiri Marten

The Nilgiri marten occurs only in the Nilgiri and adjacent hills of the Anaimalai Mountains of the Western Ghats in south India. This species was identified as one component of a regional, multispecies assessment and planning effort for what is called the Anaimalai Conservation Area, which spans some 2338 km^2 and includes Anaimalai Wildlife Sanctuary and managed forests of Dindugal and Kodaikanal districts (Sajeev et al. 2002). The Anaimalai Conservation Area contains 12 major forest types, including native evergreen and deciduous tropical forests, and forestry plantations of non-native tree species, as well as agricultural plantations of coffee, tea, rubber, banana, and other species of high economic value.

This assessment was the first of its kind in the region. It included developing wildlife-habitat relationships models, based on literature review and expert judgment; describing habitats of rare or endemic species of conservation concern; and suggesting landscape-scale guidelines for habitat conservation and restoration in the context of human habitations and resource use of the region. In this assessment, the Nilgiri marten was identified as one of the most endangered mammals in the study area, and was featured as 1 of 7 regionally endemic wildlife species, including the Nilgiri langur (*Presbytis johni*), lion-tailed macaque (*Macaca silenus*), grizzled giant squirrel (*Ratufa macroura*), dhole or Indian wild dog (*Cuon alpinus*), and Nilgiri tahr (*Hemitragus hylocrius*). The regional assessment also included consideration of habitat associations of a larger suite of species, including 51 mammals, 260 birds, 40 reptiles, 47 amphibians, and 28 fish.

The Nilgiri marten was identified as being primarily associated with old growth and late-successional stages of native evergreen and moist deciduous tropical forests. The Nilgiri marten was also considered sensitive to a wide array of human activities including clearing of underbrush, cutting of lianas, extraction of nontimber forest products, human-set ground fires, cutting of tree branches for fodder, felling of snags and hollow trees, and animal poaching. The assessment also identified old forest and large snags and logs (for dens) as key habitat features of the Nilgiri marten. The multispecies aspect of the assessment enabled resource managers to identify a wide array of other wildlife species that share the marten's forest habitat (including microhabitat elements) and are also sensitive to these human activities.

The conservation and restoration of these habitat features, and the mitigation or control of human-activity stressors, were highlighted by the assessment for inclusion in forest-management plans outside protected areas within the Anaimalai Conservation Area, thereby benefiting not only the Nilgiri marten but a host of other wildlife species as well. For example, the assessment suggested a generalized approach to antipoaching measures, including locating enforcement camps at key vantage points to intercept poachers, developing and strengthening the infrastructure of intelligence networks and coordination with other enforcement agencies, conducting security audits with local tribes, monitoring poaching patterns, providing insurance coverage for the protection staff, and offering rewards and incentives for reporting and thwarting poaching activities. General guidelines were also suggested for curtailing or regulating the removal of nontimber forest products, restoring old native forests, designating habitat corridors among protected areas, and engaging in other activities that would benefit the Nilgiri marten and many other at-risk species in the region.

American and Pacific Martens and Fisher

Much work has been done on martens and fishers in North America, especially field research studies and the development of species-specific conservation plans and harvest regulations. Four major regional assessments included martens and fishers explicitly in multispecies assessments and ecosystem-management guidelines at broad geographic scales.

Interior Columbia River Basin, Western United States

Both the Pacific marten and fisher were included in wildlife-habitat relationships databases and integrated into overall habitat-management guidelines in the U.S. Forest Service's and Bureau of Land Management's Interior Columbia Basin Ecosystem Management Project conducted from 1993 to 1996 (Quigley et al. 1996; Marcot 1997). This project covered a 580,000-km^2 area of the interior western United States—an area larger than France. The project included development of models for Pacific marten and fisher source habitats (i.e., optimal habitats that contribute to stationary or growing populations) and population carrying capacity (Wisdom et al. 2000; Raphael et al. 2001).

Although never implemented, the resulting land-use plan (U.S. Departments of Agriculture and Interior 2000) nonetheless integrated consideration of Pacific marten and fisher habitat requirements into a broad suite of multispecies management guidelines at the regional scale. This integration was accomplished partly by developing and applying several dozen Bayesian network (probability-based) models of population density outcomes at the subwatershed scale, given habitat density (proportion of the assessment area in source [optimal] habitat conditions for the species, compared with historical median habitat density) and expected trends in key resting and denning microhabitats (large snags and down logs) and anthropogenic stressors of human populations and roads (Figure 19.1). The models were run on historical, current, and several potential future-scenario conditions for each subwatershed in the assessment area, providing regional habitat projections for each species (Figure 19.2). Habitat maps were then combined among species, and areas of multispecies habitat concentrations were identified as one basis for developing conservation guidelines.

Both the Pacific marten and fisher were included in a suite of wildlife species whose old-forest, broad-elevation source habitat had declined substantially in geographic extent compared with historical amounts in the project area (Wisdom et al. 2000). The Pacific marten was selected to represent species that were positively associated with large snags and logs, and negatively associated with human disturbance. Pacific marten habitat was classified into

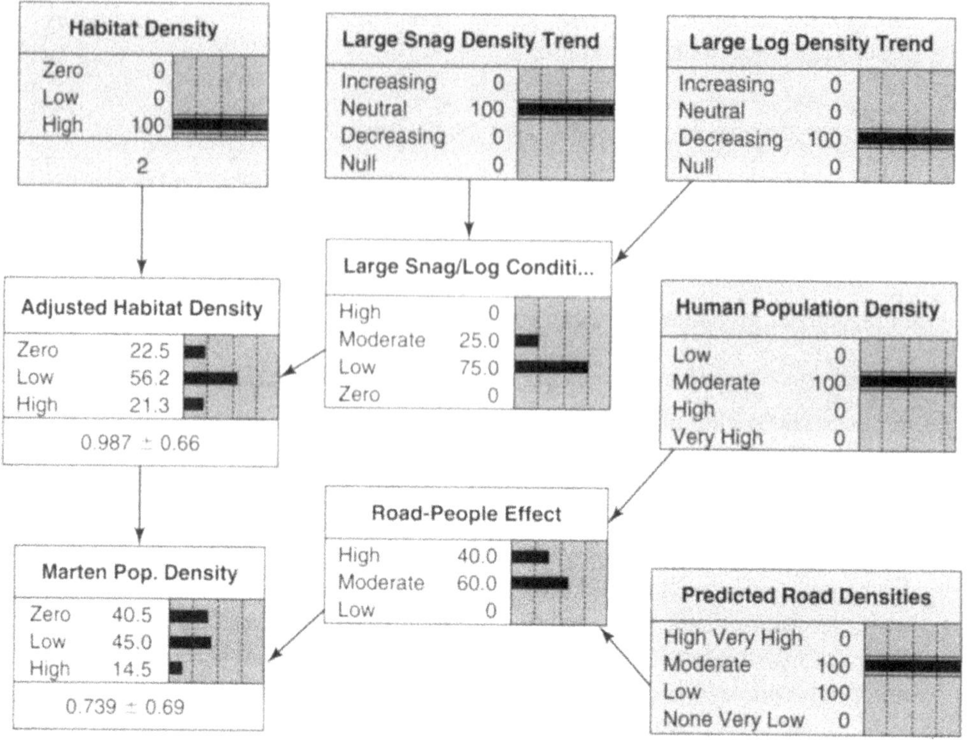

Figure 19.1. Example of a Bayesian network model for the Pacific marten from the Interior Columbia Basin Ecosystem Management Project in the western United States (Raphael et al. 2001). This example shows parameterization of the model for 1 subwatershed, resulting in a dominant probability of "low" marten population density (see text for further explanation of variables).

"environmental-index variables" that included source habitat, large-log density, road density, and human-population density (Raphael et al. 2001). Raphael et al. (2001) found that current source habitats have dropped from historical levels but were projected to return to those levels under the proposed land-management alternatives. The projected future numbers of subwatersheds with high environmental-index scores were relatively equal under all alternatives, and more than twice the current level. In response to changes in the amount of source habitat within subwatersheds, population outcome (an index of the overall likelihood of population viability) was projected to have declined strongly from historical levels, but to have improved into the future under all alternatives. Management guidelines were specified in the plan for restoring such habitat along with snags, large trees, and old-forest conditions that would collectively benefit the Pacific marten, fisher, and an array of associated wildlife species.

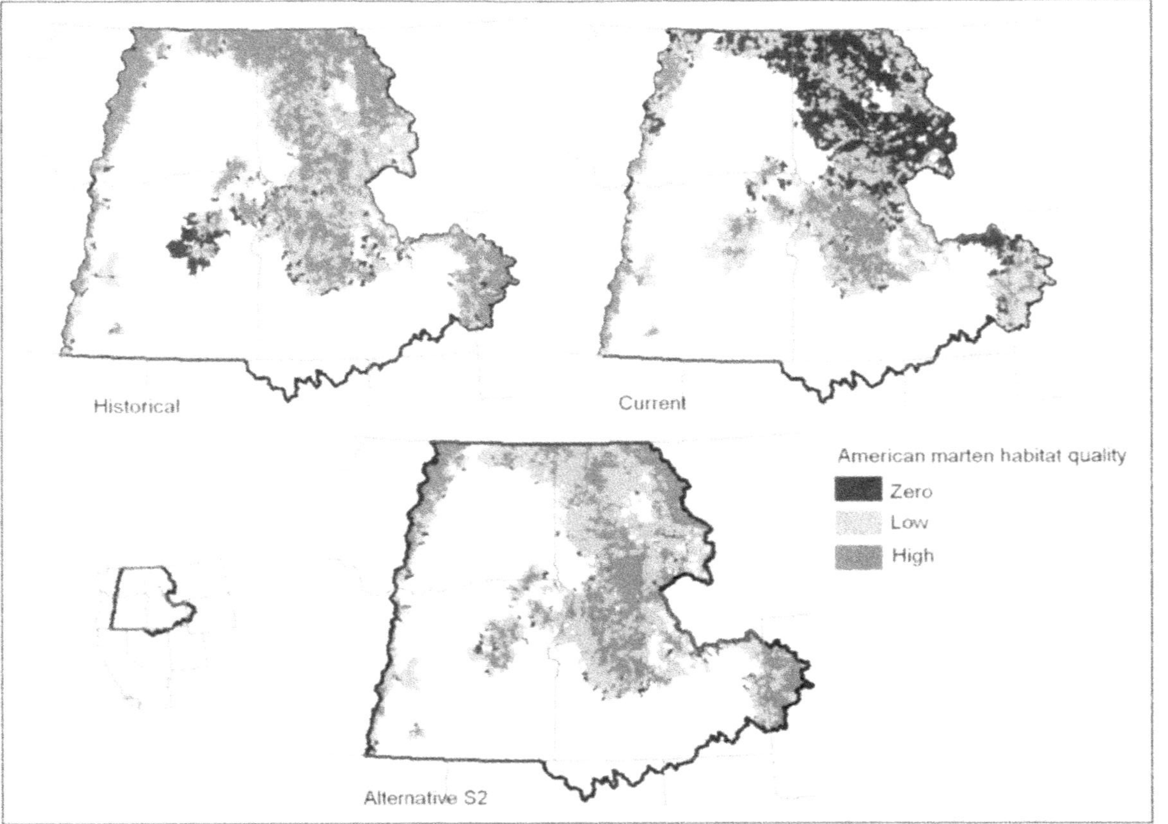

Figure 19.2. Example of using the Pacific marten model shown in Figure 19.1 to produce maps of projected historical, current, and future relative population density, from the Interior Columbia Basin Ecosystem Management Project in the western United States (Raphael et al. 2001). Several alternative future scenarios were analyzed, but only the preferred alternative (S2) is illustrated here (U.S. Departments of Agriculture and Interior 2000).

Sierra Nevada, California

Forest planning in the Sierra Nevada has included conservation of the Pacific marten and fisher as major components, particularly in the Sierra Nevada National Forest Plan Amendment of 2001, updated in 2004 and 2010 (U.S. Department of Agriculture 2010), which provided habitat guidelines for both species and gave high priority for adaptive-management studies; that is, research to reduce the areas of greatest uncertainty of the most salient stressors on these species to help guide habitat management. The plan has as its goal the assessment, development, and implementation of guidelines for managing species and ecosystems of the 11 National Forests in the Sierra Nevada bioregion. This bioregion covers some 69,560 km^2 and includes about 17% of the state of California (van Wagtendonk and Fites-Kaufman 2006). In the plan assessment, the Pacific marten and fisher were identified along with the California spotted owl (*Strix occidentalis occidentalis*), northern goshawk (*Accipiter gentilis*), Sierra Nevada red fox (*Vulpes vulpes necator*), and wolverine (*Gulo gulo*) as focal species associated with old-forest ecosystems.

Habitat needs for the Pacific marten and fisher were integrated into the broader array of forest-management guidelines for this suite of old forest-associated species. Specific guidelines for managing Pacific marten habitat included designation of 40-ha buffers for each den site, with about 5 conifer trees/ha >61 cm in dbh with suitable denning cavities, tree-canopy closure >60%, >22 metric tons/ha of coarse woody debris in intact decay classes, and an average of about 15 snags/ha west or 7 snags/ha east of the crest of the Sierra Nevada. Further guidelines pertained to minimizing habitat fragmentation. Specific guidelines for managing fisher habitat were similar to those for the Pacific marten, entailing a den-site buffer and guidelines for minimizing forest fragmentation. Additionally, a Southern Sierra Fisher Conservation Area (SSFCA) was designated to be managed for ≥50% of forested area in ≥60% canopy cover, where feasible.

Similarly, an evaluation of habitat, population status, and effects from fire and fuels management for the fisher in the Sierra Nevada has been prepared for the SSFCA area (Spencer et al. 2008). This evaluation was the first to use a spatially explicit individual-dispersal model (PATCH; Schumaker 1998; Carroll et al., this volume) to analyze effects on fisher populations from future changes under scenarios for fire and fuels reduction and vegetation management. Results included maps of microhabitat suitability, occupied sites, areas of population sources and sinks, and projections of potential habitat and population levels for several decades into the future. Implications of conserving connected habitats for the fisher also extend to consideration for other species, as well as the efficacy, scheduling, and costs of fuels management and fire containment. Thus, although this modeling effort for the fisher was not a multispecies bioregional assessment, it provides the basis for broader, ecosystem-level management guidelines and considerations.

U.S. Pacific Northwest

In this region, the Northwest Forest Plan (U.S. Departments of Agriculture and Interior 1994), which encompasses about 250,000 km^2, was instituted for the purposes of conserving and restoring mature and old-growth forest conditions for a wide array of old forest-associated plant and animal species (e.g., the northern spotted owl [*Strix occidentalis caurina*], marbled murrelet [*Brachyramphus marmoratus*], and at-risk species of anadromous fish) and for maintaining interconnected old-forest ecosystems on federal public lands. Both the Pacific marten and fisher were included in species databases and in assessments of habitat conditions and potential population viability outcomes (see below) under management guidelines that provided for a network of late-successional forest and riparian reserves, and the retention of some old-forest conditions in the intervening managed matrix lands.

In their science assessment supporting the Northwest Forest Plan, the Forest Ecosystem Management Assessment Team (FEMAT) conducted expert-panel viability assessments of 1118 plant and animal species, including 26 mammals (Forest Ecosystem Management Assessment Team 1993). For these assessments, 9 species-expert panelists were asked to spread 100 points among 1 or more of 4 possible population-viability outcomes: (1) stable and well distributed, (2) persistent, but with gaps in distribution, (3) restricted to refugia, and (4) extirpated. The Pacific marten and fisher both scored highest for outcomes 1 and 2. After the FEMAT process was completed, regional land managers developed an environmental impact statement and Record of Decision (U.S. Departments of Agriculture and Interior 1994), providing additional mitigation strategies to strengthen habitat protection for these species. These mitigations included providing greater levels of snags and down logs in the lands available for timber harvest.

Southeast Alaska

Tongass National Forest of southeast Alaska, covering about 68,000 km^2, includes some of the most extensive areas of temperate coniferous rainforest on the planet in a complex mainland and archipelago setting. Some forests of the region have been heavily clear-cut, raising concerns over the connectivity and viability of wildlife species that are closely associated with old-growth forest conditions. In the 1990s, this concern led to development of an interagency, multispecies viability assessment, and guidelines for conserving and restoring old-forest habitats (Suring et al. 1993). Both the Pacific and American martens were included in the assessment, along with the northern goshawk, Alexander archipelago wolf (*Canis lupus ligoni*), and other species. In 1995, the multiagency team broadened this assessment with proposed guidelines (based on metapopulation theory) for a network of old-growth reserves (habitat-conservation areas) to be distributed among islands of the Alexander

Archipelago and adjacent mainland. This conservation network was intended to help conserve the entire suite of old forest–associated wildlife in the region.

Resulting management guidelines were later incorporated into the Tongass National Forest Management Plan (U.S. Department of Agriculture 2008). For martens, the guidelines included identifying dens and resting sites, and providing snags and live trees with a mean dbh of 93 cm, whereby 68% of those trees with dens or resting use should be >61 cm dbh. Such criteria for conserving marten habitat were part of a broader set of forest-wide standards and guidelines for many other wildlife species. For example, a set of guidelines was provided for maintaining a 300-m (1000-ft) beach fringe of mostly unmodified forest as habitats, corridors, and connectivity areas for the bald eagle (*Haliaeetus leucocephalus*), northern goshawk, Sitka black-tailed deer (*Odocoileus hemionus sitkensis*), Pacific and American martens, river otter (*Lontra canadensis*), American black bear (*Ursus americanus*), and other wildlife species associated with the maritime-influenced environment.

Discussion

Successful conservation of *Martes* species will doubtless continue to require local, autecological studies on population distribution, abundance, trends, and responses to environmental conditions and anthropogenic stressors, including sources of mortality. Additionally, studies on taxonomy, reintroduction dynamics, life-history attributes, and innovative monitoring methods will provide critical information for conservation strategies. However, single-species management will not always solve broader conservation problems, or conflicts with land-use practices and habitat alteration. In many cases, there may be significant advantages to integrating single-species research and management into multispecies bioregional conservation assessments and strategies.

In this context, our review suggests that multispecies bioregional assessments are often initially driven by other land-management and natural-resource issues. For instance, the examples from Southeast Alaska and the U.S. Pacific Northwest resulted from concerns over excessive timber harvest, loss and fragmentation of old-growth forests, or declines in the long-term viability of other species (e.g., northern spotted owl). In such cases, the habitat and ecological requirements of martens, fishers, and other species are often integrated as mitigations for minimizing adverse effects of land-management activities on these species. Thus, some multispecies bioregional assessments were instituted less as a proactive strategy to conserve *Martes* species by restoring or conserving habitats or populations, than they were to provide mitigation measures to reduce adverse effects of other activities on particular suites of species. Nonetheless, lessons from these assessments can be synthesized into proactive approaches, which we endorse as a more efficient means of conserving *Martes* species in the context of broader goals for biodiversity conservation and ecosystem management.

Toward a Framework for a Multispecies Bioregional Assessment Approach

We have shown that multispecies bioregional assessments vary considerably in geographic scope, overall goals, information required and used, taxonomic breadth, inclusion of multiple land owners, administrative institutions and management agencies, political entities, and even international boundaries. Despite these divergent characteristics, however, the following features of a multispecies bioregional assessment seem more or less universal, and are ones that could work well for integrating *Martes* species into broader assessments:

1. Clearly articulate a set of overall planning goals, management objectives, and assumptions for the bioregional conservation strategy; for example, long-term production and use of natural resources (including wildlife populations), restoration of at-risk species, and sustainable ecosystem management.
2. Delineate the bioregion so it encompasses a significant portion of the *Martes* species' regional distribution, but also consider other important administrative and ecological boundaries.
3. Compile information on the *Martes* species of interest, including macrohabitat associations, use of microhabitat elements, life history, prey selection, key threats and stressors, responses to disturbance events, and, if available, population size, trend, distribution, and structure. Sources of information can include literature, ongoing research, and expert judgment.
4. Clarify the portion of the land base within the bioregion for which the assessment will pertain, including listing the pertinent land ownerships and management allocations. This may include specifying the set of pertinent stakeholders and interest groups potentially affected by management of this land base.
5. Compile information on an assemblage of wildlife species that are generally sympatric with the *Martes* species of interest. Such information can include habitat associations, effects of human activities on populations, and degree of concern for their conservation (e.g., IUCN Red Book vulnerability levels). Species can be selected based on focal interest for management, their degree of endemism or rarity, or other factors.
6. Identify macrohabitat conditions, use of microhabitat elements, and vulnerability to stressors for *Martes* species as well as for others being considered. Evaluate landscape patterns of key habitats for *Martes*

and other species to describe distribution, amount, condition, and connectivity. Compare with existing or projected conditions, and determine conservation needs, with management guidelines for restoration in areas where the species' needs may not be met.

7. From the previous step, compile spatially explicit management guidelines for key habitats and environments for restoration or conservation that encompass all species of concern.
8. As needed, apply the previous step to any future management scenarios, changes in human occurrence and land use, changes in climate, and disturbance events to develop mitigation and planning guidelines that may be required to meet the goals and objectives stated in step 1.

Of course, this is only a skeletal outline to provide ideas by which the conservation of *Martes* species could be embedded within broader objectives and geographic scopes. The sequence and specification of the individual steps suggested here will doubtless vary according to specific needs and interests; for example, additional consideration should be given to setting harvest levels, if any, and to improving coordination among regulatory bodies and political institutions.

Raphael et al. (2007) provided a similar but more-detailed description of procedures for considering rare or little-known species in a broader risk-assessment and conservation planning framework. Their procedures could also be used as a template for including *Martes* species in multispecies bioregional assessments. They also emphasized the critical importance of setting clear management goals, identifying short- and long-term conservation objectives, and including learning objectives to fill in knowledge gaps. In many cases, a combination of single-species and ecosystem management approaches may also help address a larger set of ecological goals and objectives (see Holthausen and Sieg 2007).

Opportunities for Incorporating Other *Martes* Species and Regions

Not every *Martes* species has been included in a bioregional assessment (Table 19.1); we could find no multispecies bioregional assessments that included the yellow-throated (*M. flavigula*), stone (*M. foina*), European pine (*M. martes*), or Japanese (*M. melampus*) martens (Table 19.1). Collectively, these 4 species range from western Europe, through Eurasia, parts of the Middle East and south Asia, to the Far East including Japan and the Korean Peninsula. They occupy a wide range of mostly north-temperate forests, including Mediterranean woodlands, Siberian conifer forests, and mixed hardwood-conifer forests of the Far East; the yellow-throated marten also occurs in subtropical and tropical forests in southeast Asia. Within the geographic and ecological ranges of these 4 species, environmental stressors and anthropogenic factors that affect habitats and populations are highly diverse. Some elements of existing single-species conservation guidelines, including ecological and life-history studies and existing management guidelines, may provide a useful foundation for building a multispecies assessment.

Some existing situations are well suited for integrating *Martes* species conservation into a broader framework. One example is the unique administrative structure available for managing American martens and fishers in the northeastern United States and eastern Canada. In that region, management guidelines for martens and fishers are established by the Northeast Furbearer Resources Technical Committee (NEFRTC), consisting of regional public representatives based on jurisdictions and serving, in part, to share information on marten populations and data collection and to help identify seasons and harvest quotas for the species. This organizational structure could facilitate the development of a cross-border, interdisciplinary bioregional assessment for these species.

Another example would be to build on the array of natural history and ecological studies of American martens and fishers in the northern and midwestern United States. Studies there have been associated with the reintroduction of both species in Wisconsin and Michigan (Williams et al. 2007), the home-range dynamics and habitat selection of martens in the Lower Peninsula of Michigan (McFadden 2007), and the use of noninvasive hair sampling and genetic tagging for both martens and fishers (Williams et al. 2009). Such research could prove useful for including martens and fishers in multispecies assessments of the full suite of carnivores that occupy the region, with the goal of developing more comprehensive conservation guidelines for this assemblage.

In the Rocky Mountain and Great Basin regions of the western United States, wildlife-habitat management on public forest lands generally consists of standards and guidelines for the conservation of old-growth forests, and for maintaining certain attributes of forest composition, forest structure, and habitat elements (e.g., snags, down logs). The Pacific marten and fisher are addressed specifically in some project-level documents in these regions, if they are recognized as a special-status species in associated National Forest land and resource-management plans, or if they are raised as a conservation issue during public scoping of projects on federally managed public land. This context seems ripe for developing broader arrays of multispecies guidelines that at least provide macrohabitat conditions and microhabitat elements for martens, fishers, and an array of other wildlife species associated with similar habitat conditions. Southwest Idaho is currently developing a Wildlife Conservation Strategy that includes the fisher as a focal species (Clint McCarthy, U.S. Forest Service, personal communication). If implemented, it would be the first bioregional assessment and conservation strategy in the region that includes the fisher, and could provide the foundation for a broader multispecies approach.

Basic research on resource use by *Martes* species and their ecological separation from other species can be central to developing effective multispecies guidelines. Examples include studies of the ecological separation of the yellow-throated marten within a guild of mesocarnivores in Thailand (Grassman et al. 2006b), the European pine marten's ecological relations with other vertebrate predators in Europe (Sidorovich et al. 2006), and the ecological relations of the American marten and fisher with other mesocarnivores in the northeastern United States (Ray 2000).

The conservation of *Martes* species could be integrated into broader biodiversity goals, guidelines for ecosystem management, conservation or restoration of old-forest conditions, and consideration of disturbance regimes. In such a context, *Martes* species conservation can be scaled up spatially from local or jurisdictional management to broader landscape, state, province, and regional scales (Proulx and Santos-Reis, this volume), as has been done on National Forests in the northeastern United States and Upper Peninsula of Michigan (USA). Management aimed at providing for the habitat needs of *Martes* species could be integrated into broader ecosystem-scale assessments and guidelines, as was done in the Northwest Forest Plan in the U.S. Pacific Northwest, and in guidelines from the Interior Columbia Basin Ecosystem Management Project in the western United States.

A number of existing species-specific assessments and conservation guidelines could also be integrated with other management objectives into more complete bioregional assessments. For example, the Rocky Mountain and Great Basin regions of the western United States are experiencing severe outbreaks of mountain pine beetle (*Dendroctonus ponderosae*) and spruce beetle (*D. rufipennis*), with massive infestations spreading in Colorado and southern Wyoming since 1996. The effects of these historic disturbances on habitat and prey of the Pacific marten could be addressed in a multispecies, multiresource, ecosystem assessment. Similarly, the effects of reduction of fuels and down wood on the Pacific marten and its mammalian prey have been evaluated in northeastern Oregon by Bull and Blumton (1999). Their analysis could be extended to multiple species of carnivores or a broader evaluation of ecosystem impacts.

One approach that is complementary to the multispecies assessments described here involves evaluating each species' key ecological functions; that is, its main ecological roles in the ecosystem. Aubry et al. (2003) evaluated the collective ecological roles of tree-dwelling mammals in western coniferous forests of the United States, including the Pacific marten and fisher, along with 4 other carnivores, 11 arboreal rodents, and 14 bats. They noted that forest carnivores can influence their ecosystems by affecting the behavior and demography of prey and competitor populations, dispersing seeds, affecting and facilitating the life cycles of pathogens and parasites, distributing nutrients through carrion feeding, and transporting nutrients and contaminants. Marcot and Aubry (2003) provided an even broader evaluation of the ecological roles of all mammal species in conifer forests of the western United States, involving an assessment of the ecological functions of entire species assemblages, including *Martes* species. In general, understanding the ecological roles of *Martes* species in the context of their ecological communities can help determine their contributions to fully functional ecosystems or their effect on the habitats and populations of other species. These findings, in turn, can provide the basis for ecosystem management guidelines that account for the collective ecological functions of entire communities.

Conclusions and Future Development

In the examples of multispecies bioregional assessments discussed in this chapter, most concerns for the future of *Martes* species pertain to loss of habitat, especially the loss of old native forests and the structural elements (e.g., large snags and down logs) needed for denning and resting sites. Not surprisingly, multispecies approaches have demonstrated that the conservation and restoration of such habitat conditions can also provide for a wide array of other wildlife species.

A less obvious benefit to these approaches is that economic and social costs associated with the conservation and restoration of old native forests can be "spread" among multiple species to counter the perception that only 1 species is to be "blamed" for such costs, as was the case with conserving mature and old-growth forests for the northern spotted owl (e.g., Beuter 1990; Montgomery et al. 1994). Multispecies approaches also address the conservation needs of many species at once, providing more efficient and economical solutions than are possible by addressing each species individually. Such multispecies approaches have been used by the U.S. Fish and Wildlife Service (U.S. Fish and Wildlife Service 1999; Clark and Harvey 2002) and other regulatory agencies, such as the Northwest Power and Conservation Council (Marcot et al. 2002).

However, there are many challenges to implementing a multispecies bioregional approach. Often, local autecological studies are lacking for *Martes* species, as well as other co-occurring species, so little is known about their distribution and abundance, demography, threats, or habitat associations. As a first step in such cases, qualitative databases on wildlife-habitat relationships can be compiled for sets of priority, at-risk, or focal species (including *Martes* species), as we described for the Nilgiri marten in south India and the Pacific marten and fisher in the U.S. Pacific Northwest. Databases of this type can be developed by expert panels or individual expert judgment, as was done for the Interior Columbia Basin Ecosystem Management Project. Maps of species' distributions can be compiled from a variety of sources and overlaid to determine hot spots of species diversity, as we described for the sable and other forest species in the Far East, or to map key locations of source habitats, as in the Interior Columbia Basin Ecosystem Management Project.

With the use of predictive tools, the future of integrating *Martes* species assessments and conservation strategies into broader bioregional, multispecies, and ecosystem-scale approaches appears bright and may prove essential for solving increasingly challenging problems of sustainable forest-resource management and land-use conflicts. Overall, including *Martes* species in more comprehensive approaches to conservation, restoration, and management of their populations and habitats provides substantial advantages for ensuring their future and that of the species assemblages, ecological communities, and ecosystems in which they reside.

Acknowledgments

We gratefully acknowledge the following individuals who provided information, documents, and helpful suggestions from many geographic areas. Northeastern United States: Wally Jakubas, Paul Jensen, Bill Krohn, Justina Ray; Alaska: Joe Cook, Rod Flynn, Wini Kessler, Chuck Parsley, Winston Smith; U.S. Rocky Mountain and Great Basin regions: Jim Claar, Greg Hayward, Clint McCarthy; U.S. Pacific Northwest: Laura Finley, Bob Naney; U.S. north-central and midwestern regions: Tim Bertram, Dan Eklund, John Erb, Dwayne Etter, Dorothy Fecske, Thomas Gehring, Jonathan Gilbert, Beth Hahn, Kim Scribner, Bronwyn Williams, Adrian P. Wydeven, Jim Woodford, Patrick Zollner; and Sierra Nevada in California: Peter Stine. We also thank Keith Aubry, Christina Vojta, and Bill Zielinski for additional information and for their general guidance on this project, and for helpful comments on the manuscript from Keith Aubry, Bill Zielinski, and 2 anonymous reviewers.

20

A Century of Change in Research and Management on the Genus *Martes*

GILBERT PROULX AND MARGARIDA SANTOS-REIS

ABSTRACT

During the 20th century, *Martes* taxonomy, evolution, biogeography, habitat, and populations were investigated extensively, and conservation practices were developed. During this period, human populations and lifestyles, socioeconomics, and cultural and ethical values changed, with a gradual shift away from traditional wildlife uses to nonconsumptive values. In this chapter, we assess trends in research and management practices from 1901 to 2010 in relation to the worldwide conservation of *Martes* species. Based on our review of 1298 publications, we identified 3 distinct time periods in the history of *Martes* research and management: (1) 1901–1960, a utilitarian period when research was focused on species status and distribution; (2) 1961–1990, a period of scientific interest in population declines and habitat deterioration; and (3) 1991–2010, a period of conservation concern for populations and habitats. We review the characteristics of each time period and show that in 110 years of research, wildlife managers and researchers have acquired extensive knowledge about the evolution, taxonomy, morphophysiology, genetics, population dynamics, habitat and predator-prey relations, nutrition and energetics, parasites, and diseases of most *Martes* species. We believe that *Martes* conservation would benefit from better integration of this knowledge into wildlife management programs. The future of *Martes* research and conservation may ultimately depend on integrating the needs of *Martes* species into multispecies management programs, testing whether habitat-management recommendations improve the fitness of target populations, and developing effective education programs.

Introduction

For centuries, the American marten (*Martes americana*), European pine marten (*M. martes*), stone marten (*M. foina*), sable (*M. zibellina*), and fisher

(*M. pennanti*) were renowned for the quality of their pelts, which have been both an article of trade and a currency (Innis 1956; Delort 1986). In the 1900s, when the scientific world became interested in this genus, *Martes* species were still widely distributed throughout Europe, Eurasia, and North America (Delort 1986). During the 20th century, *Martes* taxonomy, evolution, biogeography, habitat, and populations were investigated extensively, and conservation practices were developed. Human populations and lifestyles, socioeconomics, and cultural and ethical values also changed dramatically (Aramburu and King-Dagen 1995; Salwasser 1995). There has been a gradual shift from traditional wildlife uses to nonconsumptive values (Proulx and Barrett 1989; Heberlein 1991; Peterson and Manfredo 1993), as a result of increasing affluence, education, and urbanization (Manfredo et al. 2003).

In this chapter, we assess trends in research and management practices pertaining to the conservation and management of *Martes* species. This assessment includes a review of the ways in which societal changes have affected research themes, and our view of the continuing influence of traditional uses on conservation programs for *Martes* species today.

Methods

We reviewed an extensive body of literature where we expected most *Martes* research to have been published, including key reviews, syntheses, and compendia, such as Powell (1981), Clark et al. (1987), Buskirk et al. (1994), Proulx et al. (1997), Harrison et al. (2004), and Santos-Reis et al. (2006). We also searched all periodicals available digitally through the University of Alberta libraries, and reviewed the content of 13 technical journals: *Zoological Society of London* (since 1830), *Journal of Zoology, London* (since 1832), *Canadian Field-Naturalist* (since 1869), *American Midland Naturalist* (since 1909), *Journal of Mammalogy* (since 1919), *Journal of Animal Ecology* (since 1932), *Journal of Wildlife Management* (since 1937), *Acta Theriologica* (since 1954), *Canadian Journal of Zoology* (since 1929), *Canadian Journal of Forest Research* (since 1971), *Environmental Reviews* (since 1993), *Mammal Review* (since 1970), and *Wildlife Society Bulletin* (since 1973).

We compiled the following information for each article: date of publication, *Martes* species studied, and subject studied; we limited potential study subjects to the following:

1. Populations, with 3 themes: (a) status and distribution, (b) biology, and (c) management
2. Habitats, with 3 themes: (a) home range, habitat use and selection, and landscape characteristics, (b) food habits (food is a resource related to habitat selection, e.g., Baltrūnaitė 2006a), and (c) management programs and models

3. Taxonomy and genetics
4. Other subjects, including studies of morphology, physiology, behavior (e.g., diel activity, curiosity, scent marking), parasites, diseases, economics, and paleontology

A publication could be coded as covering 1 subject (i.e., 1 dataset corresponded to 1 subject) or >1 subject. We considered 2 groups of species based on geography, similar histories of exploitation, and habitat-conservation policies (Proulx et al. 2004): Group (1) the American marten and fisher; and Group (2) the European pine marten, stone marten, and sable. The yellow-throated (*M. flavigula*), Japanese (*M. melampus*), and Nilgiri (*M. gwatkinsii*) martens were excluded because of the paucity of publications on these species. The majority of *Martes* papers were published after 1900. We therefore focused our review on the 110-year period from 1901 to 2010. To obtain adequate sample sizes for statistical analyses, we grouped study subjects into 4 time periods of similar length: 1901–1930, 1931–1960, 1961–1990, and 1991–2010. We knew that interest in *Martes* biology and conservation had increased in the 1990s after the first International *Martes* Symposium was convened in 1991 (Buskirk et al. 1994) and the *Martes* Working Group was created in 1993, so we subdivided the first 90 years of *Martes* publications into three 30-year periods, and used a 20-year period after 1990. We used the proportion of 110 years represented by each period to determine the expected frequency of studies for each period, assuming the null hypothesis of no change in *Martes* publication activity over time. We used Chi-square statistics with Yates' correction to compare observed to expected frequencies of studies among periods (Zar 1999). When >20% of a period had an expected frequency <5, or when any expected frequency was <1, we combined consecutive periods (Cochran 1954). When Chi-square values indicated significant differences, we compared observed to expected frequencies for each period using the G test for correlated proportions (Sokal and Rohlf 1981). We considered P values ≤0.05 to be statistically significant.

Results

We located a total of 1319 *Martes* publications from 1758 to 2010, with 1298 of these published from 1901 to 2010, representing studies of 1402 different subjects. The largest numbers of publications and subjects studied were for American martens and fishers (Figure 20.1).

Taxonomy and Genetics

For the entire analysis period (1901–2010), observed frequencies of taxonomic and genetic studies for *Martes* Groups 1 and 2 differed significantly ($\chi^2 \geq 64.52$, df ≥ 1, $P < 0.001$) from expectation (Figure 20.2). Studies were

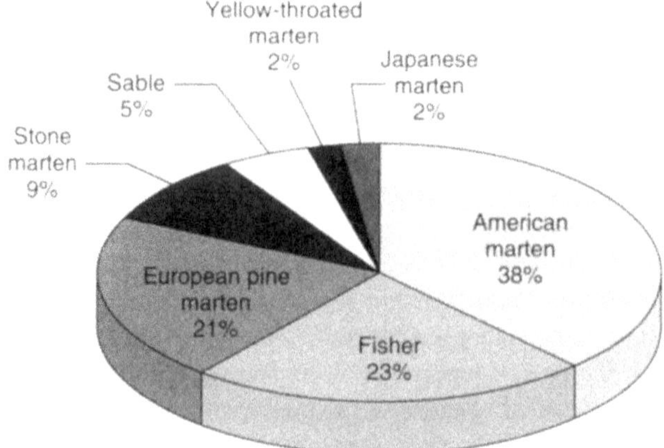

Figure 20.1. Frequency (%) of publications (n = 1298) collated in this study for each *Martes* species.

significantly less frequent than expected from 1931 to 1990 for Group 1, and from 1901 to 1990 for Group 2. Studies were more frequent than expected from 1991 to 2010 for both groups (Figure 20.2).

Population

For the entire analysis period, observed frequencies of population studies for *Martes* Groups 1 and 2 differed significantly ($\chi^2 \geq 114.58$, df ≥ 2, $P < 0.001$) from expectation (Figure 20.2). For Group 1, population studies were significantly less frequent than expected from 1901 to 1930, but more frequent than expected from 1991 to 2010 (Figure 20.2). From 1901 to 1960, population studies consisted mostly of status and distribution reports (Figure 20.3). Studies on population biology and management began from 1931 to 1960, but became more important after 1960 (Figure 20.3). In Group 2, studies were significantly less frequent than expected from 1901 to 1960, but more frequent than expected from 1961 to 1990 and 1991 to 2010 (Figure 20.2). Studies on the distribution and biology of species became more important after 1960, and population management programs, after 1990 (Figure 20.3).

Habitat

For the entire analysis period, observed frequencies of habitat studies of *Martes* Groups 1 and 2 differed significantly ($\chi^2 \geq 244.65$, df = 2, $P < 0.001$) from expectation (Figure 20.2). Studies were significantly less frequent than expected from 1901 to 1960, but more frequent than expected from 1991 to 2010 (Figure 20.2). Studies on the characteristics of habitats and landscapes increased gradually from 1931 to 2010, and habitat-management programs and models were more frequent from 1991 to 2010 (Figure 20.4).

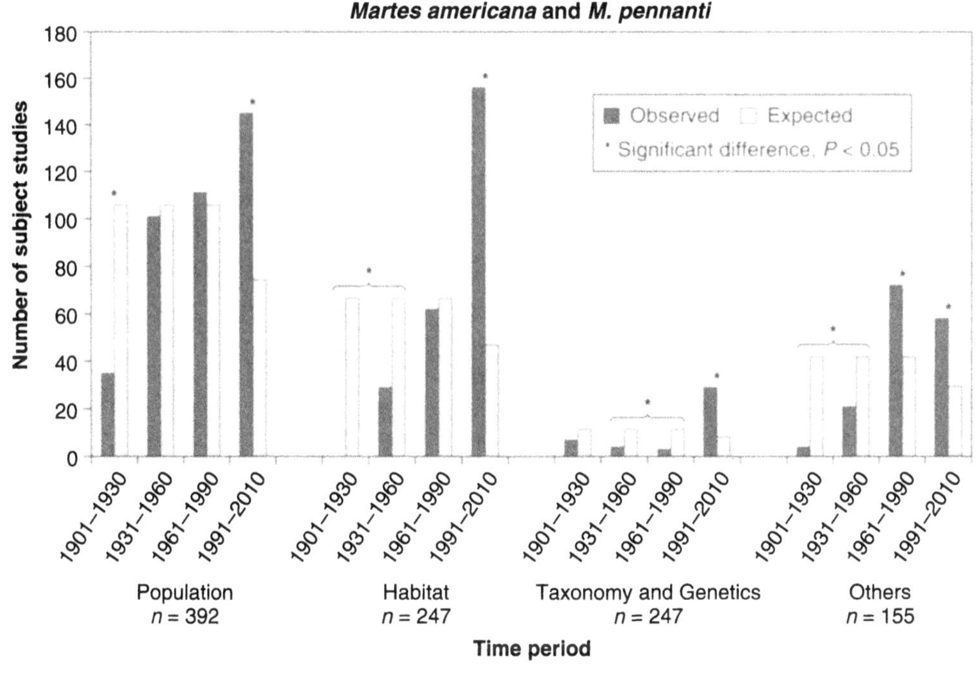

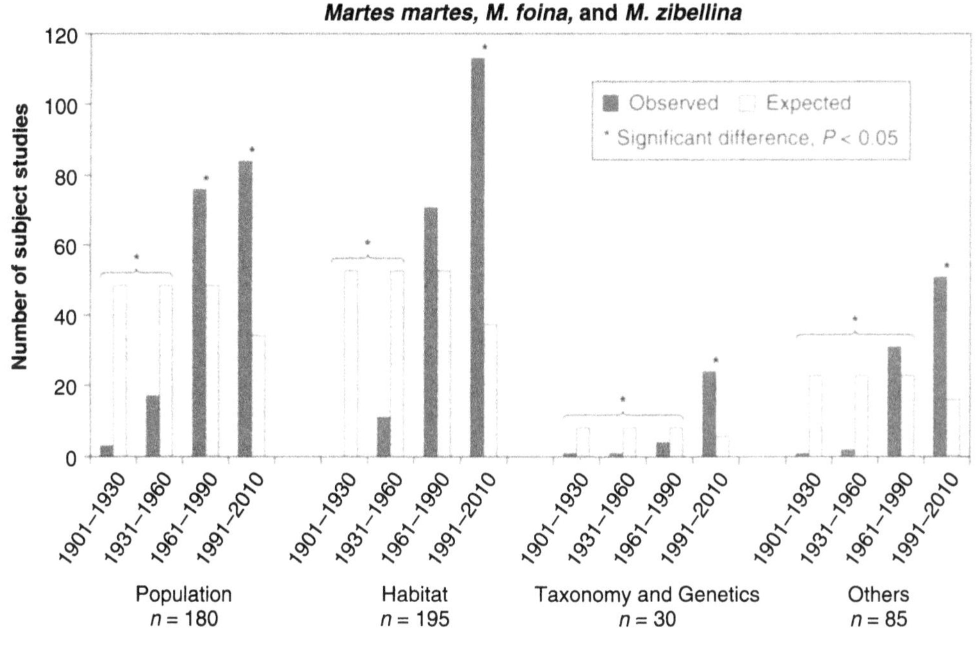

Figure 20.2. Frequency of studies by subject and time period for *Martes* species (brace brackets indicate periods that were pooled for statistical analysis).

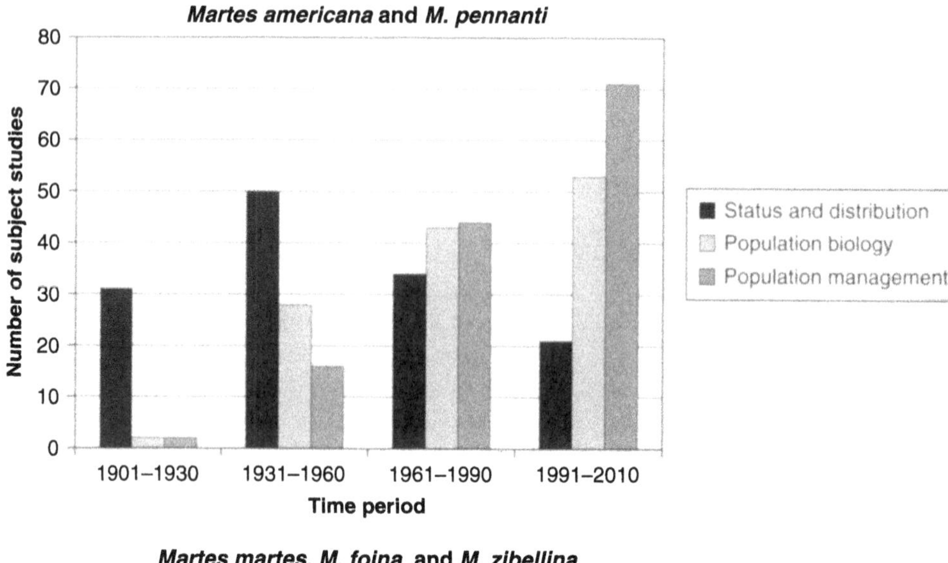

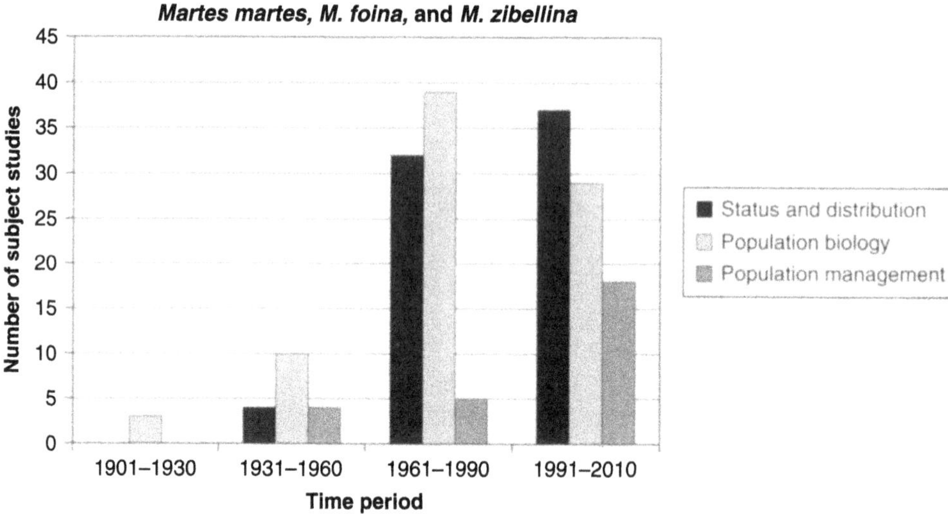

Figure 20.3. Frequency of subtopics for population studies by time period for *Martes* species.

Other Subjects

For the entire analysis period, observed frequencies of other studies for *Martes* Groups 1 and 2 differed significantly ($\chi^2 \geq 68.85$, df ≥ 1, $P < 0.001$) from expectation (Figure 20.2). Studies were significantly less frequent than expected from 1901 to 1960 and 1901 to 1990 for Groups 1 and 2, respectively (Figure 20.2). The frequency of studies was significantly higher than

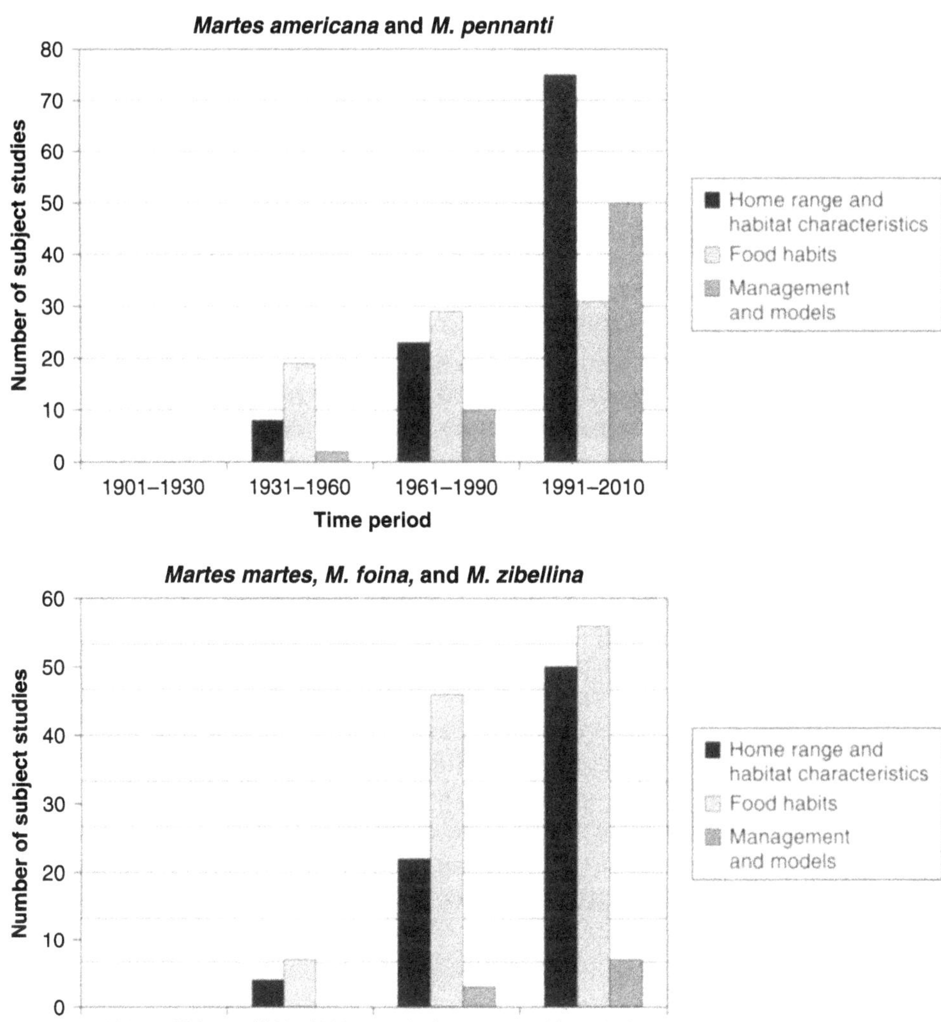

Figure 20.4. Frequency of subtopics for habitat studies by time period for *Martes* species.

expected from 1961 to 1990 and 1991 to 2010 for Group 1, and from 1991 to 2010 for Group 2 (Figure 20.2). For Group 1, the greater frequency of studies after 1960 was due largely to a marked increase in research on morphometry, physiology (e.g., thermoregulation, metabolic rate, body-fat dynamics), parasites, and diseases. For Group 2, more studies of morphometry and physiology were published after 1990.

Discussion

Changing Times

Martes research and management activities increased significantly from 1901 to 2010. We expected such a result, reflecting the evolution of the entire enterprise of science during the 20th century, that is, little scientific research, publication, and funding in the first half of the century, increased research activities and publications in the mid-1900s, when governments invested more in the advancement of science and technology, and constant annual growth in research, publications, and development programs after 1980 (Mabe and Amin 2001). Our results suggest that research themes for *Martes* species changed significantly at approximately 2 points in time: 1960 and 1990, resulting in 3 time periods during which the focus of *Martes* research and management was relatively consistent (Figure 20.5):

- 1901–1960: Utilitarian period. Research on *Martes* species was limited and included studies comprising mainly status and distribution reports.
- 1961–1990: Scientific period. This was a period of increased scientific interest in *Martes* species. In North America, studies on *Martes* populations and habitats increased, and significantly more work was conducted on the morphology, physiology, and behavior of species. In Europe and Asia, the number of publications on populations and the general biology of the species significantly increased.
- 1991–2010: Conservation period. During this period, research on *Martes* became much more management oriented. At the end of the 20th and beginning of the 21st centuries, public values and priorities for research and management were oriented toward the conservation of *Martes* populations and habitats. During this period, the number of publications on all subjects increased significantly.

Transitions from one period to the next did not occur abruptly (Figure 20.5). For example, although biologists were concerned about *Martes* populations, habitats, and harvests from 1961 to 1990 in North America, such concerns had been expressed during previous periods (e.g., Dixon 1925). Also, early thinking about wildlife management (Leopold 1933) and the factors that limit vertebrate populations (Errington 1946, 1956), along with the creation of scientific organizations such as The Wildlife Society (Bennitt et al. 1937) and the American Game Association (Allen 1985), paved the way for increased interest in research and management programs during the 1960s. In the 1950s, most biologists were interested primarily in the regional distribution of species (Dagg 1972), but some began to study population biology and habitat use. Thus, we could have set the beginning of the scientific period at

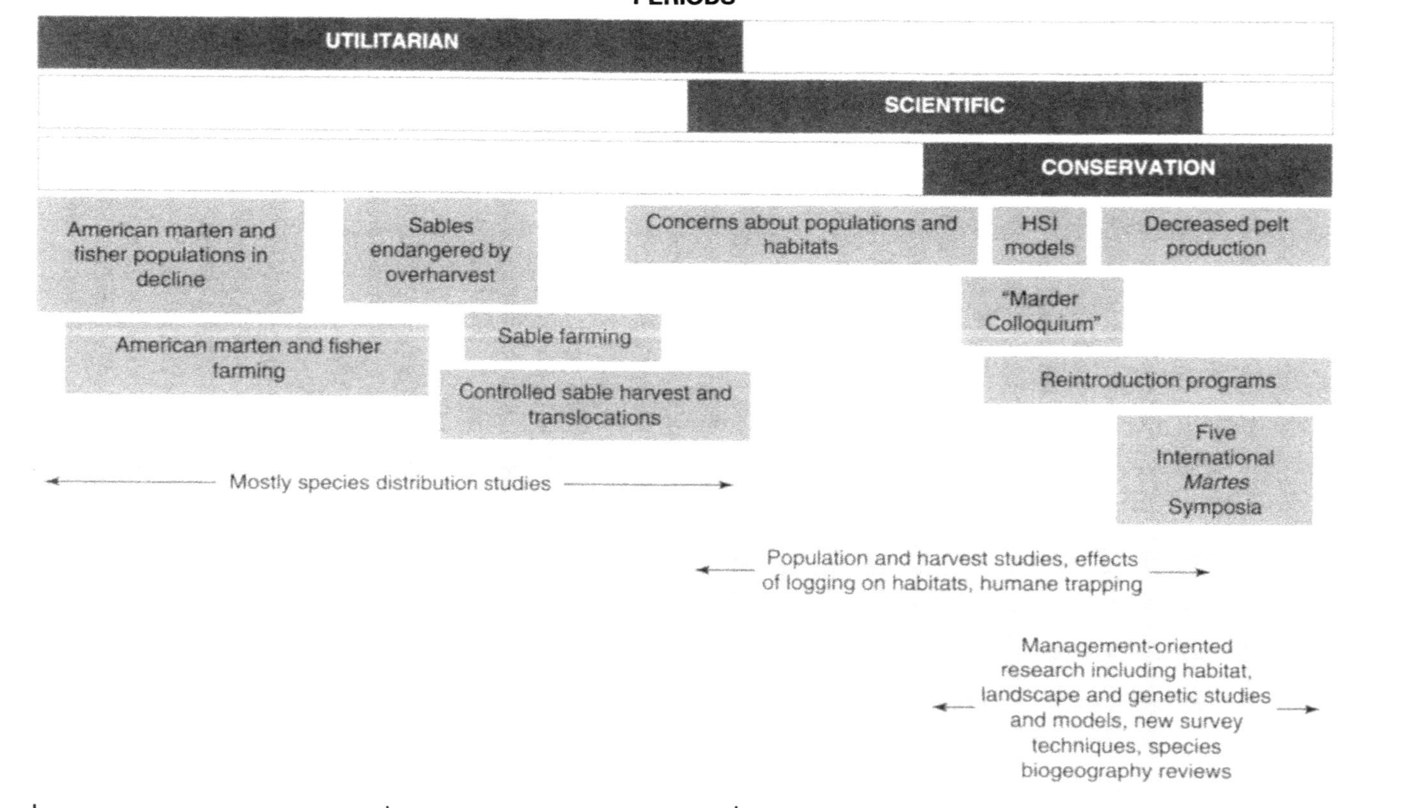

Figure 20.5. Schematic representation of chronological changes in *Martes* research and management from 1901 to 2010.

1950. It is important to realize that these themes were not exclusive to any one period, however; for example, many conservation activities began much earlier than 1991 (e.g., Anderson 1934).

1901-1960: Utilitarian Period

By the end of the 19th century, the genus *Martes* and its 3 subgenera (*Pekania*, *Charronia,* and *Martes*) had been established (Nowak 1999). At the beginning of the 20th century, the sable, the "black gold" of the early Middle Ages (Delort 1986), had been driven almost to extinction during the 1800s from overharvesting. Both European pine martens and stone martens were part of the fur trade in the early days and still being hunted at the beginning of the 20th century (Delort 1986; Bakeyev 1994; Proulx et al. 2004). In the 19th century, despite fluctuating prices (Ray 1987), pelts of American martens and fishers were consistently valuable (Innis 1956; Obbard et al. 1987). Interest in *Martes* species during the first half of the 20th century focused primarily on the value of martens, sables, and fishers as furbearing animals (Delort 1986; Proulx 2000); yet, they were trapped with little concern about resulting effects on wild stocks. At the beginning of the 20th century, the demand for raw furs was increasing, despite decreases in numbers (Gotlieb 1927). Our literature review indicated that although American martens and fishers were economically important furbearers, and their populations were in decline by the early 1900s (Dixon 1925; Dodds and Martell 1971), little field research was being conducted at that time. During the first half of the century, distributional assessments were most common (Dagg 1972).

Because of the value of *Martes* pelts and declining wild stocks, fur farming was attempted during the first half of the 20th century. Fur farming attained its greatest popularity during the 1920s as a thoroughly modern answer to the apparent and inevitable "exhaustion of nature" (Colpitts 1997). Sable farming began in the former Soviet Union in the late 1930s (Nes et al. 1988) and was successful (Korhonen et al. 2001). Fur farms for the American marten and fisher were also established in North America with some success (Jones 1913; Patton 1925; Hodgson 1937). However, their comparatively small litters, the long period before first parturition, and difficulties getting the animals to breed made farming of these species a risky business (Rand 1944; Douglas and Strickland 1987; Robitaille 2000).

Concurrent with the public demand for fur, researchers began investigating the reproductive biology of *Martes* species, including breeding (Reinhardt 1929; Ashbrook and Hanson 1930; Brassard and Bernard 1939), delayed implantation and gestation (Ashbrook and Hanson 1930; Wright 1942; Enders and Pearson 1943), and physiological and physical characteristics (Prell 1928; Brassard and Bernard 1939; Enders and Leekley 1941).

In the 1930s, a few scientists started showing some interest in the status of harvested populations. Anderson (1934: 4064) stated that "with every northern trapper after its pelt, unless the fisher becomes successfully acclimated on fur farms, this valuable species seems doomed to ultimate extinction." The status of the American marten was also in decline (Grinnell et al. 1937; Schorger 1942; Twining and Hensley 1947).

In 1935, active restoration of sable populations began with a 5-year ban on the hunting and sale of pelts in Yakutia and the Far East (Bakeyev and Sinitsyn 1994). This was followed by controlled harvests with quotas, and the development of farms where animals were bred in captivity to rebuild wild populations. From 1940 to 1965 in Russia, more than 19,000 sables were translocated into areas with low densities in the Far East and Yakutia, western Siberia and the Urals, and eastern and central Siberia (Bakeyev and Sinitsyn 1994). American martens and fishers benefited from the extensive establishment of registered traplines in Canadian provinces (Eklund 1946; Crichton 1948). In the United States, the establishment of annual harvest surveys and the closure of trapping seasons (Linhart 1985; Lewis and Zielinski 1996) also addressed conservation concerns.

By the early 1950s, wildlife managers knew that American martens and fishers were overharvested, and it had become imperative to learn more about harvested populations (Yeager 1950). Investigations showed that males and juveniles were more vulnerable to trapping than females or adults, respectively (Yeager 1950; Quick 1956); also, an equal sex ratio or one favoring females likely indicated overharvesting (Soukkala 1983; Archibald and Jessup 1984). By the mid-20th century, harvest management involved establishing quotas for registered traplines in Canada for each species, based on harvests during the preceding years. Although such an approach appeared satisfactory at the time (de Vos 1951b), later work suggested that it was unlikely to prevent overharvesting (Fortin and Cantin 1994). As Quick (1956: 271) pointed out, "the difficult problem of quantitatively measuring the effects of exploitation still demands a good appraisal of pre-season populations."

When conducting autecological studies, wildlife biologists began to explore factors other than trapping to explain declines in *Martes* populations. Researchers noticed that *Martes* species were associated with mature and old forests with dense cover, snags, and woody debris (de Vos 1952) and suggested that fire, logging, road building, mining, powerline clearing, and similar human activities were causing the disappearance of American martens and fishers from much of their range (Schmidt 1942; Yeager 1950; Miller et al. 1955; Lutz 1956), reflecting Seton's (1926: 206, 211) prediction that species "must disappear as the forest disappears."

1961–1990: Scientific Period

With the establishment of various research and management programs, the 1960s (and subsequent decades) represent a turning point for *Martes* management in North America (e.g., Anderson 1987; DiStefano 1987; Slough et al. 1987). This was also a time when public concerns about managing wildlife for consumptive uses resulted in more emphasis on environmental health and aesthetic values (Bolen 1989). In North America, American marten and fisher population levels had declined so much by 1950 (Yeager 1950; de Vos 1952) that government agencies throughout the continent initiated a number of reintroduction efforts (Table 20.1). From 1965 to 1993 in Russia, a captive-breeding population of 500 sables was established to restore the species in China (Ma and Xu 1994). European pine martens were also suffering the effects of habitat loss due to urbanization but had not yet been part of reintroduction programs.

Around 1950, biologists began to realize that there was a lack of information about factors affecting population densities and habitat carrying capacity (de Vos 1952). This realization initiated an era of studies on population dynamics (e.g., Hawley and Newby 1957), home ranges (e.g., Mech and Rogers 1977; Raine 1982; Wynne and Sherburne 1984), food habits (e.g., Cowan and Mackay 1950; Brown and Will 1979; Nagorsen et al. 1989), and the relations

Table 20.1. Reintroductions of American martens and fishers in North America from 1961 to 1990

Location	Reference
American marten	
Eastern and Central Canada	Rettie 1971; Bateman 1984; Sullivan 1984; Quann 1985; Sinclair 1986; Boss et al. 1987; Bissonette et al. 1988; Drysdale and Charlton 1988
Western Canada	Miller 1961; van Zyll de Jong 1969; Hobson et al. 1989
Canadian Territories	Slough 1994
Eastern United States	Soutiere and Coulter 1975
Midwest United States	Brander and Books 1973; Schupbach 1977; Churchill et al. 1981; Davis 1983; Gieck 1986; Fredrickson 1989
Western United States	Burris and McKnight 1973; Rognrud 1983
Fisher	
Eastern and Central Canada	Dodds and Martell 1971; Dilworth 1974
Western Canada	Davie 1984; Proulx et al. 1994
Eastern United States	Fuller 1975; Cottrell 1978; Pack and Cromer 1981; Wallace and Henry 1985
Midwest United States	Williams 1962a, 1963; Irvine et al. 1964; Petersen et al. 1977; Luque 1984; Kohn and Eckstein 1987
Western United States	Kebbe 1961a; Weckwerth and Wright 1968

among these subjects (e.g., Weckwerth and Hawley 1962; Zielinski et al. 1983; Thompson and Colgan 1987). These studies were accompanied by the development of new techniques to better determine the structure of populations (Parsons et al. 1978; Kuehn and Berg 1981; Strickland et al. 1982a). Armed with a better understanding of *Martes* populations, wildlife biologists revised harvest regulations to establish more realistic quotas (Strickland and Douglas 1981; Fortin and Cantin 1994; Garant et al. 1996) and implemented systematic trapper surveys (Lafond 1990). Despite improved harvest programs, the trapping of furbearers for economic gain and the use of archaic trapping devices became major public concerns. From the early 1980s to mid-1990s, the scientific community investigated and developed new trapping devices (Proulx 1999a) to increase capture efficiency and animal welfare. Traps that caused undue suffering were identified (Proulx et al. 1989b; Cole and Proulx 1994) and humane-trapping devices were developed according to new standards (Barrett et al. 1989; Proulx et al. 1989a; Proulx and Barrett 1994; Proulx 1997, 1999b).

From 1970 to 1990, the effects of logging on *Martes* were the subject of several investigations. Koehler and Hornocker (1977) and Pulliainen (1981b) observed limited use of openings by American martens and European pine martens, respectively. Large clear-cuts were repeatedly found to provide inadequate habitat for American martens in winter (Steventon and Major 1982; Snyder 1984). However, small clear-cuts interspersed with uncut forest stands and selectively cut stands were considered suitable for fishers (Allen 1983). Soutiere (1979) suggested that partial timber harvests that retained residual stands of 20–25 m^2/ha basal area in pole and larger trees were used by martens in winter. Similar findings were reported for fishers (Ingram 1973; Kelly 1977); however, later work would show that a change from clear-cutting to extensive partial harvesting can cause significant loss and fragmentation of American marten habitat (Fuller and Harrison 2005; Simons 2009).

In the western United States, Koehler et al. (1975) considered the American marten to be a "barometer" of the health of ecosystems that may provide clues to the effects of disturbances on other species, such as the fisher and woodland caribou (*Rangifer tarandus*). They also warned land managers that (1) during winter, martens need old-growth forests on mesic sites with canopy cover >30%; (2) martens rarely cross openings >100-m wide during the winter, and do not hunt in openings; and (3) mature-forest communities may support more martens in a given area, but a more diverse forest community will support more martens over time. They concluded that "managers could use logging to partially replace fire as an agent of diversity but mature forest communities must be maintained" (Koehler et al. 1975: 36).

In the early 1980s, new knowledge about American marten and fisher habitat use gave rise to untested habitat suitability index (HSI) models (Allen 1982, 1983). These models stressed the importance of tree-canopy closure and coniferous tree species for both American martens and fishers, whereas the successional stage of forest stands and the presence of abundant downfall (woody debris) were considered important to American martens only. HSI models were widely used by government agencies throughout North America, but because of regional variations in life-history parameters (including predation) and the poor quality of available vegetation data, they often generated questionable results (e.g., Laymon and Barrett 1986). By the end of the period from 1961 to 1990, biologists had concluded that American martens preferred dense, mature coniferous or mixed forests with high overstory cover (Francis and Stephenson 1972; Koehler et al. 1975; Douglas et al. 1983; Raine 1983; Bateman 1986). This turned out to be an overgeneralization, as later studies would show that in some areas, American martens were more closely associated with complex habitat structure than with particular forest age-classes (Potvin et al. 2000; Payer and Harrison 2003; Hearn et al. 2010). Nevertheless, although the reasons for the American marten's habitat preferences were still not clearly understood, other factors such as overhead cover from predation, prey abundance and availability, and thermoregulatory needs during winter appeared to be involved (Koehler and Hornocker 1977; Soutiere 1979; Buskirk et al. 1988; Bissonette et al. 1989).

In Europe, the 1980s were the turning point for research on *Martes*. In 1982, a "Marder Colloquium" took place in Germany, which was the first in an ongoing series of scientific meetings, including the 28th colloquium, held in The Netherlands in 2010. Originally, the goal of these meetings was to bring together biologists sharing an interest in martens, but later meetings included consideration of all mustelids, and the numbers of participants and countries represented have been growing steadily. In the 1980s, research conducted in Europe showed that the European pine marten and stone marten occupied a diversity of landscape conditions that differed from each other according to available habitat (e.g., Pulliainen 1981c; Clément and Saint-Girons 1982; O'Sullivan 1983; Marchesi and Mermod 1989) and prey (e.g., Marchesi et al. 1989). Intra- and interspecific studies (e.g., Pulliainen 1981c; Labrid 1987) were initiated to better understand population densities and distributions. Although the European pine marten was believed to be a woodland-dwelling mustelid (Pulliainen 1981c), the stone marten was associated with rural, urban, and forested habitats (Lodé 1991a). Despite a better understanding of habitat use by *Martes* species, patterns of habitat selection, the use of specific denning and resting sites, and animal movements across fragmented landscapes were still poorly understood at the end of the Scientific Period.

1991–2010: Conservation Period

During the last decade of the 20th century and the beginning of the 21st, the wildlife profession experienced changes in conservation values and public attitudes toward wildlife (Manfredo et al. 2003). As Salwasser (1995) noted, the world experienced the biodiversity crisis, the global-atmosphere crisis, the groundwater crisis, the urban-crime crisis, and the "you-name-it" crisis. As animal-rights activists raised concerns about the fur industry's practices, pelt production fell 62% in the early 1990s (Statistics Canada 2006), and the fur industry continued to struggle in the 2000s (Kozlov 2002; Trapping Today 2009). On the other hand, conservation biology and biodiversity assessment became exciting new research disciplines in the life sciences (Salafsky et al. 2002; Reid and Mace 2003).

In 1993, soon after the first International *Martes* Symposium was convened (Buskirk et al. 1994), the *Martes* Working Group (MWG) was created to facilitate communication among people with common interests in *Martes* research, conservation, and management programs. Thereafter, the MWG organized 3 additional symposia, and published the proceedings for each one (Proulx et al. 1997; Harrison et al. 2004; Santos-Reis et al. 2006).

Although our literature review suggested that research on the European pine marten, stone marten, and sable became increasingly important from 1961 to 1990, many papers on these species were published after 1991 on habitat and landscape selection (e.g., Pereboom et al. 2008; Sălek et al. 2009; Balestrieri et al. 2010; Herr et al. 2010; Vladimirova and Mozgovoy 2010), food habits (e.g., Jędrzejewski et al. 1993; Lachat-Feller 1993b; Brzeziński 1994; Buskirk et al. 1996a; Baltrūnaitė 2006a), and interspecific relations (e.g., Santos-Reis et al. 2004; Grassman et al. 2006b; Sidorovich et al. 2006), among other subjects.

As in Europe, *Martes* studies were numerous in North America from 1991 to 2010 (e.g., Coffin et al. 1997; Potvin and Breton 1997; Powell et al. 1997; Sturtevant and Bissonette 1997; Andruskiw et al. 2008). In this period of *Martes* history, biologists were faced with the challenge of understanding the conditions promoting population persistence, that is, what should be conserved, where, and how?

Habitat selection was the focus of many studies on *Martes* species from 1991 to 2010. Researchers conducted habitat-selection studies using forest-inventory data (Zielinski et al. 2006c, Proulx 2009), and both old and new techniques to detect the presence of *Martes* species (Zielinski and Kucera 1995; Aubry et al. 1997; Zielinski et al. 1997b; Poulton et al. 2006; Proulx and O'Doherty 2006; Rosellini et al. 2008). These and other studies reinforced the importance of forest structural complexity as a habitat predictor (Buskirk et al. 1996a; Miyoshi and Higashi 2005; Proulx 2006a). The prospect that fishers use cognitive maps as they make nonrandom use of space was also explored during this period (Powell 2000, 2004). Studies also stressed the potential effects of abiotic factors on *Martes* species (Krohn et al. 1997, 2004; Weir et al. 2004). This improved understanding of the habitat and landscape conditions occupied by *Martes* species allowed biologists to address a number of questions related to habitat selection and conservation. Habitat selection is, however, a complex process. All selection studies are relative, and they are especially sensitive to the habitat types (and other factors) that are present (or absent) in each study area. In addition, habitat use varies with such factors as prey type, prey abundance, forest structure, natural fragmentation, and the presence or absence of other carnivores (e.g., Hearn et al. 2010). Recognizing this complexity was a major research finding in itself. *Martes* researchers also developed spatially explicit population models to better predict trends under various environmental conditions (Carroll 2007).

To improve the management and conservation of *Martes* populations, biologists working in the Conservation Period also began to explore the genetics of species to better understand their biogeography and taxonomy, and to preserve genetic diversity (e.g., Hicks and Carr 1992; Carr and Hicks 1997; Koepfli et al. 2008). These data, together with studies on skeletal morphology (e.g., Reig 1992; Fortin et al. 1997; Monakhov 2010) led to discussions on subspecific taxonomy and the various factors that produce taxonomic variation (e.g., Nagorsen 1994; Bryant et al. 1997). Combining information about the distribution, spatial organization, and movements of animals with genotypic data enabled biologists to gain important new insights into the zoogeography and ecology of *Martes* populations (Kyle et al. 2000; Aubry et al. 2004; Pertoldi et al. 2008a). Reviews of the distribution and status of *Martes* populations became more comprehensive (Proulx et al. 2004), and biologists suggested that some populations were at risk or should be subject to further investigation to better conserve their habitats (Wisely et al. 2004; Álvares and Brito 2006; Matos and Santos-Reis 2006).

What's Next?

At the beginning of the 20th century, little was known about the populations and habitats of martens, sables, and fishers. Nearly 100 years later, researchers and wildlife managers know a great deal about the evolution, taxonomy, morphophysiology, genetics, population dynamics, habitat and predator-prey relations, food preferences, parasites, and diseases of most *Martes* species. This level of knowledge is particularly impressive when one considers that compared with large felids and canids, small carnivores receive relatively little attention from conservationists, researchers, government agencies, and funding sources (Schipper et al. 2009). Unfortunately, not all *Martes* species are well studied. Little is known about the yellow-throated marten in Southeast Asia (Proulx et al. 2004) or the Japanese marten in Japan and Korea (Saeki 2006), and almost nothing is known about the Nilgiri marten in southern India (Proulx et al. 2004). We urgently need new studies on these species that include the publication of research findings in scientific journals. It is hoped that research and development programs undertaken in North America and

Europe will be duplicated in those parts of the world where the ecological requirements of *Martes* populations are still relatively unknown.

The genera *Eira* and *Gulo* appear to form a clade with *Martes*; that is, their species possibly evolved from a common ancestor, or they may be sister lineages to a clade containing *Martes* (e.g., Koepfli et al. 2008; Hughes, this volume). Our understanding of the phylogenetic relations among the wolverine (*Gulo gulo*), tayra (*Eira barbara*), and *Martes* species will undoubtedly benefit from further investigations of mitochondrial and nuclear genes. Nevertheless, ecomorphological similarities among all these species should encourage *Martes* researchers to further investigate the ecological needs of these species when developing population and habitat conservation programs.

There is also a disparity between the amount of knowledge acquired by biologists, and the integration of such knowledge into wildlife-conservation programs. The ecology of most species is well known, and tools have been developed to identify important habitats that should be protected (Brainerd et al. 1994; Weir and Corbould 2008; Proulx 2009). Despite this knowledge, timber companies often adopt forest-management programs that do not properly address the needs of *Martes* species (Brainerd et al. 1994; Proulx 2009). For a forest-management plan to be effective for conserving *Martes* species, it must be based on spatially explicit data that relate to specific habitat requirements. Field work and improved forest-inventory data are needed to develop multiscale management programs that will address the needs of martens, sables, and fishers effectively. More work is also required to differentiate habitats that provide *Martes* species with optimal living conditions at times when survival depends on meeting unusually high energetic needs (e.g., winter, breeding season). More studies are needed at the microhabitat level to identify critical structural elements (if any), and to determine how and under which circumstances *Martes* use various portions of their home ranges. The movements and activities of predators and prey living in sympatry with *Martes* species should be studied under different environmental conditions, and at different phenological periods of *Martes* biology, to better understand habitat selection by martens, sables, and fishers within their home ranges, and why they may behave differently from one habitat type to another, or from one region to another. For example, predation by red foxes (*Vulpes vulpes*) may influence selection of resting and denning sites by European pine martens (Brainerd et al. 1995). Habitats with high densities of ground squirrels (*Spermophilus* spp.) are more important to American martens in summer, when the ground squirrels are active, than in winter when they hibernate and become inaccessible (Zielinski et al. 1983). Movements of American martens among habitats may be more linear and rapid along cut-block edges, but more tortuous in structurally complex forests (Heinemeyer 2002), where efficiencies in encountering and killing small mammals may be linked to higher abundance of coarse woody debris (Andruskiw et al. 2008).

In contrast to traditional population-genetic studies, modern landscape genetics provides a framework for investigating the influence of landscape elements and environmental features on gene flow, genetic discontinuities, and genetic population structure (Storfer et al. 2010; Schwartz et al., this volume). How dispersing martens navigate the landscape is still an unanswered question (Broekhuizen 2006). The combination of genetic and landscape-ecology tools creates an exciting field of research on *Martes* and other species that are sensitive to habitat loss and degradation (Virgós and García 2002; Gehring and Swihart 2003) and to potential dispersal barriers (e.g., roads; Grilo et al. 2009).

Martes conservation would benefit from integrating the needs of martens, sables, and fishers into multispecies conservation plans. Thomas et al. (1988) pointed out that numerous wildlife species in the western United States use old-growth forest disproportionately to its occurrence within their home ranges. These include the American marten, fisher, red tree vole (*Arborimus longicaudus*), northern flying squirrel (*Glaucomys sabrinus*), several species of bats, northern spotted owl (*Strix occidentalis caurina*), pileated woodpecker (*Dryocopus pileatus*), and many others. Thus, conserving the American marten and fisher would provide valuable habitat for many other species. Similarly, Proulx (2005a) integrated the habitat needs of the American marten and fisher with those of endangered species, such as the woodland caribou, wolverine, and grizzly bear (*Ursus arctos*), to identify areas supporting all these species where increased conservation efforts in landscapes managed for timber harvesting would benefit several species of concern. In the context of such a multispecies management program, *Martes* habitat and forest-succession models should be used to predict the location and spatial extent of suitable habitats, predict the outcomes of various forest-management scenarios, and assess the short- and long-term viability of *Martes* populations inhabiting landscapes that are vulnerable to climate change, support various predator and prey populations, and are subject to timber or fur harvests. Most importantly, models and management recommendations must be tested in replicated treatments under different conditions to learn more about *Martes* population needs and responses to various management scenarios. In particular, we need to determine whether habitat-management activities have fitness implications for *Martes* populations. Scientific research should not be separated from management activities, and management activities should not be carried out without proper scientific evaluation (Romesburg 1981; Sinclair 1991). On the other hand, the need to relate management activities to correlates of fitness should not delay the conservation of habitats shown in previous studies to be important to *Martes* species. In other words, we should not fail to implement what we know just because "we do not know it all."

Although the cumulative effects of hunting, trapping, and habitat loss may compromise the future of *Martes* populations (Banci and Proulx 1999;

Proulx and Verbisky 2001; Broekhuizen 2006), these factors have often been treated independently. In addition, despite the existence of new trapping technologies (Proulx 1999a) and standards (Proulx 1997; Powell and Proulx 2003), many wildlife researchers and trappers are unfamiliar with state-of-the-art trapping technology (Proulx et al. 2012). Older, less-humane trapping methods continue to be used; thus, *Martes* biologists must remain vigilant to ensure implementation of the best scientific and management practices in the field.

Despite several popular publications on the biology of martens and fishers (e.g., Haley 1975; Noblet 2002; Proulx 2010), we believe that most members of the general public know very little about *Martes* species. This is expected for species that are not only rare and elusive but also have a restricted geographic distribution (Wilson and Tisdell 2005; Aubry and Jagger 2006). To implement *Martes* conservation programs, however, wildlife professionals need the support of naturalist clubs, outdoor enthusiasts, and citizens from all walks of life. This need is particularly true where human development and economic growth impact biodiversity (Karmona 2007; Dawe and Mosley 2009). Agencies, companies, and individuals are often resistant to change and have difficulty adopting new findings and technologies. Change is also stymied by a lack of communication among *Martes* specialists, industry, and the public. Conserving *Martes* populations and their habitats is as much about changing human behavior as it is about collecting scientifically sound datasets. Reporting what we know to the public and finding win-win solutions between industrial interests and biologists will ultimately enhance the conservation of *Martes* populations and their habitats.

In the past, wildlife researchers have identified population- and habitat-management conflicts caused by agriculture, forestry, mining and oil exploration, and fur trapping (Proulx 2000). Unfortunately, little has been published on successful management programs and conflict resolutions, and examples of adaptive management for *Martes* conservation are scarce. Yet, *Martes* biologists in all parts of the world must deal with similar issues to conserve or restore threatened populations and habitats (Proulx et al. 2004). It is important to disseminate the results of both successful and unsuccessful management programs, so that other *Martes* biologists can improve their own programs and identify areas where more work is required to develop effective multiscale and multispecies programs for *Martes* management and conservation.

Acknowledgments

We express our gratitude to Bill Zielinski for his advice and editing work. We thank 2 anonymous referees for their critical review of an earlier version of the manuscript.

Literature Cited

Abramov, A., and C. Wozencraft. 2008. *Martes zibellina*. IUCN Red List of Threatened Species. Version 2010.4. http://www.iucnredlist.org (accessed 5 February 2011).

Abramov, A.V., S.V. Kruskop, and A.A. Lissovsky. 2006. Distribution of the stone marten *Martes foina* (Carnivora, Mustelidae) in the European part of Russia. Russian Journal of Theriology 5:37–41.

Acevedo-Whitehouse, K., and A.A. Cunningham. 2006. Is MHC enough for understanding wildlife immunogenetics? Trends in Ecology & Evolution 21:433–438.

Ackman, R.G., and S.C. Cunnane. 1992. Long-chain polyunsaturated fatty acids: sources, biochemistry and nutritional/clinical applications. Pages 161–215 *in* Advances in applied lipid research. Volume 1. F.B. Padley, editor. JAI Press, London, UK.

Adams, A.L. 1873. Field and forest rambles: with notes and observations on the natural history of eastern Canada. Henry S. King, London, UK.

Adams, L.A., P. Angulo, and K.D. Lindor. 2005. Nonalcoholic fatty liver disease. Canadian Medical Association Journal 172:899–905.

Adney, T. 1893a. Some New Brunswick traps—III [deadfall for fisher]. Forest and Stream 40:51.

Adney, T. 1893b. Some New Brunswick traps—III [deadfall for marten]. Forest and Stream 40:72.

Agosta, S.J., and J.A. Klemens. 2008. Ecological fitting by phenotypically flexible genotypes: implications for species associations, community assembly and evolution. Ecology Letters 11:1–12.

Agren, E.O., L. Nordenberg, and T. Morner. 2000. Surgical implantation of radiotelemetry transmitters in European badgers (*Meles meles*). Journal of Zoo and Wildlife Medicine 31:52–55.

Aguilar, A., G. Roemer, S. Debenham, M. Binns, D. Garcelon, and R.K. Wayne. 2004. High MHC diversity maintained by balancing selection in an otherwise genetically monomorphic mammal. Proceedings of the National Academy of Sciences, USA 101:3490–3494.

Agustí, J., A. Sanz de Siriaa, and M. Garcé. 2003. Explaining the end of the hominoid experiment in Europe. Journal of Human Evolution 45:145–153.

Ahima, R.S., D. Prabakaran, C. Mantzoros, D. Qu, B. Lowell, E. Maratos-Flier, and J.S. Flier. 1996. Role of leptin in the neuroendocrine response to fasting. Nature 382:250–252.

Ahn, S., W.B. Krohn, A.J. Plantinga, T.J. Dalton, and J.A. Hepinstall. 2002. Agricultural land changes in Maine: a compilation and brief analysis of Census of Agriculture data, 1850–1997. Maine Agricultural and Forest Experiment Station, Technical Bulletin 182. University of Maine, Orono, USA.

Alasaad, S., L. Rossi, R.C. Soriguer, L. Rambozzi, D. Soglia, J.M. Perez, and X.Q. Zhu. 2009. *Sarcoptes* mite from collection to DNA extraction: the lost realm of the neglected parasite. Parasitology Research 104:723–732.

Albrecht, N.M., C.L. Heusser, and M.K. Schwartz. 2009. Differences between fisher and marten distributions in north Idaho. Page 1 *in* Biology and conservation of martens, sables, and fishers: a new synthesis. Abstracts from the 5th International *Martes* Symposium, University of Washington, Seattle, USA.

Alcover, J.A., 1980 (1982). Note on the origin of the present mammalian fauna from the Balearic and Pityusics islands. Miscellaneous Zoology (Barcelona) 6:141–149.

Allan, A.A. 2001. Ticks (class Arachnida: order Acarina). Pages 72–106 *in* Parasitic diseases of wild mammals. W.M. Samuel, M.J. Pybus, and A.A. Kocan, editors. Iowa State University Press, Ames, USA.

Allen, A.W. 1982. Habitat suitability index models: marten. Report FWS/OBS-82/10.11, USDI Fish and Wildlife Service, Washington, D.C., USA.

Allen, A.W. 1983. Habitat suitability index models: fisher. Report FWS/OBS-82/10.45, USDI Fish and Wildlife Service, Washington, D.C., USA.

Allen, A.W. 1984. Habitat suitability index models: marten. Revised. Report FWS/OBS-82/10.11, USDI Fish and Wildlife Service, Washington, D.C., USA.

Allen, C.D., and D.D. Breshears. 1998. Drought-induced shift of a forest-woodland ecotone: rapid landscape response to climate variation. Proceedings of the National Academy of Sciences, USA 95:14839–14842.

Allen, D.L. 1985. These fifty years: the conservation record of North American Wildlife and Natural Resources Conference. Transactions of the North American Wildlife and Natural Resources Conference 50:11–67.

Allen, E.A., D.J. Morrison, and G.W. Wallis. 1996. Common tree diseases of British Columbia. Natural Resources Canada, Canadian Forest Service, Victoria, British Columbia, Canada.

Allen, J.A. 1876. The former range of some New England carnivorous mammals. American Naturalist 10:708–715.

Allendorf, F.W., and G. Luikart. 2007. Conservation and the genetics of populations. Blackwell Publishing, Malden, Massachusetts, USA.

Álvares, F., and J.C. Brito. 2006. Habitat requirements and potential areas of occurrence for the pine marten in north-western Portugal: conservation implications. Pages 29–45 *in* *Martes* in carnivore communities. M. Santos-Reis, J.D.S. Birks, E.C. O'Doherty, and G. Proulx, editors. Alpha Wildlife Publications, Sherwood Park, Alberta, Canada.

American Association of Zoo Veterinarians. 2010. Small carnivore medical management guidelines. http://www.aazv.org/displaycommon.cfm?an=1&subarticlenbr=272 (accessed January 2011).

Amori, G., F.M. Angelici, C. Prigioni, and A.V. Taglianti. 1996. The mammal fauna of Italy: a review. Hystrix 8:3–7.

An, D.J., S.H. Yoon, J.Y. Park, I.S. No, and B.K. Park. 2008. Phylogenetic characterization of canine distemper virus isolates from naturally infected dogs and a marten in Korea. Veterinary Microbiology 132:389–395.

Anderson, B. 2002. Reintroduction of fishers (*Martes pennanti*) to the Catoosa Wildlife Area in Tennessee. Tennessee Wildlife Resources Agency, Crossville, USA.

Anderson, E. 1970. Quaternary evolution of the genus *Martes* (Carnivora, Mustelidae). Acta Zoologica Fennici 130:1–132.

Anderson, E. 1994. Evolution, prehistoric distribution, and systematics of *Martes*. Pages 13–25 *in* Martens, sables and fishers: biology and conservation. S.W. Buskirk, A.S. Harestad, M.G. Raphael, and R.A. Powell, editors. Cornell University Press, Ithaca, New York, USA.

Anderson, R.C. 2000. Nematode parasites of vertebrates: their development and transmission. CABI Publishing, Cambridge, Massachusetts, USA.

Anderson, R.M. 1934. The distribution, abundance, and economic importance of the game and fur-bearing mammals of western North America. Pages 4055–4075 *in* Proceedings of the Fifth Pacific Science Congress, University of Toronto Press, Ontario, Canada.

Anderson, S.B. 1987. Wild furbearer management in eastern Canada. Pages 1040–1048 *in* Wild furbearer management and conservation in North America. M. Novak, J.A. Baker, M.E. Obbard, and B. Malloch, editors. Ontario Trappers Association and Ontario Ministry of Natural Resources, Toronto, Canada.

Andruskiw, M., J.M. Fryxell, I.D. Thompson, and J.A. Baker. 2008. Habitat-mediated variation in predation risk by the American marten. Ecology 89:2273–2280.

Anonymous. 1779. Records for Machias, Maine. Indian truck house accounts, 1776–1779. Massachusetts Archives 147:557 (microfilm).

Anonymous. 1895. Fish and game. The Industrial Journal, Bangor, Maine. May 17:6.

Aramburu, C.E, and M.E. King-Dagen. 1995. Population and the environment: perspectives, propositions and experience. Pages 99–102 *in* Integrating people and wildlife for a sustainable future. J.A. Bissonette and P.R. Krausman, editors. Wildlife Society, Bethesda, Maryland, USA.

Archibald, W.R., and R.H. Jessup. 1984. Population dynamics of the pine marten (*Martes americana*) in the Yukon Territory. Pages 81–97 *in* Northern ecology and resource management. R. Olson, R. Hastings, and F. Geddes, editors. University of Alberta Press, Edmonton, Canada.

Armstrong, D.P., and P.J. Seddon. 2008. Directions in reintroduction biology. Trends in Ecology & Evolution 23:20–25.

Arnemo J.M., A. Fahlman, K. Madslien, S. Brunberg, B. Ytrehus, and J. Swenson. 2007. Long-term evaluation of Telonics® intraperitoneal radiotransmitters in free-ranging brown bears (*Ursus arctos*). Pages 96–97 *in* Proceedings of the AAZV/AAWV/NAG Joint Conference, Knoxville, Tennessee, USA.

Arnold, M. 1997. Natural hybridization and evolution. Oxford University Press, UK.

Arthur, S.M., and W.B. Krohn. 1991. Activity patterns, movements, and reproductive ecology of fishers in southcentral Maine. Journal of Mammalogy 72:379–385.

Arthur, S.M., W.B. Krohn, and J.R. Gilbert. 1989. Home range characteristics of adult fishers. Journal of Wildlife Management 53:674–679.

Arzoumanian, L. 2003. What is hemolysis, what are the causes, what are the effects? BD Tech Talk 2:1–3.

Ashbrook, F.G., and K.B. Hanson. 1930. The normal breeding season and gestation period of martens. U.S. Department of Agriculture, Circular No. 107, Washington, D.C., USA.

Ashizawa, H., T. Murakami, D. Nosaka, S. Tateyama, and S. Habe. 1978. *Concinnum ten.* Murakami Bulletin 25:77–84.

Aubry, K.B., J.P. Hayes, B.L. Biswell, and B.G. Marcot. 2003. The ecological role of tree-dwelling mammals in western coniferous forests. Pages 405–443 *in* Mammal community dynamics: management and conservation in the coniferous forests of western North America. C.J. Zabel and R.G. Anthony, editors. Cambridge University Press, UK.

Aubry, K.B., and D.B. Houston. 1992. Distribution and status of the fisher (*Martes pennanti*) in Washington. Northwestern Naturalist 73:69–79.

Aubry, K.B., and L.A. Jagger. 2006. The importance of obtaining verifiable occurrence of data on forest carnivores and an interactive website for archiving results from standardized surveys. Pages 159–176 *in Martes* in carnivore communities. M. Santos-Reis, J.D.S. Birks, E.C. O'Doherty, and G. Proulx, editors. Alpha Wildlife Publications, Sherwood Park, Alberta, Canada.

Aubry, K.B., and J.C. Lewis. 2003. Extirpation and reintroduction of fishers (*Martes pennanti*) in Oregon: implications for their conservation in the Pacific states. Biological Conservation 114:79–90.

Aubry, K.B., K.S. McKelvey, and J.P. Copeland. 2007. Distribution and broadscale habitat relations of the wolverine in the contiguous United States. Journal of Wildlife Management, 71:2147–2158.

Aubry K.B., and C.M. Raley. 2002a. Selection of nest and roost trees by pileated woodpeckers in coastal forests of Washington. Journal of Wildlife Management 66:392–406.

Aubry, K.B., and C.M. Raley. 2002b. The pileated woodpecker as a keystone habitat modifier in the Pacific Northwest. Pages 257–274 *in* Proceedings of the symposium on the ecology and management of dead wood in western forests. W.F. Laudenslayer, Jr., P.J. Shea, B.E. Valentine, C.P. Weatherspoon, and T.E. Lisle, technical coordinators. USDA Forest Service, General Technical Report PSW-GTR-181.

Aubry, K.B., and C.M. Raley. 2006. Ecological characteristics of fishers (*Martes pennanti*) in the southern Oregon Cascade Range—update: July 2006. Unpublished report, USDA Forest Service, Pacific Northwest Research Station, Olympia, Washington, USA.

Aubry, K.B., F.E. Wahl, J. von Kienast, T.J. Catton, and S.G. Armentrout. 1997. Use of remote video cameras for the detection of forest carnivores and in radio-telemetry studies of fishers. Pages 350–361 *in Martes*: taxonomy, ecology, techniques and management. G. Proulx, H.N. Bryant, and P.M. Woodard, editors. Provincial Museum of Alberta, Edmonton, Canada.

Aubry, K., S. Wisely, C. Raley, and S. Buskirk. 2004. Zoogeography, spacing patterns, and dispersal in fishers: insights gained from combining field and genetic data. Pages 201–220 *in* Martens and fishers (*Martes*) in human-altered landscapes: an international perspective. D.J. Harrison, A.K. Fuller, and G. Proulx, editors. Springer Science+Business Media, New York, USA.

Audubon, J.J., and J. Bachman. 1852. Quadrupeds of North America. Volume 1. V.G. Aubudon, New York, USA.

Aune, K.E., and P. Schladweiler. 1997. Age, sex structure, and fecundity of the American marten in Montana. Pages 61–77 *in Martes*: taxonomy, ecology, techniques, and management. G. Proulx, H.N. Bryant, and P.M. Woodard, editors. Provincial Museum of Alberta, Edmonton, Canada.

Avise, J.C. 2000. Phylogeography: the history and formation of species. Harvard University Press, Cambridge, Massachusetts, USA.

Avise, J.C. 2004. Molecular markers, natural history, and evolution. Sinauer Associates, Sunderland, Massachusetts, USA.

Avise, J.C., J. Arnold, R.M. Ball, E. Bermingham, T. Lamb, J.E. Neigel, C.A. Reeb, and N.C. Saunders. 1987. Intraspecific phylogeography: the mitochondrial DNA bridge between pouplation genetics and systematics. Annual Review of Ecology and Systematics 18:489–522.

Azanza, B., A.M. Alonso-Zarza, M.A. Álverez-Sierra, J.P. Calvo, S. Fraile, I. García-Paredes, E. Gómez, M. Hernández-Fernández, A. Van Der Meulen, D. De Miguel, P. Montoya, J. Morales, X. Murelaga, P. Peláez-Campomanes, B. Perez, V. Quiralte, M.J. Salesa, I.M. Sánchez, A. Sánchez-Marco, and D. Soria. 2004. Los yacimientos de vertebrados continentals del Aragoniense superior (Mioceno medio) de Toril, Cuenca de Calatayd-Daroca. Geo-Temas 6:271–274.

Bach, L.A., R. Thomsen, C. Pertoldi, and V. Loeschcke. 2006. Evolution of density-dependent dispersal in an individual-based adaptive metapopulation model including explicit kin competition and demographic stochasticity. Ecological Modelling 192:658–666.

Badgley, C., J.C. Barry, M.E. Morgan, S. V. Nelson, A.K. Behrensmeyer, T.E. Cerling, and D. Pilbeam. 2008. Ecological changes in Miocene mammalian record show impact of prolong climatic forcing. Proceedings of the National Academy of Sciences, USA 105:12145–12149.

Baguette, M., and G. Mennechez. 2004. Resource and habitat patches, landscape ecology and metapopulation biology: a consensual viewpoint. Oikos 106:399–403.

Bailey, L.L., T.R. Simons, and K.H. Pollock. 2004. Estimating site occupancy and detection probability parameters for terrestrial salamanders. Ecological Applications 14:692–702.

Bailey, L.L., J.E. Hines, J.D. Nichols, and D.I. MacKenzie. 2007. Sampling design trade-offs in occupancy studies with imperfect detections: examples and software. Journal of Applied Ecology 17:281–290.

Bailey, L.L., J.A. Reid, E.D. Forsman, and J.D. Nichols. 2009. Modeling co-occurrence of northern spotted and barred owls: accounting for detection probability differences. Biological Conservation 142:2983–2989.

Baird, R., and S. Frey. 2000. Riding Mountain National Park fisher reintroduction program 1994–1995. Unpublished report, Riding Mountain National Park, Manitoba, Canada.

Bakeev, N.N., and V.V. Timofeev. 1973. Sable. Acclimatization. Pages 16–25 *in* Sable, martens, yellow-throated marten: distribution, resources, ecology, and conservation. A.A. Nasimovich, editor. Nauka, Moscow, USSR. (in Russian)

Baker, J.M. 1992. Habitat use and spatial organization of pine marten on southern Vancouver Island, British Columbia. Thesis, Simon Fraser University, Burnaby, British Columbia, Canada.

Baker, R.J., L.C. Bradley, R.D. Bradley, J.W. Dragoo, M.D. Engstrom, R.S. Hoffmann, C.A. Jones, F. Reid, D.W. Rice, and C. Jones. 2003. Revised checklist of North American mammals north of Mexico, 2003. Occasional Papers, Museum, Texas Tech University 229:1–23.

Bakeyev, N.N., and A.A. Sinitsyn. 1994. Status and conservation of sables in the Commonwealth of Independent States. Pages 246–254 *in* Martens, sables, and fishers: biology and conservation. S.W. Buskirk, A.S. Harestad, M.G. Raphael, and R.A. Powell, editors. Cornell University Press, Ithaca, New York, USA.

Bakeyev, Y.N. 1994. Stone martens in the Commonwealth of Independent States. Pages 243–245 *in* Martens, sables, and fishers: biology and conservation. S.W. Buskirk, A.S. Harestad, M.G. Raphael, and R.A. Powell, editors. Cornell University Press, Ithaca, New York, USA.

Baldwin, R.A., and L.C. Bender. 2008. Distribution, occupancy, and habitat correlates of American martens (*Martes americana*) in Rocky Mountain National Park, Colorado. Journal of Mammalogy 89:419–427.

Balestrieri, A., L. Remonti, N. Ferrari, A. Ferrari, T.L. Valvo, S. Robetto, and R. Orusa. 2006. Sarcoptic mange in wild carnivores and its co-occurrence with parasitic helminths in the Western Italian Alps. European Journal of Wildlife Research 52:196–201.

Balestrieri, A., L. Remonti, A. Ruiz-González, B.J. Gómez-Moliner, M. Vergara, and C. Prigioni. 2010. Range expansion of the pine marten (*Martes martes*) in an agricultural landscape matrix (NW Italy). Mammalian Biology 75:412–419.

Balestrieri, A., A. Ruiz-González, L. Remonti, B. Gómez-Moliner, S. Genovese, L. Gola, P.F. del Po, C. Prigioni, and others. 2008. A non-invasive genetic survey of the pine marten (*Martes martes*) in the western River Po plain (Italy): preliminary results. Hystrix 19:77–80.

Balharry, D. 1993. Social organization in martens: an inflexible system? Symposium of the Zoological Society of London 65:321–345.

Balkenhol, N., L.P. Waits, and R.J. Dezznai. 2009. Statistical approaches in landscape genetics: an evaluation of methods for linking landscape and genetic data. Ecography 32:818–830.

Ball, I.R., H.P. Possingham, and M.E. Watts. 2009. Marxan and relatives: software for spatial conservation prioritization. Pages 185–195 *in* Spatial conservation prioritization: quantitative methods and computational tools. A. Moilanen, K.A. Wilson, and H.P. Possingham, editors. Oxford University Press, UK.

Balloux, F. 2010. The worm in the fruit of the mitochondrial DNA tree. Heredity 104:419–420.

Balmysheva, N.P., and L.L Solovenchuk. 1999a. Association between mutations of mitochondrial DNA genes for cytochrome *b* and NADH dehydrogenase 5/6 in sable *Martes zibellina* L. Russian Journal of Genetics 35:1447–1451.

Balmysheva, N.P., and L.L. Solovenchuk. 1999b. Genetic variation of the mitochondrial DNA gene encoding cytochrome *b* in the Magadan population of sable *Martes zibellina* L. Russian Journal of Genetics 35:1077–1081.

Baltrūnaitė, L. 2006a. Diet and winter habitat selection of the pine marten (*Martes martes* L.) in sandy and clay plains, Lithuana. Pages 99–108 *in Martes* in carnivore communities. M. Santos-Reis, J.D.S. Birks, E.C. O'Doherty, and G. Proulx, editors. Alpha Wildlife Publications, Sherwood Park, Alberta, Canada.

Baltrūnaitė, L. 2006b. Diet and winter habitat use of the red fox, pine marten and raccoon dog in Dzūkija national park, Lithuania. Acta Zoologica Lituanica 16:46–60.

Banci, V., and G. Proulx. 1999. Resiliency of furbearers to trapping in Canada. Pages 175–203 *in* Mammal trapping. G. Proulx, editor. Alpha Wildlife Publications, Sherwood Park, Alberta, Canada.

Banfield, A.F.W. 1974. The mammals of Canada. University of Toronto Press, Ontario, Canada.

Baniandrés, N., and E. Virgós. 2009. Intraguild predation by eagle owls influences local abundance of a medium-sized carnivore, the stone marten. X International Congress of Ecology, Sydney, Australia. Poster presentation. Abstract available at http://www.intecol10.org/abstracts/pdf/0908015Abstract01260.pdf.

Barea-Azcón, J.M., E. Virgós, E. Ballesteros-Duperón, M. Moleón, and M. Chirosa. 2006. Surveying carnivores at large spatial scales: a comparison of four broad-applied methods. Biodiversity and Conservation 16:1213–1230.

Barja, I. 2005. Winter distribution of European pine marten *Martes martes* scats in a protected area of Galicia, Spain. Mammalia 69:435–438.

Barker, F.C., and J.S. Danforth. 1882. Hunting and trapping on the upper Magalloway River and Parmarchenee Lake: first winter in the wilderness. D. Lothrop and Company, Boston, Massachusetts, USA.

Barker, I., and C. Parrish. 2001. Parvovirus infections. Pages 131–146 *in* Infectious diseases of wild mammals. E.S. Williams and I.K. Barker, editors. Iowa State University Press, Ames, USA.

Barnosky, A.D., editor. 2004. Biodiversity response to climate change in the Middle Pleistocene: the Porcupine Cave fauna from Colorado. University of California Press, Berkeley, USA.

Baron, W.R. 1992. Historical climate records for the northeastern United States, 1640 to 1900. Pages 74–91 *in* Climate change since A.D. 1500. R.S. Bradley and P.D. Jones, editors. Routledge, New York, USA.

Barrett, M.W., G. Proulx, D. Hobson, D. Nelson, and J.W. Nolan. 1989. Field evaluation of the C120 Magnum trap for marten. Wildlife Society Bulletin 17:299–306.

Barrett, R.H. 1983. Smoked aluminum track plots for determining furbearer distribution and relative abundance. California Fish and Game 69:188–190.

Barrientos, R., and E.Virgós. 2006. Reduction of potential food interference in two sympatric carnivores by sequential use of shared resources. Acta Oecologica 30:107–116.

Barry, J.C., M.E. Morgan, L.J. Flynn, D. Pilbeam, A.K. Behrensmeyer, S. Mahmood Raza, I.A. Khan, C. Badgley, J. Hicks, and J. Kelley. 2002. Faunal and environmental change in the late Miocene Siwaliks of northern Pakistan. Paleobiology 28:1–71.

Barton, D.E., F.N. David, and E. Fix. 1962. Persistence in a chain of multiple events when there is simple dependence. Biometrika 49:351–357.

Baskin, J.A. 1998. Mustelidae. Pages 152–173 *in* Evolution of Tertiary mammals of North America, volume 1: Terrestrial carnivores, ungulates, and ungulatelike mammals. C.M. Janis, K.M. Scott, and L.L. Jacobs, editors. Cambridge University Press, UK.

Baskin, J.A. 2005. Carnivora from the late Miocene Love bone bed of Florida. Bulletin of the Florida Museum of Natural History 45:419–440.

Bateman, M.C. 1982. Marten re-introduction to Terra Nova National Park. Unpublished report to Parks Canada, Canadian Wildlife Service, Atlantic Region.

Bateman, M.C. 1984. Marten re-introduction to Terra Nova National Park. Unpublished report to Parks Canada, Canadian Wildlife Service, Atlantic Region.

Bateman, M.C. 1986. Winter habitat use, food habits and home range size of the marten, *Martes americana*, in western Newfoundland. Canadian Field-Naturalist 100:58–62.

Bates, K., K. Steenhof, and M.R. Fuller. 2003. Recommendations for finding PTTs on the ground without VHF telemetry. Proceedings of the Argos Animal Tracking Symposium, Annapolis, Maryland, USA. http://srfs.wr.usgs.gov/pdf/finding%20ptts.pdf (accessed 31 May 2011).

Baumgartner, M.F., and B.R. Mate. 2005. Summer and fall habitat of North Atlantic right whales (*Eubalaena glacialis*) inferred from satellite telemetry. Canadian Journal of Fisheries and Aquatic Sciences 62:527–543.

Beale, C.M., J.J. Lennon, J.M. Yearsley, M.J. Brewer, and D.A. Elston. 2010. Regression analysis of spatial data. Ecology Letters 13:246–264.

Beckmann, J.P. 2006. Carnivore conservation and search dogs: the value of a novel, noninvasive technique in the Greater Yellowstone Ecosystem. Pages 28–34 *in* Greater Yellowstone public lands: a century of discovery, hard lessons, and bright prospects. A.W. Biel, editor. Proceedings of the 8th Biennial Scientific Conference on the Greater Yellowstone Ecosystem. Yellowstone Center for Resources, Yellowstone National Park, Wyoming, USA.

Begon, M. 2008. Effects of host diversity on disease dynamics. Pages 12–29 *in* Infectious disease ecology: effects of ecosystems on disease and of diseases on ecosystems. R.S. Ostfeld, F. Keesing, and V.T. Eviner, editors. Princeton University Press, New Jersey, USA.

Behrensmeyer, A.K., and J.C. Barry. 2005. Biostratigraphic surveys in the Siwaliks of Pakistan: a method for standardized surface sampling of the vertebrate fossil record. Palaeontologia Electronica 8. http://palaeo-electronica.org/paleo/2005_1/behrens15/issue1_05.htm.

Beier, P., D.R. Majka, and W.D. Spencer. 2008. Forks in the road: choices in procedures for designing wildland linkages. Conservation Biology 22:836–851.

Beja-Pereira, A., R. Oliveira, P.C. Alves, M.K. Schwartz, and G. Luikart. 2009. Advancing ecological understandings through technological transformations in noninvasive genetics. Molecular Ecology Resources 9:1279–1301.

Belant, J.L. 2003a. A hairsnare for forest carnivores. Wildlife Society Bulletin 31:482–485.

Belant, J.L. 2003b. Comparison of 3 tracking mediums for detecting forest carnivores. Wildlife Society Bulletin 31:744–747.

Belant, J.L. 2007. Human-caused mortality and population trends of American marten and fisher in a U.S. National Park. Natural Areas Journal 27:155–160.

Ben-David, M., R.W. Flynn, and D.M. Schell. 1997. Annual and seasonal changes in diets of martens: evidence from stable isotope analysis. Oecologia 111:280–291.

Bennitt, R., J.S. Dixon, V.H. Cahalane, W.W. Chase, and W.L. McAtee. 1937. Statement of policy. Journal of Wildlife Management 1:1–2.

Benson, D.A. 1959. The fisher in Nova Scotia. Journal of Mammalogy 40:451.

Benton, A.H., and D.L. Kelly. 1975. An annotated list of New York Siphonaptera. Journal of the New York Entomological Society 83:142–156.

Berdoy, M., J.P. Webster, and D.W. Macdonald. 1995. Parasite-altered behaviour: is the effect of *Toxoplasma gondii* on *Rattus norvegicus* specific? Parasitology 111:403–409.

Berg, A.H., T.P. Combs, X. Du, M. Brownlee, and P.E. Scherer. 2001. The adipocyte-secreted protein Acrp30 enhances hepatic insulin action. Nature Medicine 7:947–953.

Berg, W.E. 1982. Reintroduction of fisher, pine marten, and river otter. Pages 159–175 *in* Midwest furbearer management. G.C. Sanderson, editor. Proceedings of the 43rd Midwest Fish and Wildlife Conference, Wichita, Kansas, USA.

Bergin, T.M. 1992. Habitat selection by the western kingbird in western Nebraska: a hierarchical analysis. Condor 94:903–911.

Bernor, R.L., V. Fahlbusch, and H.-W. Mittmann. 1996. The evolution of western Eurasian Neogene mammal faunas. Columbia University Press, New York, USA.

Beuter, J.H. 1990. Social and economic impacts of the spotted owl conservation strategy. Technical Bulletin No. 9003. American Forest Resource Alliance, Washington, D.C., USA.

Beyer, K.M., and R.T. Golightly. 1996. Distribution of Pacific fisher and other forest carnivores in coastal northwestern California—revision: April 1996. Unpublished report to California Department of Fish and Game, FG-3156-WM. Humboldt State University, Arcata, California, USA.

Biggins, D.E., J.L. Godbey, B.J. Miller, and L.R. Hanebury. 2006. Radio telemetry for black-footed ferret research and monitoring. Pages 175–190 *in* Recovery of the black-footed ferret: progress and continuing challenges. J.E. Rolle, B.J. Miller, J.L. Godbey, and D.E. Biggins, editors. U.S. Geological Survey, Scientific Investigations Report 2005–5293.

Biggins, D.E., B.J. Miller, L.R. Hanebury, and R.A. Powell. 2011. Mortality of Siberian polecats and black-footed ferrets released onto prairie dog colonies. Journal of Mammalogy 92:721–731.

Bininda-Emonds, O.R.P., M. Cardillo, K.E. Jones, R.D.E. MacPhee, R.M.D. Beck, R. Grenyer, S.A. Price, R.A. Vos, J.L. Gittleman, and A. Purvis. 2007. The delayed rise of present day mammals. Nature 446:507–512.

Bininda-Emonds, O.R.P., J.L. Gittleman, and A. Purvis. 1999. Building large trees by combining phylogenetic information: a complete phylogeny of the extant Carnivora (Mammalia). Biological Review 74:143–175.

Birks, J., J. Messenger, T. Braithwaite, A. Davison, R. Brookes, and C. Strachan. 2004. Are scat surveys a reliable method for assessing distribution and population status of pine martens? Pages 235–252 *in* Martens and fishers (*Martes*) in human-altered environments. D.J. Harrison, A.K. Fuller, and G. Proulx, editors. Springer Science+Business Media, New York, USA.

Birks, J.D.S., J.E. Messenger, and E.C. Halliwell. 2005. Diversity of den sites used by pine martens *Martes martes*: a response to the scarcity of arboreal cavities? Mammal Review 35:313–320.

Bissonette, J.A., and S. Broekhuizen. 1995. *Martes* populations as indicators of habitat spatial patterns: the need for a multiscale approach. Pages 95–121 *in* Landscape approaches in mammalian ecology and conservation. W.J. Lidicker, Jr., editor. University of Minnesota Press, Minneapolis, USA.

Bissonette, J.A., R.J. Fredrickson, and B.J. Tucker. 1988. The effects of forest harvesting on marten and small mammals in western Newfoundland. Unpublished report to Newfoundland and Labrador Wildlife Division and Corner Brook Pulp and Paper, Ltd., Utah State University, Logan, USA.

Bissonette, J.A., R.J. Frederickson, and B.J. Tucker. 1989. American marten: a case for landscape-level management. Transactions of the North American Wildlife and Natural Resources Conference 54:89–101.

Bissonette, J.A., D.J. Harrison, C.D. Hargis, and T.G. Chapin. 1997. The influence of spatial scale and scale-sensitive properties on habitat selection by American marten. Pages 368–385 *in* Wildlife and landscape ecology: effects of pattern and scale. J.A. Bissonette, editor. Springer, New York, USA.

Bjornvad, C.R., J. Elnif, and P.T. Sangild. 2004. Short-term fasting induces intra-hepatic lipid accumulation and decreases intestinal mass without reduced brush-border enzyme activity in mink (*Mustela vison*) small intestine. Journal of Comparative Physiology B 174:625–632.

Black, J.D. 1950. The rural economy of New England—a regional study. Harvard University Press, Cambridge, Massachusetts, USA.

Blix A.S., and E.S. Nordoy. 2007. Ross seal (*Ommatophoca rossii*) annual distribution, diving behaviour, breeding and moulting, off Queen Maud Land, Antarctica. Polar Biology 30:1449–1458.

Bobrov, V.V., A.A. Warshavsky, and L.A. Khlyap. 2008. Alien mammals in the ecosystems of Russia. KMK Scientific Press Ltd., Moscow, Russia. (in Russian)

Boegel, K., E. Schaal, and H. Moegle. 1977. The significance of martens as transmitters of wildlife rabies in Europe. Zentralblatt fur Bakteriol Parasitol Infektionskr Hygine Erste Abt Irug Reuge A Medical Mikrobiolgy Parasitol 283:184–190.

Boitani, L. 2003. Wolf conservation and recovery. Pages 317–340 *in* Wolves: behavior, ecology, and conservation. L.D. Mech and L. Boitani, editors. University of Chicago Press, Illinois, USA.

Bolen, E.G. 1989. Conservation biology, wildlife management, and spaceship earth. Wildlife Society Bulletin 17:351–354.

Bolen, E.G., and W.L. Robinson. 2003. Wildlife ecology and management. Prentice Hall, Upper Saddle River, New Jersey, USA.

Borgatti, S.P. 2005. Centrality and network flow. Social Networks 27:55–71.

Bornstein, S., T. Mörner, and W.M. Samuel. 2001. *Sarcoptes scabiei* and sarcoptic mange. Pages 107–119 *in* Parasitic diseases of wild mammals. W.M. Samuel, M.J. Pybus, and A.A. Kocan, editors. Iowa State University Press, Ames, USA.

Boss, J., G. Devean, and C. Drysdale. 1987. Kejimkujik National Park—American marten reintroduction program interim report, February–October, 1987. Environment Canada, Parks Canada, Natural Resources Conservervation, Kejimkujik National Park, New Brunswick, Canada.

Botzler, R.G., and S.B. Armstrong-Buck. 1985. Ethical considerations in research on wildlife diseases. Journal of Wildlife Diseases 21:341–345.

Bouchy, P., K. Theodorou, and D. Couvet. 2005. Metapopulation viability: influence of migration. Conservation Genetics 6:75–78.

Bouillant, A., and R. Hanson. 1965. Epizootiology of mink enteritis: I. Stability of the virus in feces exposed to natural environmental factors. Canadian Journal of Comparative Medicine and Veterinary Science 29:125–128.

Boulanger, J., S. Himmer, and C. Swan. 2004. Monitoring a grizzly bear population trend and demography using DNA mark-recapture methods in the Owikeno Lake area of British Columbia. Canadian Journal of Zoology 82:1267–1277.

Boulanger, J., K.C. Kendall, J.B. Stetz, D.A. Roon, L.P. Waits, and D. Paetkau. 2008. Multiple data sources improve DNA-based mark-recapture population estimates of grizzly bears. Ecological Applications 18:577–589.

Bowman, J., D. Donovan, and R.C. Rosatte. 2006. Numerical response of fishers to synchronous prey dynamics. Journal of Mammalogy 87:480–484.

Bowman, J.C., and J.-F. Robitaille. 1997. Winter habitat use of American martens *Martes americana* within second-growth forest in Ontario, Canada. Wildlife Biology 3:97–105.

Bowman, J.C., D. Sleep, G. J. Forbes, and M. Edwards. 2000. The association of small mammals with coarse woody debris at log and stand scales. Forest Ecology and Management 129:119–124.

Boyce, M.S. 2006. Scale for resource selection functions. Diversity and Distributions 12:269–276.

Boyce, M.S., and L.L. McDonald. 1999. Relating populations to habitats using resource selection functions. Trends in Ecology & Evolution 14:268–272.

Boyce, M.S., P.R. Vernier, S.E. Nielsen, and F.K.A. Schmiegelow. 2002. Evaluating resource selection functions. Ecological Modelling 157:281–300.

Bradle, B.J. 1957. The fisher returns to Wisconsin. Wisconsin Conservation Bulletin 22(11):9–11.

Bradshaw, C.J.A., M.A. Hindell, N.J. Best, K.L. Phillips, G. Wilson, and P.D. Nichols. 2003. You are what you eat: describing the foraging ecology of southern elephant seals (*Mirounga leonina*) using blubber fatty acids. Proceedings of the Royal Society B 270:1283–1292.

Brainerd, S.M., J.-O. Helldin, E. Lindström, and J. Rolstad. 1994. Eurasian pine martens and old industrial forest in southern boreal Scandinavia. Pages 343–354 *in* Martens, sables, and fishers: biology and conservation. S.W. Buskirk, A.S. Harestad, M.G. Raphael, and R.A. Powell, editors. Cornell University Press, Ithaca, New York, USA.

Brainerd, S.M., J.-O. Helldin, E.R. Lindström, E. Rolstad, J. Rolstad, and I. Storch. 1995. Pine marten (*Martes martes*) selection of resting and denning sites in Scandinavian managed forests. Annales Zoologici Fennici 32:151–157.

Brainerd, S.M., and J. Rolstad. 2002. Habitat selection by Eurasian pine martens *Martes martes* in managed forest of southern boreal Scandinavia. Wildlife Biology 8:289–297.

Braithwaite, H. 1892. Bears, wolves, blackcats and foxes: some facts about the fur bearing animals of the northern woods, by a New Brunswick trapper. The Industrial Journal, Bangor, Maine, August 19:6.

Brander, R.B., and D.J. Books. 1973. Return of the fisher. Natural History 82:52–57.

Brassard, J.A., and R. Bernard. 1939. Observations on breeding and development of marten, *Martes a. americana* (Kerr). Canadian Field-Naturalist 53:15–21.

Breiman, L. 2001. Random forests. Machine Learning 45:5–32.

Breitenmoser, U., C. Breitenmoser-Wursten, L.W. Carbyn, and S.M. Funk. 2001. Assessment of carnivore reintroductions. Pages 240–281 *in* Carnivore conservation. J.L. Gittleman, S.M. Funk, D.W. Macdonald, and R.K. Wayne, editors. Cambridge University Press, New York, USA.

Bremner-Harrison, S., S.W.R. Harrison, B.L. Cypher, J.D. Murdock, and J. Maldonado. 2006. Development of a single-sampling noninvasive hair snare. Wildlife Society Bulletin 34:456–461.

Breshears, D.D., N.S. Cobb, P.M. Rich, K.P. Price, C.D. Allen, R.G. Balice, W.H. Romme, J.H. Kastens, M.L. Floyd, J. Belnap, J.J. Anderson, O.B. Myers, and C.W. Meyer. 2005. Regional vegetation die-off in response to global change-type drought. Proceedings of the National Academy of Sciences, USA 102:15144–15148.

Bright, P.W. 2000. Lessons from lean beasts: conservation biology of the mustelids. Mammal Review 30:217–226.

Brinson, M., and J. Verhoeven. 1999. Riparian forests. Pages 265–299 *in* Maintaining biodiversity in forest ecosystems. M.L. Hunter, Jr., editor. Cambridge University Press, UK.

Britten M.W., P.L. Kennedy, and S. Ambrose. 1999. Performance and accuracy evaluation of small satellite transmitters. Journal of Wildlife Management 63:1349–1358.

Broekhuizen, S. 2006. *Martes* issues in the 21st century: lessons to learn from Europe. Pages 3–19 *in Martes* in carnivore communities. M. Santos-Reis, J.D.S. Birks, E.C. O'Doherty, and G. Proulx, editors. Alpha Wildlife Publications, Sherwood Park, Alberta, Canada.

Broekhuizen, S., and G.J.D.M. Müskens. 2000. Utilization of rural and suburban habitat by pine marten *Martes martes* and beech marten *Martes foina*: species-related potential and restrictions for adaptation. Lutra 43:223–227.

Bronson, E., M. Bush, T. Viner, S. Murray, S.M. Wisely, and S.L. Deem. 2007. Mortality of captive black-footed ferrets (*Mustela nigripes*) at Smithsonian's National Zoological Park, 1989–2004. Journal of Zoo and Wildlife Medicine 38:169–176.

Brook, B.W. 2000. Pessimistic and optimistic bias in population viability analysis. Conservation Biology 14:564–566.

Brooks, D.R., and A. Ferrao. 2005. The historical biogeography of coevolution: emerging infectious diseases are evolutionary accidents waiting to happen. Journal of Biogeography 32:1291–1299.

Brooks, D.R., and E.P. Hoberg. 2000. Triage for the biosphere: the need and rationale for taxonomic inventories and phylogenetic studies of parasites. Comparative Parasitology 67:1–25.

Brooks, D.R., and E.P. Hoberg. 2006. Systematics and emerging infectious diseases: from management to solution. Journal of Parasitology 92:426–429.

Brooks, D.R., and E.P. Hoberg. 2007. How will climate change affect host-parasite assemblages? Trends in Parasitology 23:571–574.

Brooks, D.R., V. León-Régagnon, D.A. McLennan, and D. Zelmer. 2006. Ecological fitting as a determinant of the community structure of platyhelminth parasites of anurans. Ecology 87:S76–S85.

Broquet, T., C.A. Johnson, E. Petit, I. Thompson, F. Burel, and J.M. Fryxell. 2006a. Dispersal and genetic structure in the American marten, *Martes americana*. Molecular Ecology 15:1689–1697.

Broquet, T., N. Ray, E. Petit, J.M. Fryxell, and F. Burel. 2006b. Genetic isolation by distance and landscape connectivity in the American marten (*Martes americana*). Landscape Ecology 21:877–889.

Brown, J.H., and M.V. Lomolino. 1998. Biogeography. Sinauer Associates, Sunderland, Massachusetts, USA.

Brown, J.S. 1988. Patch use as an indicator of habitat preference, predation risk, and competition. Behavioral Ecology and Sociobiology 22:37–47.

Brown, M.K., and G. Will. 1979. Food habits of the fisher in northern New York. New York Fish and Game Journal 26:87–92.

Brown, R.D., and R.O. Braaten. 1998. Spatial and temporal variability of Canadian monthly snow depths, 1946–1995. Atmosphere-Ocean 36:37–54.

Brown, R.N., M.W. Gabriel, G. Wengert, J.M. Higley, and J.E. Foley. 2006. Fecally transmitted viruses associated with Pacific fishers (*Martes pennanti*) in northwestern California. Transactions of the Western Section of The Wildlife Society 42:40–46.

Brown, R.N., M.W. Gabriel, G.M. Wengert, S. Matthews, J.M. Higley, and J.E. Foley. 2007. Pathogens associated with fishers. Pages 3–47 *in* Pathogens associated with fishers (*Martes pennanti*) and sympatric mesocarnivores in California—final draft report to the U.S. Fish and Wildlife Service for Grant # 813335G021. U.S. Fish and Wildlife Service, Yreka, California, USA.

Brumfield, R.T., P. Beerli, D.A. Nickerson, and S.V. Edwards. 2003. The utility of single nucleotide polymorphisms in inferences of population history. Trends in Ecology & Evolution 18:249–256.

Bryant, H.N., W.B. McGillivray, and W.B. Bartlett. 1997. Skeletal morphometrics of fishers from Alberta: sexual size dimorphism and age-class comparisons. Pages 40–58 *in Martes*: taxonomy, ecology, techniques and management. G. Proulx, H.N. Bryant, and P.M. Woodard, editors. Provincial Museum of Alberta, Edmonton, Canada.

Brzeziński, M. 1994. Summer diet of the sable *Martes zibellina* in the Middle Yenisei taiga, Siberia. Acta Theriologica 39:103–107.

Buck, S.G. 1982. Habitat utilization by fisher (*Martes pennanti*) near Big Bar, California. Thesis, Humboldt State University, Arcata, California, USA.

Buck, S.G., C. Mullis, and A. Mossman. 1979. A radio telemetry study of fisher in northwestern California. California-Nevada Wildlife Transactions 1979:166–172.

Buck, S.G., C. Mullis, and A.S. Mossman. 1983. Corral Bottom-Hayfork Bally fisher study: final report. Unpublished final report to USDA Forest Service. Humboldt State University, Arcata, California, USA.

Buck, S.G., C. Mullis, A.S. Mossman, I. Show, and C. Coolahan. 1994. Habitat use by fishers in adjoining heavily and lightly harvested forest. Pages 368–376 *in* Martens, sables and fishers: biology and conservation. S.W. Buskirk, A.S. Harestad, M.G. Raphael, and R.A. Powell, editors. Cornell University Press, Ithaca, New York, USA.

Bull, E.L., and A.K. Blumton. 1999. Effect of fuels reduction on American marten and their prey. USDA Forest Service, Research Note PNW-RN-539.

Bull, E.L., and T.W. Heater. 2000. Resting and denning sites of American marten in northwestern Oregon. Northwest Science 74:179–185.

Bull, E.L., and T.W. Heater. 2001. Survival, causes of mortality, and reproduction in the American marten in northwestern Oregon. Northwestern Naturalist 82:1–6.

Bull, E.L., T.W. Heater, and J.F. Shepherd. 2005. Habitat selection by the American marten in northwestern Oregon. Northwest Science 79:37–43.

Bull, E.L., R.S. Holthausen, and M.G. Henjum. 1992. Roost trees used by pileated woodpeckers in northeastern Oregon. Journal of Wildlife Management 56:786–793.

Bull, E.L., C.G. Parks, and T.R. Torgersen. 1997. Trees and logs important to wildlife in the interior Columbia River basin. USDA Forest Service, General Technical Report PNW-GTR-391.

Bulmer, M.G. 1974. A statistical analysis of the 10-year cycle in Canada. Journal of Animal Ecology 43:701–718.

Bunnell, F.L., E. Wind, and R. Wells. 2002. Dying and dead hardwoods: their implication to management. Pages 695–716 *in* Proceedings of the symposium on the ecology and management of dead wood in western forests. W.F. Laudenslayer Jr., P.J. Shea, B.E. Valentine,

C.P. Weatherspoon, and T.E. Lisle, technical coordinators. USDA Forest Service, General Technical Report PSW-GTR-18.

Burek, K. 2001. Mycotic diseases. Pages 514–533 *in* Infectious diseases of wild mammals. E.S. Williams and I.K. Barker, editors. Iowa State University Press, Ames, USA.

Burnham, K.P., and D.R. Anderson. 2002. Model selection and multi-model inference: a practical information-theoretic approach. Second edition. Springer-Verlag, New York, USA.

Burns, C.E., K.M. Johnston, and O.J. Schmitz. 2003a. Global climate change and mammalian species diversity in U.S. national parks. Proceedings of the National Academy of Sciences, USA 100:11474–11477.

Burns, R., E.S. Williams, D. O'Toole, and J.P. Dubey. 2003b. *Toxoplasma gondii* infections in captive black-footed ferrets (*Mustela nigripes*), 1992–1998: clinical signs, serology, pathology, and prevention. Journal of Wildlife Diseases 39:787–797.

Burris, O.E., and D.E. McKnight. 1973. Game transplants in Alaska. Alaska Department of Fish and Game, Game Technical Bulletin 4, Juneau, USA.

Busch, D.E., and J.C. Trexler, editors. 2003. Monitoring ecosystems: interdisciplinary approaches for evaluating ecoregional initiatives. Island Press, Covello, California, USA.

Buskirk, S.W. 1983. The ecology of marten in southcentral Alaska. Dissertation, University of Alaska, Fairbanks, USA.

Buskirk, S.W. 1984. Seasonal use of resting sites by marten in south-central Alaska. Journal of Wildlife Management 48:950–953.

Buskirk, S.W. 1994. Introduction to the genus *Martes*. Pages 1–10 *in* Martens, sables, and fishers: biology and conservation. S.W. Buskirk, A.S. Harestad, M.G. Raphael, and R.A. Powell, editors. Cornell University Press, Ithaca, New York, USA.

Buskirk, S.W., S.C. Forrest, M.G. Raphael, and H.J. Harlow. 1989. Winter resting site ecology of marten in the central Rocky Mountains. Journal of Wildlife Management 53:191–196.

Buskirk, S.W., A.S. Harestad, M.G. Raphael, and R.A. Powell, editors. 1994. Martens, sables, and fishers: biology and conservation. Cornell University Press, Ithaca, New York, USA.

Buskirk, S.W., and H.J. Harlow. 1989. Body-fat dynamics of the American marten (*Martes americana*) in winter. Journal of Mammalogy 70:191–193.

Buskirk, S.W., H.J. Harlow, and S.C. Forrest. 1988. Temperature regulation in American marten (*Martes americana*) in winter. National Geographic Research 4:208–218.

Buskirk, S.W., Y. Ma, and Z. Jiang. 1996a. Diets of, and prey selection by sables (*Martes zibellina*) in northern China. Journal of Mammalogy 77:725–730.

Buskirk, S.W., and L.L. McDonald. 1989. Analysis of variability in home range size of the American marten. Journal of Wildlife Management 53:997–1004.

Buskirk, S.W., and J.J. Millspaugh. 2006. Metrics for studies of resource selection. Journal of Wildlife Management 70:358–366.

Buskirk, S.W., and R.A. Powell. 1994. Habitat ecology of fishers and American martens. Pages 283–296 *in* Martens, sables, and fishers: biology and conservation. S.W. Buskirk, A.S. Harestad, M.G. Raphael, and R.A. Powell, editors. Cornell University Press, Ithaca, New York, USA.

Buskirk, S.W., C.M. Raley, K.B. Aubry, W.J. Zielinski, M.K. Schwartz, R.T. Golightly, K.L. Purcell, R.D. Weir, and J.S. Yaeger. 2010. Meta-analysis of resting site selection by the fisher in the Pacific coastal states and provinces—final report. Unpublished final report to the Fisher Steering Committee, USDA Forest Service, Pacific Northwest Region, Portland, Oregon, USA.

Buskirk, S.W., and L.F. Ruggiero. 1994. American marten. Pages 7–37 *in* The scientific basis for conserving forest carnivores: American marten, fisher, lynx and wolverine in the western United States. L.F. Ruggiero, K.B. Aubry, S.W. Buskirk, L.J. Lyon, and W.J. Zielinski, technical editors. USDA Forest Service, General Technical Report GTR-RM-254.

Buskirk, S., M. Yiqing, X. Li, and J. Zhaowen. 1996b. Winter habitat ecology of sables (*Martes zibellina*) in relation to forest management in China. Ecological Applications 6:318–325.

Butterworth, E.W., and M. Beverley-Burton. 1980. The taxonomy of *Capillaria* spp. (Nematoda: Trichuroidea) in carnivorous mammals from Ontario, Canada. Systematic Parasitology 1:211–236.

Cahill, G.F., Jr. 1976. Starvation in man. Clinics in Endocrinology and Metabolism 5:397–415.

Calabrese, M., and L.R. Davis. 2010. Occurrence of den trees for fisher (*Martes pennanti*), in the sub boreal pine–spruce biogeoclimatic zone in the Chilcotin area of British Columbia. Unpublished report to West Fraser Mills and BC Timber Sales, Williams Lake, British Columbia, Canada.

California Department of Fish and Game. 2010. A status review of the fisher (*Martes pennanti*) in California. http://www.dfg.ca.gov/wildlife/nongame/publications/docs/FisherStatusReviewComplete.pdf (accessed 10 March 2010).

Callas, R.L., and P. Figura. 2008. Translocation plan for the reintroduction of fishers (*Martes pennanti*) to Sierra Pacific Industries lands in the northern Sierra Nevada and southern Cascades. California Department of Fish and Game, Sacramento, USA. http://r1.dfg.ca.gov/portal/FisherTranslocation/tabid/832/Default.aspx (accessed 13 May 2011).

Campbell, G.M., J.N. Pauli, J.G. Thomas, and T. McClean. 2010. Accuracy in molecular sexing of martens (*Martes americana* and *M. caurina*) varies among sample types. Molecular Ecology Resources 10:1019–1022.

Campbell, L.A. 2004. Distribution and habitat associations of mammalian carnivores in the central and southern Sierra Nevada. Dissertation, University of California, Davis, USA.

Campbell, L.A., R.A. Long, and W.J. Zielinski. 2008. Integrating multiple methods to achieve survey objectives. Pages 223–237 *in* Noninvasive survey methods for carnivores. R.A. Long, P. MacKay, W.J. Zielinski, and J.C. Ray, editors. Island Press, Washington, D.C., USA.

Campbell, T.M. 1979. Short-term effects of timber harvests on pine marten ecology. Thesis, Colorado State University, Fort Collins, USA.

Carman, R. 1975. The complete guide to lures and baits. Spearman Publishing and Printing, Sutton, Nebraska, USA.

Carmichael, L.E. 1970. Herpesvirus canis: aspects of pathogenesis and immune response. Journal of the American Veterinary Medical Association 156:1714–1721.

Carpenter, J.W., M.J. Appel, R.C. Erickson, and M.N. Novilla. 1976. Fatal vaccine-induced canine distemper virus infection in black-footed ferrets. Journal of the American Veterinary Medical Association 169:961–964.

Carr, D., J. Bowman, C.J. Kyle, S.M. Tully, E.L. Koen, J.-F. Robitaille, and P.J. Wilson. 2007a. Rapid homogenization of multiple sources: genetic structure of a recolonizing population of fishers. Journal of Wildlife Management 71:1853–1861.

Carr, D., J. Bowman, and P.J. Wilson. 2007b. Density-dependent dispersal suggests a genetic measure of habitat suitability. Oikos 116:629–635.

Carr, S.M., and S.A. Hicks. 1997. Are there two species of marten in North America? Genetic and evolutionary relationships within *Martes*. Pages 15–28 *in Martes*: taxonomy, ecology, techniques, and management. G. Proulx, H.N. Bryant, and P.M. Woodard, editors. Provincial Museum of Alberta, Edmonton, Canada.

Carrara, P.E., T.A. Ager, and J.F. Baichtal. 2007. Possible refugia on the Alexander Archipelago of southeastern Alaska during the late Wisconsin glaciations. Canadian Journal of Earth Sciences 44:229–244.

Carrasco, M.A., B.P. Kraatz, E.B. Davis, and A.D. Barnosky. 2005. Miocene Mammalian Mapping Project (MIOMAP). University of California Museum of Paleontology. http://www.ucmp.berkeley.edu/miomap/ (accessed May 2011).

Carroll, C. 2005a. A reanalysis of regional fisher suitability including survey data from commercial forests in the redwood region. Klamath Center for Conservation Research, Final Report to USDA Forest Service, Pacific Southwest Research Station, Albany, California, USA.
Carroll, C. 2005b. Carnivore restoration in the northeastern U.S. and southeastern Canada: a regional-scale analysis of habitat and population viability for wolf, lynx, and marten (Report 2: lynx and marten viability analysis). Wildlands Project Special Paper No. 6. Richmond, Vermont, USA.
Carroll, C. 2007. Interacting effects of climate change, landscape conversion, and harvest on carnivore populations at the range margin: marten and lynx in the northern Appalachians. Conservation Biology 21:1092–1104.
Carroll, C. 2010. Connectivity Analysis Toolkit user manual, version 1.1. Klamath Center for Conservation Research, Orleans, California, USA. http://www.connectivitytools.org (accessed January 2011).
Carroll, C., D.S. Johnson, J.R. Dunk, and W.J. Zielinski. 2010. Hierarchical Bayesian spatial models for multi-species conservation planning and monitoring. Conservation Biology 24:1538–1548.
Carroll, C., B.H. McRae, and A. Brookes. 2012. Use of linkage mapping and centrality analysis across habitat gradients to conserve connectivity of gray wolf populations in western North America. Conservation Biology 26:78–87.
Carroll, C., R.F. Noss, and P.C. Paquet. 2001. Carnivores as focal species for conservation planning in the Rocky Mountain region. Ecological Applications 11:961–980.
Carroll, C., R.F. Noss, P.C. Paquet, and N.H. Schumaker. 2003a. Use of population viability analysis and reserve selection algorithms in regional conservation plans. Ecological Applications 13:1773–1789.
Carroll, C., M.K. Phillips, N.H. Schumaker, and D.W. Smith. 2003b. Impacts of landscape change on wolf restoration success: planning a reintroduction program based on static and dynamic spatial models. Conservation Biology 17:536–548.
Carroll, C., W.J. Zielinski, and R.F. Noss. 1999. Using presence-absence data to build and test spatial habitat models for the fisher in the Klamath region, U.S.A. Conservation Biology 13:1344–1359.
Caryl, F.M. 2008. Pine marten diet and habitat use within a managed coniferous forest. Dissertation, University of Stirling, Scotland, UK.
Casper, R.M. 2009. Guidelines for the instrumentation of wild birds and mammals. Animal Behaviour 78:1477–1483.
Castellini, M.A., and L.D. Rea. 1992. The biochemistry of natural fasting at its limits. Experientia 48:575–582.
Castién, E., and I. Mendiola. 1985. Atlas de los mamíferos continentales de Alava, Vizcaya y Guipúzcoa. Pages 269–335 *in* Atlas de los vertebrados continentales de Alava, Vizcaya y Guipúzcoa. A. Bea, J.M. Faus, E. Castién, and I. Mendiola, editors. Gobierno Vasco, Vitoria, Spain.
Caswell, H. 2001. Matrix population models: construction, analysis, and interpretation. Second edition. Sinauer Associates, Sunderland, Massachusetts, USA.
Caughley, G. 1994. Directions in conservation biology. Journal of Animal Ecology 63:215–244.
Caughley, G., D. Grice, R. Barker, and B. Brown. 1988. The edge of the range. Journal of Animal Ecology 57:771–785.
Cayan, D., M. Tyree, M. Dettinger, H. Hidalgo, T. Das, E. Maurer, P. Bromirski, N Graham, and R. Flick. 2009. Climate change scenarios and sea level rise estimates for the California 2008 Climate Change Scenarios Assessment. PIER Research Report, CEC-500–2009–014. California Energy Commission, Sacramento, USA.
Cerbo, A.R.D., M.T. Manfredl, M. Bregoli, N.F. Milone, and M. Cova. 2008. Wild carnivores as source of zoonotic helminths in north-eastern Italy. Helminthologia 45:13–19.

Chapin, T.G., D.J. Harrison, and D.D. Katnik. 1998. Influence of landscape pattern on habitat use by American marten in an industrial forest. Conservation Biology 12:1327–1337.

Chapin, T.G., D.J. Harrison, and D.M. Phillips. 1997a. Seasonal habitat selection by marten in an untrapped forest preserve. Journal of Wildlife Management 61:707–717.

Chapin, T.G., D.M. Phillips, D.J. Harrison, and E.C. York. 1997b. Seasonal selection of habitats by resting martens in Maine. Pages 166–181 *in Martes*: taxonomy, ecology, techniques, and management. G. Proulx, H.N. Bryant, and P.M. Woodard, editors. Provincial Museum of Alberta, Edmonton, Canada.

Charlton, K. 1994. The pathogenesis of rabies and other lyssaviral infections: recent studies. Current Topics in Microbiology and Immunology 187:95–119.

Charnov, E.L. 1976. Optimal foraging: the marginal value theorem. Theoretical Population Biology 9:129–136.

Chase, J.M., and M.A. Leibold. 2003. Ecological niches: linking classical and contemporary approaches. University of Chicago Press, Illinois, USA.

Chen, M., M.E. Tewes, K.J. Pei, and L.I. Grassman. 2009. Activity patterns and habitat use of sympatric small carnivores in southern Taiwan. Mammalia 73:20–26.

Chen, X., and P.F. Sullivan. 2003. Single nucleotide polymorphism genotyping: biochemistry, protocol, cost and throughput. The Pharmacogenomics Journal 3:77–96.

Choquenot, D., W.A. Ruscoe, and E. Murphy. 2001. Colonisation of new areas by stoats: time to establishment and requirements for detection. New Zealand Journal of Ecology 25:83–88.

Chow, L. 2009. A survey for fisher in Yosemite National Park 1992–1994. Transactions of the Western Section of The Wildlife Society 45:27–44.

Chowdhury, N., and A.A. Aguirre. 2001. Helminths of wildlife. Science Publishers, Enfield, New Hampshire, USA.

Churchill, S.J., L.A. Herman, M.F. Herman, and J.P. Ludwig. 1981. Final report on the completion of the Michigan marten reintroduction program. Ecological Research Services, Iron River, Michigan, USA.

Ciarniello, L.M., M.S. Boyce, D.R. Seip, and D.C. Heard. 2007. Grizzly bear habitat selection is scale dependent. Ecological Applications 17:1424–1440.

Clark, J.A., and E. Harvey. 2002. Assessing multi-species recovery plans under the Endangered Species Act. Ecological Applications 12:655–662.

Clark, T.W., E. Anderson, C. Douglas, and M. Strickland. 1987. *Martes americana*. Mammalian Species 289:1–8.

Cleaveland, S., T. Mlengeya, M. Kaare, D. Haydon, T. Lembo, M. Laurenson, and C. Packer. 2007. The conservation relevance of epidemiological research into carnivore viral diseases in the Serengeti. Conservation Biology 21:612–622.

Clément, R., and M.C. Saint-Girons. 1982. Le regime de la fouine *Martes foina* Erxleben 1777 dans l'agglomération nantaise et en milieu rural. Mammalia 46:550–553.

Clevenger, A.P. 1993a. Pine marten (*Martes martes* L.) home ranges and activity patterns on the island of Minorca, Spain. Zeitschrift fur Saugetierkunde 58:137–143.

Clevenger A.P. 1993b. Spring and summer food habits and habitat use of the European pine marten (*Martes martes*) on the island of Minorca, Spain. Journal of Zoology 229:153–161.

Clevenger, A.P. 1994a. Feeding ecology of Eurasian pine martens and stone martens in Europe. Pages 326–340 *in* Martens, sables, and fishers: biology and conservation. S.W. Buskirk, A.S. Harestad, M.G. Raphael, and R.A. Powell, editors. Cornell University Press, Ithaca, New York, USA.

Clevenger, A.P. 1994b. Habitat characteristics of Eurasian pine martens in an insular Mediterranean environment. Ecography 17:257–263.

Clifford, D.L., R. Woodroffe, D.K. Garcelon, S.F. Timm, and J.A.K. Mazet. 2007. Using pregnancy rates and perinatal mortality to evaluate the success of recovery strategies for endangered island foxes. Animal Conservation 10:442–451.

CLS America. 2011. Argos user's manual. http://www.argos-system.org/documents/userarea/argos_manual_en.pdf (accessed 31 May 2011).

Cobb, E.W. 2000. Physical condition of American martens, *Martes americana*, from two forest regions in northeastern Ontario. Thesis, Laurentian University, Sudbury, Ontario, Canada.

Cochran, W.G. 1954. Some methods for strengthening the common χ^2 tests. Biometrics 10:417–451.

Coffin, A.W. 2007. From roadkill to road ecology: a review of the ecological effects of roads. Journal of Transport Geography 15:396–406.

Coffin, K.W., Q.J. Kujala, R.J. Douglass, and L.R. Irby. 1997. Interactions among marten prey availability, vulnerability, and habitat structure. Pages 199–210 *in Martes*: taxonomy, ecology, techniques, and management. G. Proulx, H.N. Bryant, and P.M. Woodard, editors, Provincial Museum of Alberta, Edmonton, Canada.

Cohen, J. 1968. Weighted kappa—nominal scale agreement with provision for scaled disagreement or partial credit. Psychological Bulletin 70:213–220.

Colbert, E.H. 1935. Siwalik mammals in the American Museum of Natural History. Transactions of the American Philosophical Society, N.S. 26:1–401.

Cole, P.J., and G. Proulx. 1994. Leghold trapping: a cause of serious injuries to fishers. *Martes* Working Group Newsletter 2:14–15.

Cole, R.A., D.S. Lindsay, D.K. Howe, C.L. Roderick, J.P. Dubey, N.J. Thomas, and L.A. Baeten. 2000. Biological and molecular characterizations of *Toxoplasma gondii* strains obtained from southern sea otters (*Enhydra lutris nereis*). Journal of Parasitology 86:526–530.

Collinge, S.K., and C. Ray. 2006. Disease ecology: community structure and pathogen dynamics. Oxford University Press, London, UK.

Collinge, S.K., C. Ray, and J.F. Cully, Jr. 2008. Effects of disease on keystone species, dominant species, and their communities. Pages 129–144 *in* Infectious disease ecology: effects of ecosystems on disease and of diseases on ecosystems. R.S. Ostfeld, F. Keesing, and V.T. Eviner, editors. Princeton University Press, New Jersey, USA.

Collins, B.M., and S.L. Stephens. 2007. Managing natural wildfires in Sierra Nevada wilderness areas. Frontiers in Ecology and the Environment 5:523–527.

Colpitts, G. 1997. Conservation, science, and Canada's fur farming industry, 1913–1945. Social History 30:77–108.

Committee on the Status of Endangered Wildlife in Canada. 2007. COSEWIC assessment and update status report on the Newfoundland marten (Newfoundland population) *Martes americana atrata* in Canada. Committee on the Status of Endangered Wildlife in Canada, Ottawa, Ontario, Canada. http://www.vmunix.com/~lmayo/ (accessed 14 January 2011).

Conover, R.R., and E.T. Linder. 2009. Mud track plots: an economical, noninvasive mammal survey technique. Southeastern Naturalist 8:437–444.

Cook, J.A., E.P. Hoberg, A. Koehler, H. Henttonen, L. Wickström, V. Haukisalmi, K. Galbreath, F. Chernyavski, N. Dokuchaev, A. Lahzuhtkin, S.O. MacDonald, A. Hope, E. Waltari, A. Runck, A. Veitch, R. Popko, E. Jenkins, S. Kutz, and R. Eckerlin. 2005. Beringia: intercontinental exchange and diversification of high latitude mammals and their parasites during the Pliocene and Quaternary. Mammal Study 30:S33–S44.

Coombs, A.B., J. Bowman, and C.J. Garroway. 2010. Thermal properties of tree cavities during winter in a northern hardwood forest. Journal of Wildlife Management 74:1875–1881.

Copeland, J.P. 1996. Biology of the wolverine in central Idaho. Thesis, University of Idaho, Moscow, USA.

Corn, J.G., and M.G. Raphael. 1992. Habitat characteristics at marten subnivean access sites. Journal of Wildlife Management 56:442–448.

Cornell University Animal Health Diagnostic Center. 2010. Animal Health Diagnostic Center—Sample Submission: blood collection guide. http://diagcenter.vet.cornell.edu/pdf/bloodLabeling.pdf (accessed January 2010).

Costa, L. da F., F.A. Rodrigues, G. Travieso, and P.R. Villas Boas. 2007. Characterization of complex networks: a survey of measurements. Advances in Physics 56:167–242.

Cottrell, W. 1978. The fisher (*Martes pennanti*) in Maryland. Journal of Mammalogy 59:886.

Coues, E. 1877. Fur-bearing animals: a monograph of North American Mustelidae. U.S. Department of the Interior, Geological Survey of the Territories. Miscellaneous Publication No. 8, Washington, D.C., USA.

Coulter, M.W. 1966. Ecology and management of fishers in Maine. Dissertation, Syracuse University, New York, USA.

Cowan, I.M. 1955. An instance of scabies in the marten (*Martes americana*). Journal of Wildlife Management 19:499.

Cowan, I.M., and R.H. Mackay. 1950. Food habits of the marten (*Martes americana*) in the Rocky Mountain region of Canada. Canadian Field-Naturalist 64:100–104.

Cracraft, J. 1989. Speciation and its ontology: the empirical consequences of alternative species concepts for understanding patterns and processes of differentiation. Pages 28–59 *in* Speciation and its consequences. D. Otte and J.A. Endler, editors. Sinauer Associates, Sunderland, Massachusetts, USA.

Craig, R.E., and R.A. Borecky. 1976. Metastrongyles (Nematoda: Metastrongyloidea) of fisher (*Martes pennanti*) from Ontario. Canadian Journal of Zoology 54:806–807.

Crane, V.W. 1928. The southern frontier, 1670–1732. Duke University Press, Durham, North Carolina, USA.

Crichton, V. 1948. Registered traplines. Silva 4:3–15.

Criscione, C.D., R. Poulin, and M.S. Blouin. 2005. Molecular ecology of parasites: elucidating ecological and microevolutionary processes. Molecular Ecology 14:2247–2257.

Croiter, R., and J.-P. Brugal. 2010. Ecological and evolutionary dynamics of a carnivore community in Europe during the last 3 million years. Quaternary International 212:98–108.

Cunningham, M.W., D.B. Shindle, A.B. Allison, S.P. Terrell, D.G. Mead, and M. Owen. 2009. Canine distemper epizootic in Everglades mink. Journal of Wildlife Diseases 45:1150–1157.

Cushman, S.A., and K. McGarigal. 2004. Patterns in the species-environment relationship depend on both scale and choice of response variables. Oikos 105:117–124.

Cushman, S.A., K.S. McKelvey, J. Hayden, and M.K. Schwartz. 2006. Gene flow in complex landscapes: testing multiple hypotheses with causal modeling. American Naturalist 168:486–499.

Cutler, D.R., T.C. Edwards, Jr., K.H. Beard, A. Cutler, K.T. Hess, J. Gibson, and J.J. Lawler. 2007. Random forests for classification in ecology. Ecology 88:2783–2792.

Czyzewska, T. 1981. Natural endocranial casts of the mustelinae from Weze I near Dzialoszyn, Poland. Acta Zoologica Cracoviensia 25:261–270.

Dacheux, L., F. Larrous, A. Mailles, D. Boisseleau, O. Delmas, C. Biron, C. Bouchier, I. Capek, M. Müller, F. Ilari, T. Lefranc, F. Raffi, M. Goudal, and H. Bourhy. 2009. European bat Lyssavirus transmission among cats, Europe. Emerging Infectious Diseases 15:280–284.

Dagg, A.I. 1972. Research on Canadian mammals. Canadian Field-Naturalist 86:217–221.

Dale, V.H., L.A. Joyce, S. McNulty, R.P. Neilson, M.P. Ayres, M.D. Flannigan, P.J. Hanson, L.C. Irland, A.E. Lugo, C.J. Peterson, D. Simberloff, F.J. Swanson, B.J. Stocks, and M.B. Wotton. 2001. Climate change and forest disturbances. Bioscience 51:723–734.

Dalén, L., A. Götherström, and A. Angerbjörn. 2004. Identifying species from pieces of faeces. Conservation Genetics 5:109–111.

Dalquest, W.W. 1948. Mammals of Washington. University of Kansas Publications in Natural History 2:1–444.

Danby, R.K., and D.S. Slocombe. 2005. Regional ecology, ecosystem geography, and transboundary protected areas in the St. Elias Mountains. Ecological Applications 15:405–422.

Dark, S.J. 1997. A landscape-scale analysis of mammalian carnivore distribution and habitat use by fisher. Thesis, Humboldt State University, Arcata, California, USA.

Darsie, R.F., and G. Anastos. 1957. Geographical distribution and hosts of *Ixodes texanus* Banks (Acarina, Ixodidae). Annals of the Entomological Society of America 50:295–301.

Daszak, P., A.A. Cunningham, and A.D. Hyatt. 2000. Emerging infectious diseases of wildlife—threats to biodiversity and human health. Science 287:443–449.

Daszak, P., A.A. Cunningham, and A.D. Hyatt. 2001. Anthropogenic environmental change and the emergence of infectious diseases in wildlife. Acta Tropica 78:103–116.

Davie, J.W. 1984. Fisher reintroduction. Project completion report. Alberta Department of Forests, Lands and Wildlife, Edmonton, Canada.

Davis, F.W., C. Seo, and W.J. Zielinski. 2007. Regional variation in home-range-scale habitat models for fisher (*Martes pennanti*) in California. Ecological Applications 17:2195–2213.

Davis, L. 2003. Stand level habitat use by furbearer species in the Anahim Lake area of British Columbia. Unpublished report to Yun Ka Whu'ten Holdings Ltd. FIA Project 1023002. Williams Lake, British Columbia, Canada.

Davis, L.R. 2008. Fisher (*Martes pennanti*) habitat ecology in pine dominated habitats of the Chilcotin. Unpublished report for FIA Project Y081290 to the British Columbia Ministry of Environment, Victoria, Canada.

Davis, L.R. 2009. Denning ecology and habitat use by fisher (*Martes pennanti*) in pine dominated ecosystems of the Chilcotin Plateau. Thesis, Simon Fraser University, Burnaby, British Columbia, Canada.

Davis, M.H. 1978. Reintroduction of the pine marten into the Nicolet National Forest, Forest County, Wisconsin. Thesis, University of Wisconsin, Stevens Point, USA.

Davis, M.H. 1983. Post-release movements of introduced marten. Journal of Wildlife Management 47:59–66.

Davis, W.B. 1939. The recent mammals of Idaho. Caxton Printers, Caldwell, Idaho, USA.

Davison, A., J.D.S. Birks, R.C. Brookes, T.C. Braithwaite, and J.E. Messenger. 2002. On the origin of faeces: morphological *versus* molecular methods for surveying rare carnivores from their scats. Journal of Zoology 257:141–143.

Davison, A., J.D.S. Birks, R.C. Brookes, J.E. Messenger, and J.I. Griffiths. 2001. Mitochondrial phylogeography and population history of pine martens *Martes martes* compared with polecats *Mustela putorius*. Molecular Ecology 10:2479–2488.

Davison, R.P., W.W. Mautz, H.H. Hayes, and J.B. Holter. 1978. The efficiency of food utilization and energy requirements of captive female fishers. Journal of Wildlife Management 42:811–821.

Dawe, N.K., and G. Mosley. 2009. Striving for sustainability. How economic growth is hindering conservation. The Wildlife Professional 3(3):66–67.

Dawson, N.G. 2008. Vista norteña: tracking historical diversification and contemporary structure in high latitude mesocarnivores. Dissertation, University of New Mexico, Albuquerque, USA.

DeAngelis, D.L., and L.J. Gross, editors. 1992. Individual-based models and approaches in ecology: populations, communities, and ecosystems. Chapman and Hall, New York, USA.

de Bruijne, J.J., N. Altszuler, J. Hampshire, T.J. Visser, and W.H.L. Hackeng. 1981. Fat mobilization and plasma hormone levels in fasted dogs. Metabolism 30:190–194.

de Bruijne, J.J., and P. de Koster. 1983. Glycogenolysis in the fasting dog. Comparative Biochemistry and Physiology B 75:553–555.

Deem, S.L., L.H. Spelman, R.A. Yates, and R.J. Montali. 2000. Canine distemper in terrestrial carnivores: a review. Journal of Zoo and Wildlife Medicine 31:441–451.

DeKay, J.E. 1842. Zoology of New York, or New York fauna. Part I. Mammalia. W. and A. Whote and J. Visscher, Albany, New York, USA.

Delibes, M. 1983. Interspecific competition and the habitat of the stone marten (*Martes foina* Erxleben, 1777) in Europe. Acta Zoologica Fennica 174:229–231.

Delibes, M., and F. Amores. 1986. The stone marten *Martes foina* (Erxleben 1777) from Ibiza (Pitiusuc, Balearic islands). Miscellaneous Zoology (Barcelona) 10:335–345.

Delort, R. 1986. L'histoire de la fourrure de l'antiquité à nos jours. S.A. Edita, Lausanne, Switzerland.

D'eon, R.G., R. Serrouya, G. Smith, and C.O. Kochanny. 2002. GPS radiotelemetry error and bias in mountainous terrain. Wildlife Society Bulletin 30:430–439.

DePue, J.E., and M. Ben-David. 2007. Hair sampling techniques for river otters. Journal of Wildlife Management 71:671–674.

Deredec, A., and F. Courchamp. 2007. Importance of the Allee effect for reintroductions. Ecoscience 14:440–451.

Despopoulos, A., and S. Silbernagl. 1986. Color atlas of physiology. Third edition. Georg Thieme Verlag, Stuttgart, Germany.

de Vos, A. 1951a. Overflow and dispersal of marten and fisher in Ontario. Journal of Wildlife Management 15:164–175.

de Vos, A. 1951b. Recent findings in fisher and marten ecology and management. Transactions of the North American Wildlife Conference 16:498–505.

de Vos, A. 1951c. Tracking of fisher and marten. Sylva 7:14–18.

de Vos, A. 1952. The ecology and management of fisher and marten in Ontario. Technical Bulletin, Wildlife Series 1, Ontario Department of Lands and Forests, Toronto, Canada.

de Vos, A. 1957. Pregnancy and parasites of marten. Journal of Mammalogy 38:412.

Dexin, J., and E.I. Robbins. 2000. Quaternary palynofloras and paleoclimate of the Qaidam Basin, Qinghai Province, Northwestern China. Palynology 24:95–112.

Dhuey, B. 2009. Winter track counts, 1977–2009. Wisconsin Department of Natural Resources, Madison, USA. http://www.wnrmag.com/org/land/wildlife/harvest/reports/wntrtracks.pdf (accessed 2 April 2010).

Dick, T.A., B. Kingscote, M.A. Strickland, and C.W. Douglas. 1986. Sylvatic trichinosis in Ontario, Canada. Journal of Wildlife Diseases 22:42–47.

Dick, T.A., and R.D. Leonard. 1979. Helminth parasites of fisher *Martes pennanti* (Erxleben) from Manitoba, Canada. Journal of Wildlife Diseases 15:409–412.

Dick, T.A., and E. Pozio. 2001. *Trichinella* spp. and trichinellosis. Pages 380–396 *in* Parasitic diseases of wild mammals. W.M. Samuel, M.J. Pybus, and A.A. Kocan, editors. Iowa State University Press, Ames, USA.

Diefenbach, D.R., C.O. Kochanny, J.K. Vreeland, and B.D. Wallingford. 2003. Evaluation of an expandable, breakaway radiocollar for white-tailed deer fawns. Wildlife Society Bulletin 31:756–761.

Dilworth, T.G. 1974. Status and distribution of fisher and marten in New Brunswick. Canadian Field-Naturalist 88:495–498.

DiStefano, J.J. 1987. Wild furbearer management in the northeastern United States. Pages 1077–1090 *in* Wild furbearer management and conservation in North America. M. Novak, J.A. Baker, M.E. Obbard, and B. Malloch, editors. Ontario Trappers Association and Ontario Ministry of Natural Resources, Toronto, Canada.

Diters, R.W., and S.W. Nielsen. 1978. Toxoplasmosis, distemper, and herpesvirus infection in a skunk (*Mephitis mephitis*). Journal of Wildlife Diseases 14:132.

Divertie, G.D., M.D. Jensen, and J.M. Miles. 1991. Stimulation of lipolysis in humans by physiological hypercortisolemia. Diabetes 40:1228–1232.

Dixon, J.S. 1925. A closed season needed for fisher, marten, and wolverine. California Fish and Game 11:23–25.

Dodds, D.G., and A.M. Martell. 1971. The recent status of the fisher, *Martes pennanti pennanti* (Erxleben), in Nova Scotia. Canadian Field-Naturalist 85:62–65.

Dodge, W.E. 1977. Status of the fisher (*Martes pennanti*) in the conterminous United States. Unpublished report to the U.S. Department of the Interior, Fish and Wildlife Service, Washington, D.C., USA.

Domingo-Roura, X. 2002. Genetic distinction of marten species by fixation of a microsatellite region. Journal of Mammalogy 83:907–912.

Donadio, E., and S.W. Buskirk. 2006. Diet, morphology, and interspecific killing in Carnivora. American Naturalist 167:524–536.
Douglas, C.W., and M.A. Strickland. 1987. Fisher. Pages 510–529 *in* Wild furbearer management and conservation in North America. M. Novak, J.A. Baker, M.E. Obbard, and B. Malloch, editors. Ontario Trappers Association and Ontario Ministry of Natural Resources, Toronto, Canada.
Douglas, R.J., L.G. Fisher, and M. Mair. 1983. Habitat selection and food habits of marten, *Martes americana*, in the Northwest Territories. Canadian Field-Naturalist 97:71–74.
Dragoo, J.W., J.A. Lackey, K.E. Moore, E.P. Lessa, J.A. Cook, and T.L. Yates. 2006. Phylogeography of the deer mouse (*Peromyscus maniculatus*) provides a predictive framework for research on hantaviruses. Journal of General Virology 87:1997–2003.
Drapeau, P., A. Leduc, and Y. Bergeron. 2009. Bridging ecosystem and multiple species approaches for setting conservation targets in managed boreal forest landscapes. Pages 129–160 *in* Setting conservation targets in managed forest landscapes. M.-A. Villard and B.G. Jonnson, editors. Cambridge University Press, UK.
Drew, G.S. 1995. Winter habitat selection by American marten (*Martes americana*) in Newfoundland: why old growth? Dissertation, Utah State University, Logan, USA.
Drew, R.E., J.G. Hallett, K.B. Aubry, K.W. Cullings, S.M. Koepf, and W.J. Zielinski. 2003. Conservation genetics of the fisher (*Martes pennanti*) based on mitochondrial DNA sequencing. Molecular Ecology 12:51–62.
Drysdale, C., and R. Charlton. 1988. American marten re-introduction program progress report, Kejimkujik National Park, November 1987–October 1988. Environment Canada, Parks Canada, Natural Resources Conservation, Kejimkujik National Park, New Brunswick, Canada.
Dubey, J.P., and O.R. Hedstrom. 1993. Meningoencephalitis in mink associated with a *Sarcocystis neurona*-like organism. Journal of Veterinary Diagnostic Investigation 5:467–471.
Dubey, J.P., D.S. Lindsay, and C.A. Speer. 1998. Structures of *Toxoplasma gondii* tachyzoites, bradyzoites, and sporozoites and biology and development of tissue cysts. Clinical Microbiology Reviews 11:267.
Dubey, J.P., K. Odening, W.M. Samuel, M.J. Pybus, and A.A. Kocan. 2001. Toxoplasmosis and related infections. Pages 478–519 *in* Parasitic diseases of wild animals. W.B. Samuel, M.J. Pybus, and A.A. Kocan, editors. Iowa State University Press, Ames, USA.
Duenwald, J., J. Holland, J. Gorham, and R. Ott. 1971. Feline panleukopenia: experimental cerebellar hypoplasia produced in neonatal ferrets with live virus vaccine. Research in Veterinary Science 12:394–396.
Dumyahn, J.B., P.A. Zollner, and J.H. Gilbert. 2007. Winter home-range characteristics of American marten (*Martes americana*) in northern Wisconsin. American Midland Naturalist 158:382–394.
Duncan, R.P., D.M. Forsyth, and J. Home. 2007. Testing the metabolic theory of ecology: allometric scaling exponents in mammals. Ecology 88:324–333.
Durden, L.A. 2001. Lice (Phthiraptera). Pages 3–17 *in* Parasitic diseases of wild mammals. W.M. Samuel, M.J. Pybus, and A.A. Kocan, editors. Iowa State University Press, Ames, USA.
Durrant, S.D. 1952. Mammals of Utah: taxonomy and distribution. University of Kansas Publications in Natural History 6:1–549.
Dyke, A.S., A. Moore, and L. Robertson. 2003. Deglaciation of North America. Open File 1574. Geological Survey of Canada.
Eagle, E.C., J. Choromanski-Norris, and V.B. Kuechle. 1984. Implanting radio transmitters in mink and Franklin's ground squirrels. Wildlife Society Bulletin 12:180–184.
Earle, R.D., L.H. Mastenbrook, and T.F. Reis. 2001. Distribution and abundance of the American marten in northern Michigan. Michigan Department of Natural Resources, Wildlife Report No. 3321.
Echols, K.N., R. Vaughan, and H.D. Moll. 2004. Evaluation of subcutaneous implants for monitoring American black bear cub survival. Ursus 15:172–180.

Edmonds, M. 2001. The pleasures and pitfalls of written records. Pages 73–99 *in* The historical ecology handbook, a restorationist's guide to reference ecosystems. D. Egan and E.A. Howell, editors. Island Press, Washington, D.C., USA.

Eizirik, E., N. Yuhki, W.E. Johnson, M. Menotti-Raymond, S.S. Hannah, and S.J. O'Brien. 2003. Molecular genetics and evolution of melanism in the cat family. Current Biology 13:448–453.

Eklund, C.R. 1946. Fur resources management in British Columbia. Journal of Wildlife Management 10:29–33.

El-Badry, A.M., R. Graf, and P.-A. Clavien. 2007. Omega 3—Omega 6: what is right for the liver? Journal of Hepatology 47:718–725.

Elias, S.P., J.W. Witham, and M.L. Hunter. 2006. A cyclic red-backed vole population and seedfall over 22 years in Maine. Journal of Mammalogy 87:440–445.

Ellegren, H. 2008. Sequencing goes 454 and takes large-scale genomics into the wild. Molecular Ecology 17:1629–1635.

Ellis, L.M. 1998. Habitat use pattern of the American marten in the southern Cascade mountains of California. Thesis, Humboldt State University, Arcata, California, USA.

Elmeros, M., M.M. Birch, A.B. Madsen, H.J. Baagre, and C. Pertoldi. 2008. Skovmårens biologi og levevis i Danmark. Faglig rapport fra Danmarks Miljøundersøgelser, Aarhus Universitet nr. 692.

Emerson, K.C., and R.D. Price. 1974. A new species of *Trichodectes* (Mallophaga: Trichodectidae) from the yellow-throated marten (*Martes flavigula*). Proceedings of the Biological Society of Washington 87:77–81.

Emmons, E. 1840. Report on the quadrupeds of Massachusetts. Folsom, Wells, and Thurston, Cambridge, Massachusetts, USA.

Enders, R.K., and J.R. Leekley. 1941. Cyclic changes in the vulva of the marten (*Martes americana*). Anatomical Record 79:1–5.

Enders, R.K., and O.P. Pearson. 1943. Shortening gestation by inducing early implantation with increased light in the marten. American Fur Breeder January:18.

Environment Canada. 2009. American marten, Newfoundland population. http://www.carolynmcdademusic.com/americanmarten.html (accessed 14 January 2011).

Erickson, A.B. 1946. Incidence of worm parasites in Minnesota Mustelidae and host lists and keys to North American species. American Midland Naturalist 36:494–509.

Errington, P.L. 1946. Predation and vertebrate populations. Quarterly Review of Biology 21:144–177, 221–245.

Errington, P.L. 1956. Factors limiting higher vertebrate populations. Science 124:304–307.

[ESD] Ecologically Sustainable Development. 1996. A sustainable land use and allocation program for the Ussuri/Wusuli River watershed and adjacent territories (northeastern China and the Russian Far East). Ecologically Sustainable Development, Inc., Far Eastern Branch–Russian Academy of Sciences Institute of Aquatic and Ecological Problems, Far Eastern Branch–Russian Academy of Sciences Pacific Geographical Institute, Heilongjiang Province Territory Society, and National Committee on United States-China Relations. With map: recommended general land use allocations for sustainable development of the Ussuri River watershed and adjacent territories. [Report and map in English, Russian, and Chinese]. Ecologically Sustainable Development, Inc., Elizabethtown, New York, USA.

Eskreys-Wójcik, M., I. Wierzbowska, A. Zalewski, and H. Okarma. 2008. The land use and daily activity of urban stone marten (*Martes foina*) in Krakow, southern Poland. Page 29 *in* Abstracts, 26th Mustelid Colloquium, Eötvös Loránd University, Budapest, Hungary.

Evans, A. 1986. Feasibility study—American marten *Martes americana americana* (Turton) reintroduction to Kejimkujik National Park. Unpublished report to Parks Canada, Nesik Biological Resources, Inc.

Evans, J.S., and S.A. Cushman. 2009. Gradient modeling of conifer species using random forests. Landscape Ecology 24:673–683.

Ewald, P.W. 1995. The evolution of virulence: a unifying link between parasitology and ecology. Journal of Parasitology 81:659–669.

Excoffier, L., and G. Heckel. 2006. Computer programs for population genetics data analysis: a survival guide. Nature Reviews Genetics 7:745–758.

Fancy S.G., L.F. Pank, D.C. Douglas, C.H. Curby, G.W. Garner, S.C. Amstrup, and W.L. Regelin. 1988. Satellite telemetry: a new tool for wildlife research and management. U.S. Fish and Wildlife Service, Resource Publication No. 172, Washington D.C., USA.

Farina, A. 1998. Principles and methods in landscape ecology. Chapman & Hall, London, UK.

Faunmap Working Group. 1994. FAUNMAP: a database documenting late Quaternary distributions of mammal species in the United States. Illinois State Museum, Scientific Papers, Vol. 25, No. 1, Springfield, USA.

Fecske, D.M. 2003. Distribution and abundance of American marten and cougars in the Black Hills of South Dakota and Wyoming. Dissertation, South Dakota State University, Brookings, USA.

Fenton, A., and A.B. Pedersen. 2005. Community epidemiology framework for classifying disease threats. Emerging Infectious Diseases 11:1815–1821.

Fernández, N., S. Kramer-Schadt, and H.H. Thulke. 2006. Viability and risk assessment in species restoration: planning reintroductions for the wild boar, a potential disease reservoir. Ecology and Society 11:1–17.

Ferrière, R., F. Sarrazin, S. Legendre, and J.-P. Baron. 1996. Matrix population models applied to viability analysis and conservation: theory and practice with ULM software. Acta Oecologica 17:629–656.

Fiala, A.C.S., S.L. Garman, and A.N. Gray. 2006. Comparison of five canopy cover estimation techniques in the western Oregon Cascades. Forest Ecology and Management 232:188–197.

Fielding, A.H., and J.F. Bell. 1997. A review of methods for the assessment of prediction errors in conservation presence/absence models. Environmental Conservation 24:38–49.

Fisher, J.T., and L. Wilkinson. 2005. The response of mammals to forest fire and timber harvest in the North American boreal forest. Mammal Review 35:51–81.

Flannigan, M.D., B.J. Stocks, and B.M. Wotton. 2000. Climate change and forest fires. Science of the Total Environment 262:221–229.

Flebbe, P.A., and J.A. Herrig. 2000. Patterns of aquatic species imperilment in the southern Appalachians: an evaluation of regional databases. Environmental Management 25:681–694.

Flynn, R.W., and T.V. Schumacher. 2009. Temporal changes in population dynamics of American martens. Journal of Wildlife Management 73:1269–1281.

Fontana, A.J., I.E. Teske, K. Pritchard, and M. Evans. 1999. East Kootenay fisher reintroduction program—final report 1996–1999. Unpublished report to British Columbia Ministry of Environment, Lands, and Parks, Cranbrook, Canada.

Foran, D.R., K.R. Crooks, and S.C. Minta. 1997a. Species identification from scat: an unambiguous genetic method. Wildlife Society Bulletin 25:835–839.

Foran, D.R., S.C. Minta, and K.S. Heinemeyer. 1997b. DNA-based analysis of hair to identify species and individuals for population research and monitoring. Wildlife Society Bulletin 25:840–847.

Forest Ecosystem Management Assessment Team (FEMAT). 1993. Forest ecosystem management: an ecological, economic, and social assessment. U.S. Government Printing Office, Washington, D.C., USA.

Foreyt, W.J. 2001. Veterinary parasitology: reference manual. Fifth edition. Iowa State University Press, Ames, USA.

Foreyt, W.J., and J.E. Lagerquist. 1993. Internal parasites from the marten (*Martes americana*) in eastern Washington. Journal of the Helminthological Society of Washington 60:72–75.

Forman, R.T.T., and L.E. Alexander. 1998. Roads and their major ecological effects. Annual Review of Ecology and Systematics 29:207–231.

Forsey, O., J. Bissonette, J. Brazil, K. Curnew, L. Lemon, L. Mayo, I. Thompson, L. Bateman, and L. O'Driscoll. 1995. National recovery plan for the Newfoundland marten.

Report No. 14. Recovery of Nationally Endangered Wildlife Committee, Ottawa, Ontario, Canada.

Fortelius, M., coordinator. 2009. Neogene of the Old World database of fossil mammals (NOW). University of Helsinki. http://www.helsinki.fi/science/now/ (accessed 31 August 2009).

Fortelius, M., J. Eronen, L. Liu, D. Pushkina, A. Tesakov, I. Vislobokova, and Z. Zhang. 2006. Late Miocene and Pliocene large land mammals and climatic changes in Eurasia. Palaeogeography, Palaeoclimatology, Palaeoecology 238:219–227.

Fortin, C., and M. Cantin. 1994. The effects of trapping on a newly exploited American marten population. Pages 179–191 *in* Martens, sables, and fishers: biology and conservation. S.W. Buskirk, A.S. Harestad, M.G. Raphael, and R.A. Powell, editors. Cornell University Press, Ithaca, New York, USA.

Fortin, C., and M. Cantin. 2004. Harvest status, reproduction and mortality in a population of American martens in Québec, Canada. Pages 221–234 *in* Martens and fishers in human-altered environments: an international perspective. D.J. Harrison, A.K. Fuller, and G. Proulx, editors. Springer Science+Business Media, New York, USA.

Fortin, C., M. Julien, and Y. Leblanc. 1997. Geographic variation of American marten in central and northern Quebec. Pages 29–39 *in Martes*: taxonomy, ecology, techniques and management. G. Proulx, H.N. Bryant, and P.M. Woodard, editors. Provincial Museum of Alberta, Edmonton, Canada.

Fournier-Chambrillon, C., B. Aasted, A. Perrot, D. Pontier, F. Sauvage, M. Artois, J.M. Cassiede, X. Chauby, A. Dal Molin, C. Simon, and P. Fournier. 2004a. Antibodies to Aleutian mink disease parvovirus in free-ranging European mink (*Mustela lutreola*) and other small carnivores from southwestern France. Journal of Wildlife Diseases 40:394–402.

Fournier-Chambrillon, C., P.J. Berny, O. Coiffier, P. Barbedienne, B. Dasse, G. Delas, H. Galineau, A. Mazet, P. Pouzenc, and R. Rosoux. 2004b. Evidence of secondary poisoning of free-ranging riparian mustelids by anticoagulant rodenticides in France: implications for conservation of European mink (*Mustela lutreola*). Journal of Wildlife Diseases 40:688–695.

Fowler, C.H., and R.T. Golightly. 1994. Fisher and marten survey techniques on the Tahoe National Forest. Final report to the USDA Forest Service. Agreement No. PSW-90-0034CA. Humboldt State University, Arcata, California, USA.

Fox, J.G., R.C. Pearson, and J.R. Gorham. 1998. Viral diseases. Pages 355–374 *in* Biology and diseases of the ferret. J.G. Fox, editor. Lippincott Williams and Wilkins, Baltimore, Maryland, USA.

Frair J.L., S.E. Nielsen, E.H. Merrill, S.R. Lele, M.S. Boyce, R.H.M. Munro, G. B. Stenhouse, and H.L. Beyer. 2004. Removing GPS collar bias in habitat selection studies. Journal of Applied Ecology 41:201–212.

Francesca, V., L. Livia, M. Nadia, R. Bernardino, R. Ettore, and P. Fausto. 2004. A simple and rapid PCR-RFLP method to distinguishing *Martes martes* and *Martes foina*. Conservation Genetics 5:869–871.

Francis, G.R., and A.B. Stephenson. 1972. Marten ranges and food habits in Algonquin Provincial Park. Research Report 91, Ontario Ministry of Natural Resources, Canada.

Frank, R.K. 2001. An outbreak of toxoplasmosis in farmed mink (*Mustela vison*). Journal of Veterinary Diagnostic Investigation 13:245–249.

Frankham, R. 1995. Conservation genetics. Annual Review in Genetics 29:305–327.

Frankham, R. 2005. Genetics and extinction. Biological Conservation 126:131–140.

Franklin, J.F., and C.T. Dyrness. 1988. Natural vegetation of Oregon and Washington. Oregon State University Press, Corvallis, USA.

Fredrickson, L. 1983. Re-introduction of pine marten into the Black Hills of South Dakota, 1977–83. Pages 93–98 *in* Proceedings of the Midwest Furbearer Workshop, March 29–31, 1983, Poynette, Wisconsin, USA.

Fredrickson, L.F. 1989. Pine marten introduction in the Black Hills of South Dakota, 1979–1988. South Dakota Department of Game, Fish and Parks, Completion Report No. 90–10, Pierre, USA.

Fredrickson, R. 1990. The effects of disease, prey fluctuation, and clear-cutting on American marten in Newfoundland, Canada. Thesis, Utah State University, Logan, USA.

Freeman, L.C. 1977. A set of measures of centrality based on betweenness. Sociometry 40:35–41.

Fretwell, S.D. 1972. Populations in a seasonal environment. Princeton University Press, New Jersey, USA.

Fretwell, S.D., and H.L. Lucas. 1969. On the territorial behaviour and other factors influencing habitat distribution in birds. Acta Biotheoretica 14:16–36.

Frolich, K., O. Czupalla, L. Haas, J. Hentschke, J. Dedek, and J. Fickel. 2000. Epizootiological investigations of canine distemper virus in free-ranging carnivores from Germany. Veterinary Microbiology 74:283–292.

Frost, H.C., and W.B. Krohn. 1997. Factors affecting the reproductive success of captive female fishers. Pages 100–109 *in Martes*: taxonomy, ecology, techniques, and management. G. Proulx, H.N. Bryant, and P.M. Woodard, editors. Provincial Museum of Alberta, Edmonton, Canada.

Frost, H.C., E.C. York, W.B. Krohn, K.D. Elowe, T.A. Decker, S.M. Powell, and T.K. Fuller. 1999. An evaluation of parturition indices in fishers. Wildlife Society Bulletin 27:221–230.

Fryxell, J.M., J.B. Falls, E.A. Falls, and R.J. Brooks. 1998. Long-term dynamics of small mammal populations in Ontario. Ecology 79:213–225.

Fryxell, J.M., J.B. Falls, E.A. Falls, R.J. Brooks, L. Dix, and M.A. Strickland. 1999. Density dependence, prey dependence, and population dynamics of martens in Ontario. Ecology 80:1311–1321.

Fryxell, J., J.B. Falls, E.A. Falls, R.J. Brooks, L. Dix, and M. Strickland. 2001. Harvest dynamics of mustelid carnivores in Ontario, Canada. Wildlife Biology 7:151–159.

Fukushima, H., and M. Gomyoda. 1991. Intestinal carriage of *Yersinia pseudotuberculosis* by wild birds and mammals in Japan. Applied Environmental Microbiology 57:1152–1155.

Fuller, A.K., and D.J. Harrison. 2005. Influence of partial timber harvesting on American martens in north-central Maine. Journal of Wildlife Management 69:710–722.

Fuller, A.K., D.J. Harrison, and H.J. Lachowski. 2004. Stand scale effects of partial harvesting and clearcutting on small mammals and forest structure. Forest Ecology and Management 191:373–386.

Fuller, R.W. 1975. The 1974 fisher trapping season in Vermont. Vermont's 1975 Game Annual:23–30.

Fuller, T.K., E.C. York, S.M. Powell, T.A. Decker, and R.M. DeGraaf. 2001. An evaluation of territory mapping to estimate fisher density. Canadian Journal of Zoology 79:1691–1696.

Fulton, T.L., and C. Strobeck. 2006. Molecular phylogeny of the Arctoidea (Carnivora): effect of missing data on supertree and supermatrix analysis of multiple gene data sets. Molecular Phylogenetics and Evolution 41:165–181.

Furman, D.P., and E.C. Loomis. 1984. The ticks of California (Acari: Ixodida). Volume 25. University of California Press, Berkeley, USA.

Gabriel, M.W. 2006. Exposure to *Anaplasma phagocytophilum* and ticks in gray foxes (*Urocyon cinereoargenteus*) in northern Humboldt County, California. Thesis, Humboldt State University, Arcata, California, USA.

Gabriel, M.W., G.M. Wengert, J.E. Foley, J.M. Higley, S. Matthews, and R.N. Brown. 2007. Pathogens associated with fishers (*Martes pennanti*) and sympatric mesocarnivores in California. Final draft report to the U.S. Fish and Wildlife Service for grant #813335G021. U.S. Fish and Wildlife Service, Yreka, California, USA.

Gabriel, M.W., G.M. Wengert, J.M. Higley, S. Matthews, J.E. Foley, A. Blades, M. Sullivan, and R.N. Brown. 2010. Efectiveness of rapid diagnostic test to access pathogens of fishers and gray foxes. Journal of Wildlife Diseases 46:966–970.

Galbreath, E.C. 1953. A contribution to the Tertiary geology and paleontology of northeastern Colorado. University of Kansas Paleontological Contributions, Vertebrata 4:1–120.

Galbreath, K.E. 2009. Of pikas and parasites: historical biogeography of an alpine host-parasite assemblage. Dissertation, Cornell University, Ithaca, New York, USA.

Galbreath, K.E., D.J. Hafner, K.R. Zamudio, and K. Agnew. 2010. Isolation and introgression in the Intermountain West: contrasting gene genealogies reveal the complex biogeographic history of the American pika (*Ochotona princeps*). Journal of Biogeography 37:344–362.

Galbreath, K.E., and E.P. Hoberg. 2012. Return to Beringia: parasites reveal cryptic biogeographic history of North American pikas. Proceedings of the Royal Society B 279:371–378.

Gannon, W.L., R.S. Sikes, and the Animal Care and Use Committee of the American Society of Mammalogists. 2007. Guidelines of the American Society of Mammalogists for the use of wild mammals in research. Journal of Mammalogy 88:809–823.

Gantz, G.F., L.C. Stoddart, and F.F. Knowlton. 2006. Accuracy of aerial telemetry locations in mountainous terrain. Journal of Wildlife Management 70:1809–1812.

Garant, Y., and M. Crête. 1999. Prediction of water, fat, and protein content of fisher carcasses. Wildlife Society Bulletin 27:403–408.

Garant, Y., R. Lafond, and R. Courtois. 1996. Analyse du système de suivi de la martre d'Amérique (*Martes americana*) au Québec. Ministère de l'Environnement et de la Faune, Direction de la Faune et des Habitats, Quebec, Canada.

Gardner, B., J.A. Royle, M.T. Wegan, R.E. Rainbolt, and P.D. Curtis. 2010. Estimating black bear density using DNA data from hair snares. Journal of Wildlife Management 74:318–325.

Garroway, C.J., J. Bowman, D. Carr, and P.J. Wilson. 2008. Applications of graph theory to landscape genetics. Evolutionary Applications 1:620–630.

Garshelis, D.L. 2006. On the allure of noninvasive genetic sampling—putting a face to the name. Ursus 17:109–123.

Gaston, K.J. 2003. The structure and dynamics of geographic ranges. Oxford University Press, UK.

Gazin, C.L. 1934. Upper Pliocene mustelids from the Snake River Basin of Idaho. Journal of Mammalogy 15:137–149.

Gehring, T.M., and R.K. Swihart. 2003. Body size, niche breath, and ecologically scaled responses to habitat fragmentation: mammalian predators in an agricultural landscape. Biological Conservation 109:283–295.

Geisel, O.V., H.E. Krampitz, and A. Pospischil. 1979. Zur pathomorphologie eina Hepatozoon infektion bei musteliden. Berlin Münchener Tierärztlichen Wochenschrift 92:421–425.

Genovesi, P., and L. Boitani. 1995. Preliminary data on the social ecology of the stone marten (*Martes foina* Erxleben 1777) in Tuscany (Central Italy). Hystrix 7:159–163.

Genovesi, P., and L. Boitani. 1997a. Day resting sites of the stone marten. Hystrix 9:75–78.

Genovesi, P., and L. Boitani. 1997b. Social ecology of the stone marten in central Italy. Pages 110–120 *in Martes*: taxonomy, ecology, techniques, and management. G. Proulx, H.N. Bryant, and P.M. Woodard, editors. Provincial Museum of Alberta, Edmonton, Canada.

Genovesi, P., M. Secchi, and L. Boitani. 1996. Diet of stone martens: an example of ecological flexibility. Journal of Zoology 238:545–555.

Genovesi, P., P. Sinibaldi, and L. Boitani. 1997. Spacing patterns and territoriality of the stone marten. Canadian Journal of Zoology 75:1966–1971.

Gerhold, R.W., E.W. Howerth, and D.S. Lindsay. 2005. *Sarcocystis neurona*-associated meningoencephalitis and description of intramuscular sarcocysts in a fisher (*Martes pennanti*). Journal of Wildlife Diseases 41:224–230.

Ghaffar, A., J. A. Khan, M. Nazir, and M. Akhtar. 2004. First record of *Martes lydekkeri* from Dhok Pathan Formation of Pakistan. Punjab University Journal of Zoology 19:97–102.

Gibbard, P., and T. van Kolfschoten. 2004. The Pleistocene and Holocene epochs. Pages 441–452 *in* A geological time scale. F.M. Gradstein, J. Ogg, G. Smith, and A. Gilbert, editors. Cambridge University Press, UK.

Gibilisco, C.J. 1994. Distributional dynamics of modern *Martes* in North America. Pages 59–71 *in* Martens, sables, and fishers: biology and conservation. S.W. Buskirk, A.S. Harestad, M.G. Raphael, and R.A. Powell, editors. Cornell University Press, Ithaca, New York, USA.

Gidley, J.W. 1927. A true marten from the Madison valley (Miocene) of Montana. Journal of Mammalogy 8:239–242.

Gieck, E.M. 1986. Wisconsin pine marten recovery plan. Wisconsin Department of Natural Resources, Wisconsin Endangered Resources Report 22, Madison, USA.

Gilbert, J.H., J.L. Wright, D.J. Lauten, and J.R. Probst. 1997. Den and rest-site characteristics of American marten and fisher in northern Wisconsin. Pages 135–145 *in Martes*: taxonomy, ecology, techniques, and management. G. Proulx, H.N. Bryant, and P.M. Woodard, editors. Provincial Museum of Alberta, Edmonton, Canada.

Gilbert, J.H., P.A. Zollner, A.K. Green, J.L. Wright, and W.H. Karasov. 2009. Seasonal field metabolic rates of American martens in Wisconsin. American Midland Naturalist 162:327–334.

Gillespie, J.H. 1962. The virus of canine distemper. Annals of the New York Academy of Sciences 101:540–547.

Gilman, N. 2004. Shelter medicine for veterinarians and staff: sanitation in the animal shelter. Blackwell Publishing, Ames, Iowa, USA.

Gilmer, D.S., L.M. Cowardin, R.L. Duval, L.M. Mechlin, and C.W. Shaiffer. 1981. Procedures for the use of aircraft in wildlife biotelemetry studies. U.S. Fish and Wildlife Resource Publication 140.

Ginsburg, L. 1999. Order Carnivora. Pages 109–148 *in* The Miocene land mammals of Europe. G.E. Rössner and K. Heisseg, editors. Friedrich Pfeil, Munich, Germany.

Ginsburg, L. 2001. Les faunes de mammiferes terrestres du Miocene moyen des Faluns du basin de Savigne-sur-Lathan (France). Geodiversitas 23:381–394.

Ginsburg, L. 2002. Les carnivores fossils des sables de l'Orléanais. Annales de Paléontologie 88:115–146.

Girvetz, E., C. Zganjar, G. Raber, E.P. Maurer, P. Kareiva, and J.J. Lawler. 2009. Applied climate change analysis: the Climate Wizard tool. PLoS ONE 4:e8320.

Godbout, G., and J.-P. Ouellet. 2008. Habitat selection of American marten in a logged landscape at the southern fringe of the boreal forest. Ecoscience 15:332–342.

Goldberg, A.L., M. Tischler, G. DeMartino, and G. Griffin. 1980. Hormonal regulation of protein degradation and synthesis in skeletal muscle. Federation Proceedings 39:31–36.

Goldman, E.A. 1935. New American mustelids of the genera *Martes*, *Gulo*, and *Lutra*. Proceedings of the Biological Society of Washington 48:175–186.

Golightly, R.T., T.F. Penland, W.J. Zielinski, and J.M. Higley. 2006. Fisher diet in the Klamath/North Coast bioregion. Unpublished report, Humboldt State University, Arcata, California, USA.

Gompper, M.E., R.W. Kays, J.C. Ray, S.D. Lapoint, D.A. Bogan, and J.R. Cryan. 2006. A comparison of noninvasive techniques to survey carnivore communities in northeastern North America. Wildlife Society Bulletin 34:1142–1151.

Gompper, M.E., and E.S. Williams. 1998. Parasite conservation and the black-footed ferret recovery program. Conservation Biology 12:730–732.

Goodman, M.N., P.R. Larsen, M.M. Kaplan, T.T. Aoki, V.R. Young, and N.B. Ruderman. 1980. Starvation in the rat. II. Effect of age and obesity on protein sparing and fuel metabolism. American Journal of Physiology 239:E277–E286.

Goossens, B., L.P. Waits, and P. Taberlet. 1998. Plucked hair samples as a source of DNA: reliability of dinucleotide microsatellite genotyping. Molecular Ecology 7:1237–1241.

Gordon, J., and E. Angrick. 1986. Canine parvovirus: environmental effects on infectivity. American Journal of Veterinary Research 47:1464–1467.

Gosse, J.W., R. Cox, and S.W. Avery. 2005. Home-range characteristics and habitat use by American martens in eastern Newfoundland. Journal of Mammalogy 86:1156–1163.

Gosse, J., and B. Hearn. 2005. Seasonal diets of Newfoundland martens, *Martes americana atrata*. Canadian Field-Naturalist 119:43–47.

Goszczyński, J., M. Posłuszny, M. Pilot, and B. Gralak. 2007. Patterns of winter locomotion and foraging in two sympatric marten species: *Martes martes* and *Martes foina*. Canadian Journal of Zoology 85:239–249.

Gotlieb, A. 1927. Fur truths. The story of furs and the fur business. Harper & Brothers, New York, USA.

Graf, R.F., K. Bollmann, W. Suter, and H. Bugmann. 2005. The importance of spatial scale in habitat models: capercaillie in the Swiss Alps. Landscape Ecology 20:703–717.

Graham, R.W., and M.A. Graham. 1994. Late Quaternary distribution of *Martes* in North America. Pages 26–58 *in* Martens, sables and fishers: biology and conservation. S.W. Buskirk, A.S. Harestad, M.G. Raphael, and R.A. Powell, editors. Cornell University Press, Ithaca, New York, USA.

Grahl-Nielsen, O., M. Andersen, A.E. Derocher, C. Lydersen, Ø. Wiig, and K.M. Kovacs. 2003. Fatty acid composition of the adipose tissue of polar bears and of their prey: ringed seals, bearded seals and harp seals. Marine Ecology Progress Series 265:275–282.

Grakov, N.N. 1981. Lesnaya kunitsa [The pine marten]. Nauka, Moskva:1–109. (in Russian)

Grand, J., J. Buonaccorsi, S.A. Cushman, C.R. Griffin, and M.C. Neel. 2004. A multiscale landscape approach to predicting bird and moth rarity hotspots in a threatened pitch pine–scrub oak community. Conservation Biology 18:1063–1077.

Grassman, L.I., Jr., A.M. Haines, J.E. Janečka, and M.E. Tewes. 2006a. Activity periods of photo-captured mammals in north central Thailand. Mammalia 70:306–309.

Grassman, L.I., Jr., J.E. Janečka, and M.E. Tewes. 2006b. Ecological separation of *Martes flavigula* with five sympatric mesocarnivores in north-central Thailand. Pages 63–76 *in Martes* in carnivore communities. M. Santos-Reis, J.D.S. Birks, E.C. O'Doherty, and G. Proulx, editors. Alpha Wildlife Publications, Sherwood Park, Alberta, Canada.

Grassman, L.I., Jr., N. Sarataphan, M.E. Tewes, N.J. Silvy, and T. Nakanakrat. 2004. Ticks (Acari: Ixodidae) parasitizing wild carnivores in Phu Khieo Wildlife Sanctuary, Thailand. Journal of Parasitology 90:657–659.

Gray, J.E. 1865. Revision of the genera and species of Mustelidae contained in the British Museum. Proceedings of the Zoological Society of London 1865:100–154.

Greeley, W.B. 1925. The relation of geography to timber supply. Economic Geography 1:1–14.

Green, J.S., R.T. Golightly, S.L. Lindsey, and B.R. Leamaster. 1985. Use of radio transmitter implants in wild canids. Great Basin Naturalist 45:567–570.

Green, R.E. 2007. Distribution and habitat associations of forest carnivores and an evaluation of the California Wildlife Habitat Relationships model for the American marten in Sequoia and Kings Canyon National Parks. Thesis, Humboldt State University, Arcata, California, USA.

Greene, C.M., and J.A. Stamps. 2001. Habitat selection at low population densities. Ecology 82:2091–2100.

Grenier, M.B., D.B. McDonald, and S.W. Buskirk. 2007. Rapid population growth of a critically endangered carnivore. Science 317:779.

Griffith, B., J.M. Scott, J.W. Carpenter, and C. Reed. 1989. Translocation as a species conservation tool: status and strategy. Science 245:477–480.

Grilo, C., J. Bissonette, and M. Santos-Reis. 2009. Spatial-temporal patterns in Mediterranean carnivore road casualties: consequences for mitigation. Biological Conservation 142:301–313.

Grimm, V., and S.F. Railsback. 2005. Individual-based modeling and ecology. Princeton University Press, New Jersey, USA.

Grinnell, J., J.S. Dixon, and J.M. Linsdale. 1937. Fur-bearing mammals of California. University of California Press, Berkeley, USA.

Grivert, D., V.L. Sork, R.D. Westfall, and F.W. Davis. 2008. Conserving the evolutionary potential of California valley oak (*Quercus lobata* Nee): a multivariate genetic approach to conservation planning. Molecular Ecology 17:139–156.

Groscolas, R., and J.-P. Robin. 2001. Long-term fasting and re-feeding in penguins. Comparative Biochemistry and Physiology A 128:645–655.

Grove, J.M. 1988. The Little Ice Age. Methuen, New York, USA.

Guerin, C., and P. Mein. 1971. Les principaux gisements de mammiferes Miocènes et Pliocènes du domaine rhodanien. Pages 131–170 *in* Documents des Laboratoires de Géologie de Lyon HS. Le Néogène rhodanien—Vè congrès du Néeogène mediterranéen.

Gumtow-Farrior, D.L. 1991. Cavity resources in Oregon white oak and Douglas-fir stands in the mid-Willamette Valley, Oregon. Thesis, Oregon State University, Corvallis, USA.

Gundersen, V.S. 1995. Habitatbruk hos mår vinterstid [Habitat use by pine marten in winter]. Thesis, Agricultural University of Norway, As. (in Norwegian)

Gusev, O.K. 1971. Restoration of sable in the USSR. Priroda (Moscow, USSR) 11:68–74. (in Russian)

Guynn, D.C., J.R. Davis, and A.F. Von Recum. 1987. Pathological potential of intraperitoneal transmitter implants in beavers. Journal of Wildlife Management 51:605–606.

Haas, G.E., N. Wilson, T.O. Osborne, R.L. Zarnke, L. Johnson, and J.O. Wolff. 1989. Mammal fleas (Siphonaptera) of Alaska and Yukon Territory. Canadian Journal of Zoology 67:394–405.

Hagmeier, E.M. 1955. The genus *Martes* (Mustelidae) in North America: distribution, variation, classification, phylogeny and relationships to Old World forms. Dissertation, University of British Columbia, Vancouver, Canada.

Hagmeier, E.M. 1956. Distribution of marten and fisher in North America. Canadian Field-Naturalist 70:149–168.

Haley, D. 1975. Sleek & savage. Pacific Search, Seattle, Washington, USA.

Halfpenny, J., R.W. Thompson, S.C. Morse, T. Holden, and P. Rezendes. 1995. Snow tracking. Pages 91–163 *in* American marten, fisher, lynx, and wolverine: survey methods for their detection. W.J. Zielinski and T.E. Kucera, technical editors. USDA Forest Service, General Technical Report PSW-GTR-157.

Hall, E.R. 1931. Description of a new mustelid from the late Tertiary of Oregon, with assignment of *Parictis primaevus* to the Canidae. Journal of Mammalogy 12:156–158.

Hall, E.R. 1946. Mammals of Nevada. University of California Press, Berkeley, USA.

Hall, E.R. 1981. The mammals of North America. Second edition. John Wiley & Sons, New York, USA.

Hamilton, W.J., Jr., and A.H. Cook. 1955. The biology and management of the fisher in New York. New York Fish and Game Journal 2:13–35.

Hamm, K.A., L.V. Diller, R.R. Klug, and T.L. McDonald. 2003. Spatial independence of fisher (*Martes pennanti*) detections at track plates in northwestern California. American Midland Naturalist 149:201–210.

Hapeman, P.H. 2006. The population genetics of fishers (*Martes pennanti*) across northeastern North America. Dissertation, University of Vermont, Burlington, USA.

Hapeman, P., E.K. Latch, J.A. Fike, O.E. Rhodes, and C.W. Kilpatrick. 2011. Landscape genetics of fisher (*Martes pennanti*) in the northeast: dispersal barriers and historical influences. Journal of Heredity 102:251–259.

Hardy, C. 1869. Forest life in Acadie. Sketches of sport and natural history in the lower provinces of the Canadian dominion. Chapman & Hall, London, UK.

Hardy, M. 1903. A Maine woods walk in sixty-one. Forest and Stream 60:263–264.

Hardy, M. 1910. A fall fur hunt in Maine. Forest and Stream 74:928–929.

Hare, M.P., L. Nunney, M.K. Schwartz, D.E. Ruzzante, M. Burford, R.S. Waples, K. Ruegg, and F. Palstra. 2011. Understanding and estimating effective popualtion size for practical application in marine species management. Conservation Biology 25:438–449.

Harger, E.M., and D.F. Switzenberg. 1958. Returning the pine marten to Michigan. Report 2199, Michigan Department of Conservation, Game Division, Lansing, USA.

Hargis, C.D., J.A. Bissonette, and D.L. Turner. 1999. The influence of forest fragmentation and landscape pattern on American martens. Journal of Applied Ecology 36:157–172.

Hargis, C.D., and D.R. McCullough. 1984. Winter diet and habitat selection of marten in Yosemite National Park. Journal of Wildlife Management 48:140–146.

Hargiss, C.L.M., E.S. DeKeyser, D.R. Kirby, and M.J. Ell. 2008. Regional assessment of wetland plant communities using the index of plant community integrity. Ecological Indicators 8:303–307.

Harington, C.R. 2005. The eastern limit of Beringia: mammoth remains from Banks and Melville Islands, Northwest Territories. Arctic 58:361–369.

Harlow, H.J. 1994. Trade-offs associated with the size and shape of American martens. Pages 391–403 *in* Martens, sables, and fishers: biology and conservation. S.W. Buskirk, A.S. Harestad, M.G. Raphael, and R.A. Powell, editors. Cornell University Press, Ithaca, New York, USA.

Harlow, H.J., and S.W. Buskirk. 1991. Comparative plasma and urine chemistry of fasting white-tailed prairie dogs (*Cynomys leucurus*) and American martens (*Martes americana*): representative fat- and lean-bodied animals. Physiological Zoology 64:1262–1278.

Harlow, H.J., and S.W. Buskirk. 1996. Amino acids in plasma of fasting fat prairie dogs and lean martens. Journal of Mammalogy 77:407–411.

Harper, F. 1961. Land and fresh-water mammals of the Ungava Peninsula. University of Kansas, Lawrence, USA.

Harris, R.B., S.G. Fancy, D.C. Douglas, G.W. Garner, S.C. Amstrup, T.R. McCabe, and L.F. Pank. 1990. Tracking wildlife by satellite: current systems and performance. U.S. Fish and Wildlife Service, Technical Report 30, Washington D.C., USA.

Harrison, D.J., A.K. Fuller, and G. Proulx, editors. 2004. Martens and fishers (*Martes*) in human-altered environments: an international perspective. Springer Science+Business Media, New York, USA.

Harrison, R.L. 2006. A comparison of survey methods for detecting bobcats. Wildlife Society Bulletin 34:548–552.

Harvey, D.S., and P.J. Weatherhead. 2006. A test of the hierarchical model of habitat selection using eastern massasauga rattlesnakes (*Sistrurus c. catenatus*). Biological Conservation 130:206–216.

Hastie, T.J., and R.J. Tibshirani. 1990. Generalized additive models. Chapman and Hall/CRC, Boca Raton, Florida, USA.

Haukisalmi, V., H. Henttonen, and L. Hardman. 2006. Taxonomy, diversity, and zoogeography of *Paranoplocephala* spp. (Cestoda: Anoplocephalidae) in voles and lemmings of Beringia, with a description of three new species. Biological Journal of the Linnean Society 89:277–299.

Haukisalmi, V., L. Wickström, J. Hantula, and H. Henttonen. 2001. Taxonomy, genetic differentiation, and Holarctic biogeography of *Paranoplocephala* spp. (Cestoda: Anoplocephalidae) in collared lemmings (*Dicrostonyx*; Arvicolinae). Biological Journal of the Linnean Society 74:171–196.

Hauptman, T.N. 1979. Spatial and temporal distribution and feeding ecology of the pine marten. Thesis, Idaho State University, Moscow, USA.

Hawley, V.D., and F.E. Newby. 1957. Marten home ranges and population fluctuations. Journal of Mammalogy 38:174–184.

Haydon, D.T., and J.M. Fryxell. 2003. Using knowledge of recruitment to manage harvesting. Ecology 85:78–85.

Haydon, D.T., D.A. Randall, L. Matthews, D.L. Knobel, L.A. Tallents, M.B. Gravenor, S.D. Williams, J.P. Pollinger, S. Cleaveland, and M.E.J. Woolhouse. 2006. Low-coverage vaccination strategies for the conservation of endangered species. Nature 443:692–695.

Hayhoe, K., D. Cayan, C.B. Field, P.C. Frumhoff, E.P. Maurer, N.L. Miller, S.C. Moser, S.H. Schneider, K.N. Cahill, E.E. Cleland, L. Dale, R. Drapek, R.M. Hanemann, L.S. Kalkstein, J. Lenihan, C.K. Lunch, R.P. Neilson, S.C. Sheridan, and J.H. Verville.

2004. Emissions pathways, climate change, and impacts on California. Proceedings of the National Academy of Sciences, USA 101:12422–12427.
Haywood, D. (aka D.E. Heywood). 1897. Parmachenee Lake, Maine, notes. Shooting and Fishing 22:107.
Hearn, B.J. 2007. Factors affecting habitat selection and population characteristics of American marten (*Martes americana atrata*) in Newfoundland. Dissertation, University of Maine, Orono, USA.
Hearn, B.J., D.J. Harrison, A.K. Fuller, C.G. Lundrigan, and W.J. Curran. 2010. Paradigm shifts in habitat ecology of threatened Newfoundland martens. Journal of Wildlife Management 74:719–728.
Heberlein, T.A. 1991. Changing attitudes and future for wildlife—preserving the sport hunter. Wildlife Society Bulletin 19:528–534.
Heinemeyer, K.S. 1993. Temporal dynamics in the movements, habitat use, activity, and spacing of reintroduced fishers in northwestern Montana. Thesis, University of Montana, Missoula, USA.
Heinemeyer, K.S. 2002. Translating individual movements into population patterns: American marten in fragmented forested landscapes. Dissertation, University of California, Santa Cruz, USA.
Heinemeyer, K.S., T.J. Ulizio, and R.L. Harrison. 2008. Natural sign: tracks and scats. Pages 45–74 *in* Noninvasive survey methods for carnivores. R.A. Long, P. MacKay, W.J. Zielinski, and J.C. Ray, editors. Island Press, Washington, D.C., USA.
Heinrichs, J.A., D.J. Bender, D.L. Gummer, and N.H. Schumaker. 2010. Assessing critical habitat: evaluating the relative contribution of habitats to population persistence. Biological Conservation 143:2229–2237.
Hejlíček, K., I. Literák, and J. Nezval. 1997. Toxoplasmosis in wild mammals from the Czech Republic. Journal of Wildlife Diseases 33:480–485.
Helldin, J.-O. 1999. Diet, body condition, and reproduction of Eurasian pine martens *Martes martes* during cycles in microtine density. Ecography 22:324–339.
Helldin, J.-O. 2000. Population trends and harvest management of pine marten *Martes martes* in Scandinavia. Wildlife Biology 6:111–120.
Helle, E., P. Helle, H. Linden, and M. Wikman. 1996. Wildlife populations in Finland during 1990–1995, based on wildlife triangle data. Finnish Game Research 49:12–17.
Helle, P., and A. Nikula. 1996. Usage of geographic information systems (GIS) in analyses of wildlife triangle data. Finnish Game Research 49:26–36.
Hellstedt, P., J. Sundell, P. Helle, and H. Henttonen. 2006. Large-scale spatial and temporal patterns in population dynamics of the stoat, *Mustela erminea*, and the least weasel, *M. nivalis*, in Finland. Oikos 115:286–298.
Herbst, L. 1991. Pathological and reproductive effects in intraperitoneal telemetry devices on female armadillos. Journal of Wildlife Management 55:628–631.
Herman, T., and K. Fuller. 1974. Observations of the marten, *Martes americana*, in the Mackenzie District, Northwest Territories. Canadian Field-Naturalist 88:501–503.
Herr, J. 2008. Ecology and behaviour of urban stone martens (*Martes foina*) in Luxembourg. Dissertation, University of Sussex, Brighton, UK.
Herr, J., L. Schley, E. Engel, and T.J. Roper. 2010. Den preferences and denning behaviour in urban stone martens. Mammalian Biology 75:138–145.
Herr, J., L. Schley, and T.J. Roper. 2008. Fate of translocated wild-caught and captive-reared stone martens (*Martes foina*). European Journal of Wildlife Research 54:511–514.
Herr, J., L. Schley, and T.J. Roper. 2009. Socio-spatial organization of urban stone martens. Journal of Zoology 277:54–62.
Herrmann, M. 1994. Habitat use and spatial organization by the stone marten. Pages 122–136 *in* Martens, sables, and fishers: biology and conservation. S.W. Buskirk, A.S. Harestad, M.G. Raphael, and R.A. Powell, editors. Cornell University Press, Ithaca, New York, USA.

Herzog, C.J., R.W. Kays, J.C. Ray, M.E. Gompper, W.J. Zielinski, R. Higgins, and M. Tymeson. 2007. Using patterns in track-plate footprints to identify individual fishers. Journal of Wildlife Management 71:955–963.

Hewitt, G.M. 1989. The subdivision of species by hybrid zones. Pages 85–110 *in* Speciation and its consequences. D. Otte and J.A. Endler, editors. Sinauer Associates, Sunderland, Massachusetts, USA.

Hewitt, G.M. 1996. Some genetic consequences of ice ages, and their role in divergence and speciation. Biological Journal of the Linnean Society 58:247–276.

Hewitt, G.M. 1999. Post-glacial re-colonization of European biota. Biological Journal of the Linnean Society 68:87–112.

Hibbard, W.C., and E.S. Riggs. 1949. Upper Pliocene vertebrates from Keefe Canyon, Meade County, Kansas. The Geological Society of America Bulletin 60:829–860.

Hicks, S.A., and S.M. Carr. 1992. Genetic analysis of a threatened subspecies, the Newfoundland pine marten (*Martes americana atrata*). Pages 287–290 *in* Science and management in protected areas. J.H.M. Willison, S. Boundrup-Nielsen, C. Drysdale, T.B. Herman, N.W.P. Munro, and T.L. Pollock, editors. Elsevier, Amsterdam, The Netherlands.

Higgins, J.V., M.T. Bryer, M.L. Khoury, and T.W. Firtzhugh. 2005. A freshwater classification approach for biodiversity conservation planning. Conservation Biology 19:432–445.

Higley, J.M., and S.M. Matthews. 2009. Fisher habitat use and population monitoring on the Hoopa Valley Reservation, California. Unpublished final report to U.S. Fish and Wildlife Service, USFWS TWG U-12-NA-1, Yreka, California. Hoopa Tribal Forestry and Wildlife Conservation Society, Hoopa, California, USA.

Higuchi, H., Y. Shibaev, J. Minton, K. Ozaki, S. Surmach, G. Fujita, K. Momose, Y. Momose, M. Ueta, V. Andronov, N. Mita, and Y. Kanai. 1998. Satellite tracking of the migration of the red-crowned crane *Grus japonensis*. Ecological Research 13:273–282.

Hillyer, E.V., and K.E. Quesenberry, editors. 1997. Ferrets, rabbits and rodents: clinical medicine and surgery. Second edition. W.B. Saunders Company, Philadelphia, Pennsylvania, USA.

Hines, J., J.D. Nichols, J.A. Royle, D.I. MacKenzie, A.M. Gopalaswamy, N.S. Kumar, and K.U. Karanth. 2010. Tigers on trails: occupancy modeling for cluster sampling. Ecological Applications 20:1456–1466.

Hobbs, N.T., and T.A. Hanley. 1990. Habitat evaluation: do use/availability data reflect carrying capacity? Journal of Wildlife Management 54:515–522.

Hoberg, E.P. 1997a. Parasite biodiversity and emerging pathogens: a role for systematics in limiting impacts on genetic resources. Pages 71–83 *in* Global genetic resources: access, ownership and intellectual property rights. K.E. Hoaglund and A.Y. Rossman, editors. Association of Systematics Collections, Washington, D.C., USA.

Hoberg, E.P. 1997b. Phylogeny and historical reconstruction: host-parasite systems as keystones in biogeography and ecology. Pages 243–261 *in* Biodiversity II: understanding and protecting our biological resources. M. Reaka-Kudla, D.E. Wilson, and E.O. Wilson, editors. Joseph Henry Press, Washington, D.C., USA.

Hoberg, E.P. 2005a. Coevolution and biogeography among Nematodirinae (Nematoda: Trichostrongylina), Lagomorpha and Artiodactyla (Mammalia): exploring determinants of history and structure for the northern fauna across the Holarctic. Journal of Parasitology 91:358–369.

Hoberg, E.P. 2005b. Coevolution in marine systems. Pages 327–339 *in* Marine parasitology. K. Rohde, editor. CSIRO, Sydney, Australia.

Hoberg, E.P. 2006. Phylogeny of *Taenia*: defining species and origins of human parasites. Parasitology International 50:S23–S30.

Hoberg, E.P. 2010. Invasive processes, mosaics and the structure of helminth parasite faunas. Revue Scientifique et Technique Office International des Épizooties 29:255–272.

Hoberg, E.P., K.B. Aubry, and J.D. Brittell. 1990. Helminth parasitism in martens (*Martes americana*) and ermines (*Mustela erminea*) from Washington, with comments on the distribution of *Trichinella spiralis*. Journal of Wildlife Diseases 26:447–452.

Hoberg, E.P., and D.R. Brooks. 2008. A macroevolutionary mosaic: episodic host-switching, geographic colonization, and diversification in complex host-parasite systems. Journal of Biogeography 35:1533–1550.

Hoberg, E.P., and D.R. Brooks. 2010. Beyond vicariance: integrating taxon pulses, ecological fitting and oscillation in historical biogeography and evolution. Pages 7–20 *in* The biogeography of host-parasite interactions. S. Morand and B. Krasnov, editors. Oxford University Press, New York, USA.

Hoberg, E.P., C.J. Henny, O.R. Hedstrom, and R.A. Grove. 1997. Intestinal helminths of river otters (*Lutra canadensis*) from the Pacific Northwest. Journal of Parasitology 83:105–110.

Hoberg, E.P., S.J. Kutz, K.E. Galbreath, and J. Cook. 2003. Arctic biodiversity: from discovery to faunal baselines—revealing the history of a dynamic ecosystem. Journal of Parasitology 89:S84–S95.

Hoberg, E.P., and S.G. McGee. 1982. Helminth parasitism in raccoons, *Procyon lotor hirtus* Nelson and Goldman, in Saskatchewan. Canadian Journal of Zoology 60:53–57.

Hoberg, E.P., P. Pilitt, and K.E. Galbreath. 2009. Why museums matter: a tale of pinworms (Oxyuroidea: Heteroxynematidae) among pikas (*Ochotona princeps* and *O. collaris*) in the American west. Journal of Parasitology 95:450–501.

Hoberg, E.P., L. Polley, E.J. Jenkins, S.J. Kutz, A. Veitch, and B. Elkin. 2008. Integrated approaches and empirical models for investigation of parasitic diseases in northern wildlife. Emerging Infectious Diseases 14:10–17.

Hobson, D.P., G. Proulx, and B.L. Dew. 1989. Initial post-release behavior of marten, *Martes americana*, introduced in Cypress Hills Provincial Park, Saskatchewan. Canadian Field-Naturalist 103:398–400.

Hodgman, T.P., D.J. Harrison, D.D. Katnik, and K.D. Elowe. 1994. Survival in an intensively trapped marten population in Maine. Journal of Wildlife Management 58:593–600.

Hodgman, T.P., D.J. Harrison, D.M. Phillips, and K.D. Elowe. 1997. Survival of American marten in an untrapped forest preserve in Maine. Pages 86–99 *in Martes*: taxonomy, ecology, techniques, and management. G. Proulx, H.N. Bryant, and P.M. Woodard, editors. Provincial Museum of Alberta, Edmonton, Canada.

Hodgson, R.G. 1937. Fisher farming. Fur Trade Journal of Canada, Toronto, Ontario, Canada.

Hoffman, J.A. 1983. Progress report, marten re-introduction to Terra Nova National Park. Environment Canada, Parks Canada, Natural Resources Conservation, Terra Nova National Park, Newfoundland, Canada.

Hoffmann, A.A., and C.M. Sgrò. 2011. Climate change and evolutionary adaptation. Nature 470:479–485.

Holland, G.P. 1985. The fleas of Canada, Alaska and Greenland (Siphonaptera). Memoirs of the Entomological Society of Canada 130:1–631.

Holmes, E.E., J.L. Sabo, and S.V. Viscido. 2007. A statistical approach to quasi-extinction forecasting. Ecology Letters 10:1182–1198.

Holmes, J.C. 1963. Helminth parasites of pine marten, *Martes americana,* from the District of Mackenzie. Canadian Journal of Zoology 41:333.

Holmes, T., and R.A. Powell. 1994. Morphology, ecology, and the evolution of sexual dimorphism in North American *Martes*. Pages 72–84 *in* Martens, sables, and fishers: biology and conservation. S.W. Buskirk, A.S. Harestad, M.G. Raphael, and R.A. Powell, editors. Cornell University Press, Ithaca, New York, USA.

Holsinger, K.E., and B.S. Weir. 2009. Genetics in geographically structured populations: defining, estimating, and interpreting F_{ST}. Nature Review Genetics 10:639–650.

Holthausen, R.S., and C.H. Sieg. 2007. Effectiveness of alternative management strategies in meeting conservation objectives. Pages 187–235 *in* Conservation of rare or little-known species: biological, social, and economic considerations. M.G. Raphael and R. Molina, editors. Island Press, Washington, D.C., USA.

Hooper, J., and R.V. Rea. 2009. The use of an orthotic casting foam as a track-plate medium for wildlife research and monitoring. Wildlife Biology 15:106–112.

Hoover, J.P., C.R. Root, and M.A. Zimmer. 1984. Clinical evaluation of American river otters in a reintroduction study. Journal of the American Veterinary Medical Association 185:1321–1326.

Hopkins, D.M. 1959. Cenozoic history of the Bering Land Bridge. Science 129:1519–1528.

Hosoda, T., H. Suzuki, M. Harada, K. Tsuchiya, S-H. Han, Y.P. Zhang, A.P. Kryukov, and L.K. Lin. 2000. Evolutionary trends of the mitochondrial lineage differentiation in species of genera *Martes* and *Mustela*. Genes & Genetic Systems 75:259–267.

Hosoda, T., H. Suzuki, M.A. Iwasa, M. Hayashida, S. Watanabe, M. Tatara, and K. Tsushiya. 1999. Genetic relationships within and between the Japanese marten *Martes melampus* and the sable *M. zibellina*, based on variation of mitochondrial DNA and nuclear ribosomal DNA. Mammal Study 24:25–33.

Hosoda, T., H. Suzuki, K. Tsuchiya, H. Lan, L. Shi, and A.P. Kryukov. 1997. Phylogenetic relationships within *Martes* based on nuclear ribosomal DNA and mitochondrial DNA. Pages 3–14 *in Martes*: taxonomy, ecology, techniques, and management. G. Proulx, H.N. Bryant, and P.M. Woodard, editors. Provincial Museum of Alberta, Edmonton, Canada.

Hoving, C.L., D.J. Harrison, W.B. Krohn, R.A. Joseph, and M. O'Brien. 2005. Broad-scale predictors of Canada lynx occurrence in eastern North America. Journal of Wildlife Management 69:739–751.

Hoving, C.L., R.A. Joseph, and W.B. Krohn. 2003. Recent and historical distributions of Canada lynx in Maine and the northeast. Northeastern Naturalist 10:363–382.

Hubbard, C.A. 1947. Fleas of western North America. Iowa State College Press, Ames, USA.

Hudson, P.J., A.P. Dobson, and K.D. Lafferty. 2006. Is a healthy ecosystem one that is rich in parasites? Trends in Ecology & Evolution 21:381–385.

Hudson, P.J., A. Rizzoli, B.T. Grenfell, H. Heesterbeek, and A.P. Dobson. 2004. The ecology of wildlife diseases. Oxford University Press, London, UK.

Hughes, S.S. 2009. Noble marten (*Martes americana nobilis*) revisited: its adaptation and extinction. Journal of Mammalogy 90:74–92.

Hunt, R.M., Jr. 1996. Biogeography of the Order Carnivora. Pages 485–541 *in* Carnivore behavior, ecology, and evolution. Volume 2. J.L. Gittleman, editor. Cornell University Press, Ithaca, New York, USA.

Hurt, A., B. Davenport, and E. Greene. 2000. Training dogs to distinguish between black bear (*Ursus americanus*) and grizzly bear (*Ursus arctos*) feces. University of Montana Under-Graduate Biology Journal. http://Ibscore.dbs.umt.edu/journal/Articles_all/2000/Hurt.htm (accessed 2 April 2010).

Idaho Department of Lands. 2006. Idaho Department of Lands GIS. Boise, Idaho, USA. http://gis1.idl.idaho.gov/index.htm (accessed November 2006).

Imbrie, J., and K.P. Imbrie. 1986. Ice ages: solving the mystery. Harvard University Press, Cambridge, Massachusetts, USA.

Ims, R.A. 1995. Movement patterns in relation to landscape structures. Pages 89–109 *in* Mosaic landscapes and ecological processes. L. Hansson., L. Fajrig, and G. Merriam, editors. Springer-Verlag, Berlin, Germany.

Ingram, R. 1973. Wolverine, fisher and marten in central Oregon. Oregon State Game Commission, Central Region Administration, Report 73-2.

Inman, R.M., R.R. Wigglesworth, K.H. Inman, M.K. Schwartz, B.L. Brock, and J.D. Rieck. 2004. Wolverine makes extensive movements in the Greater Yellowstone Ecosystem. Northwest Science 78:261–266.

Innis, H.A. 1956. The fur trade in Canada. University of Toronto Press, Ontario, Canada.

Inoue, T., T. Murakami, A.V. Abramov, and R. Masuda. 2010. Mitochondrial DNA control region variations in the sable *Martes zibellina* of Hokkaido Island and the Eurasian continent, compared with the Japanese marten *M. melampus*. Mammal Study 35:145–155.

[IPCC] Intergovernmental Panel on Climate Change. 2007a. Climate change 2007: impacts, adaptation and vulnerability. Contribution of Working Group II to the Fourth Assessment Report of the Intergovernmental Panel on Climate Change. Cambridge University Press, UK.

[IPCC] Intergovernmental Panel on Climate Change. 2007b. Climate change 2007: the physical science basis. Contribution of Working Group I to the Fourth Assessment Report of the Intergovernmental Panel on Climate Change. Cambridge University Press, UK.

Irvine, G.W., L.T. Magnus, and B.J. Bradle. 1964. The restocking of fisher in Lake State forests. Transactions of the North American Wildlife and Natural Resources Conference 29:307–315.

Irving, L., K. Schmidt-Nielsen, and N.S.B. Abrahamsen. 1957. On the melting points of animal fats in cold climates. Physiological Zoölogy 30:93–105.

Isaac, N.J., S.T. Turvey, B. Collen, C. Waterman, and J.E. Baillie. 2007. Mammals on the EDGE: conservation priorities based on threat and phylogeny. PLoS ONE 2:e296.

Ito, T.Y., N. Miura, B. Lhagvasuren, D. Enkhbileg, S. Takatsuki, A. Tsunekawa, and Z. Jiang. 2006. Satellite tracking of Mongolian gazelles (*Procapra gutturosa*) and habitat shifts in their seasonal ranges. Journal of Zoology 269:291–298.

[IUCN] International Union for Conservation of Nature. 1987. IUCN position statement on translocation of living organisms: introductions, reintroductions, and re-stocking. http://www.iucn.org/themes/ssc/pubs/policy/transe.

[IUCN] International Union for Conservation of Nature. 1994. Guidelines for protected area management categories. IUCN, Gland, Switzerland and Cambridge, UK.

[IUCN] International Union for Conservation of Nature. 1995. IUCN/SSC Guidelines for re-introductions. Forty-first meeting of the IUCN Council, Gland, Switzerland. http://www.iucn.org/themes/ssc/pubs/policy/reinte.

[IUCN] International Union for Conservation of Nature. 1998. Guidelines for re-introductions. International Union for Conservation of Nature, Gland, Switzerland.

Ivan, J.S. 2000. Effectiveness of covered track plates for detecting American marten. Thesis, University of Montana, Missoula, USA.

Iversen, J.A. 1972. Basal energy metabolism of mustelids. Journal of Comparative Physiology 81:341–344.

Iverson, S.J., K.J. Frost, and L.F. Lowry. 1997. Fatty acid signatures reveal fine scale structure of foraging distribution of harbor seals and their prey in Prince William Sound, Alaska. Marine Ecology Progress Series 151:255–271.

Jacob, G., R. Debrunner, F. Gugerli, B. Schmid, and K. Bollmann. 2010. Field surveys of capercaillie (*Tetrao urogallus*) in the Swiss Alps underestimated local abundance of the species as revealed by genetic analyses of non-invasive samples. Conservation Genetics 11:33–44.

Jacobson, G.L., I.J. Fernandez, P.A. Mayewski, and C.V. Schmitt, editors. 2009. Maine's climate future: an initial assessment. A report for the Governor of Maine. University of Maine, Orono, USA.

Janeway, C.A., P. Travers, and M. Walport. 2007. Immunobiology. Seventh edition. Garland Science, New York, USA.

Jansman, H., and S. Broekhuizen. 2000. Do the local Dutch pine marten populations form parts of a metapopulation? Potentials of molecular analysis. Lutra 43:101–107.

Jędrzejewski, W., A. Zalewski, and B. Jędrzejewska. 1993. Foraging by pine marten *Martes martes* in relation to food resources in Bialowieza National Park, Poland. Acta Theriologica 38:405–426.

Jennings, S.B., N.D. Brown, and D. Sheil. 1999. Assessing forest canopies and understory illumination: canopy closure, canopy cover and other measures. Forestry 72:59–73.

Jensen, A., and B. Jensen. 1970. Husmaaren (*Martes foina*) og maarjagten i Danmark. Danske Vildtundersþgelser, Hefte 15. Vildtbiologisk Station. (in Danish)

Jensen, O., and G. Schmidt. 2002. American marten monitoring project. Final report to Parks Canada, Riding Mountain National Park, Manitoba, Canada.

Jessup, D.A., M.A. Miller, C. Kreuder-Johnson, P.A. Conrad, M.T. Tinker, J. Estes, and J.A. Mazet. 2007. Sea otters in a dirty ocean. Journal of the American Veterinary Medical Association 231:1648–1652.

Jessup, D., M. Murray, D. Casper, D. Brownstein, and C. Kreuder-Johnson. 2009. Canine distemper vaccination is a safe and useful preventive procedure for southern sea otters (*Enhydra lutra nereis*). Journal of Zoo and Wildlife Medicine 40:705–710.

Jiang, Z., L. Xu, Y. Ma, Y. Wang, Y. Li, and S. Buskirk. 1998. The winter habitat selection of sables in Daxinganling Mountains. Acta Theriologica Sinica 18:112–119.

Johnson, C.A., J.M. Fryxell, I.D. Thompson, and J.A. Baker. 2009. Mortality risk increases with natal dispersal distance in American martens. Proceedings of the Royal Zoological Society of London B 276:3361–3367.

Johnson, C.J., S.E. Nielsen, E.H. Merrill, T.L. McDonald, and M.S. Boyce. 2006. Resource selection functions based on use-availability data: theoretical motivation and evaluation methods. Journal of Wildlife Management 70:347–357.

Johnson, D.H. 1980. The comparison of usage and availability measurements for evaluating resource preference. Ecology 61:65–71.

Johnson, K.N., F. Swanson, M. Herring, and S. Greene, editors. 1999. Bioregional assessments: science at the crossroads of management and policy. Island Press, Washington, D.C., USA.

Johnson, S.A. 1984. Home range, movements, and habitat use of fishers in Wisconsin. Thesis, University of Wisconsin, Stevens Point, USA.

Johnson, T. 2004. Shelter medicine for veterinarians and staff: the animal shelter building: design and maintenance of a healthy and efficient facility. Blackwell Publishing, Ames, Iowa, USA.

Jones, A., and M.J. Pybus. 2001. Taeniasis and echinococcosis. Pages 150–192 *in* Parasitic diseases of wild mammals. W.M. Samuel, M.J. Pybus, and A.A. Kocan, editors. Iowa State University Press, Ames, USA.

Jones, J.L. 1991. Habitat use of fisher in northcentral Idaho. Thesis, University of Idaho, Moscow, USA.

Jones, J.L., and E.O. Garton. 1994. Selection of successional stages by fishers in northcentral Idaho. Pages 377–387 *in* Martens, sables and fishers: biology and conservation. S.W. Buskirk, A.S. Harestad, M.G. Raphael, and R.A. Powell, editors. Cornell University Press, Ithaca, New York, USA.

Jones, J.W. 1913. Fur-farming in Canada. Committee on Fisheries, Game and Fur-bearing Animals, Canada Commission of Conservation, The Mortimer Co. Ltd., Ottawa, Ontario, Canada.

Jones, L.L.C., and M.G. Raphael. 1993. Inexpensive camera systems for detecting martens, fishers, and other animals: guidelines for use and standardization. USDA Forest Service, General Technical Report PNW-GTR-306.

Jones, L.L.C., L.F. Ruggiero, and J.K. Swingle. 1991. Ecology of marten in the Pacific Northwest: technique evaluation. Page 532 *in* Wildlife and vegetation of unmanaged Douglas-fir forests. L.F. Ruggiero, K.B. Aubry, A.B. Carey, and M.H. Huff, technical coordinators. USDA Forest Service, General Technical Report PNW-GTR-285.

Jones, T.C., R.D. Hunt, and N.W. King. 1997. Diseases due to protozoa. Pages 549–600 *in* Veterinary pathology. Sixth edition. C. Cann and S. Hunsberger, editors. Williams and Wilkins, Baltimore, Maryland, USA.

Jones, Y.L., S.D. Fitzgerald, J.G. Sikarske, A. Murphy, N. Grosjean, and M. Kiupel. 2006. Toxoplasmosis in a free-ranging mink. Journal of Wildlife Diseases 42:865–869.

Jordan, M.J. 2007. Fisher ecology in the Sierra National Forest, California. Dissertation, University of California, Berkeley, USA.

Jordan, M., J. Higley, S. Matthews, O. Rhodes, M. Schwartz, R. Barrett, and P. Palsboll. 2007. Development of 22 new microsatellite loci for fishers (*Martes pennanti*) with variability results from across their range. Molecular Ecology Notes 2007:1–5.

Kadosaki, M. 1981. Zoology, natural history and cultural history of the Nopporo Hill and its environs, Hokkaido, Japan. Research Bulletin of the Historical Museum of Hokkaido 6:25–38. (in Japanese)

Kaloyianni, M., and R.A. Freedland. 1990. Effect of diabetes and time after *in vivo* insulin administration on ketogenesis and gluconeogenesis in isolated rat hepatocytes. International Journal of Biochemistry 22:159–164.

Kamiya, H., and K. Ishigaki. 1972. Helminths of mustelidae in Hokkaido. Japanese Journal of Veterinary Research 20:117–128.

Karanth, K.P. 2003. Evolution of disjunct distributions among wet-zone species of the Indian subcontinent: testing various hypotheses using a phylogenetic approach. Current Science 85:1276–1283.

Karmanova, E.M. 1986. Dioctophymidea of animals and man and diseases caused by them. Fundamentals of nematodology, vol. 20. English translation. Amerind Publishing, New York, USA.

Karmona, J. 2007. Unduly constrained: implementing conservation areas under British Columbia's Forest and Range Practices Act. Thesis, University of British Columbia, Vancouver, Canada.

Karpenko, S.V., N.E. Dokuchaev, and E.P. Hoberg. 2007. Nearctic shrews, *Sorex* spp., as paratenic hosts of *Soboliphyme baturini* (Nematoda: Soboliphymidae). Comparative Parasitology 74:81–87.

Kauffman, J. 1996. Parasitic infections of domestic animals: a diagnostic manual. Birkhauser, Basel, Germany.

Kaya, T., D. Geraads, and V. Tuna. 2005. A new Late Miocene mammalian fauna in the Karaburun Peninsula (W Turkey). Neues Jahrbuch für Geologie und Palaeontologie, Abhandlungen 236:321–349.

Kays, R.W., B. Kranstauber, P. Jansen, C. Carbone, M. Rowcliffe, T. Fountain, and S. Tilak. 2009. Camera traps as sensor networks for monitoring animal communities. The 34th Annual IEEE Conference on Local Computer Networks, Zurich, Switzerland.

Kays, R.W., and K.M. Slauson. 2008. Remote cameras. Pages 110–140 *in* Noninvasive survey methods for carnivores. R.A. Long, P. MacKay, W.J. Zielinski, and J.C. Ray, editors. Island Press, Washington, D.C., USA.

Keating, K.A., W.G. Brewster, and C.H. Key. 1991. Satellite telemetry—performance of animal-tracking systems. Journal of Wildlife Management 55:160–171.

Keay, F.E. 1901. The animals our fathers found in New England. New England Magazine 24:535–545.

Kebbe, C.E. 1961a. Return of the fisher. Oregon State Game Commission Bulletin 16:3–7.

Kebbe, C.E. 1961b. Transplanting fisher. Western Association of State Game and Fish Commissioners 41:165–167.

Keith, D.A., H.R. Akçakaya, W. Thuiller, G.F. Midgely, R.G. Pearson, S.J. Phillips, H.M. Regan, M.B. Araújo, and T.G. Rebelo. 2008. Predicting extinction risks under climate change: coupling stochastic population models with dynamic bioclimatic habitat models. Biology Letters 4:560–563.

Kelly, D.W., R.A. Paterson, C.R. Townsend, R. Poulin, and D.M. Tompkins. 2009a. Parasite spillback: a neglected concept in invasion ecology? Ecology 90:2047–2056.

Kelly, G.M. 1977. Fisher (*Martes pennanti*) biology in the White Mountain National Forest and adjacent areas. Dissertation, University of Massachusetts, Amherst, USA.

Kelly, J.R. 2005. Recent distribution and population characteristics of American marten in New Hampshire and potential factors affecting their occurrence. Thesis, University of Massachusetts, Amherst, USA.

Kelly, J.R., T.K. Fuller, and J.J. Kanter. 2009b. Records of recovering American marten, *Martes americana*, in New Hampshire. Canadian Field-Naturalist 123:1–6.

Kelly, M.J. 2008. Design, evaluate, refine: camera trap studies for elusive species. Animal Conservation 11:182–184.

Kendall, K.C., and K.S. McKelvey. 2008. Hair collection. Pages 141–182 *in* Noninvasive survey methods for carnivores. R.A. Long, P. MacKay, W.J. Zielinski, and J.C. Ray, editors. Island Press, Washington, D.C., USA.

Kendall, K.C., J.B. Stetz, D.A. Roon, L.P. Waits, J.B. Boulanger, and D. Paetkau. 2008. Grizzly bear density in Glacier National Park, Montana. Journal of Wildlife Management 72:1693–1705.

Kenward, R.E. 2001. A manual for wildlife radio tagging. Academic Press, San Diego, California, USA.

Kenyon, A., B. Kenyon, and E. Hahn. 1978. Protides of the Mustelidae: immunoresponse of mustelids to Aleutian mink disease virus. American Journal of Veterinary Research 39:1011–1015.

Kerley, L.L., and G.P. Salkina. 2007. Using scent-matching dogs to identify individual Amur tigers from scat. Wildlife Society Bulletin 71:1341–1356.

Kerry, M., J.A. Royle, and H.Schmidt. 2005. Modeling avian abundance from replicated counts using binomial mixture models. Ecological Applications 15:1450–1461.

Kilham, L., G. Margolis, and E.D. Colby. 1967. Congenital infections of cats and ferrets by feline panleukopenia virus manifested by cerebellar hypoplasia. Laboratory Investigation 17:465–480.

Kimber, K.R., G.V. Kollias, and E.J. Dubovi. 2000. Serologic survey of selected viral agents in recently captured wild North American river otters (*Lontra canadensis*). Journal of Zoo and Wildlife Medicine 31:168–175.

King, A.W. 1997. Hierarchy theory: a guide to system structure for wildlife biologists. Pages 185–212 *in* Wildlife and landscape ecology: effects of pattern and scale. J.A. Bissonette, editor. Springer-Verlag, New York, USA.

King, C.M., and R.L. Edgar. 1977. Techniques for trapping and tracking stoats (*Mustela erminea*): a review, and a new system. New Zealand Journal of Zoology 4:193–212.

Kingscote, B.F. 1986. Leptospirosis in red foxes in Ontario. Journal of Wildlife Diseases 22:475–478.

Kirk, T.A., and W.J. Zielinski. 2009. Developing and testing a landscape habitat suitability model for the American marten (*Martes americana*) in the Cascades mountains of California. Landscape Ecology 24:759–773.

Kirsch, P.E., S.J. Iverson, and W.D. Bowen. 2000. Effect of a low-fat diet on body composition and blubber fatty acids of captive juvenile harp seals (*Phoca groenlandica*). Physiological and Biochemical Zoology 73:45–59.

Klopfleisch, R., P.U. Wolf, C. Wolf, T. Harder, E. Starick, M. Niebuhr, T.C. Mettenleiter, and J.P. Teifke. 2007. Encephalitis in a stone marten (*Martes foina*) after natural infection with highly pathogenic avian influenza virus subtype H5N1. Journal of Comparative Pathology 137:155–159.

Klug, R.R. 1997. Occurrence of the Pacific fisher (*Martes pennanti*) in the redwood zone of northern California and the habitat attributes associated with their detections. Thesis, Humboldt State University, Arcata, California, USA.

Knaus, B.J., R. Cronn, A. Liston, K. Pilgrim, and M.K. Schwartz. 2011. Mitochondrial genome sequences illuminate maternal lineages of conservation concern in a rare carnivore. BMC Ecology 11. doi:10.1186/1472-6785-11-10.

Knobel, D., A. Fooks, S. Brookes, D. Randall, S. Williams, K. Argaw, F. Shiferaw, L. Tallents, and M. Laurenson. 2007. Trapping and vaccination of endangered Ethiopian wolves to control an outbreak of rabies. Journal of Applied Ecology 45:109–116.

Kochanny, C.O., G.D. Delgiudice, and J. Fieberg. 2009. Comparing global positioning system and very high frequency telemetry home ranges of white-tailed deer. Journal of Wildlife Management 73:779–787.

Koehler, A.V.A. 2006. Systematics, phylogeography, distribution and lifecycle of *Soboliphyme baturini*. Thesis, University of New Mexico, Albuquerque, USA.

Koehler, A.V.A., E.P. Hoberg, N.E. Dokuchaev, and J.A. Cook. 2007. Geographic and host range of the nematode *Soboliphyme baturini* across Beringia. Journal of Parasitology 93:1070–1083.

Koehler, A.V.A., E.P. Hoberg, N.E. Dokuchaev, N.A. Trabenkova, J.S. Whitman, D.W. Nagorsen, and J.A. Cook. 2009. Phylogeography of a Holarctic nematode, *Soboliphyme baturini*, among mustelids: climate change episodic colonization, and diversification in a complex host-parasite system. Biological Journal of the Linnean Society 96:651–663.

Koehler, G.M., P.B. Hall, M.H. Norton, and D.J. Pierce. 2001. Implant- versus collar-transmitter use on black bears. Wildlife Society Bulletin 29:600–605.

Koehler, G.M., and M.G. Hornocker. 1977. Fire effects on marten habitat in the Selway-Bitterroot Wilderness. Journal of Wildlife Management 41:500–505.

Koehler, G.M., W.R. Moore, and A.R. Taylor. 1975. Preserving the pine marten. Management guidelines for western forests. Western Wildlands 2:31–36.

Koen, E. 2005. Home range, population density, habitat preference, and survival of fishers *(Martes pennanti)* in eastern Ontario. Thesis, Carleton Institute of Biology, Ottawa, Ontario, Canada.

Koen, E.L., J. Bowman, and C.S. Findlay. 2007. Fisher survival in eastern Ontario. Journal of Wildlife Management 71:1214–1219.

Koepfli, K.-P., K.A. Deere, G.J. Slater, C. Begg, K. Begg, L. Grassman, M. Lucherini, G. Veron, and R.K. Wayne. 2008. Multigene phylogeny of the Mustelidae: resolving relationships tempo and biogeographic history of a mammalian adaptive radiation. BMC Biology 6:10.

Koepfli, K.-P., and R.K. Wayne. 2003. Type ISTS markers are more informative than cytochrome *b* in phylogenetic reconstruction of the Mustelidae (Mammalia: Carnivora). Systematic Biology 52:571–593.

Kohn, B.E. 1991. Minnesota pine martens introduced to Wisconsin. Niche—Newsletter of the Wisconsin Bureau of Endangered Resources 5:4.

Kohn, B.E., and R.G. Eckstein. 1987. Status of marten in Wisconsin, 1985. Wisconsin Department of Natural Resources, Research Report 143, Madison, USA.

Kohn, M.H., and R.K. Wayne. 1997. Facts from feces revisited. Trends in Ecology & Evolution 12:223–227.

Kontrimavichus, V.L. 1985. Helminths of mustelids and trends in their evolution. English Translation. Amerind Publishing Company, New Delhi, India.

Kopylov, I.P. 1958. The sable. Pages 20–33 *in* Guidelines for the resettlement of fur-bearing animals. N.P. Lavrov, editor. Centrosoyuz Publishing House, Moscow, USSR. (in Russian)

Korhonen, K., P. Niemelä, and P. Siirilä. 2001. Temperament and reproductive performance in farmed sable. Agricultural and Food Science in Finland 10:91–98.

Koubek, P., V. Baru, and B. Koubkova. 2004. *Troglotrema acutum* (Digenea) from carnivores in the Czech Republic. Helminthologia 41:25–31.

Koufos, G.D. 2006. The Neogene mammal localities of Greece: faunas, chronology, and biostratigraphy. Hellenistic Journal of Geosciences 4:183–214.

Koufos, G.D. 2008. Carnivores from the early/middle Miocene locality of Antonios (Chalkidiki, Macedonia, Greece). Geobios 41:365–380.

Koufos, G.D., D.S. Kostopoulos, and T.D. Vlachou. 2005. Neogene/Quaternary mammalian migrations in eastern Mediterranean. Belgian Journal of Zoology 135:181–190.

Koussoroplis, A.-M., C. Lemarchand, A. Bec, C. Desvilettes, C. Amblard, C. Fournier, P. Berny, and G. Bourdier. 2008. From aquatic to terrestrial food webs: decrease of the docosahexaenoic acid/linoleic acid ratio. Lipids 43:461–466.

Kovach, A.I., and R.A. Powell. 2003. Reproductive success of male black bears. Canadian Journal of Zoology 81:1257–1268.

Kozlov, V. 2002. Russian fur industry struggles for survival. The Russia Journal, Issue 149. http://www.russiajournal.com/node/5901 (accessed 21 March 2011).

Kranz, A., A. Tikhonov, J. Conroy, P. Cavallini, J. Herrero, M. Stubbe, T. Maran, and A. Abramov. 2008. *Martes martes*. IUCN Red List of Threatened Species. Version 2010.4. http://www.iucnredlist.org (accessed 21 April 2011).

Kraus, T. 2005. Ovulation rates in harvested fisher, *Martes pennanti*, populations of central Ontario, with special reference to habitat conditions. Thesis, Laurentian University, Sudbury, Ontario, Canada.

Krebs, J.R. 1978. Optimal foraging: decision rules for predators. Pages 23–63 *in* Behavioural ecology: an evolutionary approach. J.R. Krebs and N.B. Davies, editors. Sinaeur Associates Sunderland, Massachusetts, USA.

Krebs, J., S. Williams, J. Smith, C. Rupprecht, and J. Childs. 2003. Rabies among infrequently reported mammalian carnivores in the United States, 1960–2000. Journal of Wildlife Diseases 39:253–261.

Kreuder, C., M. Miller, D. Jessup, L. Lowenstine, M. Harris, J. Ames, T. Carpenter, P. Conrad, and J. Mazet. 2003. Patterns of mortality in southern sea otters (*Enhydra lutris nereis*) from 1998–2001. Journal of Wildlife Diseases 39:495–509.

Krohn, W.B. 2005. A fall fur-hunt from Maine to New Brunswick, Canada: the 1858 journal of Manly Hardy. Northeastern Naturalist 12:509–540.

Krohn, W.B., S.M. Arthur, and T.F. Paragi. 1994. Mortality and vulnerability of a heavily trapped fisher population. Pages 137–145 *in* Martens, sables, and fishers: biology and conservation. S.W. Buskirk, A.S. Harestad, M.G. Raphael, and R.A. Powell, editors. Cornell University Press, Ithaca, New York, USA.

Krohn, W.B., K.D. Elowe, and R.B. Boone. 1995. Relations among fishers, snow, and martens: development and evaluation of two hypotheses. Forestry Chronicle 71:97–105.

Krohn, W.B., and C.L. Hoving. 2010. Early Maine wildlife: historical accounts of Canada lynx, moose, mountain lion, white-tailed deer, wolverine, wolves, and woodland caribou, 1603–1930. University of Maine Press, Orono, USA.

Krohn, W.B., C. Hoving, D. Harrison, D. Philips, and H. Frost. 2004. *Martes* foot-loading and snowfall patterns in eastern North America: implications to broad-scale distributions and interactions of mesocarnivores. Page 115–131 *in* Martens and fishers (*Martes*) in human-altered landscapes: an international perspective. D.J. Harrison, A.K. Fuller, and G. Proulx, editors. Springer Science+Business Media, New York, USA.

Krohn, W.B., W.J. Zielinski, and R.B. Boone. 1997. Relations among fishers, snow, and martens in California: results from small-scale spatial comparisons. Pages 211–232 *in* *Martes*: taxonomy, ecology, techniques, and management. G. Proulx, H.N. Bryant, and P.M. Woodard, editors. Provincial Museum of Alberta, Edmonton, Canada.

Kruger, H.H. 1990. Home ranges and patterns of distribution of stone and pine martens. Transactions of the XIXth Congress of the International Union of Game Biologists, Trondheim, Norway.

Kucera, T.E., A.M. Soukkala, and W.J. Zielinski. 1995. Photographic bait stations. Pages 25–65 *in* American marten, fisher, lynx, and wolverine: survey methods for their detection. W.J. Zielinski and T.E. Kucera, technical editors. USDA Forest Service, General Technical Report PSW-GTR-157.

Kuehn, D.W., and W.E. Berg. 1981. Use of radiographs to identify age-classes of fisher. Journal of Wildlife Management 45:1009–1010.

Kukekova, A.V., L.N. Trut, I.N. Oskina, A.V. Kharlamova, S.G. Shikhevich, E.F. Kirkness, G.D. Aguirre, and G.M. Acland. 2004. A marker set for construction of a genetic map of the silver fox (*Vulpes vulpes*). Journal of Heredity 95:185–194.

Kurki, S., A. Nikula, P. Helle, and H. Lindén. 1998. Abundances of red fox and pine marten in relation to the composition of boreal forest landscapes. Journal of Animal Ecology 67:874–886.

Kurose, N., R. Masuda, B. Siriaroonrat, and M.C. Yoshido. 1999. Intraspecific variation of mitochondrial cytochrome *b* gene sequences of the Japanese marten *Martes melampus* and the sable *Martes zibellina* (Mustelidae, Carnivora, Mammalia) in Japan. Zoological Science 16:693–700.

Kurose, N., R. Masuda, and M. Tatara. 2005. Fecal DNA analysis for identifying species and sex of sympatric carnivores: a noninvasive method for conservation on the Tsushima islands, Japan. Journal of Heredity 96:688–697.
Kurtén, B. 1968. Pleistocene mammals of Europe. Aldine Publishing, New York, USA.
Kyle, C.J., C.S. Davis, and C. Strobeck. 2000. Microsatellite analysis of North American pine marten (*Martes americana*) populations from the Yukon and Northwest Territories. Canadian Journal of Zoology 78:1150–1157.
Kyle, C.J., A. Davison, and C. Strobeck. 2003. Genetic structure of European pine martens (*Martes martes*), and evidence for introgression with *M. americana* in England. Conservation Genetics 4:179–188.
Kyle, C.J., J.-F. Robitaille, and C. Strobeck. 2001. Genetic variation and structure of fisher (*Martes pennanti*) populations across North America. Molecular Ecology 10:2341–2347.
Kyle, C.J., and C. Strobeck. 2003. Genetic homogeneity of Canadian mainland marten populations underscores the distinctiveness of Newfoundland pine martens (*Martes americana atrata*). Canadian Journal of Zoology 81:57–66.
Labrid, M. 1986. La martre (*Martes martes* Linnaeus, 1758). *In* Encyclopédie des carnivores de France. Volume 9. Société Française pour l'Etude et la Protection des Mammifères, Paris, France.
Labrid, M. 1987. La martre *Martes martes* et la fouine *Martes foina*: utilisation de l'espace et du temps et régime alimentaire de deux Mustélidés sympatriques en milieu forestier. Dissertation, University of Paris, France.
Lachat-Feller, N. 1993a. Eco-éthologie de la fouine (*Martes foina* Erxleben, 1777) dans le Jura Suisse. Dissertation, Université de Neuchâtel, Switzerland.
Lachat-Feller, N. 1993b. Régime alimentaire de la fouine (*Martes foina*) durant un cycle de pullulation du campagnol terrestre (*Arvicola terrestris scherman*) dans le Jura Suisse. Mammalian Biology 58:275–280.
Lacy, R.C., M. Borbat, and J.P. Pollak. 2005. VORTEX: a stochastic simulation of the extinction process. Version 9.50. Chicago Zoological Society, Brookfield, Illinois, USA.
Lacy, R.C., and T.W. Clark. 1993. Simulation modeling of American marten (*Martes americana*) populations: vulnerability to extinction. Great Basin Naturalist 53:282–292.
Lafferty, K.D. 2008. Effects of disease on community interactions and food web structure. Pages 205–222 *in* Infectious disease ecology: effects of ecosystems on disease and of diseases on ecosystems. R.S. Ostfeld, F. Keesing, and V.T. Eviner, editors. Princeton University Press, New Jersey, USA.
Lafferty, K.D. 2009. The ecology of climate change and infectious diseases. Ecology 90:888–900.
Lafond, R. 1990. Analyse du système de suivi des animaux à fourrure. Québec Ministère du Loisir, de la Chasse et de la Pêche, Direction de la Gestion des Espèces et des Habitats, Canada.
Laliberte, A.S., and W.J. Ripple. 2004. Range contraction of North American carnivores and ungulates. Bioscience 54:123–138.
Lamberson, R.H., B.R. Noon, C. Voss, and K.S. McKelvey. 1994. Reserve design for territorial species: the effects of patch size and spacing on the viability of the northern spotted owl. Conservation Biology 8:185–195.
Landram, M. 2002. Canopy cover study—implications for Sierra Nevada Forest Plan Amendment implementation. Unpublished report. USDA Forest Service, Pacific Southwest Region, Vallejo, California.
Lane, R.S. 1984. New host records of ticks (Acari: Argasidae and Ixodidae) parasitizing wildlife in California and a case of tick paralysis in a deer. California Fish and Game 70:11–17.
Langlois, I. 2005. Viral diseases of ferrets. Veterinary Clinics of North America: Exotic Animal Practice 8:139–160.
Latimer, A.M., S. Wu, A.E. Gelfand, and J.A. Silander, Jr. 2006. Building statistical models to analyze species distributions. Ecological Applications 16:33–50.

Latour, P.B., N. Maclean, and K.G. Poole. 1994. Movements of martens, *Martes americana*, in burned and unburned taiga in the Mackenzie Valley, Northwest Territories. Canadian Field-Naturalist 108:351–354.

Lawler, J.J., S.L. Shafer, D. White, P. Karieva, E.P. Maurer, A.R. Blaustein, and P.J. Bartlein. 2009. Projected climate-induced faunal change in the Western Hemisphere. Ecology 90:588–597.

Lawler, J.J., D. White, R.P. Neilson, and A.R. Blaustein. 2006. Predicting climate-induced range shifts: model differences and model reliability. Global Change Biology 12:1568–1584.

Lawton, J.H. 1993. Range, population abundance and conservation. Trends in Ecology & Evolution 8:409–413.

Laymon, S.A., and R.H. Barrett. 1986. Developing and testing habitat-capability models: pitfalls and recommendations. Pages 87–91 *in* Wildlife 2000: modeling habitat relationships of terrestrial vertebrates. J. Verner, M.L. Morrison, and C.J. Ralph, editors. University of Wisconsin Press, Madison, USA.

Lefkovitch, L.P. 1965. The study of population growth in organisms grouped by stages. Biometrika 35:183–212.

Leiby, P.D., and W.G. Dyer. 1971. Cyclophyllidean tapeworms of wild Carnivora. Pages 174–234 *in* Parasitic diseases of wild mammals. J.W. Davis and R.C. Anderson, editors. Iowa State University Press, Ames, USA.

Le Maho, Y., H. Vu Van Kha, H. Koubi, G. Dewasmes, J. Girard, P. Ferré, and M. Cagnard. 1981. Body composition, energy expenditure, and plasma metabolites in long-term fasting geese. American Journal of Physiology 241:E342–E354.

Lenihan, J.M., D. Bachelet, R.P. Neilson, and R. Drapek. 2008. Response of vegetation distribution, ecosystem productivity, and fire to climate change scenarios for California. Climatic Change 87:S215–S230.

Lenihan, J.M., R. Drapek, D. Blachelet, and R.P. Neilson. 2003a. Climate change effects on vegetation distribution, carbon, and fire in California. Ecological Applications 13:1667–1681.

Lenihan, J.M., R. Drapek, R.P. Neilson, and D. Bachelet. 2003b. Climate change effects on vegetation distribution, carbon stocks, and fire regimes in California. Appendix IV *in* Global climate change and California: potential implications for ecosystems, health, and the economy. Electric Power Research Institute, Palo Alto, California, USA.

Lenormand, T. 2002. Gene flow and the limits to natural selection. Trends in Ecology & Evolution 17:183–189.

Leopold, A. 1933. Game management. Charles Scribner's Sons, New York, USA.

Lessa, E.P., J.A. Cook, and J.L. Patton. 2003. Genetic footprints of demographic expansion in North America, but not Amazonia, following the Late Pleistocene. Proceedings of the National Academy of Sciences, USA 100:10331–10334.

Lessa, E.P., G. D'Elia, and U.F.J. Pardiñas. 2010. Genetic footprints of late Quaternary climate change in the diversity of Patagonian-Fuegian rodents. Molecular Ecology 19:3013–3037.

Levin, S.A. 1992. The problem of pattern and scale in ecology. Ecology 73:1943–1967.

Lewis, J.C. 2006. Implementation plan for reintroducing fishers to Olympic National Park. Washington Department of Fish and Wildlife, Olympia, USA. http://wdfw.wa.gov/wlm/diversty/soc/fisher/ (accessed June 2011).

Lewis, J.C., and P.J. Happe. 2008. Olympic fisher reintroduction project: 2008 progress report. Washington Department of Fish and Wildlife, Olympia, USA. http://wdfw.wa.gov/publications/pub.php?id=00226 (accessed June 2011).

Lewis, J.C., P.J. Happe, K.J. Jenkins, and D.J. Manson. 2010. Olympic fisher reintroduction project: 2009 progress report. Washington Department of Fish and Wildlife, Olympia, USA. http://wdfw.wa.gov/publications/pub.php?id=00812 (accessed June 2011).

Lewis, J.C., P.J. Happe, K.J. Jenkins, and D.J. Manson. 2011. Olympic fisher reintroduction project: 2010 progress report. Washington Department of Fish and Wildlife, Olympia, USA. http://wdfw.wa.gov/publications/pub.php?id=01186 (accessed June 2011).

Lewis, J.C., and G.E. Hayes. 2004. Feasibility assessment for reintroducing fishers to Washington. Washington Department of Fish and Wildlife, Olympia, USA. http://wdfw.wa.gov/publications/pub.php?id=00231 (accessed June 2011).

Lewis, J.C., R.A. Powell, and W.J. Zielinski. 2012. Carnivore translocations and conservation: insights from population models and field data for fishers (*Martes pennanti*). PLoS ONE 7:e32726.

Lewis, J.C., and W.J. Zielinski. 1996. Historical harvest and incidental capture of fishers in California. Northwest Science 70:291–297.

Lewis, J.S., J.L. Rachlow, E.O. Garton, and L.A. Vierling. 2007. Effects of habitat on GPS collar performance: using data screening to reduce location error. Journal of Applied Ecology 44:663–671.

Lewis, R.E., J.H. Lewis, and C. Maser. 1988. The fleas of the Pacific Northwest. Oregon State University Press, Corvallis, USA.

Libois, R., and A. Waechter. 1991. La fouine (*Martes foina*). *In* Encyclopédie des carnivores de France. Volume 10. Societé Française pour l'Étude et Protection des Mammifères, Paris, France.

Liebenberg, L. 1990. The art of tracking: the origin of science. David Philip Publishers, Cape Town, South Africa.

Lima, S.L., and L.M. Dill. 1990. Behavioural decisions made under the risk of predation: a review and prospectus. Canadian Journal of Zoology 68:619–640.

Lindenfors, P., L. Dálen, and A. Angerbjörn. 2003. The monophyletic origin of delayed implantation in carnivores and its implications. Evolution 57:1952–1956.

Lindquist, E.E., K.W. Wu, and J.H. Redner. 1999. A new species of the tick genus *Ixodes* (Acari: Ixodidae) parasitic on mustelids (Mammalia: Carnivora) in Canada. Canadian Entomologist 131:151–170.

Lindström, E.R. 1989. The role of medium-sized carnivores in the Nordic boreal forest. Finnish Game Research 46:53–63.

Lindström, E.R., S.M. Brainerd, J.O. Helldin, and K. Overskaug. 1995. Pine marten-red fox interactions: a case of intraguild predation? Annales Zoologici Fennici 32:123–130.

Linhart, S.B. 1985. Furbearer management and the steel foothold trap. Pages 52–63 *in* Proceedings of the Great Plains Wildlife Damage Control Workshop. D.B. Fagre, editor. San Antonio, Texas, USA.

Lipsey, M.K., and M.F. Child. 2007. Combining the fields of reintroduction biology and restoration ecology. Conservation Biology 21:1387–1388.

Lister, A. 2004. The impact of Quaternary ice ages on mammalian evolution. Philosophical Transactions of the Royal Society B: Biological Sciences 359:221–241.

Liu, Y.-M., J.-M. Lacorte, N. Viguerie, C. Poitou, V. Pelloux, B. Guy-Grand, C. Coussieu, D. Langin, A. Basdevant, and K. Clément. 2003. Adiponectin gene expression in subcutaneous adipose tissue of obese women in response to short-term very low calorie diet and refeeding. Journal of Clinical Endocrinology and Metabolism 88:5881–5886.

Livia, L., V. Francesca, P. Antonella, P. Fausto, and R. Bernardino. 2006. A PCR-RFLP method on faecal samples to distinguish *Martes martes*, *Martes foina*, *Mustela putorius* and *Vulpes vulpes*. Conservation Genetics 8:757–759.

Livingston, T.R., P.S. Gipson, W.B. Ballard, D.M. Sanchez, and P.R. Krausman. 2005. Scat removal: a source of bias in feces-related studies. Wildlife Society Bulletin 33:172–178.

Lledo, L., C. Gimenez-Pardo, G. Domınguez-Penafiel, R. Sousa, M.I. Gegundez, N. Casado, and A. Criado. 2010. Molecular detection of hemoprotozoa and rickettsia species in arthropods collected from wild animals in the Burgos Province, Spain. Vector-borne and Zoonotic Diseases 10:735–738.

Lodé, T. 1991a. Conspecific recognition and mating in stone marten *Martes foina*. Acta Theriologica 36:275–283.

Lodé, T. 1991b. Exploitation des milieux et organisation de l'espace chez deux mustélides européens: la fouine et le putois. Vie et Milieu 41:29–38.

Lofroth, E.C., J.M. Higley, R.H. Naney, C.M. Raley, J.S. Yaeger, S.A. Livingston, and R.L. Truex. 2011. Conservation of fishers (*Martes pennanti*) in south-central British Columbia, western Washington, western Oregon, and California—volume II: key findings from fisher habitat studies in British Columbia, Montana, Idaho, Oregon, and California. USDI Bureau of Land Management, Denver, Colorado, USA.

Lofroth, E.C., C.M. Raley, J.M. Higley, R.L. Truex, J.S. Yaeger, J.C. Lewis, P.J. Happe, L.L. Finley, R.H. Naney, L.J. Hale, A.L. Krause, S.A. Livingston, A.M. Myers, and R.N. Brown. 2010. Conservation of fishers (*Martes pennanti*) in south-central British Columbia, western Washington, western Oregon, and California—volume I: conservation assessment. USDI Bureau of Land Management, Denver, Colorado, USA.

Long, C. 2008. The wild mammals of Wisconsin. Pensoft Series Faunistica 68, Sofia-Moscow, Russia.

Long, R.A., T.M. Donovan, P. MacKay, W.J. Zielinski, and J.S. Buzas. 2007a. Comparing scat detection dogs, cameras, and hair snares for surveying carnivores. Journal of Wildlife Management 71:2018–2025.

Long, R.A., T.M. Donovan, P. MacKay, W.J. Zielinski, and J.S. Buzas. 2007b. Effectiveness of scat detection dogs for detecting forest carnivores. Journal of Wildlife Management 71:2007–2017.

Long, R.A., T.M. Donovan, P. MacKay, W.J. Zielinski, and J.S. Buzas. 2011. Predicting carnivore occurrence with noninvasive surveys and occupancy modeling. Landscape Ecology 26:327–340.

Long, R.A., P. MacKay, W.J. Zielinski, and J.C. Ray, editors. 2008. Noninvasive survey methods for carnivores. Island Press, Washington, D.C., USA.

Long, R.A., and W.J. Zielinski. 2008. Designing effective noninvasive carnivore surveys. Pages 8–44 *in* Noninvasive survey methods for carnivores. R.A. Long, P. MacKay, W.J. Zielinski, and J.C. Ray, editors. Island Press, Washington, D.C., USA.

Longpré, M.H., Y. Bergeron, D Paré, and M. Béland. 1994. Effect of companion species on the growth of jack pine. Canadian Journal of Forest Research 24:1846–1853.

Loos-Frank, B. 2000. An update of Verster's (1969) "Taxonomic revision of the genus *Taenia* Linnaeus" (Cestoda) in table format. Systematic Parasitology 45:155–183.

Loos-Frank, B., and E. Zeyhle. 1982. The intestinal helminths of the red fox and some other carnivores in southwest Germany. Zeitschrift für Parasitenkunde 67:99–113.

Lubelczyk, C.B., T. Hanson, E.H. Lacombe, M.S. Holman, and J.E. Keirans. 2007. First U.S. record of the hard tick *Ixodes* (*Pholeoixodes*) *gregsoni* Lindquist, Wu, and Redner. Journal of Parasitology 93:718–719.

Lucchesi, P.M.A., G.H. Arroyo, A.I. Etcheverría, A.E. Parma, and A.C. Seijo. 2004. Recommendations for the detection of *Leptospira* in urine by PCR. Revista da Sociedade Brasileira de Medicina Tropical 37:131–134.

Luikart, G., N. Ryman, D.A. Tallmon, M.K. Schwartz, and F.W. Allendorf. 2010. Estimation of census and effective population sizes: the increasing usefulness of DNA-based approaches. Conservation Genetics 11:355–373.

Lunn, D.J., A. Thomas, N. Best, and D. Spiegelhalter. 2000. WinBUGS—a Bayesian modeling framework: concepts, structure, and extensibility. Statistics and Computing 10:325–337.

Luque, M. 1984. The fisher: Idaho's forgotten furbearer. Idaho Wildlife 4:12–15.

Lutz, H.J. 1956. Ecological effects of forest fires in the interior of Alaska. USDA Technical Bulletin 1133, Washington, D.C., USA.

Lyman, R.L. 2011. Paleoecological and biogeographical implications of late Pleistocene noble marten (*Martes americana nobilis*) in eastern Washington state, USA. Quaternary Research 75:176–182.

Lynch, A.B. 2006. An investigation into the ecology of the pine marten (*Martes martes*) in Killarney National Park. Dissertation, Trinity College, Dublin, Ireland.

Lynch, A.B., and M.J.F. Brown. 2006. Molecular sexing of pine marten (*Martes martes*): how many replicates? Molecular Ecology Notes 6:631–633.

Lynch, A.B., M.J.F. Brown, and J.M. Rochford. 2006. Fur snagging as a method of evaluating the presence and abundance of a small carnivore, the pine marten (*Martes martes*). Journal of Zoology (London) 270:330–339.

Lynch, M. 1996. A quantitative-genetic perspective on conservation issues. Pages 471–501 *in* Conservation genetics: case histories from nature. J.C. Avise and J.L. Hamrick, editors. Chapman and Hall, New York, USA.

Ma, J., L. Xu, H. Zhang, and X. Bao. 1999. Activity patterns of sables (*Martes zibellina*) in Daxinganling Mountains, China. Acta Theriologica Sinica 19:95–100.

Ma, Y., and L. Xu. 1994. Distribution and conservation of sables in China. Pages 255–261 *in* Martens, sables, and fishers: biology and conservation. S.W. Buskirk, A.S. Harestad, M.G. Raphael, and R.A. Powell, editors. Cornell University Press, Ithaca, New York, USA.

Mabe, M., and M. Amin. 2001. Growth dynamics of scholarly and scientific journals. Scientometrics 51:147–162.

MacDonald, S.O., and J.A. Cook. 1996. The land mammal fauna of southeast Alaska. Canadian Field-Naturalist 110:571–598.

MacDonald, S.O., and J.A. Cook. 2007. Mammals and amphibians of Southeast Alaska. Special Publication No. 8, Museum of Southwestern Biology, University of New Mexico, Albuquerque, USA.

MacKay, P., D.A. Smith, R.A. Long, and M. Parker. 2008a. Scat detection dogs. Pages 183–222 *in* Noninvasive survey methods for carnivores. R.A. Long, P. MacKay, W.J. Zielinski, and J.C. Ray, editors. Island Press, Washington, D.C., USA.

MacKay, P., W.J. Zielinski, R.A. Long, and J.C. Ray. 2008b. Noninvasive research and carnivore conservation. Pages 1–7 *in* Noninvasive survey methods for carnivores. R.A. Long, P. MacKay, W.J. Zielinski, and J.C. Ray, editors. Island Press, Washington, D.C., USA.

MacKenzie, D.I., and L.R. Bailey. 2004. Assessing the fit of site-occupancy models. Journal of Agricultural, Biological & Environmental Statistics 9:300–318.

MacKenzie, D.I., L.L. Bailey, and J.D. Nichols. 2004. Investigating species co-occurrence patterns when species are detected imperfectly. Journal of Animal Ecology 73:546–555.

MacKenzie, D.I., and J.E. Hines. 2006. Program Presence, version 2.0. http://www.mbr-pwrc.usgs.gov/software/doc/presence/presence.html (accessed 20 May 2011).

MacKenzie, D.I., and J.D. Nichols. 2004. Occupancy as a surrogate for abundance estimation. Animal Biodiversity and Conservation 27:461–467.

MacKenzie, D.I., J.D. Nichols, J.E. Hines, M.G. Knutson, and A.B. Franklin. 2003. Estimating site occupancy, colonization, and extinction when a species is detected imperfectly. Ecology 84:2200–2207.

MacKenzie, D.I., J.D. Nichols, G.B. Lachman, S. Droege, J.A. Royle, and C.A. Langtimm. 2002. Estimating site occupancy rates when detection probabilities are less than one. Ecology 83:2248–2255.

MacKenzie, D.I., J.D. Nichols, J.A. Royle, K.H. Pollock, L.L. Bailey, and J.E. Hines. 2006. Occupancy estimation and modeling: inferring patterns and dynamics of species occurrence. Academic Press, Burlington, Massachusetts, USA.

MacKenzie, D.I., J.D. Nichols, M.E. Seamans, and R.J. Gutiérrez. 2009. Modeling species occurrence dynamics with multiple states and imperfect detection. Ecology 90:823–835.

MacKenzie, D.I., and J.A. Royle. 2005. Designing occupancy studies: general advice and allocating of survey effort. Journal of Applied Ecology 42:1105–1114.

MacNulty, D.R., G.E. Plumb, and D.W. Smith. 2008. Validation of a new video and telemetry system for remotely monitoring wildlife. Journal of Wildlife Management 72:1834–1844.

Maffei, M., J. Halaas, E. Ravussin, R.E. Pratley, G.H. Lee, Y. Zhang, H. Fei, S. Kim, R. Lallone, S. Ranganathan, P.A. Kern, and J.M. Friedman. 1995. Leptin levels in human and rodent: measurement of plasma leptin and ob RNA in obese and weight-reduced subjects. Nature Medicine 1:1155–1161.

Magoun, A.J., C.D. Long, M.K. Schwartz, K.L. Pilgrim, R.E. Lowell, and P. Valkenburg. 2011a. Integrating motion-detection cameras and hair snags for wolverine identification. Journal of Wildlife Management 75:731–739.

Magoun, A.J., P. Valkenburg, D.N. Peterson, C.D. Long, and R.E. Lowell. 2011b. Wolverine images—using motion-detection cameras for photographing, identifying, and monitoring wolverines. Blurb http://www.blurb.com/bookstore/detail/1914572 (accessed 23 May 2011).

Main, W.A., D.M. Paananen, and R.E. Burgan. 1990. Fire Family Plus. USDA Forest Service General Technical Report, GTR-NC-138.

Malyarchuk, B.A., A.V. Petrovskaya, and M.V. Derenko. 2010. Intraspecific structure of sable *Martes zibellina* L. inferred from nucleotide variation of the mitochondrial DNA cytochrome *b* gene. Russian Journal of Genetics 46:64–68.

Manel, S., M.K. Schwartz, G. Luikart, and P. Taberlet. 2003. Landscape genetics: combining landscape ecology and population genetics. Trends in Ecology & Evolution 18:189–197.

Manel, S., O.E. Gaggiotii, and R.S. Waples. 2005. Assignment methods: matching biological questions wth appropriate techniques. Trends in Ecology & Evolution 20:136–142.

Manfredo, M.J., T.L. Teel, and A.D. Bright. 2003. Why are public values toward wildlife changing? Human Dimensions of Wildlife 8:287–306.

Mangas, J.G., M. Carboles, L.H. Alcázar, D. Bellón, and E. Virgós. 2007. Aproximación al estudio de la ecología espacial de especies simpátricas: la garduña (*Martes foina*) y la gineta (*Genetta genetta*). Galemys 19:61–71.

Manion, P.D. 1991. Tree disease concepts. Second edition. Prentice Hall Career and Technology, New Jersey, USA.

Manley, P.N., B. Van Horne, J.K. Roth, W.J. Zielinski, M.M. McKenzie, T.J. Weller, F.W. Weckerly, and C. Hargis. 2006. Multiple species inventory and monitoring technical guide. USDA Forest Service, General Technical Report WO-GTR-73.

Manly, B.F.J., L.L. McDonald, D.L. Thomas, T.L. McDonald, and W.P. Erickson. 2002. Resource selection by animals: statistical design and analysis for field studies. Second edition. Kluwer Academic, Dordrecht, The Netherlands.

Mann, M.E. 2002. Little Ice Age. Pages 504–509 *in* Encyclopedia of global environmental change. M.C. MacCracken and J. S. Perry, editors. John Wiley & Sons, Chichester, UK.

Marchesi, P. 1989. Ecologie et comportement de la martre (*Martes martes* L.) dans le Jura Suisse. Dissertation, Universite Neuchatel, Switzerland. (in French)

Marchesi, P., N. Lachat, R. Lienhard, Ph. Debiève, and C. Mermod. 1989. Comparaison des régimes alimentaires de la fouine (*Martes foina* Erxl.) et de la martre (*Martes martes* L.) dans une région du Jura Suisse. Revue Suisse de Zoologie 96:281–296.

Marchesi, P., and C. Mermod. 1989. Régime alimentaire de la martre (*Martes martes* L.) dans le Jura Suisse (Mammalia: Mustelidae). Revue Suisse de Zoologie 96:127–146.

Marcogliese, D.J. 2005. Parasites of the superorganism: are they indicators of ecosystem health? International Journal of Parasitology 35:705–716.

Marcot, B.G. 1997. Species-environment relations (SER) database. Interior Columbia Basin Ecosystem Management Project, USDA Forest Service and USDI Bureau of Land Management. Database available at http://www.icbemp.gov/spatial/metadata/databases/dbase.html.

Marcot, B.G., and K.B. Aubry. 2003. The functional diversity of mammals in coniferous forests of western North America. Pages 631–664 *in* Mammal community dynamics: management and conservation in the coniferous forests of western North America. C.J. Zabel and R.G. Anthony, editors. Cambridge University Press, UK.

Marcot, B.G., S.S. Ganzei, T. Zhang, and B.A. Voronov. 1997. A sustainable plan for conserving forest biodiversity in Far East Russia and northeast China. Forestry Chronicle 73:565–571.

Marcot, B.G., W.E. McConnaha, P.H. Whitney, T.A. O'Neil, P.J. Paquet, L.E. Mobrand, G.R. Blair, L.C. Lestelle, K.M. Malone, and K.I. Jenkins. 2002. A multi-species framework approach for the Columbia River Basin: integrating fish, wildlife, and ecological functions. Available on CD-ROM and at http://www.nwcouncil.org/edt/framework/. Northwest Power Planning Council, Portland, Oregon, USA.

Marincovich, L., Jr., and A.Y. Gladenkov. 1999. Evidence for an early opening of the Bering Strait. Nature 397:149–151.

Markham, A.C., and A.J. Wilkinson. 2008. EcoLocate: a heterogeneous wireless network system for wildlife tracking. Pages 293–298 *in* Novel algorithms and techniques in telecommunications, automation and industrial electronics. T. Sobh, K. Elleithy, A. Mahmood, and M.A. Kari, editors. Springer Science+Business Media, Dordrecht, The Netherlands.

Marmi, J., J.F. Lopez-Giraldez, and X. Domingo-Roura. 2004. Phylogeny, evolutionary history and taxonomy of the Mustelidae based on sequences of the cytochrome *b* gene and a complex repetitive flanking region. Zoological Scripta 33:481–499.

Martinek, K., L. Kolarova, and J. Cerveny. 2001. *Echinococcus multilocularis* in carnivores from the Klatovy district of the Czech Republic. Journal of Helminthology 75:61–66.

Marucco, F., L. Boitani, D.H. Pletscher, and M.K. Schwartz. 2011. Bridging the gaps between non-invasive genetic sampling and population parameter estimation. European Journal of Wildlife Research 57:1–13.

Mascarenhas, A., P. Coelho, E. Subtil, and T.B. Ramos. 2010. The role of common local indicators in regional sustainability assessment. Ecological Indicators 10:646–656.

Masseti, M. 1995. Quaternary biogeography of the Mustelidae family on the Mediterranean islands. Hystrix 7:17–34.

Masuda, R. 2009. *Martes melampus* (Wagner, 1840). Page 250–251 *in* The wild mammals of Japan. S.D. Ohdachi, Y. Ishibashi, M.A. Iwasa, and T. Saitoh, editors. Shoukadoh, Kyoto, Japan.

Mathews, F., D. Moro, R. Strachan, M. Gelling, and N. Buller. 2006. Health surveillance in wildlife reintroductions. Biological Conservation 131:338–347.

Matos, H.M., M.J. Santos, F. Palomares, and M. Santos-Reis. 2009. Does riparian habitat condition influence mammalian carnivore abundance in Mediterranean ecosystems? Biodiversity and Conservation 18:373–386.

Matos, H., and M. Santos-Reis. 2006. Distribution and status of the pine marten *Martes martes* in Portugal. Pages 47–61 *in Martes* in carnivore communities. M. Santos-Reis, J.D.S. Birks, E.C. O'Doherty, and G. Proulx, editors. Alpha Wildlife Publications, Sherwood Park, Alberta, Canada.

Matsubara, M., S. Maruoka, and S. Katayose. 2002. Inverse relationship between plasma adiponectin and leptin concentrations in normal-weight and obese women. European Journal of Endocrinology 147:173–180.

Matthew, W.D. 1924. Third contribution to the Snake Creek fauna. Bulletin of the American Museum of Natural History 50:59–210.

Mauritzen, M., A.E. Derocher, O. Wiig, S.E. Belikov, A.N. Boltunov, E. Hansen, and G.W. Garner. 2002. Using satellite telemetry to define spatial population structure in polar bears in the Norwegian and western Russian Arctic. Journal of Applied Ecology 39:79–90.

Mayer, K.A., M.D. Dailey, and M.A. Miller. 2003. Helminth parasites of the southern sea otter *Enhydra lutris nereis* in central California: abundance, distribution and pathology. Diseases of Aquatic Organisms 53:77–88.

Mayor, S.J., J.A. Schaefer, D.C. Schneider, and S.P. Mahoney. 2007. Spectrum of selection: new approaches to detecting the scale-dependent response to habitat. Ecology 88:1634–1640.

Mazzoni, A.K. 2002. Habitat use by fishers (*Martes pennanti*) in the southern Sierra Nevada, California. Thesis, California State University, Fresno, USA.

McAllister, M.M. 2005. A decade of discoveries in veterinary protozoology changes our concept of "subclinical" toxoplasmosis. Veterinary Parasitology 132:241–247.

McCann, N.P., P.A. Zollner, and J.H. Gilbert. 2010. Survival of adult martens in northern Wisconsin. Journal of Wildlife Management 74:1502–1507.

McCarthy, A.J., M.A. Shaw, and S.J. Goodman. 2007. Pathogen evolution and disease emergence in carnivores. Proceedings of the Royal Society B 274:3165–3174.

McCarthy, K.P., T.K. Fuller, M. Ming, T.M. McCarthy, L. Waits, and K. Jumabaev. 2008. Assessing estimators of snow leopard abundance. Journal of Wildlife Management 72:1826–1833.

McCaw, D., and J. Hoskins. 2006. Canine viral enteritis. Pages 63–73 *in* Infectious diseases of the dog and cat. Third edition. C.E. Green, editor. W.B. Saunders, Philadelphia, Pennsylvania, USA.

McClelland, B.R., and P.T. McClelland. 1999. Pileated woodpecker nest and roost trees in Montana: links with old-growth and forest "health." Wildlife Society Bulletin 27:846–857.

McComb, B., B. Zuckerberg, D. Vesely, and C. Jordan. 2010. Monitoring animal populations and their habitats: a practitioner's guide. CRC Press, Boca Raton, Florida, USA.

McCullagh, P., and J.A. Nelder. 1989. Generalized linear models. Second Edition. Chapman and Hall/CRC, Boca Raton, Florida, USA.

McDaniel, G.W., K.S. McKelvey, J.R. Squires, and L.F. Ruggiero. 2000. Efficacy of lures and hair snares to detect lynx. Wildlife Society Bulletin 28:119–123.

McDonald, R.A., and S. Larivière. 2001. Diseases and pathogens of *Mustela* spp., with special reference to the biological control of introduced stoat *Mustela erminea* populations in New Zealand. Journal of the Royal Society of New Zealand 31:721–744.

McFadden, L.M. 2007. Home-range dynamics and habitat selection of American martens (*Martes americana*) in Michigan's northern Lower Peninsula. Thesis, Central Michigan University, Mount Pleasant, USA.

McGarigal, K., S.A. Cushman, M.C. Neel, and E. Ene. 2002. FRAGSTATS: spatial pattern analysis program for categorical maps. University of Massachusetts, Amherst, USA.

McGavin, D. 1987. Inactivation of canine parvovirus by disinfectants and heat. Journal of Small Animal Practice 28:523–535.

McGowan, C., L.A. Howes, and W.S. Davidson. 1999. Genetic analysis of an endangered pine marten (*Martes americana*) population from Newfoundland using randomly amplified polymorphic DNA markers. Canadian Journal of Zoology 77:661–666.

McGuill, M.W., and A.N. Rowan. 1989. Biological effects of blood loss: implications for sampling volumes and techniques. ILAR News 31:5–18.

McIntyre, N.E. 1997. Scale-dependent habitat selection by the darkling beetle *Eleodes hispilabris* (Coleoptera: Tenebrionidae). American Midland Naturalist 138:230–235.

McKague, C.I. 2007. Winter resource selection by the American marten (*Martes americana*): the effect of model resolution. Thesis, University of Guelph, Ontario, Canada.

McKelvey, K.S., K.B. Aubry, and M.K. Schwartz. 2008. Using anecdotal occurrence data for rare or elusive species: the illusion of reality and a call for evidentiary standards. BioScience 58:549–555.

McKelvey, K.S., and J.D. Johnston. 1992. Historical perspectives on forests of the Sierra Nevada and Transverse Ranges of southern California: forest conditions at the turn of the century. Pages 225–246 *in* The California spotted owl: a technical assessment of its current status. J. Verner, K.S. McKelvey, B.R. Noon, R.J. Gutiérrez, G.I. Gould, Jr., and T.W. Beck, technical coordinators. USDA Forest Service, General Technical Report PSW-GTR-133.

McKelvey, K.S, J. von Kienast, K.B. Aubry, G.M. Koehler, B.T. Maletzke, J.R. Squires, E.L. Lindquist, S. Loch, and M.K. Schwartz. 2006. DNA analysis of hair and scat collected along snow tracks to document the presence of Canada lynx. Wildlife Society Bulletin 34:451–455.

McKenzie, D., Z. Gedalof, D.L. Peterson, and P. Mote. 2004. Climate change, wildfire, and conservation. Conservation Biology 18:890–902.

McNeely, J.A. 2002. The role of taxonomy in conserving biodiversity. Journal for Nature Conservation 10:145–153.

McRae, B.H., and P. Beier. 2007. Circuit theory predicts gene flow in plant and animal populations. Proceedings of the National Academy of Sciences, USA 104:19885–19890.

McRae, B.H., B.G. Dickson, T.H. Keitt, and V.B. Shah. 2008. Using circuit theory to model connectivity in ecology and conservation. Ecology 10:2712–2724.

McRae, B.H., and V.B. Shah. 2009. Circuitscape user guide. http://www.circuitscape.org/Circuitscape/Welcome.html (accessed 23 April 2010).

Mead, J.F., R.B. Alfin-Slater, D.R. Howton, and G. Popják. 1986. Lipids: chemistry, biochemistry, and nutrition. Plenum, New York, USA.

Mead, R.A. 1994. Reproduction in *Martes*. Pages 404–422 *in* Martens, sables, and fishers: biology and conservation. S.W. Buskirk, A.S. Harestad, M.G. Raphael and R.A. Powell, editors. Cornell University Press, Ithaca, New York, USA.

Mech, L.D. 1974. Current techniques in the study of elusive wilderness carnivores. Pages 315–322 *in* XI International Congress of Game Biologists, National Swedish Environmental Protection Board. I. Kjerner and P. Bjurholm, editors. Swedish National Environment Protection Board, Stockholm, Sweden.

Mech, L.D., and S.M. Barber. 2002. A critique of wildlife radio-tracking and its use in national parks: report to the US National Park Service. U.S. National Park Service, Fort Collins, Colorado, USA.

Mech, L.D., and L.L. Rogers. 1977. Status, distribution and movements of marten in northeastern Minnesota. USDA Forest Service, Research Paper NC-143.

Meidinger, D., and J. Pojar. 1991. Ecosystems of British Columbia. British Columbia Ministry of Forests, Victoria, Canada. http://www.for.gov.bc.ca/hfd/pubs/Docs/Srs/Srs06.htm (accessed 15 April 2011).

Mergey, M. 2007. Réponses des populations de martres d'Europe (*Martes martes*) à la fragmentation de l'habitat: mécanismes comportementaux et consequences. Dissertation, University of Reims, Campagne-Ardenne, France.

Mergey, M., R. Helder, and J.-J. Roeder. 2011. Effect of forest fragmentation on space-use patterns in the European pine marten (*Martes martes*). Journal of Mammalogy 92:328–335.

Merriam, C.H. 1882. The vertebrates of the Adirondack region, northeastern New York. Transactions of the Linnaean Society of New York 1:5–168.

Merriam, C.H. 1890. Description of a new marten (*Mustela caurina*) from the north-west coast region of the United States. North American Fauna 4:27–29.

Merrill, G.P. 1920. Contributions to a history of American state geological and natural history surveys. U.S. National Museum Bulletin 109. Government Printing Office, Washington, D.C., USA.

Meshcherskii, I.G., V.V. Rozhnov, and S.V. Naidenko. 2003. On certain properties of water and energy metabolism in representatives of *Martes* and *Mustela* genera (Mammalia: Mustelidae). Biology Bulletin 30:406–410.

Meyer, C.B., and W. Thuiller. 2006. Accuracy of resource selection functions across spatial scales. Diversity and Distributions 12:288–297.

Michelat, D., J.P. Quéré, and P. Giraudoux. 2001. Charactéristiques des gîtes utilisés par la fouine (*Martes foina*, Erxleben, 1777) dans le Haut-Doubs. Revue Suisse de Zoologie 108:263–274.

Michigan Department of Natural Resources. 1970. 25th biennial report of the Department of Natural Resources of the State of Michigan, 1969–1970. Michigan Department of Natural Resources, Lansing, USA.

Miguel, J., J.C. Casanova, F. Tenora, C. Filiu, and J. Torres. 1995. A scanning electron-microscope study of some Rictulariidae (Nematoda) parasites of Iberian mammals. Helminthologia 32:3–14.

Millan, J., and E. Ferroglio. 2001. Helminth parasites in stone martens (*Martes foina*) from Italy. Zeitschrift für Jagdwissenschaft 47:229–231.

Miller, B., D. Biggins, C. Wemmer, R. Powell, L. Calvo, and T. Wharton. 1990a. Development of survival skills in captive-raised Siberian polecats (*Mustela eversmanni*). II. Predator avoidance. Ethology 8:95–104.

Miller, B., D. Biggins, C. Wemmer, R. Powell, L. Hanebury, D. Horn, and A. Vargas. 1990b. Development of survival skills in captive-raised Siberian polecats (*Mustela eversmanni*). I. Locating prey. Ethology 8:89–94.

Miller, B., K. Ralls, R.P. Reading, J.M. Scott, and J. Estes. 1999. Biological and technical considerations of carnivore translocation: a review. Animal Conservation 2:59–68.

Miller, C., and D. Urban. 1999. Forest pattern, fire and climatic change in the Sierra Nevada. Ecosystems 2:76–87.

Miller, D.R. 1961. Marten transplanting in northern Manitoba. Biological Report, Manitoba Game Branch, Winnepeg, Canada.

Miller, J.D., H.D. Safford, M. Crimmins, and A.E. Thode. 2009. Quantitative evidence for increasing forest fire severity in the Sierra Nevada and southern Cascade Mountains, California and Nevada, USA. Ecosystems 12:16–32.

Miller, J.D., C.N. Skinner, H.D. Safford, E.E. Knapp, and C.M. Ramirez. 2012. Trends and causes of severity, size, and number of fires in northwestern California, USA. Ecological Applications 22:184–203.

Miller, P.S., and R.C. Lacy. 2005. VORTEX: a stochastic simulation of the extinction process. Version 9.50 User's Manual. Conservation Breeding Specialist Group (SSC/IUCN), Apple Valley, Minnesota, USA.

Miller, R.G., R.W. Ritcey, and R.Y. Edwards. 1955. Live-trapping marten in British Columbia. Murrelet 36:1–8.

Millspaugh, J.J., and J.M. Marzluff. 2001. Radio tracking and animal populations. Academic Press, San Diego, California, USA.

Minta, S.C., P.M. Kareiva, and A.P. Curlee. 1999. Carnivore research and conservation: learning from history and theory. Pages 323–404 *in* Carnivores in ecosystems: the Yellowstone experience. T.W. Clark, A.P. Curlee, S.C. Minta, and P.M. Kareiva, editors. Yale University Press, New Haven, Connecticut, USA.

Mitchell, M.S., and R.A. Powell. 2003. Response of black bears to forest management in the southern Appalachian Mountains. Journal of Wildlife Management 67:692–705.

Mitchell, M.S., J.W. Zimmerman, and R.A. Powell. 2002. Test of a habitat suitability index for black bears in the southern Appalachians. Wildlife Society Bulletin 30:794–808.

Mitchell, T.D., and P.D. Jones. 2005. An improved method of constructing a database of monthly climate observations and associated high-resolution grids. International Journal of Climatology 25:693–712.

Mitchell-Jones, A.J., G. Amori, W. Bogdanowicz, B. Krystufek, P.J.H. Reijnders, F. Spitzenberger, M. Stubbe, J.B.M. Thissen, V. Vohralik, and J. Zima. 1999. The atlas of European mammals. Academic Press, London, UK.

Mitcheltree, D.H., T.L. Serfass, M.T. Whary, W.M. Tzilkowski, R.P. Brooks, and R.L. Peper. 1997. Captive care and clinical evaluation of fishers during the first year of a reintroduction project. Pages 317–328 *in Martes*: taxonomy, ecology, techniques, and management. G. Proulx, H.N. Bryant, and P.M. Woodard, editors. Provincial Museum of Alberta, Edmonton, Canada.

Miyoshi, K., and S. Higashi. 2005. Home range and habitat use by the sable *Martes zibellina brachyura* in a Japanese cool-temperate mixed forest. Ecological Research 20:95–101.

Mladenoff, D.J. 2004. LANDIS and forest landscape models. Ecological Modelling 180:7–19.

Möhl, K., K. Grosse, A. Hamedy, T. Wüste, P. Kabelitz, and E. Lücker. 2009. Biology of *Alaria* spp. and human exposition risk to *Alaria mesocercariae*—a review. Parasitological Research 105:1–15.

Moilanen, A., H. Kujala, and J. Leathwick. 2009. The Zonation framework and software for conservation prioritization. Pages 196–210 *in* Spatial conservation prioritization:

quantitative methods and computational tools. A. Moilanen, K.A. Wilson, and H.P. Possingham, editors. Oxford University Press, UK.

Moloney, F.X. 1931. The fur trade in New England, 1620–1676. Harvard University Press, Cambridge, Massachusetts, USA.

Monakhov, G.I. 1976. Geographic variability and taxonomic structure of sable in the USSR. Pages 54–86 *in* Proceedings for the Research Institute for Game Farming and Animal Breeding, Kirov, Russia.

Monakhov, G.I., and N.N. Bakeev. 1981. The sable. Lesnaja Promishlennost Publishing House, Moscow, USSR. (in Russian)

Monakhov, G.I., V.S. Kriuchkov, V.G. Monakhov, and V.V. Shurigin. 1982. Results of East-Siberian sables introduction in Yenisei Siberia and Vasugan River basin. Pages 136–148 *in* Game theriology. D.I. Bibikov and N.N. Grakov, editors. Nauka Publishing House, Moscow, USSR. (in Russian)

Monakhov, G.I., and V.G. Monakhov. 1978. Patterns of acclimatization in sable populations restored through the introduction. Pages 185–187 *in* Acclimatization of game animals in USSR. Y.N. Chichikin, editor. Urojhai Publishing House, Minsk, USSR. (in Russian)

Monakhov, V.G. 1978. Economic effectiveness of reacclimatization works in sable at Yenisei Siberia. Pages 114–116 *in* Acclimatization of game animals in USSR. Y.N. Chichikin, editor. Urojhai Publishing House, Minsk, USSR. (in Russian)

Monakhov, V.G. 1982. Economic efficiency of sable reintroduction works. Pages 94–95 *in* Enrichment of the fauna and breeding of game animals. V.G. Safonov, editor. VNIIOZ, Kirov, USSR. (in Russian)

Monakhov, V.G. 1984. Morphological changes of sables in Middle Siberia and Ob River basin under influence of Baikal immigrants. Dissertation, Institute of Plant and Animal Ecology, Russian Academy of Sciences, Yekaterinburg, USSR. (in Russian)

Monakhov, V.G. 1995. Sable in the Urals, Ob Region, and Yenisei Siberia: results of acclimatization. Bank of Cultural Information Publishing House, Yekaterinburg, USSR. (in Russian)

Monakhov, V.G. 1999. The prevalence of filaroidiasis (*Filaroides martis*) in Russian sables (*Martes zibellina*) of different sex and age. Russian Journal of Ecology 30:420–427. (in Russian)

Monakhov, V.G. 2001. Phenetic analysis of aboriginal and introduced populations of sable (*Martes zibellina*) in Russia. Russian Journal of Genetics 37:1074–1081.

Monakhov, V.G. 2006. Dynamics of size and phenetic structure of sable in specific area. Ural Division of Russian Academy of Science, Bank of Cultural Information Publishing House, Ekaterinburg, USSR. (in Russian)

Monakhov, V.G. 2009. Is sexual size dimorphism variable? Data on species of the genus *Martes* in the Urals. Biology Bulletin 36:45–52.

Monakhov, V.G. 2010. Phenogeography of a cranial trait of the sable *Martes zibellina* L. in the species area. Doklady Biological Sciences 431:94–99.

Monfort, S.L., C.C. Schwartz, and S.K. Wasser. 1993. Monitoring reproduction in moose using urinary and fecal steroid metabolites. Journal of Wildlife Management 57:400–407.

Montgomery, C.A., G.M. Brown, Jr., and D.M. Adams. 1994. The marginal cost of species preservation: the northern spotted owl. Journal of Environmental Economics and Management 26:111–128.

Moore, T.D., L.E. Spence, C.E. Dugnolle, and W.G. Hepworth. 1974. Identification of the dorsal guard hairs of some mammals of Wyoming. Wyoming Game and Fish Department, Bulletin No. 14, Cheyenne, USA.

Morin, P.A., F.I. Archer, A.D. Foote, J. Vilstrup, E.E. Allen, P. Wade, J.W. Durban, K. Parsons, R.L. Pitman, L. Li, P. Bouffard, S.C. Abel Nielsen, M. Rasmussen, E. Willerslev, M. Thomas, P. Gilbert, and T. Harkins. 2010. Complete mitochondrial genome phylogeographic analysis of killer whales (*Orcinus orca*) indicates multiple species. Genome Research 20:908–916.

Morin, P.A., and D.S. Woodruff. 1992. Paternity exclusion using multiple hypervariable microsatellite loci amplified from nuclear DNA of hair cells. Pages 63–81 *in* Paternity in primates: genetic tests and theories. R.D. Martin, A.F. Dixson, and E.J. Wickings, editors. Karger, Basel, Switzerland.

Moritz, C., J.L. Patton, C.J. Conroy, J.L. Parra, G.C. White, and S.R. Beissinger. 2008. Impact of a century of climate change on small-mammal communities in Yosemite National Park, USA. Science 322:261–264.

Morlo, M., G.F. Gunnell, and D. Nagel. 2010. Ecomorphological analysis of carnivore guilds in the Eocene through Miocene of Laurasia. Pages 269–310 *in* Carnivoran evolution: new views on phylogeny, form, and function. A. Goswami and A. Friscia, editors. Cambridge University Press, UK.

Mörner, T. 1992. Sarcoptic mange in Swedish wildlife. Revue Scientifique et Technique 11:1115–1121.

Morris, D.W. 1987. Spatial scale and the cost of density-dependent habitat selection. Evolutionary Ecology 1:379–388.

Morris, D.W. 1989. Habitat-dependent estimates of competitive interaction. Oikos 55:111–120.

Morris, D.W. 1992. Scales and cost of habitat selection in heterogeneous environments. Evolutionary Ecology 6:412–432.

Morris, D.W. 2002. Measuring the Alee effect: positive density dependence in small mammals. Ecology 83:14–20.

Morrison, M.L., and L.S. Hall. 2002. Standard terminology: toward a common language to advance ecological understanding and application. Pages 43–52 *in* Predicting species occurrences: issues of accuracy and scale. J.M. Scott, P.J. Heglund, M.L. Morrison, J.B. Haufler, M.G. Raphael, W.A. Wall, and F.B. Samson, editors. Island Press, Seattle, Washington, USA.

Morrison, T.A., J. Yoshizaki, J.D. Nichols, and D.T. Bolger. 2011. Estimating survival in photographic capture-recapture studies: overcoming misidentification error. Methods in Ecology and Evolution: published online March 25, 2011. doi:10.1111/j.2041–210X.2011.00106.x.

Mortelliti, A., and L. Boitani. 2008. Interaction of food resources and landscape structure in determining the probability of patch use by carnivores in fragmented landscapes. Landscape Ecology 23:285–298.

Morton, D.B., D. Abbot, R. Barclay, B.S. Close, R. Ewbank, D. Gask, M. Heath, S. Mattic, T. Poole, and J. Seamer. 1993. Removal of blood from laboratory mammals and birds. Laboratory Animals 27:1–22.

Moruzzi, T.L., K.J. Royar, C. Grove, R.T. Brooks, C. Bernier, F.L. Thompson, Jr., R.M. DeGraaf, and T.K. Fuller. 2003. Assessing an American marten, *Martes americana*, reintroduction in Vermont. Canadian Field-Naturalist 117:190–195.

Mos, L., P.S. Ross, D. McIntosh, and S. Raverty. 2003. Canine distemper virus in river otters in British Columbia as an emergent risk for coastal pinnipeds. Veterinary Record 152:237–239.

Moser, B.W., and E.O. Garton. 2007. Effects of telemetry location error on space-use estimates using a fixed-kernel density estimator. Journal of Wildlife Management 71:2421–2426.

Mowat, G. 2006. Winter habitat associations of American martens *Martes americana* in interior wet-belt forests. Wildlife Biology 12:51–61.

Mowat, G., and D. Paetkau. 2002. Estimating marten *Martes americana* population size using hair capture and genetic tagging. Wildlife Biology 8:201–209.

Mowat, G., C. Shurgot, and K. Poole. 2000. Using track plates and remote cameras to detect marten and short-tailed weasels in coastal cedar hemlock forests. Northwestern Naturalist 81:113–121.

Müller, T., J. Cox, W. Peter, R. Schafer, N. Johnson, L.M. McElhinney, J.L. Geue, K. Tjornehoj, and A.R. Fooks. 2004. Spill-over of European bat lyssavirus type 1 into a stone marten (*Martes foina*) in Germany. Journal of Veterinary Medicine, Series B 51:49–54.

Mullins, J., M.J. Statham, T. Roche, P.D. Turner, and C. O'Reilly. 2010. Remotely plucked hair genotyping: a reliable and non-invasive method for censusing pine marten (*Martes martes*, L. 1758) populations. European Journal of Wildlife Research 56:443–453.

Mullis, C. 1985. Habitat utilization by fisher (*Martes pennanti)* near Hayfork Bally, California. Thesis, Humboldt State University, Arcata, California, USA.

Munson, L. 2003. Scope and magnitude of disease in species conservation. Pages 19–21 *in* Animal movements and disease risk: a workbook. D. Armstrong, R. Jakob-Hoff, and U.S. Seal, editors. Conservation Breeding Specialist Group (SSC/IUCN).

Munson, L. 2004. Necropsy of wild animals. http://www.vetmed.ucdavis.edu/whc/pdfs/munsonnecropsy.pdf.

Munson, L., and W.B. Karesh. 2002. Conservation medicine: disease monitoring for the conservation of terrestrial animals. Oxford University Press, New York, USA.

Murakami, T. 2009. *Martes zibellina* (Linnaeus, 1758). Pages 252–253 *in* The wild mammals of Japan. S.D. Ohdachi, Y. Ishibashi, M.A. Iwasa, and T. Saitoh, editors. Shoukadoh, Kyoto, Japan.

Murakami, T., M. Asano, and N. Ohtaishi. 2004. Mitochondrial DNA variation in the Japanese marten *Martes melampus* and Japanese sable, *Martes zibellina*. Japanese Journal of Veterinary Research 51:135–142.

Murakami, T., and N. Ohtaishi. 2000. Current distribution of the endemic sable and introduced Japanese marten in Hokkaido. Mammal Study 25:149–152.

Murie, A. 1961. Some food habits of the marten. Journal of Mammalogy 42:516–521.

Murphy, D.D., and B.R. Noon. 1992. Integrating scientific methods with habitat planning: reserve design for northern spotted owls. Ecological Applications 2:3–17.

Murphy, M.A., L.P. Waits, and K.C. Kendall. 2000. Quantitative evaluation of fecal drying methods for brown bear DNA analysis. Wildlife Society Bulletin 28:951–957.

Murray, D.L., C.A. Kapke, J.F. Evermann, and T.K. Fuller. 1999. Infectious disease and the conservation of free-ranging large carnivores. Animal Conservation 2:241–254.

Mustonen, A.-M., J. Asikainen, J. Aho, and P. Nieminen. 2007a. Selective seasonal fatty acid accumulation and mobilization in the wild raccoon dog (*Nyctereutes procyonoides*). Lipids 42:1155–1167.

Mustonen, A.-M., J. Asikainen, K. Kauhala, T. Paakkonen, and P. Nieminen. 2007b. Seasonal rhythms of body temperature in the free-ranging raccoon dog (*Nyctereutes procyonoides*) with special emphasis on winter sleep. Chronobiology International 24:1095–1107.

Mustonen, A.-M., R. Käkelä, and P. Nieminen. 2007c. Different fatty acid composition in central and peripheral adipose tissues of the American mink (*Mustela vison*). Comparative Biochemistry and Physiology A 147:903–910.

Mustonen, A.-M., and P. Nieminen. 2006. Fatty acid composition in the central and peripheral adipose tissues of the sable (*Martes zibellina*). Journal of Thermal Biology 31:617–625.

Mustonen, A.-M., M. Puukka, K. Rouvinen-Watt, J. Aho, J. Asikainen, and P. Nieminen. 2009. Response to fasting in an unnaturally obese carnivore, the captive European polecat *Mustela putorius*. Experimental Biology and Medicine 234:1287–1295.

Mustonen, A.-M., M. Puukka, S. Saarela, T. Paakkonen, J. Aho, and P. Nieminen. 2006a. Adaptations to fasting in a terrestrial mustelid, the sable (*Martes zibellina*). Comparative Biochemistry and Physiology A 144:444–450.

Mustonen, A.-M., T. Pyykönen, J. Aho, and P. Nieminen. 2006b. Hyperthermia and increased physical activity in the fasting American mink *Mustela vison*. Journal of Experimental Zoology A 305:489–498.

Mustonen, A.-M., T. Pyykönen, T. Paakkonen, A. Ryökkynen, J. Asikainen, J. Aho, J. Mononen, and P. Nieminen. 2005a. Adaptations to fasting in the American mink (*Mustela vison*): carbohydrate and lipid metabolism. Comparative Biochemistry and Physiology A 140:195–202.

Mustonen, A.-M., S. Saarela, and P. Nieminen. 2008. Food deprivation in the common vole (*Microtus arvalis*) and the tundra vole (*Microtus oeconomus*). Journal of Comparative Physiology B 178:199–208.

Mustonen, A.-M., S. Saarela, T. Pyykönen, and P. Nieminen. 2005b. Endocrinologic adaptations to wintertime fasting in the male American mink (*Mustela vison*). Experimental Biology and Medicine 230:612–620.

Nagel, D., C. Stefen, and M. Morlo. 2009. The carnivoran community from the Miocene of Sandelzhausen (Germany). Paläontologische Zeitschrift 83:151–174.

Nagorsen, D.W. 1994. Body weight variation among insular and mainland American martens. Pages 85–97 *in* Martens, sables, and fishers: biology and conservation. S.W. Buskirk, A.S. Harestad, M.G. Raphael, and R.A. Powell, editors. Cornell University Press, Ithaca, New York, USA.

Nagorsen, D.W., K.F. Morrison, and J.E. Forsberg. 1989. Winter diet of Vancouver Island marten (*Martes americana*). Canadian Journal of Zoology 67:1394–1400.

Nakaya, H. 1994. Faunal change of late Miocene Africa and Eurasia: mammalian fauna from the Namurungule Formation, Samburu Hills, northern Kenya. African Study Monographs 20:1–112.

Nakicenovic, N., J. Alcamo, G. Davis, B. de Vries, J. Fenhann, S. Gaffin, K. Gregory, A. Grübler, T.Y. Jung, T. Kram, E.L.L. Rovere, L. Michaelis, S. Mori, T. Morita, W. Pepper, H. Pitcher, L. Price, K. Riahi, A. Roehrl, H.-H. Rogner, A. Sankovski, M. Schlesinger, P. Shukla, S. Smith, R. Swart, S. van Rooijen, N. Victor, and Z. Dadi. 2000. Special report on emissions scenarios. A Special Report of Working Group III of the Intergovernmental Panel on Climate Change. Cambridge University Press, UK.

Nams, V.O., and M. Bourgeois. 2004. Fractal analysis measures habitat use at different spatial scales: an example with American marten. Canadian Journal of Zoology 82:1738–1747.

Nasimovich, A.A., editor. 1973. Sable, martens, yellow-throated marten. Nauka Publishing House, Moscow, USSR.

National Landcover Database. 2001. Multi-resolution Land Characteristics Consortium. http://www.mrlc.gov/ (accessed November 2006).

National Parks and Wildlife Service. 2007. *Martes martes* (1357) conservation status assessment report. Unpublished report to the National Parks and Wildlife Service, Department of the Environment, Heritage and Local Government, Dublin, Ireland.

Nelson, K.A. 1979. The occurrence of wolverine (*Gulo luscus*) and other mammals by baited hair traps and snow transects in Six Rivers National Forest. Technical Report, Six Rivers National Forest, Eureka, California, USA.

Nelson, R.A., D.L. Steiger, and T.D.I. Beck. 1983. Neuroendocrine and metabolic interactions in the hibernating black bear. Acta Zoologica Fennica 174:137–141.

Nes, N., E.J. Einarsson, O. Lohi, and G. Jorgensen. 1988. Beautiful fur animals and their colour genetics. Scientifur, Glostrup, Denmark.

New, M., M. Hulme, and P.D. Jones. 1999. Representing twentieth century space-time climate variability. Part 1: development of a 1961–90 mean monthly terrestrial climatology. Journal of Climate 12:829–856.

New, M., D. Lister, M. Hulme, and I. Makin. 2002. A high-resolution data set of surface climate over global land areas. Climate Research 21:1–25.

Newman, M.E.J. 2010. Networks: an introduction. Oxford University Press, UK.

Newsom, W.M. 1937. Mammals on Anticosti Island. Journal of Mammalogy 18:435–442.

Nichols, J.D., L.L. Bailey, A.F. O'Connell, Jr., N.W. Talancy, E.H.C. Grant, A.T. Gilbert, E.M. Annand, T.P. Husband, and J.E. Hines. 2008. Multi-scale occupancy estimation and modeling using multiple detection methods. Journal of Applied Ecology 45:1321–1329.

Nichols, J.D., J.E. Hines, D.I. MacKenzie, M.E. Seamans, and R.J. Gutiérrez. 2007. Occupancy estimation with multiple states and state uncertainty. Ecology 88:1395–1400.

Nichols, J.D., and B.K. Williams. 2006. Monitoring for conservation. Trends in Ecology & Evolution 21:668–673.

Nieberding, C.M., M.-C. Durette-Desset, A. Vanderpooten, J.C. Casanova, A. Ribas, V. Deffontaine, C. Feliu, S. Morand, R. Libois, and J.R. Michaux. 2008. Geography and host biogeography matter for understanding the phylogeography of a parasite. Molecular Phylogenetics and Evolution 47:538–554.

Nieberding, C., S. Morand, R. Libois, and J.R. Michaux. 2004. A parasite reveals cryptic phylogeographic history of its hosts. Proceedings of the Royal Society B 271:2559–2568.

Nieberding, C.M., and I. Olivieri. 2007. Parasites: proxies for host geneaology and ecology? Trends in Ecology & Evolution 22:156–165.

Nielsen, L. 1988. Definitions, considerations, and guidelines for translocation of wild animals. Pages 12–51 *in* Translocation of wild animals. L. Nielson and R.D. Brown, editors. Wisconsin Humane Society and Caesar Kleberg Wildlife Research Institute, Milwaukee, USA.

Nieminen, P., J. Asikainen, and H. Hyvärinen. 2001. Effects of seasonality and fasting on the plasma leptin and thyroxin levels of the raccoon dog (*Nyctereutes procyonoides*) and the blue fox (*Alopex lagopus*). Journal of Experimental Zoology 289:109–118.

Nieminen, P., R. Käkelä, T. Pyykönen, and A.-M. Mustonen. 2006a. Selective fatty acid mobilization in the American mink (*Mustela vison*) during food deprivation. Comparative Biochemistry and Physiology B 145:81–93.

Nieminen, P., and A.-M. Mustonen. 2007. Uniform fatty acid mobilization from anatomically distinct fat depots in the sable (*Martes zibellina*). Lipids 42:659–669.

Nieminen, P., and A.-M. Mustonen. 2008. A preliminary study on the seasonal body temperature rhythms of the captive mountain hare (*Lepus timidus*). Cryobiology 56:163–167.

Nieminen, P., A.-M. Mustonen, J. Asikainen, and H. Hyvärinen. 2002. Seasonal weight regulation of the raccoon dog (*Nyctereutes procyonoides*): interactions between melatonin, leptin, ghrelin, and growth hormone. Journal of Biological Rhythms 17:155–163.

Nieminen, P., A.-M. Mustonen, V. Kärjä, J. Asikainen, and K. Rouvinen-Watt. 2009. Fatty acid composition and development of hepatic lipidosis during food deprivation—mustelids as a potential animal model for liver steatosis. Experimental Biology and Medicine 234:278–286.

Nieminen, P., K. Rouvinen-Watt, D. Collins, J. Grant, and A.-M. Mustonen. 2006b. Fatty acid profiles and relative mobilization during fasting in adipose tissue depots of the American marten (*Martes americana*). Lipids 41:231–240.

Nieminen, P., K. Rouvinen-Watt, S. Saarela, and A.-M. Mustonen. 2007. Fasting in the American marten (*Martes americana*): a physiological model of the adaptations of a lean-bodied animal. Journal of Comparative Physiology B 177:787–795.

Nikolova, S., Y. Tzvetkov, H. Najdenski, and A. Vesselinova. 2001. Isolation of pathogenic yersiniae from wild animals in Bulgaria. Journal of Veterinary Medicine, Series B 48:203–209.

Noblet, J.-F. 2002. La martre. Éveil Nature, Saint-Yrieix-sur-Charente, France.

Nomura, T. 2005. Methods for minimizing the loss of genetic diversity in conserved populations with overlapping generations. Conservation Genetics 6:655–663.

Novak, M., M.E. Obbard, J.G. Jones, R. Newman, A. Booth, A.J. Satterthwaite, and G. Linscombe. 1987. Furbearer harvests in North America, 1600–1984. Ontario Trappers Association and Ontario Ministry of Natural Resources, Toronto, Canada.

Nowak, R.M. 1999. Walker's mammals of the world. Sixth edition. John Hopkins University Press, Baltimore, Maryland, USA.

Nowak, R.M. 2005. Walker's carnivores of the world. Johns Hopkins University Press, Baltimore, Maryland, USA.

Nukerbaeva, K.K. 1981. Coccida of sables (*Martes zibellina*). Izvestiia Akademaii Nauk Kazachskoi SSR, seriya Biologicheskia Nauk 1:30–33.

Nyholm, E.S. 1961. Havaintoja näätäeläinten käyttäytymisestä (Observations on mustelid behavior). Luonnon Tutkija 65:50–56. (in Finnish)

Oatway, M.L., and D.W. Morris. 2007. Do animals select habitat at small or large scales? An experiment with meadow voles. Canadian Journal of Zoology 85:479–487.

Obbard, M.E., J.G. Jones, R. Newman, A. Booth, A.J. Stterthwaite, and G. Linscombe. 1987. Furbearer harvests in North America. Pages 1007–1034 *in* Wild furbearer management and conservation in North America. M. Novak, J.A. Baker, M.E. Obbard, and B. Malloch, editors. Ontario Trappers Association and Ontario Ministry of Natural Resources, Toronto, Canada.

O'Connell, A.F., Jr., N. Talancy, L.L. Bailey, J. Sauer, R. Cook, and A.T. Gilbert. 2006. Estimating site occupancy and detection probability parameters for mammals in a coastal ecosystem. Journal of Wildlife Management 70:1625–1633.

Ognev, S.I. 1925. A systematical review of the Russian sables. Journal of Mammalogy 6:276–280.

Oksanen, T., L. Oksanen, and S.D. Fretwell. 1985. Surplus killing in the hunting strategy of small predators. American Naturalist 126:328–346.

Olsen, O.W. 1952. *Crenosoma coloradoensis*, n. sp. (Nematoda: Metastrongylidae) from the lungs of martens, *Martes caurina origenes* (Rhoads). Journal of Parasitology 38:207–209.

O'Mahony, D., C. O'Reilly, and P. Turner. 2006. National pine marten survey of Ireland 2005. COFORD Connects 7.

O'Meara, D.C., D.D. Payne, and J.F. Witter. 1960. *Sarcoptes* infestation of a fisher. Journal of Wildlife Management 24:339.

Ontario Ministry of Natural Resources. 1988. Timber management guidelines for the provision of moose habitat. Queen's Printer for Ontario, Toronto, Canada.

Orrlock, J.L., J.F. Pagels, W.J. McShea, and E.K. Harper. 2000. Predicting presence and abundance of a small mammal species: the effect of scale and resolution. Ecological Applications 10:1356–1366.

Oshima, K. 1990. The history of straits around the Japanese islands in the Late-Quaternary. Quaternary Research 29:193–208.

O'Sullivan, P.J. 1983. The distribution of the pine marten (*Martes martes*) in the Republic of Ireland. Mammal Review 13:39–44.

Otranto, D., R.P. Lia, C. Cantacessi, E. Brianti, D. Traversa, and S. Giannetto. 2007. *Filaria martis* Gmelin 1790 (Spirurida, Filariidae) affecting beech marten (*Martes foina*): morphological description and molecular characterisation of the cytochrome oxidase c subunit I. Parasitological Research 101:877–883.

Ouborg J., C. Pertoldi, V. Loeschcke, R. Bijlsma, and P.W. Hedrick. 2010a. Conservation genetics in transition to conservation genomics. Trends in Genetics 26:177–187.

Ouborg, N.J., F. Angeloni, and P. Vergeer. 2010b. An essay on the necessity and feasibility of conservation genomics. Conservation Genetics 11:643–653.

Pack, J.C., and J.I. Cromer. 1981. Reintroduction of fisher in West Virginia. Pages 1431–1442 *in* Worldwide furbearer conference proceedings. J.A. Chapman and D. Pursley, editors. Worldwide Furbearer Conference, Frostburg, Maryland, USA.

Page, R.J., and S.D. Langton. 1996. The occurrence of ixodid ticks on wild mink *Mustela vision* in England and Wales. Medical and Veterinary Entomology 10:359–364.

Palmer, R.S., editor. 1949. Rufus B. Philbrook, trapper. New England Quarterly 22:452–474.

Palomares, F., and T.M. Caro. 1999. Interspecific killing among carnivores. American Naturalist 153:492–508.

Palstra, F.P., and D.E. Ruzzante. 2010. A temporal perspective on population structure and gene flow in Atlantic salmon (*Salmo salar*) in Newfoundland, Canada. Canadian Journal of Fisheries and Aquatic Sciences 67:225–242.

Paragi, T.F. 1990. Reproductive biology of female fishers in southcentral Maine. Thesis, University of Maine, Orono, USA.

Paragi, T.F., W.N. Johnson, D.D. Katnik, and A.J. Magoun. 1996. Marten selection of postfire seres in the Alaska taiga. Canadian Journal of Zoology 74:2226–2237.

Paragi, T.F., W.B. Krohn, and S.M. Arthur. 1994. Using estimates of fisher recruitment and survival to evaluate population trend. Northeast Wildlife 51:1–11.

Parks, C.G., and E.L. Bull. 1997. American marten use of rust and dwarf mistletoe brooms in northeastern Oregon. Western Journal of Applied Forestry 12:131–133.

Parmesan, C. 2006. Ecological and evolutionary responses to recent climate change. Annual Review of Ecology and Systematics 37:637–669.

Parmesan, C., and G. Yohe. 2003. A globally coherent fingerprint of climate change impacts across natural systems. Nature 412:37–42.

Parrish, C., P. Have, W. Foreyt, J. Evermann, M. Senda, and L. Carmichael. 1988. The global spread and replacement of canine parvovirus strains. Journal of General Virology 69:1111–1116.

Parrish, C., and Y. Kawaoka. 2005. The origins of new pandemic viruses: the acquisition of new host ranges by canine parvovirus and influenza A viruses. Annual Review of Microbiology 59:553–586.

Parrish, S.-V. 1972. The fur and skin trade of Colonial Virginia. Thesis, Old Dominion University, Norfolk, Virginia, USA.

Parsons, G.R., M.K. Brown, and G.B. Will. 1978. Determining the sex of fisher from the lower canine teeth. New York Fish and Game Journal 25:42–44.

Pasitschniak-Arts, M., and S. Larivière. 1995. *Gulo gulo*. Mammalian Species 499:1–10.

Patterson, B.D., G. Ceballos, W. Sechrest, M.F. Tognelli, T. Brooks, L. Luna, P. Ortega, I. Salazar, and B.E. Young. 2003. Digital distribution maps of the mammals of the Western Hemisphere, version 1.0. NatureServe, Arlington, Virginia, USA.

Patton, H. 1925. Raising fur-bearing animals. Wheldon & Wesley, London, UK.

Paul, T.W. 2009. Game transplants in Alaska. Second edition. Technical Bulletin No. 4, Alaska Department of Fish and Game, Juneau, USA.

Pauli, J.N. 2010. Ecological studies of the American marten (*Martes americana*): quantifying cryptic processes in an elusive species. Dissertation, University of Wyoming, Laramie, USA.

Pauli, J.N., M. Ben-David, S.W. Buskirk, J.E. DePue, and W.P. Smith. 2009a. An isotopic technique to mark mid-size vertebrates non-invasively. Journal of Zoology 278:141–148.

Pauli, J.N., M.B. Hamilton, E.B. Crain, and S.W. Buskirk. 2008. A single-sampling hair trap for mesocarnivores. Journal of Wildlife Management 72:1650–1652.

Pauli, J.N., J.P. Whiteman, M.D. Riley, and A.D. Middleton. 2009b. Defining noninvasive approaches for sampling of vertebrates. Conservation Biology 24:349–352.

Pavlinov, I.Y., and O.L. Rossolimo. 1979. Geographic variability and intraspecies systematics of sable (*Martes zibellina* L.) in the USSR. Pages 241–256 *in* Proceedings of the Zoological Museum, Moscow State University, USSR.

Pavlov, M.P. 1973a. Resettlement of sables in the USSR. Pages 24–32 *in* Acclimatization and reacclimatization of game animals. B.A. Galaka, editor. USSR Agricultural Ministry, Kiev. (in Russian)

Pavlov, M.P. 1973b. The pine marten. Pages 106–107 *in* Acclimatization of game mammals in the Soviet Union 1. I.D. Kiris, editor. Volgo-Vyatskoe Publishing House, Kirov, USSR. (in Russian)

Pavlov, M.P. 1973c. The stone marten. Pages 108–109 *in* Acclimatization of game mammals in the Soviet Union 1. I.D. Kiris, editor. Volgo-Vyatskoe Publishing House, Kirov, USSR. (in Russian)

Payer, D.C. 1999. Influences of timber harvesting and trapping on habitat selection and demographic characteristics of marten. Dissertation, University of Maine, Orono, USA.

Payer, D.C., and D.J. Harrison. 1999. Influences of timber harvesting and trapping on habitat selection and demographic characteristics of marten. Final Report, Cooperative Forestry Research Unit, University of Maine, Orono, USA.

Payer, D., and D. Harrison. 2000. Structural differences between forests regenerating following spruce budworm defoliation and clear-cut harvesting: implications for marten. Canadian Journal of Forest Research 30:1965–1972.

Payer, D.C., and D.J. Harrison. 2003. Influence of forest structure on habitat use by American marten in an industrial forest. Forest Ecology and Management 179:145–156.

Payer, D., and D. Harrison. 2004. Relationships between forest structure and habitat use by American marten in Maine, USA. Pages 173–186 *in* Martens and fishers (*Martes*) in human-altered environments: an international perspective. D.J. Harrison, A.K. Fuller, and G. Proulx, editors. Springer Science+Business Media, New York, USA.

Payer, D., D. Harrison, and D. Phillips. 2004. Territoriality and home-range fidelity of American martens in relation to timber harvesting and trapping. Pages 99–114 *in* Martens and fishers (*Martes*) in human-altered environments: an international perspective. D.J. Harrison, A.K. Fuller, and G. Proulx, editors. Springer Science+Business Media, New York, USA.

Pearson, R.G., and T.P. Dawson. 2004. Bioclimate envelope models: what they detect and what they hide; response to Hampe (2004). Global Ecology and Biogeography 13:471–473.

Pedersen, A.B., K.E. Jones, C.L. Nunn, and S. Altizer. 2007. Infectious diseases and extinction risk in wild mammals. Conservation Biology 21:1269–1279.

Peigne, S., M.J. Salesa, M. Anton, and J. Morales. 2006. New data on carnivores from the Middle Miocene (Upper Aragonian, MN 6) of Arroyo del Val area (Villafeliche, Zaragoza Province, Spain). Estudio Geologicos 62:359–374.

Pelletier, A.-M. 2005. Préférences d'habitat estival, structure d'âge et reproduction d'une population de martres d'Amérique (*Martes americana*) dans un secteur forestier aménagé de manière intensive au Nord-Ouest du Nouveau-Brunswick. Thesis, University of Moncton, New Brunswick, Canada.

Pellikka, J., H. Rita, and H. Linden. 2005. Monitoring wildlife richness—Finnish applications based on wildlife triangle censuses. Annales Zoologici Fennici 42:123–134.

Pereboom, V., M. Mergey, N. Villerette, R. Helder, J.-F. Gerard, and T. Lodé. 2008. Movement patterns, habitat selection, and corridor use of a typical woodland-dweller species, the European pine marten (*Martes martes*), in fragmented landscape. Canadian Journal of Zoology 86:983–991.

Perez, J.M., and R.L. Palma. 2001. A new species of *Felicola* (Phthiraptera: Trichodectidae) from the endangered Iberian lynx: another reason to ensure its survival. Biodiversity and Conservation 10:929–937.

Perkins, S.E., S. Altizer, O. Bjornstad, J.J. Burdon, K. Clay, L. Gomez-Aparicio, J.M. Jesche, P.T.J. Johnson, K.D. Lafferty, C.M. Malstrom, P. Martin, A. Power, D.L. Strayer, P.H. Thrall, and M. Uriarte. 2008. Invasion biology and parasitic infections. Pages 179–204 *in* Infectious disease ecology: effects of ecosystems on disease and of diseases on ecosystems. R.S. Ostfeld, F. Keesing, and V.T. Eviner, editors. Princeton University Press, New Jersey, USA.

Pertoldi, C., S.F. Barker, A.B. Madsen, H. Jørgensen, E. Randi, J. Muñoz, H.J. Baagoe, and V. Loeschcke. 2008a. Spatio-temporal population genetics of the Danish pine marten (*Martes martes*). Biological Journal of the Linnean Society 93:457–464.

Pertoldi, C., R. Bijlsma, and V. Loeschcke. 2007. Conservation genetics in a globally changing environment: present problems, paradoxes and future challenges. Biodiversity and Conservation 16:4147–4163.

Pertoldi, C., M.M. Hansen, V. Loeschcke, A.B. Madsen, L. Jacobsen, and H. Baagoe. 2001. Genetic consequences of population decline in European otter *Lutra lutra*: an assessment of microsatellite DNA variation in Danish otters from 1883 to 1993. Proceedings of the Royal Society B 268:1775–1781.

Pertoldi, C., J. Munoz, A.B. Madsen, J.S.F. Barker, D.H.H. Andersen, J. Baagøe, M. Birch, and V. Loeschcke. 2008b. Genetic variability in the mitochondrial DNA of the Danish pine marten. Journal of Zoology 276:168–175.

Pertoldi, C., and C. Topping. 2004. Impact assessment predicted by means of genetic agent-based modelling. Critical Reviews in Toxicology 34:487–498.

Petersen, L.R., M.A. Martin, and C.M. Fils. 1977. Status of fishers in Wisconsin, 1975. Report 92, Wisconsin Department Natural Resources, Madison, USA.

Peterson, M., and M.J. Manfredo. 1993. Social science and the evolving conservation philosophy. Pages 292–304 *in* Conservation and resource management. S.K. Majumder, E.W. Miler, D.E. Baker, E.K. Brown, J.R. Pratt, and R.T. Schmalz, editors. Pennsylvania Academy of Sciences, Pittsburgh, USA.

Petit, E., and N. Valiere. 2006. Estimating population size with noninvasive capture-mark-recapture data. Conservation Biology 20:1062–1073.

Petrovskaya, A.V. 2007. Genetic structure of the sable *Martes zibellina* L. populations from Magadan Oblast as inferred from mitochondrial DNA variation. Russian Journal of Genetics 43:424–429.

Petter, G. 1965. Deux mustelids nouveaux du pontien d'espagne orientale. Bulletin du Muséum National d'Histoire Naturelle, 2ème série 36:270–278.

Philippa, J., C. Fournier-Chambrillon, P. Fournier, W. Schaftenaar, M. van de Bildt, R. van Herweijnen, T. Kuiken, M. Liabeuf, S. Ditcharry, L. Joubert, M. Begnier, and A. Osterhaus. 2008. Serologic survey for selected viral pathogens in free-ranging endangered European mink (*Mustela lutreola*) and other mustelids from south-western France. Journal of Wildlife Diseases 44:791–801.

Philippa, J.D., F.A. Leighton, P.Y. Daoust, O. Nielsen, M. Pagliarulo, H. Schwantje, T. Shury, R. Van Herwijnen, B.E. Martina, T. Kuiken, M.W. Van de Bildt, and A.D. Osterhaus. 2004. Antibodies to selected pathogens in free-ranging terrestrial carnivores and marine mammals in Canada. Veterinary Record 155:135–140.

Phillips, D.M., D.J. Harrison, and D.C. Payer. 1998. Seasonal changes in home range area and fidelity of martens. Journal of Mammalogy 79:180–190.

Phillips, M.K., and D.W. Smith. 1998. The wolves of Yellowstone. Voyageur Press, Minneapolis, Minnesota, USA.

Phillips, S.J., and M. Dudik. 2008. Modeling of species distributions with Maxent: new extensions and a comprehensive evaluation. Ecography 31:161–175.

Pielou, E.C. 1991. After the Ice Age: the return of life to glaciated North America. University of Chicago Press, Illinois, USA.

Piggott, M.P. 2004. Effect of sample age and season of collection on the reliability of microsatellite genotyping of faecal DNA. Wildlife Research 31:485–493.

Piggott, M.P., and A.C. Taylor. 2003. Remote collection of animal DNA and its applications in conservation management and understanding the population biology of rare and cryptic species. Wildlife Research 30:1–13.

Pilot, M., B. Gralak, J. Goszczyński, and M. Posłuszny. 2007. A method of genetic identification of pine marten (*Martes martes*) and stone marten (*Martes foina*) and its application to faecal samples. Journal of Zoology 271:140–147.

Poizat, G., and D. Pont. 1996. Multi-scale approach to species-habitat relationships: juvenile fish in a large river section. Freshwater Biology 36:611–622.

Pollock, K.H., J.D. Nichols, T.R. Simon, G.L. Farnsworth, L.L. Bailey, and J.R. Sauer. 2002. Large scale wildlife monitoring studies: statistical methods for design and analysis. Envirometrics 13:105–119.

Poole, B.C., K. Chadee, and T.A. Dick. 1983. Helminth parasites of pine marten, *Martes americana* (Turton), from Manitoba, Canada. Journal of Wildlife Diseases 19:10–13.

Poole, K.G., A.D. Porter, A. de Vries, C. Maundrell, S.D. Grindal, and C.C. St. Clair. 2004. Suitability of a young deciduous-dominated forest for American marten and the effects of forest removal. Canadian Journal of Zoology 82:423–435.

Porter, A.S., C.C. St. Clair, and A. de Vries. 2005. Fine-scale selection by marten during winter in a young deciduous forest. Canadian Journal of Forest Research 35:901–909.

Posillico, M., P. Serafini, and S. Lovari. 1995. Activity patterns of the stone marten *Martes foina* Erxleben, 1777, in relation to some environmental factors. Hystrix 7:79–97.

Posluszny, M., M. Pilot, J. Goszczynski, and B. Gralak. 2007. Diet of sympatric pine marten (*Martes martes*) and stone marten (*Martes foina*) identified by genotyping of DNA from faeces. Annales Zoologici Fennici 44:269–284.

Postma, E., and A.J. van Noordwijk. 2005. Gene flow maintains a large genetic difference in clutch size at a small spatial scale. Nature 433:65–68.

Potter, D. 2002. Modelling fisher (*Martes pennanti*) habitat associations in Nova Scotia. Thesis, Acadia University, Wolfville, Nova Scotia, Canada.

Potvin, F. 1998. La martre d'amerique (*Martes americana*) et la coupe a blanc en foret boreal: une approche telemetrique et geomatique. Dissertation, Universite Laval, Quebec, Canada.

Potvin, F., L. Bélanger, and K. Lowell. 2000. Marten habitat selection in a clearcut boreal landscape. Conservation Biology 14:844–857.

Potvin, F., and L. Breton. 1997. Short-term effects of clearcutting on martens and their prey in the boreal forest of western Quebec. Pages 452–474 *in Martes*: taxonomy, ecology, techniques, and management. G. Proulx, H.N. Bryant, and P.M. Woodard, editors. Provincial Museum of Alberta, Edmonton, Canada.

Poulton, S., J.D.S. Birks, J.E. Messenger, and D.J. Jefferies. 2006. A quality-scoring system for using sightings data to assess pine marten distribution at low densities. Pages 177–202 *in Martes* in carnivore communities. M. Santos-Reis, J.D.S. Birks, E.C. O'Doherty, and G. Proulx, editors. Alpha Wildlife Publications, Sherwood Park, Alberta, Canada.

Powell, R.A. 1977. Hunting behaviour, ecological energetics and predator-prey community stability of the fisher (*Martes pennanti*). Dissertation, University of Chicago, Illinois, USA.

Powell, R.A. 1979a. Ecological energetics and foraging strategies of the fisher (*Martes pennanti*). Journal of Animal Ecology 48:195–212.

Powell, R.A. 1979b. Mustelid spacing patterns: variations on a theme by *Mustela*. Zietschrift für Tierpsychologie 50:153–165.

Powell, R.A. 1981. *Martes pennanti*. Mammalian Species 156:1–6.

Powell, R.A. 1982. The fisher: life history, ecology, and behavior. University of Minnesota Press, Minneapolis, USA.

Powell, R.A. 1993. The fisher: life history, ecology and behavior. Second edition. University of Minnesota Press, Minneapolis, USA.

Powell, R.A. 1994a. Evolution and biogeography: introduction. Pages 11–12 *in* Martens, sables, and fishers: biology and conservation. S.W. Buskirk, A.S. Harestad, M.G. Raphael, and R.A. Powell, editors. Cornell University Press, Ithaca, New York, USA.

Powell, R.A. 1994b. Structure and spacing of *Martes* populations. Pages 101–121 *in* Martens, sables, and fishers: biology and conservation. S.W. Buskirk, A.S. Harestad, M.G. Raphael, and R.A. Powell, editors. Cornell University Press, Ithaca, New York, USA.

Powell, R.A. 2000. Animal home ranges and territories and home range estimators. Pages 65–110 *in* Research techniques in animal ecology: controversies and consequences. L. Boitani and T.K. Fuller, editors. Columbia University Press, New York, USA.

Powell, R.A. 2004. Home ranges, cognitive maps, habitat models and fitness landscapes for *Martes*. Pages 135–146 *in* Martens and fishers (*Martes*) in human-altered landscapes: an international perspective. D.J. Harrison, A.K. Fuller, and G. Proulx, editors. Springer Science+Business Media, New York, USA.

Powell, R.A., S.W. Buskirk, and W.J. Zielinski. 2003. Fisher and marten: *Martes pennanti* and *Martes americana*. Pages 635–649 *in* Wild mammals of North America: biology, management and conservation. Second edition. G.A. Feldhamer, B.C. Thompson, and J.A. Chapman, editors. Johns Hopkins University Press, Baltimore, Maryland, USA.

Powell, R.A., and G. Proulx. 2003. Trapping and marking terrestrial mammals for research: integrating ethics, standards, techniques, and common sense. Institute of Laboratory Animal Research Journal 44:259–276.

Powell, R.A., and W.J. Zielinski. 1994. Fisher. Pages 38–73 *in* The scientific basis for conserving forest carnivores: American marten, fisher, lynx, and wolverine in the western

United States. L.F. Ruggiero, K.B. Aubry, S.W. Buskirk, L.J. Lyon, and W.J. Zielinski, technical editors. USDA Forest Service, General Technical Report RM-254.

Powell, S.M., E.C. York, and T.K. Fuller. 1997. Seasonal food habits of fishers in central New England. Pages 279–305 *in Martes*: taxonomy, ecology, techniques and management. G.Proulx, H.N. Bryant, and P.M. Woodard, editors. Provincial Museum of Alberta, Edmonton, Canada.

Pozio, E. 2000. Factors affecting the flow among domestic, synanthropic and sylvatic cycles of *Trichinella*. Veterinary Parasitology 93:241–262.

Pozio, E. 2005. The broad spectrum of *Trichinella* hosts: from cold to warm-blooded animals. Veterinary Parasitology 132:3–11.

Pozio, E., G. La Rosa, D.S. Zarlenga, and E.P. Hoberg. 2009a. Molecular taxonomy and phylogeny of nematodes of the genus *Trichinella*. Infection Genetics and Evolution 9:606–616.

Pozio, E., L. Rinaldi, G. Marucci, V. Musella, F. Galati, G. Cringoli, P. Boireau, and G. La Rosa. 2009b. Hosts and habitats of *Trichinella spiralis* and *Trichinella britovi* in Europe. International Journal for Parasitology 39:71–79.

Prasad, A.M., L.R. Iverson, and A. Liaw. 2006. Newer classification and regression tree techniques: bagging and random forests for ecological prediction. Ecosystems 9:181–199.

Preble, E.A. 1902. A biological investigation of the Hudson Bay Region. North American Fauna 22:1–140.

Prell, H. 1928. Die fortpflanzungsbiologie des Amerikanischen fichtenmardes *Martes americana* Turt. Die Pelztierzucht 4, Germany.

Presley, S.G. 2000. *Eira barbara*. Mammalian Species 636:1–6.

Pridham, T.J., and J. Belcher. 1958. Toxoplasmosis in mink. Canadian Journal of Comparative Medicine and Veterinary Science 22:99–106.

Prigioni, C., A. Balestrieri, L. Remonti, and L. Cavada. 2008. Differential use of food and habitat by sympatric carnivores in the eastern Italian Alps. Italian Journal of Zoology 75:173–184.

Proctor, M.F., B.N. McLellan, C. Strobeck, and R.M.R. Barclay. 2005. Genetic analysis reveals demographic fragmentation of grizzly bears yielding vulnerably small populations. Proceedings of the Royal Society B 272:2409–2416.

Promislow, D.E.L., and P.H. Harvey. 1990. Living fast and dying young: a comparative analysis of life-history variation among mammals. Journal of Zoology 220:417–437.

Proulx, G. 1997. Improved trapping standards for marten and fisher. Pages 362–371 *in Martes*: taxonomy, ecology, techniques and management. G. Proulx, H.N. Bryant, and P.M. Woodard, editors. Provincial Museum of Alberta, Edmonton, Canada.

Proulx, G. 1999a. Review of current mammal trap technology in North America. Pages 1–46 *in* Mammal trapping. G. Proulx, editor. Alpha Wildlife Publications, Sherwood Park, Alberta, Canada.

Proulx, G. 1999b. The Bionic: an effective marten trap. Pages 79–87 *in* Mammal trapping. G. Proulx, editor, Alpha Wildlife Publications, Sherwood Park, Alberta, Canada.

Proulx, G. 2000. The impact of human activities on North American mustelids. Pages 53–75 *in* Mustelids in a modern world: management and conservation aspects of small carnivore:human interactions. H.I. Griffiths, editor. Backhuys Publishers, Leiden, The Netherlands.

Proulx, G. 2005a. Integrating the habitat needs of fine- and coarse-filter species in landscape planning. *In* Proceedings of the Species at Risk 2004: Pathways to Recovery Conference. T.D. Hooper, editor. Species at Risk 2004: Pathways to Recovery Conference Organizing Committee, Victoria, British Columbia, Canada. http://www.llbc.leg.bc.ca/Public/PubDocs/bcdocs/400484/proulx_edited_final_march_17.pdf (accessed 21 March 2011).

Proulx, G. 2005b. Predicting the distribution of mountain caribou, wolverine, fisher, and grizzly bear in Prince George Forest District: III. Field verification of fisher predictive distribution maps, and observations on other species at risk in Tree Farm Licence 30. Final Report to Canadian Forest Products Ltd., Prince George, British Columbia, Canada.

FIA Agreement No. 2003-FIA-11. http://www.env.gov.bc.ca/wildlife/wsi/reports/3754_WSI_3754_RPT.PDF (accessed 2 April 2010).

Proulx, G. 2006a. Using forest inventory data to predict winter habitat use by fisher *Martes pennanti* in British Columbia, Canada. Acta Theriologica 51:275–282.

Proulx, G. 2006b. Winter habitat use by American marten in western Alberta boreal forests. Canadian Field-Naturalist 120:100–105.

Proulx, G. 2009. Conserving American marten *Martes americana* winter habitat in sub-boreal spruce forests affected by mountain pine beetle *Dendroctonus ponderosae* infestations and logging in British Columbia, Canada. Small Carnivore Conservation 41:51–57.

Proulx, G. 2010. Fisher. The Canadian Encyclopedia, Historica Foundation of Canada. http://www.thecanadianencyclopedia.com/PrinterFriendly.cfm?Para ms=A1ARTA0002826 (accessed 21 March 2011).

Proulx, G., K. Aubry, J. Birks, S. Buskirk, C. Fortin, H. Frost, W. Krohn, L. Mayo, V. Monakhov, D. Payer, M. Saeki, M. Santos-Reis, R. Weir, and W. Zielinski. 2004. World distribution and status of the genus *Martes* in 2000. Pages 21–76 *in* Martens and fishers (*Martes*) in human-altered environments: an international perspective. D.J. Harrison, A.K. Fuller, and G. Proulx, editors. Springer Science+Business Media, New York, USA.

Proulx, G., and M.W. Barrett. 1989. Animal welfare concerns and wildlife trapping: ethics, standards and commitments. Transactions of the Western Section of the Wildlife Society 25:1–6.

Proulx, G., and M.W. Barrett. 1994. Ethical considerations in the selection of traps to harvest American martens and fishers. Pages 192–196 *in* Martens, sables, and fishers: biology and conservation. S.W. Buskirk, A.S. Harestad, M.G. Raphael, and R.A. Powell, editors. Cornell University Press, Ithaca, New York, USA.

Proulx, G., M.W. Barrett, and S.R. Cook. 1989a. The C120 Magnum: an effective quick-kill trap for marten. Wildlife Society Bulletin 17:294–298.

Proulx, G., H.N. Bryant, and P.M. Woodard, editors. 1997. *Martes* taxonomy, ecology, techniques, and management. Provincial Museum of Alberta, Edmonton, Canada.

Proulx, G., M.R.L. Cattet, and R.A. Powell. 2012. Humane and efficient capture and handling methods for carnivores. Pages 70–129 *in* Carnivore ecology and conservation: a handbook of techniques. L. Boitani and R.A. Powell, editors. Oxford University Press, London, UK.

Proulx, G., S.R. Cook, and M.W. Barrett. 1989b. Assessment and preliminary development of the rotating-jaw Conibear 120 trap to effectively kill marten (*Martes americana*). Canadian Journal of Zoology 67:1074–1079.

Proulx, G., and B. Genereux. 2009. Persistence of a reintroduced fisher, *Martes pennanti*, population in Cooking Lake-Blackfoot Provincial Recreation Area, Central Alberta. Canadian Field-Naturalist 123:178–181.

Proulx, G., A.J. Kolenosky, M.J. Badry, R.K. Drescher, K. Seidel, and P.J. Cole. 1994. Post-release movements of translocated fishers. Pages 197–203 *in* Martens, sables, and fishers: biology and conservation. S.W. Buskirk, A.S. Harestad, M.G. Raphael, and R.A. Powell, editors. Cornell University Press, Ithaca, New York, USA.

Proulx, G., and E.C. O'Doherty. 2006. Snow-tracking to determine *Martes* winter distribution and habitat use. Pages 211–224 *in Martes* in carnivore communities. M. Santos-Reis, J.D.S. Birks, E.C. O'Doherty, and G. Proulx, editors. Alpha Wildlife Publications, Sherwood Park, Alberta, Canada.

Proulx, G., and R. Verbisky. 2001. Occurrence of American marten within a connectivity corridor. *Martes* Working Group Newsletter 9:4–5.

Pulliainen, E. 1981a. A transect survey of small land carnivore and red fox populations on a subarctic fell in Finnish Forest Lapland over 13 winters. Annales Zoologici Fennici 18:270–278.

Pulliainen, E. 1981b. Food and feeding habits of the pine marten in Finnish Forest Lapland in winter. Pages 580–598 *in* Worldwide furbearer conference proceedings. J.A. Chapman and D. Pursley, editors. Worldwide Furbearer Conference, Frostburg, Maryland, USA.

Pulliainen, E. 1981c. Winter habitat selection, home range, and movements of the pine marten (*Martes martes*) in a Finnish Lapland forest. Pages 1068–1087 *in* Worldwide furbearer conference proceedings. J.A. Chapman and D. Pursley, editors. Worldwide Furbearer Conference, Frostburg, Maryland, USA.

Pulliainen, E. 1984. Use of the home range by pine martens (*Martes martes* L.). Acta Zoologica Fennica 171:271–274.

Purcell, K.L., A.K. Mazzoni, S.R. Mori, and B.B. Boroski. 2009. Resting structures and resting habitat of fishers in the southern Sierra Nevada, California. Forest Ecology and Management 258:2696–2706.

Putnam, R.J. 1984. Facts from faeces. Mammal Review 14:79–97.

Quann, J.D. 1985. Fundy National Park American marten reintroduction program, 1985. Unpublished report to Environment Canada, Parks Canada, Natural Resources Conservation, Fundy National Park, New Brunswick, Canada.

Quick, H.F. 1956. Effects of exploitation on a marten population. Journal of Wildlife Management 20:267–274.

Quigley, H.B., and P.G. Crawshaw. 1989. Use of ultralight aircraft in wildlife radio telemetry. Wildlife Society Bulletin 17:330–340.

Quigley, T.M., R.W. Haynes, and R.T. Graham, editors. 1996. Integrated scientific assessment for ecosystem management in the interior Columbia Basin. USDA Forest Service, General Technical Report PNW-GTR-382.

Quinn, D.B., editor. 1979. New American world: a documentary history of North America to 1612. Volume 1. Macmillan Press, London, UK.

Quinn, P.J., M.E. Carter, B. Markey, and G.R. Carter. 1999. Clinical veterinary microbiology. Elsevier, Philadelphia, Pennsylvania, USA.

R Development Core Team. 2009. R: a language and environment for statistical computing. R Foundation for Statistical Computing, Vienna, Austria. ISBN 3–900051–07–0. http://www.R-project.org (accessed May 2011).

Raclot, T., and R. Groscolas. 1995. Selective mobilization of adipose tissue fatty acids during energy depletion in the rat. Journal of Lipid Research 36:2164–2173.

Raclot, T., E. Mioskowski, A.C. Bach, and R. Groscolas. 1995. Selectivity of fatty acid mobilization: a general metabolic feature of adipose tissue. American Journal of Physiology 269:R1060–R1067.

Rademacher, U., W. Jakob, and I. Bockhardt. 1999. *Cryptosporidium* infection in beech martens (*Martes foina*). Journal of Zoo and Wildlife Medicine 30:421–422.

Rafinesque, C.S. 1819. Description of a new species of North American marten (*Mustela vulpina*). American Journal of Science 1:82–84.

Raine, R.M. 1981. Winter food habits, responses to snow cover and movements of fisher (*Martes pennanti*) and marten (*Martes americana*) in southeastern Manitoba. Thesis, University of Manitoba, Winnipeg, Canada.

Raine, R.M. 1982. Ranges of juvenile fisher, *Martes pennanti*, and marten, *Martes americana*, in southeastern Manitoba. Canadian Field-Naturalist 96:431–438.

Raine, R.M. 1983. Winter habitat use and responses to snow cover of fisher (*Martes pennanti*) and marten (*Martes americana*) in southeastern Manitoba. Canadian Journal of Zoology 61:25–34.

Raine, R.M. 1987. Winter food habits and foraging behaviour of fishers (*Martes pennanti*) and martens (*Martes americana*) in southeastern Manitoba. Canadian Journal of Zoology 65:745–747.

Ralls, K., D.B. Siniff, T.D. Williams, and V.B. Kuechle. 2006. An intraperitoneal radio transmitter for sea otters. Marine Mammal Science 5:376–381.

Ramsden, R., and D. Johnston. 1975. Studies on the oral infectivity of rabies virus in Carnivora. Journal of Wildlife Diseases 11:318–324.

Rand, A.L. 1944. The status of the fisher *Martes pennanti* (Erxleben), in Canada. Canadian Field-Naturalist 58:77–81.

Randall, D.A., J. Marino, D.T. Haydon, C. Sillero-Zubiri, D.L. Knobel, L.A. Tallents, D.W. Macdonald, and M.K. Laurenson. 2006. An integrated disease management strategy for the control of rabies in Ethiopian wolves. Biological Conservation 131:151–162.

Raphael, M.G. 1984. Wildlife populations in relation to stand age and area in Douglas-fir forests of northwestern California. Pages 259–274 *in* Fish and wildlife relationships in old-growth forests: proceedings of a symposium. W.R. Meehan, T.R. Merrell, Jr., and T.A. Hanley, editors. American Institute of Fisheries Research Biologists, Juneau, Alaska, USA.

Raphael, M.G. 1994. Techniques for monitoring populations of fishers and American martens. Pages 224–240 *in* Martens, sables, and fishers: biology and conservation. S.W. Buskirk, A.S. Harestad, M.G. Raphael, and R.A. Powell, editors. Cornell University Press, Ithaca, New York, USA.

Raphael, M.G., and L.L.C. Jones. 1997. Characteristics of resting and denning sites of American martens in central Oregon and western Washington. Pages 146–165 *in Martes*: taxonomy, ecology, techniques, and management. G. Proulx, H.N. Bryant, and P.M. Woodard, editors. Provincial Museum of Alberta, Edmonton, Canada.

Raphael, M.G., R. Molina, C.H. Flather, R. Holthausen, R.L. Johnson, B.G. Marcot, D.H. Olson, J.D. Peine, C.H. Sieg, and C.S. Swanson. 2007. A process for selection and implementation of conservation approaches. Pages 334–362 *in* Conservation of rare or little-known species: biological, social, and economic considerations. M.G. Raphael and R. Molina, editors. Island Press, Washington, D.C., USA.

Raphael, M.G., M.J. Wisdom, M.M. Rowland, R.S. Holthausen, B.C. Wales, B.G. Marcot, and T.D. Rich. 2001. Status and trends of habitats of terrestrial vertebrates in relation to land management in the interior Columbia River Basin. Forest Ecology and Management 153:63–87.

Rasmussen, A.M., A.B. Madsen, T. Asferg, B. Jensen, and M. Rosengaard. 1986. Investigations of the stone marten (*Martes foina*) in Denmark. Danske Vildtundersoegelser 41:1–39.

Rausch, R.L. 1977. The specific distinction of *Taenia twitchelli* Schwartz, 1924 from *T. martis* (Zeder, 1803) (Cestoda: Taeniidae). Pages 357–366 *in* Excerta parasitological en memoria del Doctor Eduardo Caballero y Caballero. Universidad Nacional Autónoma de México. Instituto Biología, Publicaciones especiales No. 4. México.

Rausch, R.L. 1994. Transberingian dispersal of cestodes in mammals. International Journal for Parasitology 24:1203–1212.

Rausch, R.L. 2003. *Taenia pencei* n. sp. from the ringtail, *Bassariscus astutus* (Carnivora: Procyonidae), in Texas, U.S.A. Comparative Parasitology 70:1–10.

Rausch, R.L., F.H. Fay, and F.S.L. Williamson. 1990. The ecology of *Echinococcus multilocularis* (Cestoda, Taeniidae) on St. Lawrence Island, Alaska. II. Helminth populations in the definitive host. Annales de Parasitologie Humaine et Comparee 65:131–140.

Ray, A.J. 1987. The fur trade in North America: an overview from a historical geographical perspective. Pages 21–30 *in* Wild furbearer management and conservation in North America. M. Novak, J.A. Baker, M.E. Obbard, and B. Malloch, editors. Ontario Trappers Association and Ontario Ministry of Natural Resources, Toronto, Canada.

Ray, J.C. 2000. Mesocarnivores of northeastern North America: status and conservation issues. WCS Working Paper No. 15. Wildlife Conservation Society, Bronx, New York, USA.

Ray, J.C., and W.J. Zielinski. 2008. Track stations. Pages 75–109 *in* Noninvasive survey methods for carnivores. R.A. Long, P. MacKay, W.J. Zielinski, and J.C. Ray, editors. Island Press, Washington, D.C., USA.

Rayfield, B., A. Moilanen, and M.-J. Fortin. 2009. Incorporating consumer-resource spatial interactions in reserve design. Ecological Modelling 220:725–733.

Raymond, W.O. 1950. The River St. John, its physical features, legends and history from 1604 to 1784. Second edition. J.C. Webster, editor. Tribune Press, Sackville, New Brunswick, Canada. First published 1910 by J.A. Bowes, St. John, New Brunswick.

Reading, R.P., and T.W. Clark. 1996. Carnivore reintroductions: an interdisciplinary examination. Pages 296–336 *in* Carnivore behavior, ecology, and evolution. Volume 2. J.L. Gittleman, editor. Cornell University Press, Ithaca, New York, USA.

Rego, P.W. 1989. Wildlife investigation: fisher reintroduction, 10/1/88–9/30/89. Unpublished report, Project number W-49-R-14, Connecticut Department of Environmental Protection, Wildlife Division, Burlington, USA.

Rego, P.W. 1990. Wildlife investigation: fisher reintroduction, 10/1/89–9/30/90. Unpublished report, Project number W-49-R-14, Connecticut Department of Environmental Protection, Wildlife Division, Burlington, USA.

Rego, P.W. 1991. Wildlife investigation: fisher reintroduction, 10/1/90–9/30/91. Unpublished report, Project number W-49-R-14, Connecticut Department of Environmental Protection, Wildlife Division, Burlington, USA.

Reid, D.G., W.E. Melquist, J.D. Woolington, and J.M. Noll. 1986. Reproductive effects of intraperitoneal transmitter implants in river otters. Journal of Wildlife Management 50:92–94.

Reid, F., and K. Helgen. 2008. *Martes pennanti*. IUCN Red List of Threatened Species. Version 2010.4. http://www.iucnredlist.org (accessed 24 February 2011).

Reid, W.V., and G.M. Mace. 2003. Taking conservation biology to new levels in environmental decision-making. Conservation Biology 17:943–945.

Reig, S. 1992. Geographic variation in pine marten (*Martes martes*) and beech marten (*M. foina*) in Europe. Journal of Mammalogy 73:744–769.

Reinhardt, E., and N.L. Crookston. 2003. The fire and fuels extension to the forest vegetation simulator. USDA Forest Service, General Technical Report RMRS-GTR-116.

Reinhardt, H. 1929. Beitrag zur zucht und jugendentwicklung des steinmarders. Deutsche Pelztierzuchter 15:445–448.

Renard, A., M. Lavoie, and S. Larivière. 2008. Differential footload of male and female fisher, *Martes pennanti*, in Quebec. Canadian Field-Naturalist 122:269–270.

Reno, M.A., K.R. Rulon, and C.E. James. 2008. Fisher monitoring within two industrially managed forests of northern California. Unpublished progress report to California Department of Fish and Game. Sierra Pacific Industries, Anderson, California, USA.

Rettie, A. 1971. Summary of marten and fisher transplanting in Parry Sound Forest District in 1956–63. Unpublished report to Ontario Ministry of Natural Resources, Toronto, Canada.

Rettie, W.J., and F. Messier. 2000. Hierarchical habitat selection by woodland caribou: its relationship to limiting factors. Ecography 23:466–478.

Rezendes, P. 1992. Tracking and the art of seeing: how to read animal tracks and sign. Camden House Publishing, Charlotte, Vermont, USA.

Rhoads, S.N. 1902. Synopsis of the American martens. Proceedings of the Academy of Natural Sciences, Philadelphia 54:443–460.

Rhoads, S.N. 1903. The mammals of Pennsylvania and New Jersey: a biographic, historic and descriptive account. Privately printed, Philadelphia, Pennsylvania, USA.

Ribas, A., C. Milazzo, P. Foronda, and J.C. Casanova. 2004. New data on helminths of stone marten, *Martes foina* (Carnivora, Mustelidae), in Italy. Helminthologia 41:59–61.

Riddle, A.E., K.L. Pilgrim, L.S. Mills, K.S. McKelvey, and L.F. Ruggiero. 2003. Identification of mustelids using mitochondrial DNA and non-invasive sampling. Conservation Genetics 4:241–243.

Rissler, L.J., R.J. Hijmans, C.H. Graham, C. Moritz, and D.B. Wake. 2006. Phylogeographic lineages and species comparisons in conservation analyses: a case study of California herpetofauna. American Naturalist 167:655–666.

Robitaille, J.-F. 2000. Management and conservation of mustelids in Ontario and eastern Canada. Pages 77–96 *in* Mustelids in a modern world: management and conservation aspects of small carnivore:human interactions. H.I. Griffiths, editor. Backhuys Publishers, Leiden, The Netherlands.

Robitaille, J.-F., and K. Aubry. 2000. Occurrence and activity of American martens *Martes americana* in relation to roads and other routes. Acta Theriologica 45:137–143.

Robitaille, J.-F., and E.W. Cobb. 2003. Indices to estimate fat depots in American marten *Martes americana*. Wildlife Biology 9:113–121.

Robitaille, J.-F., and K. Jensen. 2005. Additional indices to estimate fat contents in fisher *Martes pennanti* populations. Wildlife Biology 11:263–269.

Roche, T.R. 2008. The use of baited hair traps and genetic analysis to determine the presence of pine marten. Thesis, Waterford Institute of Technology, Ireland.

Rodgers, A.R. 2001. Tracking animals with GPS: the first 10 years. Pages 1–10 *in* Tracking animals with GPS. A.M. Sibbald and I.J. Gordon, editors. The Macaulay Institute, Aberdeen, Scotland.

Rodgers, A.R., R.S. Rempel, and K.F. Abraham. 1996. A GPS-based telemetry system. Wildlife Society Bulletin 24:559–566.

Rodriguez, A., and E. Carbonell. 1998. Gastrointestinal parasites of the Iberian lynx and other wild carnivores from central Spain. Acta Parasitologica 43:128–136.

Roemer, G.W., M.E. Gompper, and B.V. Valkenburgh. 2009. The ecological role of the mammalian mesocarnivore. BioScience 59:165–173.

Rognrud, M. 1983. General wildlife restocking in Montana, 1941–1982. Final report, Federal Aid in Wildlife Restoration Project W-5-D, Montana Department of Fish, Wildlife, and Parks, Helena, USA.

Romashov, B.V. 2001. Three capillariid species (Nematoda, Capillariidae) from carnivores (Carnivora) and discussion of system and evolution of the nematode family Capillariidae. 2. *Eucoleus trophimenkovi* sp. n. from the marten *Martes martes* and discussion of system and evolution of the nematode family Capillariidae. Zoologichesky Zhurnal 80:145–154.

Romesburg, H.C. 1981. Wildlife science: gaining reliable knowledge. Journal of Wildlife Management 45:293–313.

Rondinini, C., and L. Boitani. 2002. Habitat use by beech martens in a fragmented landscape. Ecography 25:257–264.

Roon, D.A., M.E. Thomas, K.C. Kendall, and L.P. Waits. 2005. Evaluating mixed samples as a source of error in non-invasive genetic studies using microsatellites. Molecular Ecology 14:195–201.

Roon, D.A., L.P. Waits, and K.C. Kendall. 2003. A quantitative evaluation of two methods for preserving hair samples. Molecular Ecology Notes 3:163–166.

Root, T.L., J.T. Price, K.R. Hall, S.H. Schneider, C. Rosenzweig, and J.A. Pounds. 2003. Fingerprints of global warming on wild animals and plants. Nature 421:57–60.

Rosalino, L.M., J. Rosário, and M. Santos-Reis. 2009. The role of habitat patches on mammalian diversity in cork oak agroforestry systems. Acta Oecologica 35:507–512.

Rosalino, L.M., and M. Santos-Reis. 2009. Fruit consumption by carnivores in Mediterranean Europe. Mammal Review 39:67–78.

Rosatte, R., D. Donovan, M. Allan, L. Bruce, T. Buchanan, K. Sobey, B. Stevenson, M. Gibson, T. MacDonald, and M. Whalen. 2009. The control of raccoon rabies in Ontario Canada: proactive and reactive tactics, 1994–2007. Journal of Wildlife Diseases 45:772.

Roscoe, D. 1993. Epizootiology of canine distemper in New Jersey raccoons. Journal of Wildlife Diseases 29:390–395.

Rosellini, S., E. Osorio, A. Ruiz-González, A. Piñeiro, and I. Barja. 2008. Monitoring the small-scale distribution of sympatric European pine martens (*Martes martes*) and stone martens (*Martes foina*): a multievidence approach using faecal DNA analysis and camera-traps. Wildlife Research 35:434–440.

Rosenberg, K.V., and M.G. Raphael. 1986. Effects of forest fragmentation on vertebrates in Douglas-fir forests. Pages 263–272 *in* Wildlife 2000: modeling habitat relationships of terrestrial vertebrates. J. Verner, M.L. Morrison, and C.J. Ralph, editors. University of Wisconsin Press, Madison, USA.

Rosenburg, D.K., K.A. Swindle, and R.G. Anthony. 1994. Habitat associations of California red-backed voles in young and old-growth forests in western Oregon. Northwest Science 68:266–272.

Rosenzweig, M.L. 1981. A theory of habitat selection. Ecology 62:327–335.

Roughton, R.D., and M.W. Sweeny. 1982. Refinements in scent-station methodology for assessing trends in carnivore populations. Journal of Wildlife Management 46:217–229.

Roussiakis, S.J. 2002. Musteloids and feloids (Mammalia, Carnivora) from the late Miocene locality of Pikermi (Attica, Greece). Geobios 35:699–719.

Rouvinen-Watt, K., A.-M. Mustonen, R. Conway, C. Pal, L. Harris, S. Saarela, U. Strandberg, and P. Nieminen. 2010. Rapid development of fasting-induced hepatic lipidosis in the American mink (*Neovison vison*): effects of food deprivation and re-alimentation on body fat depots, tissue fatty acid profiles, hematology and endocrinology. Lipids 45:111–128.

Rovero, F., and A.R. Marshall. 2009. Camera trapping photographic rate as an index of density in forest ungulates. Journal of Applied Ecology 46:1011–1017.

Rowcliffe, J.M., C. Carbone, P.A. Jansen, R. Kays, and B. Kranstauber. 2011. Quantifying the sensitivity of camera traps: an adapted distance sampling approach. Methods in Ecology and Evolution: published online March 1, 2011. doi:10.1111/j.2041–210X.2011.00094.x.

Rowcliffe, J.M., J. Field, S.T. Turvey, and C. Carbone. 2008. Estimating animal density using camera traps without the need for individual recognition. Journal of Applied Ecology 45:1228–1236.

Roy, K.D. 1991. Ecology of reintroduced fishers in the Cabinet Mountains of northwest Montana. Thesis, University of Montana, Missoula, USA.

Royle, J.A., and J.D. Nichols. 2003. Estimating abundance from repeated presence-absence data or point counts. Ecology 84:777–790.

Royle, J.A., J.D. Nichols, K.U. Karanth, and A. Gopalaswamy. 2009. A hierarchical model for estimating density in camera-trap studies. Journal of Applied Ecology 46:118–127.

Royle, J.A., J.R. Stanley, and P.M. Lukacs. 2008. Statistical modeling and inference from carnivore survey data. Pages 293–312 *in* Noninvasive survey methods for carnivores. R.A. Long, P. MacKay, W.J. Zielinski, and J.C. Ray, editors. Island Press, Washington, D.C., USA.

Rozhnov, V.V., I.G. Meschersky, S.L. Pishchulina, and L.V. Simakin. 2010. Genetic analysis of sable (*Martes zibellina*) and pine marten (*M. martes*) populations in sympatric part of distribution area in the Northern Urals. Russian Journal of Genetics 46:488–492.

Ruette, S., J.-M. Vandel, G. Gayet, and M. Alberet. 2005. Caractéristiques paysagères et habitat diurne de la martre dans une zone de Bresse. ONCFS Rapport Scientifique.

Ruggiero, L.F., R.F. Holthausen, B.G. Marcot, K.B. Aubry, J.W. Thomas, and E.C. Meslow. 1988. Ecological dependency: the concept and its implications for research and management. Transactions of the North American Wildlife and Natural Resources Conference 53:115–126.

Ruggiero, L.F., D.E. Pearson, and S.E. Henry. 1998. Characteristics of American marten den sites in Wyoming. Journal of Wildlife Management 62:663–673.

Ruiz-González, A. 2011. Phylogeography and non-invasive landscape genetics of the Euopean pine marten (*Martes martes* L. 1758): insights into ancient and contemporary processes shaping genetic variation. Dissertation, Universidad del Pais, Vasco, Spain.

Ruiz-González, A., J. Rubines, O. Berdión, and B.J. Gómez-Moliner. 2008. A non-invasive genetic method to identify the sympatric mustelids pine marten (*Martes martes*) and stone marten (*Martes foina*): preliminary distribution survey on the northern Iberian Peninsula. European Journal of Wildlife Research 54:253–261.

Ruiz-Olmo, J., and J.M. López-Martín. 1995. Marta. Pages 88–92 *in* Els grans mamífers de Catalunya i Andorra. J. Ruíz-Olmo and À. Aguilar, editors. Lynx Editions, Barcelona, Spain.

Ruiz-Olmo, J., X. Parellada, and J. Porta. 1988. Sobre la distribución y el hábitat de la marta (*Martes martes* L., 1758) en Cataluña. Pirineos 131:85–93.

Rupprecht, C., K. Stöhr, and C. Meredith. 2001. Rabies. Pages 3–36 *in* Infectious diseases of wild mammals. E.S. Williams and I.K. Barker, editors. Iowa State University Press, Ames, USA.

Russell, Y., P. Evans, and G.H. Dodd. 1989. Characterization of the total lipid and fatty acid composition of rat olfactory mucosa. Journal of Lipid Research 30:877–884.

Rutledge, L.Y., J.J. Holloway, B.R. Patterson, and B.N. White. 2009. An improved field method to obtain DNA for individual identification from wolf scat. Journal of Wildlife Management 73:1430–1435.

Ryynänen, H.J., A. Tonteri, A. Vasemägi, and C.R. Primmer. 2007. A comparison of biallelic markers and microsatellites for the estimation of population and conservation genetic parameters in Atlantic salmon (*Salmo salar*). Journal of Heredity 7:692–704.

Sacchi, O., and A. Meriggi. 1995. Habitat requirements of the stone marten (*Martes foina*) on the Tyrrhenian slopes of the northern Apennines. Hystrix 7:99–104.

Saeki, M. 2006. *Martes* issues in the 21st century: lessons to learn from Asia. Pages 21–26 *in Martes* in carnivore communities. M. Santos-Reis, J.D.S. Birks, E.C. O'Doherty, and G. Proulx, editors. Alpha Wildlife Publications, Sherwood Park, Alberta, Canada.

Safronov, V.M., and R.K. Anikin. 2000. Ecology of the sable, *Martes zibellina* (Carnivora, Mustelidae), in northeastern Yakutia. Zoologichesky Zhurnal 79:471–479. (in Russian with English abstract)

Safronov V.M., E.W.S. Zakharov, and N.N. Smetanin. 2009. Numbers and ecology of the sable in the south Yakutia. International Union of Game Biologists, Abstracts from the 29th Annual Meeting, Moscow, Russia. http://www.new-staryi.ru/cd/docs/s4/Safonov %20V.M.pdf (accessed 21 May 2011).

Sagarin, R. 2002. Historical studies of species' responses to climate change: promise and pitfalls. Pages 127–163 *in* Wildlife responses to climate change: North American case studies. S.H. Schneider and T.L. Root, editors. Island Press, Washington, D.C., USA.

Sajeev, T.K., S.K. Srivastava, M.G. Raphael, S. Dutt, N.K. Ramachandran, and P.C. Tyagi. 2002. Management of forests in India for biological diversity and forest productivity, a new perspective. Volume III: Anaimalai Conservation Area (ACA). Wildlife Institute of India-USDA Forest Service collaborative project report. Wildlife Institute of India, Dehra Dun.

Salafsky, N., R. Margoluis, K.H. Redford, and J.G. Robinson. 2002. Improving the practice of conservation: a conceptual framework and research agenda for conservation science. Conservation Biology 16:1469–1479.

Sălek, M., J. Hreisinger, F. Sedláček, and T. Albrecht. 2009. Corridor vs. hayfield matrix use by mammalian predators in an agricultural landscape. Agriculture, Ecosystems and Environment 134:8–13.

Salwasser, H. 1995. Changing roles for wildlife professionals: where we've been and where we're headed. Transactions of the Western Section of the Wildlife Society 31:1–6.

Samoilov, E.B. 2005. The new sable population in Transbaikalia. Pages 300–303 *in* Conservation and rational use of plant and animal resources. S.M. Muzika, editor. Agricultural Academy Press, Irkutsk, Russia.

Samuel, W.M., M.J. Pybus, and A.A. Kocan. 2001. Parasitic diseases of wild mammals. Second edition. Iowa State University Press, Ames, USA.

Sanchez, D.M., P.R. Krausman, T.R. Livingston, and P.S. Gipson. 2004. Persistence of carnivore scat in the Sonoran Desert. Wildlife Society Bulletin 32:366–372.

Sandell, M. 1990. The evolution of seasonal delayed implantation. Quarterly Review of Biology 65:23–42.

Sano, M., K. Araki, A. Ishii, M. Maeda, and K. Arisaka. 1978. Epidemiological survey of lung fluke *Paragonimus miyazakii* in Shizuoka Prefecture Japan. International Journal of Zoonoses 5:75–79.

Santella, R.M., and S.E. Harkinson. 2008. Biosampling methods. Pages 53–62 *in* Molecular epidemiology: applications in cancer and other human diseases. T.R. Rebbeck, C.B. Ambrosone, and P.G. Shields, editors. Informa Healthcare, New York, USA.

Santos, M.J., and M. Santos-Reis. 2010. Stone marten (*Martes foina*) habitat in a Mediterranean ecosystem: effects of scale, sex and interspecific interactions. European Journal of Wildlife Research 56:275–286.

Santos, N., C. Almendra, and L. Tavares. 2009. Serologic survey for canine distemper virus and canine parvovirus in free-ranging wild carnivores from Portugal. Journal of Wildlife Diseases 45:221–226.

Santos-Reis, M., J.D.S. Birks, E.C. O'Doherty, and G. Proulx, editors. 2006. *Martes* in carnivore communities. Alpha Wildlife Publications, Sherwood Park, Alberta, Canada.

Santos-Reis, M., M.J. Santos, S. Lourenço, J.T. Marques, I. Pereira, and B. Pinto. 2004. Relationships between stone martens, genets and cork oak woodlands in Portugal. Pages 147–172 *in* Martens and fishers (*Martes*) in human-altered environments: an international perspective. D.J. Harrison, A.K. Fuller, and G. Proulx, editors. Springer Science+Business Media, New York, USA.

Sarmento, P., J. Cruz, C. Eira, and C. Fonseca. 2009. Evaluation of camera trapping for estimating red fox abundance. Journal of Wildlife Management 73:1207–1212.

SAS. 2008. SAS, version 9.2. SAS Institute, 100 SAS Campus Drive, Cary, North Carolina, USA 27513–2414. http://www.sas.com (accessed May 2011).

Sato, H., Y. Ihama, T. Inaba, M. Yagisawa, and H. Kamiya. 1999a. Helminth fauna of carnivores distributed in north-western Tohoku, Japan, with special reference to *Mesocestoides paucitesticulus* and *Brachylaima tokudai*. Journal of Veterinary Science 61:1339–1342.

Sato, H., T. Inaba, Y. Ihama, and H. Kamiya. 1999b. Parasitological survey on wild carnivora in north-western Tohoku, Japan. Journal of Veterinary Medical Science 61:1023–1026.

Sato, J.J., T. Hosoda, M. Wolsan, K. Tsuchiya, Y. Yamamoto, and H. Suzuki. 2003. Phylogenetic relationships and divergence times among mustelids (Mammalia: Carnivora) based on nucleotide sequences of the nuclear interphotoreceptor retinoid binding protein and mitochondrial cytochrome *b* genes. Zoological Science 20:243–264.

Sato, J.J., M. Wolsan, S. Minami, T. Hosoda, M.H. Sinaga, K. Hiyuma, Y. Yamaguchi, and H. Suzuki. 2009a. Deciphering and dating the red panda's ancestry and early adaptive radiation of Musteloidea. Molecular Phylogenetics and Evolution 53:907–922.

Sato, J.J., S.P. Yasuda, and T. Hosoda. 2009b. Genetic diversity of the Japanese marten (*Martes melampus*) and its implications for the conservation unit. Zoological Science 26:457–466.

Schaefer, J.A., and F. Messier. 1995. Habitat selection as a hierarchy: the spatial scales of winter foraging by muskoxen. Ecography 18:333–344.

Scharpf, R.F., editor. 1993. Diseases of Pacific Coast conifers. USDA Forest Service, Agricultural Handbook 521.

Scheller, R.M., J.B. Domingo, B.R. Sturtevant, J.S. Williams, A. Rudy, E.J. Gustafson, and D.J. Mladenoff. 2007. Design, development, and application of LANDIS-II, a spatial landscape simulation model with flexible temporal and spatial resolution. Ecological Modelling 201:409–419.

Scheller, R.M, W.D. Spencer, H. Rustigian-Romsos, A.D. Syphard, B.C. Ward, and J.R. Strittholt. 2011. Using stochastic simulation to evaluate competing risks of wildfires and fuels management on an isolated forest carnivore. Landscape Ecology 26:1491–1504.

Scherer, P.E., S. Williams, M. Fogliano, G. Baldini, and H.F. Lodish. 1995. A novel serum protein similar to C1q, produced exclusively in adipocytes. Journal of Biological Chemistry 270:26746–26749.

Schipper, J., K.M. Helgen, J.L. Belant, J. González-Maya, E. Eizirik, and M. Tsuchiya-Jerep. 2009. Small carnivores in the Americas: reflections, future research and conservation priorities. Small Carnivore Conservation 41:1–2.

Schlegel, M.W., S.E. Knapp, and R.E. Millemann. 1968. "Salmon poisoning" disease. V. Definitive hosts of the trematode vector, *Nanophyetus salmincola*. Journal of Parasitology 54:770–774.

Schlexer, F.V. 2008. Attracting animals to detection devices. Pages 262–292 *in* Noninvasive survey methods for carnivores. R.A. Long, P. MacKay, W.J. Zielinski, and J.C. Ray, editors. Island Press, Washington, D.C., USA.

Schmidt, F. 1942. Naturgeschichte des baum-und des steinmarders. Mono. Der Wildsaugeterre. Baud 10:1–258. (in German)

Schmitt, N., J.M. Saville, L. Friis, and P.L. Stovell. 1976. Trichinosis in British Columbia wildlife. Canadian Journal of Public Health 67:21–24.

Schneider, R. 1997. Simulated spatial dynamics of martens in response to habitat succession in the Western Newfoundland Model Forest. Pages 419–436 *in Martes*: taxonomy, ecology, techniques, and management. G. Proulx, H.N. Bryant, and P.M. Woodard, editors. Provincial Museum of Alberta, Edmonton, Canada.

Schneider, R.R., and P. Yodzis. 1994. Extinction dynamics in the American marten (*Martes americana*). Conservation Biology 8:1058–1068.

Schneider, S.H., and T.L. Root, editors. 2002. Wildlife responses to climate change: North American case studies. Island Press, Washington, D.C., USA.

Scholander, P.F., V. Walters, R. Hock, and L. Irving. 1950. Body insulation of some arctic and tropical mammals and birds. Biological Bulletin 99:225–236.

Schoo, G., K. Pohlmeyer, and M. Stoye. 1994. A contribution to the helminth fauna of the stone marten (*Martes foina* Erxleben 1777). Zeitschrift für Jagdwissenschaft 40:84–90.

Schorger, A.W. 1942. Extinct and endangered mammals and birds of the Upper Great Lake region. Transactions of the Wisconsin Academy of Sciences, Arts and Letters 34:23–44.

Schröpfer, R., W. Biedermann, and H. Szczesniak. 1989. Saisonale aktionsraumveränderungen beim baummarder *Martes martes* L. 1758. Pages 433–442 *in* Populationsökologie marderartiger säugetiere 2: wissenschaftliche beiträge 1989/37 (P39). M. Stubbe, editor. Martin-Luther-Universität, Halle-Wittenberg, Deutschland.

Schröpfer, R., P. Wiegand, and H.H. Hogrefe. 1997. The implications of territoriality for the social system of the European pine marten *Martes martes* (L., 1758). Zeitschrift fur Saugetierkunde 62:209–218.

Schulte, R., and U. Fielitz. 2001. High performance GPS collars, use of the latest available technology. Pages 97–101 *in* Tracking animals with GPS. A.M. Sibbald and I.J. Gordon, editors. The Macaulay Institute, Aberdeen, Scotland, UK.

Schumaker, N. 1998. A users guide to the PATCH model. U.S. Environmental Protection Agency, Office of Research and Development EPA/600/R-98/135. Washington, D.C., USA.

Schumaker, N.H., T. Ernst, D. White, J. Baker, and P. Haggerty. 2004. Projecting wildlife responses to alternative future landscapes in Oregon's Willamette basin. Ecological Applications 14:381–400.

Schupbach, T.A. 1977. History, status, and management of the pine marten in the upper peninsula of Michigan. Thesis, Michigan Technological University, Houghton, USA.

Schwartz, M.K. 2007. Ancient DNA confirms native Rocky Mountain fisher (*Martes pennanti*) avoided early 20th century extinction. Journal of Mammalogy 88:921–925.

Schwartz, M.K., J.P. Copeland, N.J. Anderson, J.R. Squires, R.M. Inman, K.S. McKelvey, K.L. Pilgrim, L.P. Waits, and S.A. Cushman. 2009. Wolverine gene flow across a narrow climatic niche. Ecology 90:3222–3232.

Schwartz, M.K., G. Luikart, and R.S. Waples. 2007. Genetic monitoring as a promising tool for conservation and management. Trends in Ecology & Evolution 22:25–33.

Schwartz, M.K., and S.L. Monfort. 2008. Genetic and endocrine tools for carnivore surveys. Pages 238–262 *in* Noninvasive survey methods for carnivores. R.A. Long, P. MacKay, W.J. Zielinski, and J.C. Ray, editors. Island Press, Washington, D.C., USA.

Schwartz, M.K., D.A. Tallmon, and G. Luikart. 1998. Review of DNA-based census and effective population size estimators. Animal Conservation 1:293–299.

Schwartz, M.K., D.A. Tallmon, and G. Luikart. 1999. Using genetics to estimate the size of wild populations: many methods, much potential, uncertain utility. Animal Conservation 2:321–323.

Scott, W.B., and H.F. Osborn. 1887. Preliminary account of the fossil mammals from the White River Formation. Bulletin of the Cambridge Museum of Comparative Zoology, Vol. 13.

Seddon, J.M., H.G. Parker, E.A. Ostrander, and H. Ellegren. 2005. SNPs in ecological and conservation studies: a test in the Scandinavian wolf population. Molecular Ecology 14:503–511.

Seddon, P.J., D.P. Armstrong, and R.F. Maloney. 2007. Developing the science of reintroduction biology. Conservation Biology 21:303–312.

Seddon, P.J., and R.F. Maloney. 2004. Tracking wildlife radio-tag signals by light fixed-wing aircraft. Department of Conservation, Technical Series 30, Wellington, New Zealand.

Sedgeley, J.A. 2001. Quality of cavity microclimate as a factor influencing selection of maternity roosts by a tree-dwelling bat, *Chalinolobus tuberculatus*, in New Zealand. Journal of Applied Ecology 38:425–438.

Seglund, A.E. 1995. The use of rest sites by the Pacific fisher. Thesis, Humboldt State University, Arcata, California, USA.

Segovia, J.-M., J. Torres, J. Miquel, E. Sospedra, R. Guerrero, and C. Feliu. 2007. Analysis of helminth communities of the pine marten, *Martes martes*, in Spain: mainland and insular data. Acta Parasitologica 52:156–164.

Seiler, A., H.H. Kruger, and A. Festetics. 1994. Reaction of a male stone marten (*Martes foina* Erxleben, 1977) to foreign feces within its territory—a field experiment. Zeitschrift fur Saugetierkunde 59:58–60.

Self, S., and S. Kerns. 2001. Pacific fisher use of a managed forest landscape in northern California. Unpublished Wildlife Research Paper No. 6, Sierra Pacific Industries, Redding, California, USA.

Serfass, T.L., R.P. Brooks, and W.M. Tzilkowski. 2001. Fisher reintroduction in Pennsylvania. Final report of the Pennsylvania fisher reintroduction project, Frostburg State University, Maryland, USA.

Seton, E.T. 1909. Life-histories of northern animals, an account of the mammals of Manitoba. Volume 2. Flesh-eaters. Charles Scribner's Sons, New York, USA.

Seton, E.T. 1926. Animals. Doubleday, Doran & Company, Garden City, New York, USA.

Seton, E.T. 1929. Lives of game animals: an account of those land animals in America, north of the Mexican border, which are considered "game." Volume 2, Part 2: bears, cats, badgers, skunks, and weasels. Doubleday, Doran & Company, Garden City, New York, USA.

Seville, R.S., and E.M. Addison. 1995. Nongastrointestinal helminths in marten (*Martes americana*) from Ontario, Canada. Journal of Wildlife Diseases 31:529–533.

Shafer, A.B.A., C.I. Cullingham, S.D. Côté, and D.W. Coltman. 2010. Of glaciers and refugia: a decade of study sheds new light on the phylogeography of northwestern North America. Molecular Ecology 19:4589–4621.

Shaffer, K.E., and W.F. Laudenslayer. 2006. Fire and animal interactions. Pages 118–144 *in* Fire in California ecosystems. N.G. Sugihara, J.W. van Wagtendonk, K.E. Shaffer, J. Fites-Kaufman, and A.E. Thode, editors. University of California Press, Berkeley, USA.

Shahan, A.E., R.L. Knudson, H.R. Seibold, and C.N. Dale. 1947. Aujuszky's disease (pseudorabies): a review with notes on two strains of the virus. North American Veterinarian 28:440–449.

Shattuck, M.R., and S.A. Williams. 2010. Arboreality has allowed for the evolution of increased longevity in mammals. Proceedings of the National Academy of Sciences, USA 107:4635–4639.

Shaw, G., and J. Livingstone. 1992. The pine marten—its reintroduction and subsequent history in the Galloway Forest Park. Transactions of the Dumfriesshire and Galloway Natural History and Antiquarian Society, 3rd Series, 67:1–7.

Shaw, M. 1987. Mount Carleton wilderness: New Brunswick's unknown north. Fiddlehead Poetry and Books and Goose Lane Editions, Fredericton, New Brunswick, Canada.

Sher, A. 1999. Traffic lights at the Beringian crossroads. Nature 397:103–104.

Shimshony, A. 1997. Epidemiology of emerging zoonoses in Israel. Emerging Infectious Diseases 3:229–238.

Shotwell, J.A. 1970. Pliocene mammals of southeast Oregon and adjacent Idaho. University of Oregon, Bulletin of the Museum of Natural History 17:1–103.

Sidorovich, V.E., D.A. Krasko, and A.A. Dyman. 2005. Landscape-related differences in diet, food supply and distribution pattern of the pine marten, *Martes martes* in the transitional mixed forest of northern Belarus. Folia Zoologica 54:39–52.

Sidorovich, V.E., D.A. Krasko, A.A. Sidorovich, I.A. Solovej, and A.A. Dyman. 2006. The pine marten's *Martes martes* ecological niche and its relationships with other vertebrate predators in the transitional mixed forest ecosystems of northern Belarus. Pages 109–126 *in Martes* in carnivore communities. M. Santos-Reis, J.D.S. Birks, E.C. O'Doherty, and G. Proulx, editors. Alpha Wildlife Publications, Sherwood Park, Alberta, Canada.

Siivonen, L., and S. Sulkava. 1994. Pohjolan nisäkkäät (Mammals of northern Europe). Otava, Helsinki, Finland. (in Finnish)

Silver, H. 1957. A history of New Hampshire game and furbearers. New Hampshire Fish and Game Department, Survey Report No. 6, Concord, USA.

Silverman, J., M. Suckow, and S. Murthy. 2006. The IACUC handbook. Second edition. CRC Press, Boca Raton, Florida, USA.

Simon, O., and J. Lang. 2007. Mit Hühneri auf marderfang—methods und fangerfolge in den wäldern um Frankfurt. Natur und Museum 137:1–11.

Simon, T.L. 1980. An ecological study of the marten in the Tahoe National Forest, California. Thesis, California State University, Sacramento, USA.

Simons, E.M. 2009. Influences of past and future forest management on the spatiotemporal dynamics of habitat supply for Canada lynx and American martens in northern Maine. Dissertation, University of Maine, Orono, USA.

Simpson, V.R. 2001. Aleutian disease, mink and otters. Veterinary Record 149:720.

Simpson, V.R., R.J. Panciera, J. Hargreaves, J.W. McGarry, S.F. Scholes, K.J. Bown, and R.J. Birtles. 2005. Myocarditis and myositis due to infection with Hepatozoon species in pine martens (*Martes martes*) in Scotland. Veterinary Record 156:442–446.

Sims, R.A., W.D. Towill, K.A. Baldwin, and G.M. Wickware. 1989. Forest ecosystem classification for northwestern Ontario. Forestry Canada and Ontario Ministry of Natural Resources, OMNR NW Technical Development Unit, Thunder Bay, Canada.

Sinclair, A.R.E. 1991. Science and the practice of wildlife management. Journal of Wildlife Management 55:767–773.

Sinclair, G. 1986. American marten live trapping results for re-introductions to Fundy and Kejimkujik National Parks. Environment Canada, Parks Canada, Natural Resources Conservation, Fundy National Park, New Brunswick, Canada.

Sinclair, G. 1987. Plan update for the reintroduction of the American marten, Fundy National Park. Environment Canada, Parks Canada, Natural Resources Conservation, Fundy National Park, New Brunswick, Canada.

Singleton, P.H., W.L. Gaines, and J.F. Lehmkuhl. 2002. Landscape permeability for large carnivores in Washington: a geographic information system weighted-distance and least-cost corridor assessment. USDA Forest Service, Pacific Northwest Research Station, Research Paper PNW-RP-549.

Sinitsyn, A.A. 2001. The sable. Pages 94–109 *in* Acclimatization and biotechnical measures in game animal population management. N.N. Grakov, editor. All-Russian Research Institute of Hunting and Fur Farming, Kirov, Russia. (in Russian)

Skalski, T., and I. Wierzbowska. 2008. Variation of insect assemblages in fox and marten faeces collected in southern Poland. Annales Zoologici Fennici 45:308–316.

Skirnisson, K. 1986. Untersuchungen zum-raum-zeit-system freilebender steinmarder (*Martes foina* Erxleben 1777). Beiträge zur Wildbiologie 6:1–200.

Skirnisson, K., and D. Feddersen. 1984. Experiences with implantations of transmitters in beech marten (*Martes foina*) living in the wild. Zeitschrift fuer Jagdwissenschaft 30:228–235.

Slauson, K.M., R.L. Truex, and W.J. Zielinski. 2008. Determining the sex of American martens and fishers at track plate stations. Northwest Science 83:185–198.

Slauson, K.M., and W.J. Zielinski. 2001. Distribution and habitat ecology of American martens and Pacific fishers in southwestern Oregon: progress report I, July 1–November 15, 2001. Unpublished report, USDA Forest Service, Pacific Southwest Research Station, Arcata, California, USA.

Slauson, K.M., and W.J. Zielinski. 2003. Distribution and habitat associations of the Humboldt marten (*Martes americana humboldtensis*), and Pacific fisher (*Martes pennanti pacifica*) in Redwood National and State Parks—final report. Unpublished report, USDA Forest Service, Pacific Southwest Research Station, Arcata, California, USA.

Slauson, K.M., and W.J. Zielinski. 2007. Strategic surveys for *Martes* populations in northwestern California: Mendocino National Forest. Unpublished report, USDA Forest Service, Pacific Southwest Research Station, Arcata, California, USA.

Slauson, K.M., W.J. Zielinski, and J.P. Hayes. 2007. Habitat selection by American martens in coastal California. Journal of Wildlife Management 71:458–468.

Slauson, K.M., W.J. Zielinski, and K.D. Stone. 2009. Characterizing the molecular variation among American marten (*Martes americana*) subspecies from Oregon and California. Conservation Genetics 10:1337–1341.

Slough, B.G. 1989. Movements and habitat use by transplanted marten in the Yukon Territory. Journal of Wildlife Management 53:991–997.

Slough, B.G. 1994. Translocations of American martens: an evaluation of factors in success. Pages 165–178 *in* Martens, sables, and fishers: biology and conservation. S.W. Buskirk, A.S. Harestad, M.G. Raphael, and R.A. Powell, editors. Cornell University Press, Ithaca, New York, USA.

Slough, B.G., R.H. Jessup, D.I. McKay, and A.B. Stephenson. 1987. Wild furbearer management in western and northern Canada. Pages 1062–1076 *in* Wild furbearer management and conservation in North America. M. Novak, J.A. Baker, M.E. Obbard, and B. Malloch, editors. Ontario Trappers Association and Ontario Ministry of Natural Resources, Toronto, Canada.

Small, M.P., K.D. Stone, and J.A. Cook. 2003. American marten (*Martes americana*) in the Pacific Northwest: population differentiation across a landscape fragmented in time and space. Molecular Ecology 12:89–103.

Smith, A.C., and J.A. Schaefer. 2002. Home range size and habitat selection by American marten in Labrador. Canadian Journal of Zoology 80:1602–1609.

Smith, B.L. W.P. Burger, and F.J. Singer. 1998. An expandable radiocollar for elk calves. Wildlife Society Bulletin 26:113–117.

Smith, D.A., K. Ralls, B.L. Cypher, and J.E. Maldonado. 2005. Assessment of scat-detection dog surveys to determine kit fox distribution. Wildlife Society Bulletin 33:897–904.

Smith, D.A., K. Ralls, B. Davenport, B. Adams, and J.E. Maldonado. 2001. Canine assistants for conservationists. Science 291:435.

Smith, D.A., K. Ralls, A. Hurt, B. Adams, M. Parker, B. Davenport, M.C. Smith, and J.E. Maldonado. 2003. Detection and accuracy rates of dogs trained to find scats of San Joaquin kit foxes (*Vulpes macrotis mutica*). Animal Conservation 6:339–346.

Smith, D.W., R.O. Peterson, T.D. Drummer, and D.S. Sheputis. 1991. Over-winter activity and body temperature patterns in northern beavers. Canadian Journal of Zoology 69:2178–2182.

Smith, J.B., J.A. Jenks, and R.W. Klaver. 2007. Evaluating detection probabilities for American marten in the Black Hills, South Dakota. Journal of Wildlife Management 71:2412–2416.

Smith, K.F., D.F. Sax, and K.D. Lafferty. 2006. Evidence for the role of infectious disease in species extinction and endangerment. Conservation Biology 20:1349–1357.

Smith, S., N. Aitken, C. Schwarz, and A. Morin. 2004. Characterization of 15 single nucleotide markers for chimpanzees (*Pan troglodytes*). Molecular Ecology Notes 4:348–351.
Snyder, J.E. 1984. Marten use of clearcuts and residual forest stands in western Newfoundland. Thesis, University of Maine, Orono, USA.
Snyder, J. 1986. Updated status report on the marten (Newfoundland population) *Martes americana atrata*. Committee on the Status of Endangered Wildlife in Canada, Ottawa, Ontario, Canada.
Snyder, J., and J.A. Bissonette. 1987. Marten use of clearcuttings and residual forest stands in western Newfoundland. Canadian Journal of Zoology 65:169–174.
Sobrino, R., J.P. Dubey, M. Pabon, N. Linarez, O.C. Kwok, J. Millan, M.C. Arnal, D.F. Luco, F. Lopez-Gatius, P. Thulliez, C. Gortazar, and S. Almeria. 2008. *Neospora caninum* antibodies in wild carnivores from Spain. Veterinary Parasitology 155:190–197.
Sokal, R.R., and F.J. Rohlf. 1981. Biometry. Second edition. W.H. Freeman and Co., San Francisco, California, USA.
Sommer, R., and N. Benecke. 2004. Late- and post-glacial history of the Mustelidae in Europe. Mammal Review 34:249–284.
Sommer, S. 2005. The importance of immune gene variability (MHC) in evolutionary ecology and conservation. Frontiers in Zoology 2:1742–9994.
Sørensen, P.G., I.M. Petersen, and O. Sand. 1995. Activities of carbohydrate and amino acid metabolizing enzymes from liver of mink (*Mustela vison*) and preliminary observations on steady state kinetics of the enzymes. Comparative Biochemistry and Physiology B 112:59–64.
Soukkala, A.M. 1983. The effects of trapping on marten populations in Maine. Thesis, University of Maine, Orono, USA.
Soutiere, E.C. 1979. Effects of timber harvesting on marten in Maine. Journal of Wildlife Management 43:850–860.
Soutiere, E.C., and M.W. Coulter. 1975. Re-introduction of marten to the White Mountain National Forest, New Hampshire, 1975. Unpublished report, University of Maine, Orono, USA.
Soutullo, A., V. Urios, M. Ferrer, and S.G. Penarrubia. 2006. Post-fledging behavior in golden eagles *Aquilla chrysaetos*: onset of juvenile dispersal and progressive distancing from the nest. Ibis 148:307–312.
Spencer, W.D., R.H. Barrett, and W.J. Zielinski. 1983. Marten habitat preferences in the northern Sierra Nevada. Journal of Wildlife Management 47:1181–1186.
Spencer, W.D., H.L. Rustigian, R.M. Scheller, A. Syphard, J. Strittholt, and B. Ward. 2008. Baseline evaluation of fisher habitat and population status, and effects of fires and fuels management on fishers in the southern Sierra Nevada. Conservation Biology Institute, Final Report to USDA Forest Service, Pacific Southwest Region, Vallejo, California, USA.
Spencer, W., H. Rustigian-Romsos, J. Strittholt, R. Scheller, W. Zielinski, and R. Truex. 2011. Using occupancy and population models to assess habitat conservation opportunities for an isolated carnivore population. Biological Conservation 144:788–803.
Spies, T.A. 1998. Forest structure: a key to the ecosystem. Northwest Science 72:34–39.
Stallknecht, D.E. 2007. Impediments to wildlife disease surveillance, research, and diagnostics. Current Topics in Microbiology and Immunology 315:445–461.
Statistics Canada. 2006. The fur industry: changing with the times. http://www41.statcan.ca/2006/1664/ceb1664_004-eng.htm (accessed 21 March 2011).
Statistics Canada. 2009. Fur statistics 2008. Statistics Canada, Agriculture Division, Catalogue 23–013-X, Vol. 6, No. 1.
Steck, F., and A. Wandeler. 1980. The epidemiology of fox rabies in Europe. Epidemiologic Reviews 2:71–96.
Steinel, A., L. Munson, M. Van Vuuren, and U. Truyen. 2000. Genetic characterization of feline parvovirus sequences from various carnivores. Journal of General Virology 81:345–350.
Steinel, A., C.R. Parrish, M.E. Bloom, and U. Truyen. 2001. Parvovirus infections in wild carnivores. Journal of Wildlife Diseases 37:594–607.

Steinhagen, P., and W. Nebel. 1985. Distemper in stone martens (*Martes foina*, Erxl.) in Schleswig-Holstein. A contribution to the epidemiology of distemper. Dtsch Tierarztl Wochenschr 92:178–181.

Steininger, F.F., and G. Wessely. 1999. From the Tethyan Ocean to the Paratethys Sea: Oligocene to Neogene stratigraphy, palegeography and paleobiogeography of the circum-Mediterranean region and the Oligocene to Neogene basin evolution in Austria. Mitteilungen der Österreichischen Geologischen 92:95–116.

Stephens, D.W., and J.R. Krebs. 1986. Foraging theory. Princeton University Press, New Jersey, USA.

Stephens, S.L., R.E. Martin, and N.E. Clinton. 2007. Prehistoric fire area and emissions from California's forests, woodlands, shrublands and grasslands. Forest Ecology and Management 251:205–216.

Stevens, M.S., and J.B. Stevens. 2003. Carnivora (Mammalia, Felidae, Canidae, and Mustelidae) from the earliest Hemphillian Screw Bean local fauna, Big Bend National Park, Brewster County, Texas. Pages 177–211 *in* Vertebrate fossils and their context: contributions in honor of Richard H. Tedford. Bulletin of the American Museum of Natural History 279, New York, USA.

Stevens, S.S., and T.L. Serfass. 2005. Sliding behavior in nearctic river otters: locomotion or play? Northeast Naturalist 12:241–244.

Steventon, J.D. 1979. Influence of timber harvesting upon winter habitat use by marten. Thesis, University of Maine, Orono, USA.

Steventon, J.D., and J.T. Major. 1982. Marten use of habitat in a commercially-clearcut forest. Journal of Wildlife Management 46:175–182.

Stier, N. 2000. Tagesverstecke des baummarders (*Martes martes* L.) in Südwest-Mecklenburg. Beiträge zur Jagd- und Wildforschung 25:165–182.

Stone, K.D., and J.A. Cook. 2002. Molecular evolution of Holarctic martens (genus *Martes*, Mammalia: Carnivora: Mustelidae). Molecular Phylogenetics and Evolution 24:169–179.

Stone, K.D., R.W. Flynn, and J.A. Cook. 2002. Post-glacial colonization of northwestern North America by the forest-associated American marten (*Martes americana*, Mammalia: Carnivora: Mustelidae). Molecular Ecology 11:2049–2063.

Storch, I. 1988. Zur raumnutzung von baummardern. Zeitschrift Fur Jagdwissenschaft 34:115–119.

Storch, I., E. Lindström, and J. de Jounge. 1990. Habitat selection and food habitats of the pine marten in relation to competition with the red fox. Acta Theriologica 35:311–320.

Storck, P., D.P. Lettenmaier, and S.M. Bolton. 2002. Measurement of snow interception and canopy effects on snow accumulation and melt in a mountainous maritime climate, Oregon, United States. Water Resources Research 38. http://www.agu.org/journals/wr/wr0211/2002WR001281/index.html (accessed 3 October 2010).

Storfer, A., M.A. Murphy, S.F. Spear, R. Holderegger, and L.P. Waits. 2010. Landscape genetics: where are we now? Molecular Ecology 19:3496–3514.

Strickland, M.A., and C.W. Douglas. 1981. The status of fisher in North America and its management in southern Ontario. Pages 1443–1458 *in* Worldwide furbearer conference proceedings. J.A. Chapman and D. Pursley, editors. Worldwide Furbearer Conference, Frostburg, Maryland, USA.

Strickland, M.A., and C.W. Douglas. 1987. Marten. Pages 531–546 *in* Wild furbearer management and conservation in North America. M. Novak, J.A. Baker, M.E. Obbard, and B. Malloch, editors. Ontario Trappers Association and Ontario Ministry of Natural Resources, Toronto, Canada.

Strickland, M.A., C.W. Douglas, M.K. Brown, and G.R. Parsons. 1982a. Determining the age of fisher from cementum annuli of teeth. New York Fish and Game Journal 29:90–94.

Strickland, M.A., C.W. Douglas, M. Novak, and M.P. Hunziger. 1982b. Fisher. Pages 586–598 *in* Wild mammals of North America: biology, management, and economics. J.A. Chapman and G.A. Feldhamer, editors. Johns Hopkins University Press, Baltimore, Maryland, USA.

Strickland, M.A., C.W. Douglas, M. Novak, and M.P. Hunziger. 1982c. Marten. Pages 599–612 *in* Wild mammals of North America: biology, management, economics. J.A. Chapman and G.A. Feldhamer, editors. Johns Hopkins University Press, Baltimore, Maryland, USA.

Sturtevant, B.R., and J.A. Bissonette. 1997. Stand structure and microtine abundance in Newfoundland: implications for marten. Pages 182–198 *in Martes*: taxonomy, ecology, techniques and management. G. Proulx, H.N. Bryant, and P.M. Woodard, editors. Provincial Museum of Alberta, Edmonton, Canada.

Sugihara, N.G., J.W. van Wagtendonk, K.E. Shaffer, J. Fites-Kaufman, and A.E. Thode. 2006. Fire in California's ecosystems. University of California Press, Berkeley, USA.

Sugimoto, T., K. Miyoshi, D. Sakata, K. Nomoto, and S. Higashi. 2009. Fecal DNA-based discrimination between indigenous *Martes zibellina* and non-indigenous *Martes melampus* in Hokkaido, Japan. Mammal Study 34:155–159.

Sullivan, M.J. 1984. American marten re-introduction progress report, Fundy National Park, 1984. Environment Canada, Parks Canada, Natural Resource Conservation, Fundy National Park, New Brunswick, Canada.

Suring, L.H., D.C. Crocker-Bedford, R.W. Flynn, C.L. Hale, G.C. Iverson, M.D. Kirchhoff, T.E. Schenck, L.C. Shea, and K. Titus. 1993. A proposed strategy for maintaining well-distributed, viable populations of wildlife associated with old-growth forests in southeast Alaska. USDA Forest Service, Alaska Region, Juneau, USA.

Swanson, B.J., and C.J. Kyle. 2007. Relative influence of temporal and geographic separation of source populations in a successful marten reintroduction. Journal of Mammalogy 88:1346–1348.

Swanson, B.J., L.R. Peters, and C.J. Kyle. 2006. Demographic and genetic evaluation of an American marten reintroduction. Journal of Mammalogy 87:272–280.

Swarbrick, M.M., and P.J. Havel. 2008. Physiological, pharmacological, and nutritional regulation of circulating adiponectin concentrations in humans. Metabolic Syndrome and Related Disorders 6:87–102.

Taberlet, P., S. Griffin, B. Goossens, S. Questiau, V. Manceau, N. Escaravage, L.P. Waits, and J. Bouvet. 1996. Reliable genotyping of samples with very low DNA quantities using PCR. Nucleic Acids Research 26:3189–3194.

Taberlet, P., L.P. Waits, and G. Luikart. 1999. Noninvasive genetic sampling: look before you leap. Trends in Ecology & Evolution 14:323–327.

Talancy, N.W. 2005. Effects of habitat fragmentation and landscape context on medium-sized mammalian predators in northeastern national parks. Thesis, University of Rhode Island, Kingston, USA.

Tallmon, D.A., D. Gregovich, R. Waples, C.S. Baker, J. Jackson, B. Taylor, F. Archer, F.W. Allendorf, and M.K. Schwartz. 2010. When are genetic methods useful for estimating contemporary abundance and detecting population trends? Molecular Ecology Resources. doi:10.1111/j.1755–0998.2010.02831.x.

Tallmon, D.A., G. Luikart, and R.S. Waples. 2004. The alluring simplicity and complex reality of genetic rescue. Trends in Ecology & Evolution 16:330–342.

Tatara, M. 1994. Ecology and conservation status of the Tsushima marten. Pages 272–279 *in* Martens, sables, and fishers: biology and conservation. S.W. Buskirk, A.S. Harestad, M.G. Raphael, and R.A. Powell editors. Cornell University Press, Ithaca, New York, USA.

Tatsumi, N., S. Miwa, and S.M. Lewis. 2002. Specimen collection, storage, and transmission to the laboratory for hematological tests. International Journal of Hematology 75:261–268.

Taylor, M.E., and N. Abrey. 1982. Marten, *Martes americana*, movements and habitat use in Algonquin Provincial Park, Ontario. Canadian Field-Naturalist 96:439–447.

Taylor, S.L., and S.W. Buskirk. 1994. Forest microenvironments and resting energetics of the American marten *Martes americana*. Ecography 17:249–256.

Taylor, S.L., and S.W. Buskirk. 1996. Dynamics of subnivean temperature and wind speed in subalpine forests of the Rocky Mountains. Journal of Thermal Biology 21:91–99.

Tedford, R.H., M.F. Skinner, R.W. Fields, J.M. Rensberger, D.P. Whistler, T. Galusha, B.E. Taylor, J.R. Macdonald, and S.D. Webb. 1987. Faunal succession and biochronology of the Arikareean through Hemphillian interval (late Oligocene through earliest Pliocene epochs) in North America. Pages 153–210 *in* Cenozoic mammals of North America: geochronology and biostratigraphy. M.O. Woodburne, editor. University of California Press, Berkeley, USA.

Tenter, A.M., A.R. Heckeroth, and L.M. Weiss. 2000. *Toxoplasma gondii*: from animals to humans. International Journal for Parasitology 30:1217–1258.

Theobald, D.M. 2006. Exploring the functional connectivity of landscapes using landscape networks. Pages 416–443 *in* Connectivity conservation: maintaining connections for nature. K.R. Crooks and M.A. Sanjayan, editors. Cambridge University Press, UK.

Thomas, J.W., L.F. Ruggiero, R.W. Mannan, J.W. Schen, and R.A. Lancia. 1988. Management and conservation of old-growth forests in the United States. Wildlife Society Bulletin 16:252–262.

Thomasma, L.E., T.D. Drummer, and R.O. Peterson. 1991. Testing the habitat suitability index model for the fisher. Wildlife Society Bulletin 19:291–297.

Thomasma, L.E., T.D. Drummer, and R.O. Peterson. 1994. Modeling habitat selection by fishers. Pages 316–325 *in* Martens, sables and fishers: biology and conservation. S.W. Buskirk, A.S. Harestad, M.G. Raphael, and R.A. Powell, editors. Cornell University Press, Ithaca, New York, USA.

Thompson, C.M., and K. McGarigal. 2002. The influence of research scale on bald eagle habitat selection along the lower Hudson River, New York (USA). Landscape Ecology 17:569–586.

Thompson, C., K. Purcell, J. Garner, and R. Green. 2010. Kings River Fisher Project: progress report 2007–2009. Unpublished report to the USDA Forest Service, Pacific Southwest Region, Vallejo, California, USA.

Thompson, C.M., W.J. Zielinski, and K.L. Purcell. 2011. Evaluating management risks using landscape trajectory analysis: a case study of California fisher. Journal of Wildlife Management 75:1164–1176.

Thompson, I.D. 1994. Marten populations in uncut and logged boreal forests in Ontario. Journal of Wildlife Management 58:272–280.

Thompson, I.D., J.A. Baker, C. Jastrebski, J. Dacosta, J. Fryxell, and D. Corbett. 2008. Effects of post-harvest silviculture on use of boreal forest stands by amphibians and marten in Ontario. Forestry Chronicle 84:741–747.

Thompson, I.D., and P.W. Colgan. 1987. Numerical responses of marten to a food shortage in northcentral Ontario. Journal of Wildlife Management 51:824–835.

Thompson, I.D., and P.W. Colgan. 1994. Marten activity in uncut and logged boreal forests in Ontario. Journal of Wildlife Management 58:280–288.

Thompson, I.D., and W.J. Curran. 1995. Habitat capability of second-growth balsam fir forests for marten in Newfoundland. Canadian Journal of Zoology 73:2059–2064.

Thompson, I.D., I.J. Davidson, S. O'Donnell, and F. Brazeau. 1989. Use of track transects to measure the relative occurrence of some boreal mammals in uncut forest and regeneration stands. Canadian Journal of Zoology 67:1816–1823.

Thompson, I.D., and A.S. Harestad. 1994. Effects of logging on American martens, and models for habitat management. Pages 355–367 *in* Martens, sables, and fishers: biology and conservation. S.W. Buskirk, A.S. Harestad, M.G. Raphael, and R.A. Powell, editors. Cornell University Press, Ithaca, New York, USA.

Thompson, J., L. Diller, R. Golightly, and R. Klug. 2007. Fisher (*Martes pennanti*) use of a managed forest in coastal northwest California. Pages 245–246 *in* Proceedings of the Redwood Region forest science symposium: what does the future hold? R.B. Standiford, G.A. Giusti, Y. Valachovic, W.J. Zielinski, and M.J. Furniss, editors. USDA Forest Service, General Technical Report PSW-GTR-194.

Thompson, W.D. 1991. Could marten become the spotted owl of Canada? Forestry Chronicle 67:136–140.

Thompson, W.K. 1949. A study of martens in Montana. Pages 181–188 *in* Proceedings of the 29th Annual Conference of the Western Association of State Game and Fish Commissioners, Seattle, Washington, USA.

Thompson, Z. 1842. History of Vermont, natural, civil and statistical. Part I. Privately printed, Burlington, Vermont, USA.

Thorne, E.T., and E.S. Williams. 1988. Disease and endangered species: the black-footed ferret as a recent example. Conservation Biology 2:66–74.

Timofeev, V.V., and V.N. Nadeev. 1955. The sable. Zagotizdat Publishing House, Moscow, USSR. (in Russian)

Timofeev, V.V., and M.P. Pavlov. 1973. The sable. Pages 51–105 *in* Acclimatization of game mammals in the Soviet Union 1. I.D. Kiris, editor. Volgo-Vyatskoe Publishing House, Kirov, USSR. (in Russian)

Tizard, I.R., J.B. Billett, and R.O. Ramsden. 1976. The prevalence of antibodies against *Toxoplasma gondii* in some Ontario mammals. Journal of Wildlife Diseases 12:322–325.

Toma, S., and L. Lafleur. 1974. Survey on the incidence of *Yersinia enterocolitica* infection in Canada. Applied Enviromental Microbiology 28:469–473.

Torchin, M.E., K.D. Lafferty, A.P. Dobson, V.J. Mackenzie, and A.N. Kuris. 2003. Introduced species and their missing parasites. Nature 412:628–629.

Tóth, M. 2008. A new non-invasive method for detecting mammals from birds' nests. Journal of Wildlife Management 72:1237–1240.

Tóth, M., A. Bárány, and R. Kis. 2009. An evaluation of stone marten (*Martes foina*) records in the city of Budapest, Hungary. Acta Zoologica Academiae Scientiarum Hungaricae 55:199–209.

Trapping Today. 2009. NAFA promises to rebuild wild fur market. http://trappingtoday.com/index.php/2009/03/03/nafa-promises-to-rebuild-wild-fur-market/ (accessed 21 March 2011).

Trouessart, D.E. 1893. Mammiferes. Extrait de L'Annuaire Geologique Universel, Tome X.

Truex, R.L., and W.J. Zielinski. 2005. Short-term effects of fire and fire surrogate treatments on fisher habitat in the Sierra Nevada. Unpublished final report to Joint Fire Science Program Project JFSP 01C-3-3-02. USDA Forest Service, Sequoia National Forest, Porterville, California, and USDA Forest Service, Pacific Southwest Research Station, Arcata, California, USA.

Truex, R.L., W.J. Zielinski, R.T. Golightly, R.H. Barrett, and S.M. Wisely. 1998. A meta-analysis of regional variation in fisher morphology, demography, and habitat ecology in California. Unpublished draft report to California Department of Fish and Game, Wildlife Management Division, Nongame Bird and Mammal Section, Sacramento, USA.

Tschöp, M., D.L. Smiley, and M.L. Heiman. 2000. Ghrelin induces adiposity in rodents. Nature 407:908–913.

Tully, S.M. 2006. Habitat selection of fishers (*Martes pennanti*) in an untrapped refugium: Algonquin Provincial Park. Thesis, Trent University, Peterborough, Ontario, Canada.

Turner, M.G. 1989. Landscape ecology: the effect of pattern on process. Annual Review of Ecology and Systematics 20:171–197.

Turton, W. 1806. A general system of nature, through the three grand kingdoms of animals, vegetables and minerals. Lackington and Allen, London, UK.

Twining, H., and A. Hensley. 1947. The status of pine martens in California. California Fish and Game 33:133–137.

Urban, D.L. 1987. Landscape ecology. Bioscience 37:119–127.

U.S. Department of Agriculture. 2004. Sierra Nevada Forest Plan Amendment Final Supplemental Environmental Impact Statement. USDA Forest Service, Pacific Southwest Region, Vallejo, California, USA.

U.S. Department of Agriculture. 2008. Land and resource management plan, Tongass National Forest. R10-MB-603b. USDA Forest Service, Tongass National Forest, Juneau, Alaska, USA.

U.S. Department of Agriculture. 2010. Sierra Nevada Forest Plan amendment draft supplemental environmental impact statement. R5-MB-213. USDA Forest Service, Pacific Southwest Region, Vallejo, California, USA. http://www.fs.fed.us/r5/snfpa/.

U.S. Departments of Agriculture and Interior. 1994. Record of decision for amendments to Forest Service and Bureau of Land Management planning documents within the range of the northern spotted owl and the standards and guidelines for management of habitat for late-successional and old-growth related species within the range of the northern spotted owl. USDA Forest Service and USDI Bureau of Land Management, Portland, Oregon, USA.

U.S. Departments of Agriculture and Interior. 2000. Interior Columbia Basin final environmental impact statement proposed decision. Interior Columbia Basin Ecosystem Management Project. USDA Forest Service and USDI Bureau of Land Management, Washington, D.C., USA.

U.S. Fish and Wildlife Service. 1999. South Florida multi-species recovery plan: a species plan and ecosystem approach. U.S. Fish and Wildlife Service, Region IV, Atlanta, Georgia, USA.

U.S. Fish and Wildlife Service. 2004. Notice of 12-month finding for a petition to list the West Coast Distinct Population Segment of the fisher (*Martes pennanti*). Federal Register 69:18770–18792.

U.S. Fish and Wildlife Service. 2006. Species assessment and listing priority assignment form: fisher, West Coast Distinct Population Segment. U.S. Fish and Wildlife Service, Portland, Oregon, USA.

U.S. Fish and Wildlife Service. 2010. 90-day finding on a petition to list a distinct population segment of the fisher in its United States northern Rocky Mountain range as endangered or threatened with critical habitat. Federal Register 75:19925–19935.

U.S. Geological Survey. 1997. Forum on wildlife telemetry: innovations, evaluations, and research needs. Northern Prairie Wildlife Research Center Online, U.S. Geological Survey and The Wildlife Society, Jamestown, North Dakota, USA. http://www.npwrc.usgs.gov/resource/wildlife/telemtry/index.htm (accessed 25 June 1998).

Vadillo, J.M., J. Reija, and C. Vilà. 1997. Distribución y selección de hábitat de la garduña (*Martes foina*, Erxleben, 1777) en Vizcaya y Sierra Salvada (Burgos). Doñana Acta Vertebrata 24:39–49.

Våge, D.I., E.B. Stavdal, J. Beheim, and H. Klungland. 2003. Why certain silver fox genotypes develop red hairs in their coat. Scientifur 27:79–83.

Valentsev, A.S. 1996. Helminth invasion of sable. *Martes* Working Group Newsletter 4:22–24.

Van der Lans, H.E., J. Tonckens, C.E. van der Ziel, and J.L. Mulder. 2006. Kansen voor de boommarter in Noord-Brabant (Opportunity for the pine marten in Noord-Brabant). Ecoplan-Natuurontwikkeling & Mulder Natuurlijk, De Bilt, Netherlands. (in Dutch)

van Diggelen, F. 2010. Expert advice: are we there yet? The state of the consumer industry. GPS World, March 1, 2010.

Van Horne, B. 1983. Density as a misleading indicator of habitat quality. Journal of Wildlife Management 47:893–901.

Van Moll, P., S. Alldinger, W. Baumgärtner, and M. Adami. 1995. Distemper in wild carnivores: an epidemiological, histological and immunocytochemical study. Veterinary Microbiology 44:193–199.

Van Vuren, D. 1989. Effects of intraperitoneal transmitter implants on yellow-bellied marmots. Journal of Wildlife Management 53:320–323.

van Wagtendonk, J.W., and J.A. Fites-Kaufman. 2006. Sierra Nevada bioregion. Pages 264–294 *in* Fire in California's ecosystems. N.G. Sugihara, J.W. van Wagtendonk, K.E. Shaffer, J.A. Fites-Kaufman, and A.E. Thode, editors. University of California Press, Berkeley, USA.

Van Why, K.R., and W.M. Giuliano. 2001. Fall food habits and reproductive condition of fishers, *Martes pennanti*, in Vermont. Canadian Field-Naturalist 115:52–56.

van Zyll de Jong, C.G. 1969. The restoration of marten in Manitoba, an evaluation. Manitoba Wildlife Branch, Biological Report, Winnipeg, Canada.

Vaughan, T.A., J.M. Ryan, and N.J. Czaplewski. 2000. Mammalogy. Fourth edition. Thomson Learning, Stamford, Connecticut, USA.

Velander, K.A. 1983. Pine marten survey of Scotland, England and Wales 1980–1982. The Vincent Wildlife Trust, London, UK.

Venables, W.N., and D.M. Smith. 2008. An introduction to R: a programming environment for data analysis and graphics. Version 2. R Development Core Team. http://cran.r-project.org/.

Videla, L.A., R. Rodrigo, J. Araya, and J. Poniachik. 2004. Oxidative stress and depletion of hepatic long-chain polyunsaturated fatty acids may contribute to nonalcoholic fatty liver disease. Free Radical Biology & Medicine 37:1499–1507.

Vignal, A., D. Milan, M. SanCristobal, and A. Eggen. 2002. A review on SNP and other types of molecular markers and their use in animal genetics. Genetics Selection Evolution 34:275–305.

Vinkey, R.S. 2003. An evaluation of fisher (*Martes pennanti*) introductions in Montana. Thesis, Univeristy of Montana, Missoula, USA.

Vinkey, R.S., M.K. Schwartz, K.S. McKelvey, K.R. Foresman, K.L. Pilgrim, B.J. Giddings, and E.C. Lofroth. 2006. When reintroductions are augmentations: the genetic legacy of fishers (*Martes pennanti*) in Montana. Journal of Mammalogy 87:265–271.

Virgós, E. 2001. Relative value of riparian woodlands in landscapes with different forest cover for medium-sized Iberian carnivores. Biodiversity and Conservation 10:1039–1049.

Virgós, E., S. Cabezas-Díaz, J.G. Mangas, and J. Lozano. 2010. Spatial distribution models in a frugivorous carnivore, the stone marten (*Martes foina*): is the fleshy-fruits availability a useful predictor? Animal Biology 60:423–436.

Virgós, E., and J.G. Casanovas. 1998. Distribution patterns of the stone marten (*Martes foina*) in Mediterranean mountains of central Spain. Zeitschrift für Säugetierkunde 63:193–199.

Virgós, E., and F.J. García. 2002. Patch occupancy by stone martens *Martes foina* in fragmented landscapes of central Spain: the role of fragment size, isolation and habitat structure. Acta Oecologia 23:231–237.

Virgós, E., M.R. Recio, and Y. Cortés. 2000. Stone marten (*Martes foina*) use of different landscape types in the mountains of central Spain. Zeitschrift für Säugetierkunde 65:375–379.

Vladimirova, E.J., and J.P. Mozgovoy. 2010. Winter ecology of the pine marten (*Martes martes* L.) in the Volga Floodplain opposite Samara. Russian Journal of Ecology 41:333–339.

Voorhies, M.R. 1990. Vertebrate biostratigraphy of the Ogallala group in Nebraska. Pages 115–151 *in* Geologic framework and regional hydrology: upper Cenozoic Blackwater Draw and Ogallala Formation, Great Plains. T.C. Gustavson, editor. University of Texas, Austin, USA.

Wabakken, P. 1985. Vinternæring, habitatbruk og jaktadferd hos mår (*Martes martes*) i sørøst-norsk barskog. Thesis, University of Oslo, Norway. (in Norwegian)

Waechter, A. 1975. Écologie de la fouine en Alsace. Revue d'Ecologie (Terre et Vie) 29:399–457.

Wagener, W.W., and R.W. Davidson. 1954. Heart rots in living trees. Botanical Review 20:61–134.

Waits, L.P., and D. Paetkau. 2005. Noninvasive sampling tools for wildlife biologists: a review of applications and recommendations for accurate data collection. Journal of Wildlife Management 64:1419–1433.

Walker, F.A. 1872. Ninth census: Volume I, the statistics of the population of the United States, embracing the tables of race, nationality, sex, selected ages, and occupations. U.S. Department of the Interior, Government Printing Office, Washington, D.C., USA.

Wallace, K., and R. Henry. 1985. Return of a Catskill native. The Conservationist 40(3):16–19.

Waltari, E., E.P. Hoberg, E.P. Lessa, and J.A. Cook. 2007. Eastward Ho: phylogeographical perspectives on colonization of hosts and parasites across the Beringian nexus. Journal of Biogeography 34:561–574.

Walton, L.R., H.D. Cluff, P.C. Paquet, and M.A. Ramsay. 2001. Movement patterns of barren-ground wolves in the central Canadian Arctic. Journal of Mammalogy 82:867–876.

Wan-bo, H. 1979. On the age of the cave faunas of south China. Vertebrata PalAsiatica 17:327–343.

Wang, J., M. Lin, A. Crenshaw, A. Hutchinson, B. Hicks, M. Yeager, S. Berndt, W.-Y. Huang, R.B. Hayes, S.J. Chanock, R.C. Jones, and R. Ramakrishnan. 2009. High-throughput single nucleotide polymorphism genotyping using nanofluidic dynamic arrays. BMC Genomics 10:561–574.

Warheit, K.I. 2004. Fisher (*Martes pennanti*) control region sequences from Alberta, and re-analysis of sequences from other regions in North America: recommendations for the reintroduction of fishers into Washington. Draft report, Washington Department of Fish and Wildlife, Olympia, USA.

Wasser, S.K., B. Davenport, E.R. Ramage, K.E. Hunt, M. Parker, C. Clarke, and G. Stenhouse. 2004. Scat detection dogs in wildlife research and management: application to grizzly and black bears in the Yellowhead ecosystem, Alberta, Canada. Canadian Journal of Zoology 82:475–492.

Wasser, S.K., H. Smith, L. Madden, N. Marks, and C. Vynne. 2009. Scent-matching dogs determine number of unique individuals from scat. Journal of Wildlife Management 73:1233–1240.

Wasserman, T.N. 2008. Habitat relationships and landscape genetics of *Martes americana* in northern Idaho. Thesis, Western Washington University, Bellingham, USA.

Wasserman, T.N., S.A. Cushman, M.K. Schwartz, and D.O. Wallin. 2010. Spatial scaling and multi-model inference in landscape genetics: *Martes americana* in northern Idaho. Landscape Ecology 25:1601–1612.

Wayne, R.K., and D.M. Brown. 2001. Hybridization and conservation of carnivores. Pages 145–162 *in* Carnivore conservation. J.L. Gittleman, S.M. Funk, D.W. Macdonald, and R.K. Wayne, editors. Cambridge University Press, London, UK.

Webster, J.A. 2001. A review of the historical evidence of the habitat of the pine marten in Cumbria. Mammal Review 31:17–31.

Webster, J.P. 2007. The effect of *Toxoplasma gondii* on animal behavior: playing cat and mouse. Schizophrenia Bulletin 33:752–756.

Webster, J.P., C.F. Brunton, and D.W. Macdonald. 1994. Effect of *Toxoplasma gondii* upon neophobic behaviour in wild brown rats, *Rattus norvegicus*. Parasitology 109:37–43.

Weckwerth, R.P., and V.D. Hawley. 1962. Marten food habits and population fluctuations in Montana. Journal of Wildlife Management 26:55–74.

Weckwerth, R.P., and P.L. Wright. 1968. Results of transplanting fishers in Montana. Journal of Wildlife Management 32:977–980.

Weir, R.D. 1995. Diet, spatial organization, and habitat relationships of fishers in south-central British Columbia. Thesis, Simon Fraser University, Burnaby, British Columbia, Canada.

Weir, R.D. 1999. Inventory of fishers in the sub-boreal forests of north-central British Columbia Year III (1998/99): radiotelemetry monitoring. Peace/Williston Fish and Wildlife Compensation Program, Report No. 206, Prince George, British Columbia, Canada.

Weir, R.D. 2009. Fisher ecology in the Kiskatinaw Plateau Ecosystem. Unpublished report to the British Columbia Ministry of Environment and Louisiana-Pacific Canada Ltd., Vancouver, Canada.

Weir, R.D., I.T. Adams, G. Mowat, and A.J. Fontana. 2003. East Kootenay fisher assessment. Unpublished report to British Columbia Ministry of Water, Land, and Air Protection, Cranbrook, Canada.

Weir, R.D., and F.B. Corbould. 2006. Density of fishers in the sub-boreal spruce biogeoclimatic zone of British Columbia. Northwestern Naturalist 87:118–127.

Weir, R.D., and F.B. Corbould. 2007. Factors affecting diurnal activity of fishers in north-central British Columbia. Journal of Mammalogy 88:1508–1514.

Weir, R.D., and F.B. Corbould. 2008. Ecology of fishers in the sub-boreal forests of north-central British Columbia. Peace/Williston Fish and Wildlife Compensation Program, Report No. 315, Prince George, British Columbia, Canada.

Weir, R.D., and F.B. Corbould. 2010. Factors affecting landscape occupancy by fishers in north-central British Columbia. Journal of Wildlife Management 74:405–410.

Weir, R., F. Corbould, and A. Harestad. 2004. Effect of ambient temperature on the selection of rest structures by fishers. Pages 187–197 *in* Martens and fishers (*Martes*) in human-altered environments: an international perspective. D.J. Harrison, A.K. Fuller and G. Proulx, editors. Springer Science+Business Media, New York, USA.

Weir, R.D., and A.S. Harestad. 1997. Landscape-level selectivity by fishers in south-central British Columbia. Pages 252–264 *in Martes*: taxonomy, ecology, techniques, and management. G. Proulx, H.N. Bryant, and P.M. Woodard, editors. Provincial Museum of Alberta, Edmonton, Canada.

Weir, R.D., and A.S. Harestad. 2003. Scale-dependent habitat selectivity by fishers in south-central British Columbia. Journal of Wildlife Management 67:73–82.

Weir, R.D., A.S. Harestad, and R.C. Wright. 2005. Winter diet of fishers in British Columbia. Northwestern Naturalist 86:12–19.

Wengert, G.M., M.W. Gabriel, and D.L. Clifford. 2012. Investigating cause-specific mortality and diseases in carnivores. Pages 294–313 *in* Carnivore ecology and conservation: a handbook of techniques. L. Boitani and R.A. Powell, editors. Oxford University Press, London, UK.

Werdelin, L. 1996. Carnivores, exclusive of Hyaenidae, from the later Miocene of Europe and western Asia. Pages 271–289 *in* The evolution of western Eurasian Neogene mammal faunas. R.L. Bernor, V. Fahlbusch, and H.-W. Mittmann, editors. Columbia University Press, New York, USA.

Westerling, A.L., B.P. Bryant, H.K. Preisler, H.G. Hidalgo, and T. Das. 2009. Climate change, growth and California wildfire. CEC-500–2009–046-F. California Energy Commission, Sacramento, USA.

Westerling, A.L., H.G. Hidalgo, D.R. Cayan, and T.W. Swetnam. 2006. Warming and earlier spring increase western U.S. forest wildfire activity. Science 313:940–943.

Whitaker, J.O., Jr., and W.J. Hamilton, Jr. 1998. Mammals of the eastern United States. Cornell University Press, Ithaca, New York, USA.

White, G.C., and R.A. Garrott. 1990. Analysis of wildlife and radio-tracking data. Academic Press, New York, USA.

Wickström, L., V. Haukisalmi, S. Varis, J. Hantula, V.B. Federov, and H. Henttonen. 2003. Phylogeography of circumpolar *Paranoplocephala arctica* species complex (Cestoda; Anoplocephalidae) parasitizing collared lemmings. Molecular Ecology 12:3359–3371.

Wickstrom, M., and C. Eason. 1999. Literature search for mustelid-specific toxicants. Science for Conservation 127:57–65.

Wiebe, K.L. 2001. Microclimate of tree cavity nests: is it important for reproductive success in northern flickers? Auk 118:412–421.

Wiens, J.A. 1976. Population responses to patchy environments. Annual Review of Ecology and Systematics 7:81–120.

Wiens, J.A., N.C. Stenseth, B. Van Horne, and R.A. Ims. 1993. Ecological mechanisms and landscape ecology. Oikos 66:369–380.

Wilbert, C.J., S.W. Buskirk, and K.G. Gerow. 2000. Effects of weather and snow on habitat selection by American martens (*Martes americana*). Canadian Journal of Zoology 78:1691–1696.

Williams, B.K., J.D. Nichols, and M.J. Conroy. 2002. Analysis and management of animal populations. Academic Press, San Diego, California, USA.

Williams, B.W., D.R. Etter, D.W. Linden, K.F. Millenbah, S.R. Winterstein, and K.T. Scribner. 2009. Noninvasive hair sampling and genetic tagging of co-distributed fishers and American martens. Journal of Wildlife Management 73:26–34.

Williams, B.W., J.H. Gilbert, and P.A. Zollner. 2007. Historical perspective on the reintroduction of the fisher and American marten in Wisconsin and Michigan. USDA Forest Service, General Technical Report GTR-NRS-5.
Williams, B.W., and K.T. Scribner. 2007. Demographic and genetic evaluation of an American marten reintroduction: a comment on Swanson et al. Journal of Mammalogy 88:1342–1345.
Williams, B.W., and K.T. Scribner. 2010. Effects of multiple founder populations on spatial genetic structure of reintroduced American martens. Molecular Ecology 19:227–240.
Williams, E.S. 2001. Canine distemper. Pages 50–59 *in* Infectious diseases of wild mammals. E.S. Williams and I.K. Barker, editors. Iowa State University Press, Ames, USA.
Williams, E.S., and I.K. Barker, editors. 2001. Infectious diseases of wild mammals. Third edition. Iowa State University Press, Ames, USA.
Williams, E.S., K. Mills, D.R. Kwiatkowski, E.T. Thorne, and A. Boerger-Fields. 1994. Plague in a black-footed ferret (*Mustela nigripes*). Journal of Wildlife Diseases 30:581–585.
Williams, E.S., and E.T. Thorne. 1996. Infectious and parasitic diseases of captive carnivores, with special emphasis on the black-footed ferret (*Mustela nigripes*). Review of Science and Technology 15:91–114.
Williams, R.M. 1962a. The fisher returns to Idaho. Idaho Wildlife Review 15:8–9.
Williams, R.M. 1962b. Trapping and transplant project, fisher transplant segment. Project W 75-D-9, completion report. Idaho Department of Fish and Game, Boise, USA.
Williams, R.M. 1963. Trapping and transplanting. Part I—fisher. Federal Aid in Wildlife Retoration, Final Segment Report, Project W 75-D-10. Idaho Department of Fish and Game, Boise, USA.
Williams, R.N., O.E. Rhodes, Jr., and T.L. Serfass. 2000. Assessment of genetic variance among source and reintroduced fisher populations. Journal of Mammalogy 81:895–907.
Wilson, C., and C. Tisdell. 2005. What roles does knowledge of wildlife play in providing support for species conservation? Journal of Social Sciences 1:47–51.
Wilson, D.E., and D.M. Reeder. 2005. Mammal species of the world: a taxonomic and geographic reference. Third edition. Johns Hopkins University Press, Baltimore, Maryland, USA.
Wilson, R.L. 1968. Systematics and faunal analysis of a Lower Pliocene vertebrate assemblage. University of Michigan Contributions from the Museum of Paleontology 22:75–126.
Wimsatt, J., D.E. Biggins, E.S. Williams, and V.M. Becerra. 2006. The quest for a safe and effective canine distemper virus vaccine for black-footed ferrets. Pages 248–266 *in* Recovery of the black-footed ferret: progress and continuing challenges. J.E. Roelle, B.J. Miller, J.L. Godbey, and D.E. Biggins, editors. U.S. Geological Survey, Fort Collins, Colorado, USA.
Wisdom, M.J., R.S. Holthausen, B.C. Wales, C.D. Hargis, V.A. Saab, D.C. Lee, W.J. Hann, T.D. Rich, M.M. Rowland, W.J. Murphy, and M.R. Eames. 2000. Source habitats for terrestrial vertebrates of focus in the Interior Columbia Basin: broad-scale trends and management implications. 3 volumes. USDA Forest Service, General Technical Report PNW-GTR-485.
Wisely, S.M., S.W. Buskirk, G.A. Russell, K.B. Aubry, and W.J. Zielinski. 2004. Genetic diversity and structure of the fisher (*Martes pennanti*) in a peninsular and peripheral metapopulation. Journal of Mammalogy 85:640–648.
Wobeser, G.A. 2006. Essentials of disease in wild animals. Blackwell Publishing, Ames, Iowa, USA.
Wobeser, G.A. 2007. Disease in wild animals: investigation and management. Springer-Verlag, Berlin, Germany.
Wolsan, M. 1988. Morphological variations of the first upper molar in the Genus *Martes* (Carnivora, Mustelidae). Pages 241–254 *in* Teeth revisited: proceedings of the seventh international symposium on dental morphology. D.E. Russell, J.-P. Santoro, and D. Sigog-

neau-Russell, editors. Mémoires du Musée National d'Histoire Naturelle, Paris (serie C), 53.

Wolsan, M. 1989. Dental polymorphism in the genus *Martes* (Carnivora: Mustelidae) and its evolutionary significance. Acta Theriologica 34:545–593.

Wolsan, M. 1993a. Evolution des carnivores quaternaires en Europe centrale dans leur contexte stratigraphique et paleoclimatique. L'Anthropologie (Paris), Tome 97, nos 2/3:203–222.

Wolsan, M. 1993b. Phylogeny and classification of early European mustelida (Mammalia: Carnivora). Acta Theriologica 38:345–384.

Wolsan, M. 1999. Oldest mephitine cranium and its implications for the origin of skunks. Acta Palaeontological Polonica 44:223–229.

Wolsan, M., A.L. Ruprecht, and T. Buchalczyk. 1985. Variation and asymmetry in the dentition of the pine and stone martens (*Martes martes* and *M. foina*) from Poland. Acta Theriologica 30:79–114.

Wood, D.J.A., J.L. Koprowski, and P.W.W. Lurz. 2007. Tree squirrel introduction: a theoretical approach with population viability analysis. Journal of Mammalogy 88:1271–1279.

Wood, J. 1977. The fisher is:. . . . National Wildlife 15(3):18–21.

Wood, J.E. 1959. Relative estimates of fox population levels. Journal of Wildlife Management 23:53–63.

Woodford, J.E. 2009a. Stocking and monitoring American marten in Wisconsin. Final Project Report, Wisconsin Department of Natural Resources, Madison, USA.

Woodford, J.E. 2009b. Winter track surveys for American marten in northern Wisconsin, 2008–09. Wisconsin Natural Resources, Madison, USA. http://dnr.wi.gov/org/land/wildlife/harvest/reports/09martensurv.pdf (accessed 2 April 2010).

Woodroffe, R. 1999. Managing disease threats to wild mammals. Animal Conservation 2:185–193.

Woodroffe, R. 2001. Assessing the risks of intervention: immobilization, radio-collaring and vaccination of African wild dogs. Oryx 35:234–244.

Worthen, G.L., and D.L. Kilgore, Jr. 1981. Metabolic rate of pine marten in relation to air temperature. Journal of Mammalogy 62:624–628.

Wozencraft, W.C. 2005. Order Carnivora. Pages 532–628 *in* Mammal species of the world. Third edition. D.E. Wilson and D.M. Reeder, editors. Johns Hopkins University Press, Baltimore, Maryland, USA.

Wright, P.L. 1942. Delayed implantation in the long-tailed weasel (*Mustela frenata*), the short-tailed weasel (*Mustela cicognani*), and the marten (*Martes americana*). Anatomical Research 83:341–349.

Wright, P.L. 1953. Intergradation between *Martes americana* and *Martes caurina* in western Montana. Journal of Mammalogy 34:74–86.

Wynne, K.M., and J.A. Sherburne. 1984. Summer home range used by adult marten in northwestern Maine. Canadian Journal of Zoology 62:941–943.

Yaeger, J.S. 2005. Habitat at fisher resting sites in the Klamath Province of Northern California. Thesis, Humboldt State University, Arcata, California, USA.

Yanai, T., A. Tomita, T. Masegi, K. Ishikawa, T. Iwasaki, K. Yamazoe, and K. Ueda. 1995. Histopathologic features of naturally occurring hepatozoonosis in wild martens (*Martes melampus*) in Japan. Journal of Wildlife Diseases 31:233–237.

Yeager, L.E. 1950. Implications of some harvest and habitat factors on pine marten management. Transactions of the North American Wildlife Conference 15:319–334.

Yonezawa, T., M. Nikaido, N. Kohno, Y. Fukumoto, N. Okada, and M. Hasegawa. 2007. Molecular phylogenetic study on the origin and evolution of Mustelidae. Gene 396:1–12.

York, E.C. 1996. Fisher population dynamics in north-central Massachusetts. Thesis, University of Massachusetts, Amherst, USA.

Youngman, P.M., and F.W. Schueler. 1991. *Martes nobilis* is a synonym of *Martes americana*, not an extinct Pleistocene-Holocene species. Journal of Mammalogy 72:567–577.

Yu, L., Q.W. Li, O.A. Ryder, and Y.P. Zhang. 2004. Phylogenetic relationships within mammalian order Carnivora indicated by sequences of two nuclear DNA genes. Molecular Phylogenetics and Evolution 33:694–705.

Zabala, J., I. Zuberogoitia, and J.A. Martinez-Climent. 2009. Testing for niche segregation between two abundant carnivores using presence-only data. Folia Zoologica 58:385–395.

Zachos, J., M. Pagani, L. Sloan, E. Thomas, and K. Billups. 2001. Trends, rhythms, and aberrations in global climate 65 Mya to present. Science 292:686–693.

Zajac, A.M., and G.A. Conboy. 2006. Veterinary clinical parasitology. Seventh edition. Blackwell Publishing, Ames, Iowa, USA.

Zalewski, A. 1997a. Factors affecting selection of resting site type by pine marten in primeval deciduous forests (Bialowieza National Park, Poland). Acta Theriologica 42:271–288.

Zalewski, A. 1997b. Patterns of resting site use by pine marten, *Martes martes*, in Bialowieza National Park (Poland). Acta Theriologica 42:153–168.

Zalewski, A. 1999. Identifying sex and individuals of pine marten using snow track measurements. Wildlife Society Bulletin 27:28–31.

Zalewski, A. 2000. Factors affecting the duration of activity by pine martens (*Martes martes*) in the Białowieża National Park, Poland. Journal of Zoology 251:439–447.

Zalewski, A. 2007. Does size dimorphism reduce competition between sexes? The diet of male and female pine martens at local and wider geographical scales. Acta Theriologica 52:237–250.

Zalewski, A., and W. Jędrzejewski. 2006. Spatial organisation and dynamics of the pine marten *Martes martes* population in Bialowieza Forest (E. Poland) compared with other European woodlands. Ecography 29:31–43.

Zalewski, A., W. Jędrzejewski, and B. Jędrzejewska. 1995. Pine marten home ranges, numbers and predation on vertebrates in a deciduous forest (Bialowieza National Park, Poland). Annales Zoologici Fennici 32:131–144.

Zar, J.H. 1999. Biostatistical analysis. Fourth edition. Prentice Hall, Englewood Cliffs, New Jersey, USA.

Zarlenga, D.S., B. Rosenthal, E. Pozio, G. La Rosa, and E.P. Hoberg. 2006. Post-Miocene expansion, colonization, and host switching drove speciation among extant nematodes of the archaic genus *Trichinella*. Proceedings of the National Academy of Sciences, USA 103:7354–7359.

Zarnke, R.L., J.S. Whitman, R.W. Flynn, and J.M. Ver Hoef. 2004. Prevalence of *Soboliphyme baturini* in marten (*Martes americana*) populations from three regions of Alaska, 1990–1998. Journal of Wildlife Diseases 40:452–455.

Zhang, Y., R. Proenca, M. Maffei, M. Barone, L. Leopold, and J.M. Friedman. 1994. Positional cloning of the mouse *obese* gene and its human homologue. Nature 372:425–432.

Zielinski, G.A., and B.D. Keim. 2003. New England weather, New England climate. University Press of New England, Lebanon, New Hampshire, USA.

Zielinski, W.J. 1984. Plague in pine martens and the fleas associated with its occurrence. Great Basin Naturalist 41:170–175.

Zielinski, W.J. 1995. Track plates. Pages 67–89 *in* American marten, fisher, lynx, and wolverine: survey methods for their detection. W.J. Zielinski and T.E. Kucera, technical editors. USDA Forest Service, General Technical Report PSW-GTR-157.

Zielinski, W.J., R.H. Barrett, and R.L. Truex. 1997a. Southern Sierra Nevada fisher and marten study: progress report IV. Unpublished report to the USDA Forest Service, Pacific Southwest Region, Vallejo, California, USA.

Zielinski, W.J., C. Carroll, and J.R. Dunk. 2006a. Using landscape suitability models to reconcile conservation planning for two key forest predators. Biological Conservation 133:409–430.

Zielinski, W.J., N.P. Duncan, E.C. Farmer, R.L. Truex, A.P. Clevenger, and R.H. Barrett. 1999. Diet of fishers (*Martes pennanti*) at the southernmost extent of their range. Journal of Mammalogy 80:961–971.

Zielinski, W.J., J.R. Dunk, J.S. Yaeger, and D.W. LaPlante. 2010. Developing and testing a landscape-scale habitat suitability model for fisher (*Martes pennanti*) in forests of interior northern California. Forest Ecology and Management 260:1579–1591.

Zielinski, W.J., and T.E. Kucera, technical editors. 1995. American marten, fisher, lynx, and wolverine: survey methods for their detection. USDA Forest Service, General Technical Report PSW-GTR-157.

Zielinski, W.J., T.E. Kucera, and R.H. Barrett. 1995. Current distribution of the fisher, *Martes pennanti*, in California. California Fish and Game 81:104–112.

Zielinski, W.J., and F.V. Schlexer. 2009. Inter-observer variation in identifying mammals from their tracks at enclosed track plate stations. Northwest Science 83:299–307.

Zielinski, W.J., F.V. Schlexer, K.L. Pilgrim, and M.K. Schwartz. 2006b. The efficacy of wire and glue hair snares in identifying mesocarnivores. Wildlife Society Bulletin 34:1152–1161.

Zielinski, W.J., W.D. Spencer, and R.H. Barrett. 1983. Relationship between food habits and activity patterns of pine martens. Journal of Mammalogy 64:387–396.

Zielinski, W.J., and H.B. Stauffer. 1996. Monitoring *Martes* populations in California: survey design and power analysis. Ecological Applications 6:1254–1267.

Zielinski, W.J., and R.L. Truex. 1995. Distinguishing tracks of marten and fisher at track-plate stations. Journal of Wildlife Management 59:571–579.

Zielinski, W.J., R.L. Truex, J.R. Dunk, and T. Gaman. 2006c. Using forest inventory data to assess fisher resting habitat suitability in California. Ecological Applications 16:1010–1025.

Zielinski, W.J., R.L. Truex, C.V. Ogan, and K. Busse. 1997b. Detection surveys for fishers and American martens in California, 1989–1994: summary and interpretations. Pages 372–392 *in Martes*: taxonomy, ecology, techniques and management. G. Proulx, H.N. Bryant, and P.M. Woodard, editors. Provincial Museum of Alberta, Edmonton, Canada.

Zielinski, W.J., R.L. Truex, F.V. Schlexer, L.A. Campbell, and C. Carroll. 2005. Historical and contemporary distributions of carnivores in forests of the Sierra Nevada, California, USA. Journal of Biogeography 32:1385–1407.

Zielinski, W.J., R.L. Truex, G.A. Schmidt, F.V. Schlexer, K.N. Schmidt, and R.H. Barrett. 2004a. Home range characteristics of fishers in California. Journal of Mammalogy 85:649–657.

Zielinski, W.J., R.L. Truex, G.A. Schmidt, F.V. Schlexer, K.N. Schmidt, and R.H. Barrett. 2004b. Resting habitat selection by fishers in California. Journal of Wildlife Management 68:475–492.

Zschille, J., N. Stier, and M. Roth. 2008. Radio tagging American mink (*Mustela vison*)—experience with collar and intraperitoneal-implanted transmitters. European Journal of Wildlife Research 54:263–268.

Index

Milton Keynes UK
Ingram Content Group UK Ltd.
UKHW051912090824
446632UK00008B/77

9 780801 450884